Lecture Notes in Networks and Systems

Volume 738

The series “Lecture Notes in Networks and Systems” publishes the latest developments in Networks and Systems—quickly, informally and with high quality. Original research reported in proceedings and post-proceedings represents the core of LNNS.

Volumes published in LNNS embrace all aspects and subfields of, as well as new challenges in, Networks and Systems.

The series contains proceedings and edited volumes in systems and networks, spanning the areas of Cyber-Physical Systems, Autonomous Systems, Sensor Networks, Control Systems, Energy Systems, Automotive Systems, Biological Systems, Vehicular Networking and Connected Vehicles, Aerospace Systems, Automation, Manufacturing, Smart Grids, Nonlinear Systems, Power Systems, Robotics, Social Systems, Economic Systems and other. Of particular value to both the contributors and the readership are the short publication timeframe and the world-wide distribution and exposure which enable both a wide and rapid dissemination of research output.

The series covers the theory, applications, and perspectives on the state of the art and future developments relevant to systems and networks, decision making, control, complex processes and related areas, as embedded in the fields of interdisciplinary and applied sciences, engineering, computer science, physics, economics, social, and life sciences, as well as the paradigms and methodologies behind them.

Indexed by SCOPUS, INSPEC, WTI Frankfurt eG, zbMATH, SCImago.

All books published in the series are submitted for consideration in Web of Science.

For proposals from Asia please contact Aninda Bose (aninda.bose@springer.com).

Jyotsna Kumar Mandal · Biswapati Jana ·
Tzu-Chuen Lu · Debashis De
Editors

Proceedings of International Conference on Network Security and Blockchain Technology

ICNSBT 2023

Springer

Editors
Jyotsna Kumar Mandal
Department of Computer Science and Engineering
Kalyani University
Kalyani, West Bengal, India

Raiganj University
Raiganj, West Bengal, India

Tzu-Chuen Lu
Department of Information Management
Chaoyang University of Technology
Taichung, Taiwan

Biswapati Jana
Department of Computer Science
Vidyasagar University
Midnapore, West Bengal, India

Debashis De
Department of Computer Science and Engineering
Maulana Abul Kalam Azad University of Technology
Kolkata, West Bengal, India

ISSN 2367-3370 ISSN 2367-3389 (electronic)
Lecture Notes in Networks and Systems
ISBN 978-981-99-4432-3 ISBN 978-981-99-4433-0 (eBook)
https://doi.org/10.1007/978-981-99-4433-0

This Springer imprint is published by the registered company Springer Nature Singapore Pte Ltd.
The registered company address is: 152 Beach Road, #21-01/04 Gateway East, Singapore 189721, Singapore

Paper in this product is recyclable.

Foreword

Since the days of the Greek empire, steganography has existed in a variety of ways. Steganography itself derives from the Greek words steganos which represents concealed or covered and graphia means to write something. The Greek historian Herodotus wrote about how a slave served as the conduit for secret communication. After shaving the slave's head and tattooing the message on the exposed skull, the hair was then permitted to grow back. The slave was delivered to the recipient of the message, who shaved the slave's head to disclose the contents. The communication was definitely not time-sensitive! As a scout for the British during the Boer War, Lord Robert Baden-Powell marked the locations of Boer artillery sites by incorporating maps into drawings of butterflies. Anonymization, for example, may be construed to be a result of such a steganographic approach toward data.

Materials can be classified into identifiable and non-identifiable categories based on their identifiability. Information is created by adding value to data, which are facts and claims. Applications may not be lincar, but data is jokingly said to be the smallest building block of information. Informally, the following formula can be used to distinguish between information and data: The phrase "you provided me with valuable information" is often used because information plus value equals data. As a result, value is created in the part that transforms data into information, which can be bought and sold.

A name cannot be unique as it is open to be adopted by one and all. That is why, a name is never bought or sold, i.e., transacted against consideration. While that very name, when laced with other attributes making it unique, creates value. The following depiction will give a clear idea. It is becoming more and more crucial for individuals and organizations to be able to send and receive data anonymously over the Internet in a time when much of the information sent over the Internet is stored and analyzed by various companies, governments, and online criminals. It has also become crucial for individuals and organizations to be able to conceal their locations. An adversary can determine the location or even the identity of the individual behind the communication by knowing the sender's or receiver's original IP address.

The encryption had always served to safeguard the confidentiality between the sender and the intended recipient. To add a stronger layer of protection to the hidden data, steganography methods are now being used more frequently in addition to cryptography. This study addressed how to embed a message into a JPEG image by controlling JPEG quantization tables, and it demonstrated how the quality factor in a JPEG image can be an embedding space (QTs). This method can be utilized as a permutation algorithm combination.

Of course, using social networks which are conducting their *lossless data transfers* over these networks cannot be guaranteed due to internal picture compression. Nevertheless, these social networks can be used to distribute connections to amusing pictures on other websites. Websites that offer visitors original (undisturbed) media material may be the best candidates for using it as a channel for steganography data flows. It is often concluded that there should be a way to either check all incoming and outgoing Internet traffic (check for steganographic data on a firewall, proxy, etc.) or to directly check Internet traffic at workplaces in a web browser. A corporate web browser might be the best option to avoid steganographic data flow.

The common name in the first slot is a valueless fact, while it acquires value when laced with a mobile number because it becomes unique. Lastly, adding another unique attribute viz. AADHAR, etc. will make it invaluable. Sharing one's name is an etiquette while sharing identifiable information that creates unique attributes is frightening. Therefore, a casual share of data can become a devastating disclosure of identity. Thus, information is classified into personally identifiable information and sensitive personally identifiable information.

The empty locker is the one that does not need a lock. Otherwise, nothing of value is contained there. Similar to this, if information loses any of its valuable qualities, it loses all of its value. There is no need for a lock or encryption. Otherwise, the likelihood of losing the information decreases significantly if it is shared or known because of a need or necessity. A piece of information is typically collected for promotion, verification, and identification purposes. Until one discloses other characteristics, such as a residence or an organization they are a member of, their name does not identify them. Second, one's identity can be verified before accessing some restricted services thanks to unique identity proof like AADHAR, etc. Everyone has the right to know why being identified, the limits of identification, and the authority of the identifier. These three correlate to one's fundamental right of privacy enshrined in MY INFORMATION—MY RIGHT and RIGHT TO BE FORGOTTEN.

Documents for clinical trials can number in the thousands of pages. They are submitted for approval to regulatory agencies like the FDA. Clinical research organizations make the laborious effort of manually cleaning up the paperwork to make sure that these documents do not contain the participants' private information.

Privacy AI shortens the turnaround time from days to hours by employing an automated process that includes entity detection, extraction, relationship management, and anonymization.

There is a definitive role for steganography in security. It is meant to support cryptography rather than supplant it. Steganography techniques lessen the likelihood that communication will be picked up. However, if that communication is also encrypted, it must also be decrypted if it is found (yet another layer of protection). Applications for steganography are virtually limitless. A very small portion of the field of steganography is covered in this essay. It involves much more than just including words in an image. In addition to digital images, steganography also applies to other types of files (such as voice, text, and binary ones) as well as other types of media like contact channels.

The road ahead lies in technological intervention involving steganographic cryptography and/or other methods, coupled with process reform and it is called Data Anonymiser. Anonymiser is a strainer that strains attributable data from personally identifiable information making it lose value but not loses context. It is neither just redacting nor masking of data. Because both redacting and masking processes are retractable. Anonymiser can be an application, a process, or a hybrid of both. For example, a person keys in one's login credentials oblivious to the CCTV behind one's back. It is clear that the feed stored via that CCTV has the person's SPII, and using that, the person's privacy can easily be violated. How to avoid that? The answer could be repositioning that CCTV. But is it possible on the fly? No. The simplest solution is to reposition the person himself. Similarly, Anonymiser could be a process reform too. Hence, a single Anonymiser may not be the solution for various problem statements. There could be many Anonymiser Solutions.

The State Government of West Bengal in the Department of Information Technology and Electronics has rolled out the Anonymiser hackathon through which any individual can contribute to the ever-growing repository of problem statements via Ideathon and thereafter come together to solve each of such problem statements by innovative methods comprising the application, process reforms et al which will be known as the Anonymiser. These solutions will then be taken through the live transactional data reposited with the government to find out whether they can really strain the valuable attributes laced in the PII/SPII and create an anonymised data lake.

Various market study suggests that the anonymized data has a market value of $ 220 Billion in 2021 that will grow to a whopping $ 343 Billion in 2030, and the tons of data generated by public services on a daily basis in this country could be a game changer for the academia, industry as well as the professionals. They will create affordable and accessible solutions for the mass. The 2nd ICNSBT 2023 is one of the platforms that brings academicians, researchers, technological students, and industry people with common interests and interact issues and innovations relating to the information and system security, cryptographic issues, and blockchain.

Sanjay Kumar Das
West Bengal Civil Service (Exe.,'96)
Joint Secretary and State Information
Security Officer
IT&E Department
State Information Security
Government of West Bengal
Kolkata, West Bengal, India

Preface

This ICNSBT 2023 is focused on network security, privacy, and blockchain technology and its applications. The conference was held during March 25–26, 2023, at the Vidyasagar University, West Bengal, India. This volume consists of the privacy, authentication, digital watermarking, cyber-security, access control, security modeling, security in social networks, digital rights management, blockchain in the Internet of things (IoT), blockchain in cyber-physical systems, blockchain in social networking, blockchain in supply chain management, and cryptocurrency.

This volume contains a total of 44 distinct chapters authored by various researchers worldwide. This volume contains the research articles which are classified into three thrust areas, such as security and privacy, network security and its applications, and blockchain technology and IoT.

The first part "Security and Privacy" consists of the nineteen (19) research articles, contributed by the allied researchers, principally illustrates the machine learning-based identification of DDoS flood attack, phishing E-mail detection, improvement of a mutual authentication, collision avoidance and drowsiness detection of vehicles, face mask detection using CNN, and finds out the IoT-DDoS attack.

The second part "Network Security and Its Applications" consists of nine (09) articles mainly focused on the classification of brute-force attacks using CNN, malware analysis, asymptotic diffusion analysis, enhancement of data security for cloud computing, and increases the network security using genetic algorithm.

The third part "Blockchain Technology and IoT" consists of sixteen (16) original research articles and typically highlights the food supply chain for IoT, development of IoT-based biometric attendance system, protecting the privacy of IoMT, blockchain-based crowdfunding platform, and blockchain-based transparent solution for achieving investment for farming.

The conference received articles from more than ten countries with more than ten foreign authors. We are grateful to the Springer Nature for publishing the accepted and presented original research papers of ICNSBT 2023 in the "Lecture Notes in Networks and Systems" (LNNS), Springer Nature. On behalf of the organizing committee, we are grateful to the inaugurators, keynote presenters, and eminent experts who delivered the expert talks. Our sincere gratitude to the esteemed authors and reviewers for extending support and cooperation. Hope this volume will be a valuable document for the researchers and budding engineers.

Raiganj/Kalyani, India — Jyotsna Kumar Mandal
Midnapore, India — Biswapati Jana
Taichung, Taiwan — Tzu-Chuen Lu
Kolkata, India — Debashis De

Contents

Network Security and Its Applications

Blockchain Technology and IoT

Editors and Contributors

About the Editors

Jyotsna Kumar Mandal completed M.Tech. (Computer Science, University of Calcutta) and Ph.D. (Engg., Jadavpur University) in the field of Data Compression and Error Correction Techniques and is Professor of Computer Science and Engineering, University of Kalyani, India; Former Director, IQAC, Kalyani University; Life Member of CSI, CRSI; Associate Member ACM, IEEE; Fellow Member of IETE; Former Dean Faculty of Engineering, Technology and Management, working in the field of Network Security, Steganography, Remote Sensing and GIS Application, and Image Processing; 35 years of teaching and research experiences; twenty-eight scholars were awarded Ph.D., four submitted, and eight are pursuing; the total number of publications is more than four hundred in addition of publication of twelve books from LAP Lambert, Germany, and IGI Global; organized 55 conferences of Springer Nature, Elsevier, IEEE, etc.; edited more than 55 proceedings as corresponding editor/editors; Member of NAAC Peer Team and AICTE Expert Team of EVC; Academic auditor for universities and institutes; delivered more than 100 expert lectures across the globe; Former Member of Governing Council, IETE, Delhi; and received Siksha Ratna Award from Government of West Bengal for outstanding performance of teaching and research.

Biswapati Jana completed M.Tech. (Computer Science and Engineering, University of Calcutta) and Ph.D. (Computer Science, Vidyasagar University) in the field of Data Hiding Techniques; Associate Professor in Computer Science, Vidyasagar University, India; 20 years of teaching and research experiences; four scholars awarded Ph.D. and six are pursuing; the total number of publications is more than hundred; and delivered more than 25 expert lectures across the globe. He served as Reviewer for a good number of international journals and conferences. His research interest includes Data Hiding, Image Processing, Data Security, Steganography, and Watermarking.

Tzu-Chuen Lu received the B.M. degree (1999) and MSIM degree (2001) in Information Management from Chaoyang University of Technology, Taiwan. She received her Ph.D. degree (2006) in Computer Engineering from National Chung Cheng University. Her current title is Professor in Department of Information Management at Chaoyang University of Technology.

Debashis De is Professor in Department of Computer Science and Engineering at Maulana Abul Kalam Azad University of Technology, West Bengal, India. He received M.Tech. from the University of Calcutta, 2002, and a Ph.D. from Jadavpur University in 2005. He is Senior Member, IEEE; Fellow, IETE; and Life Member, CSI. He was awarded the prestigious Boyscast Fellowship by the Department of Science and Technology, Government of India, to work at the Heriot-Watt University, Scotland, UK. He received the Endeavour Fellowship Award from 2008–2009 by DEST Australia to work at the University of Western Australia. He received the Young Scientist Award both in 2005 at New Delhi and in 2011 in Istanbul, Turkey, from the International Union of Radio Science, Belgium. In 2016, he received the J. C. Bose Research Award by IETE, New Delhi. In 2019, he received Shiksha Ratna Award by the Government of West Bengal. He established the Center of Mobile Cloud Computing (CMCC) for IoT applications. He is Vice Chair of Dew Computing STC of IEEE Computer Society. He published in 320 journals and 200 conference papers, fifteen books, and filed ten patents. His h-index is 36; citation 6300; and listed in Top 2% Scientist List of the world, Stanford University, USA. His research interest is Cloud, IoT, and Quantum Computing.

Contributors

Subhajit Adhikari Assistant Professor, BSH Department, Institute of Engineering and Management, University of Engineering and Management, Kolkata, India;
Research Scholar, Department of Computer Science, University of Burdwan, Burdwan, India

C. Ambika Department of Mathematics, Ethiraj College for Women, Chennai, India

Fatima Mohammad Amin Vellore Institute of Technology, Vellore, India

N. Anitha Head-Talent Development, 1.Ms.N.Anitha, Head-TalPixelExpert Technology & Services Pvt. Ltd., Chennai, India

S. Anitha Department of Information Technology, Kongu Engineering College, Erode, Tamilnadu, India

O. N. Arunkumar Symbiosis Centre for Management Studies (SCMS), Symbiosis International (Deemed University) (SIU), Bengaluru, Karantaka, India;
Symbiosis Institute of Business Management (SIBM), Bengaluru, India

N. Ashok Department of Information Technology, Kongu Engineering College, Erode, Tamilnadu, India

Arpan Baiagi Department of Computer Science, Ramakrishna Mission Residential College, Narendrapur, Calcutta University, Kolkata, India

B. Balamurugan Velammal Institute of Technology, Chennai, Tamilnadu, India

Siddhartha Banerjee Department of Computer Science, Ramakrishna Mission Residential College, Narendrapur, Calcutta University, Kolkata, India

Arkojeet Bera National Institute of Technology Karnataka, Surathkal, Mangalore, Karnataka, India

Debkumar Bera Department of Computer Science, Vidyasagar University, West Midnapore, West Bengal, India

Rahul Deb Bhandari Department of Computer Science, Ramakrishna Mission Residential College, Narendrapur, Calcutta University, Kolkata, India

K. Bharadwaj Department of CSE, SCSVMV (Deemed to be University), Tamilnadu, India

Medhavi Bhargava School of Engineering and Technology, SAGE University, Bhopal, India

Debnandan Bhattacharyya Department of Electronics and Communication Engineering, St. Thomas' College of Engineering and Technology, Kolkata, India

Srikakulapu Bhavitha Department of Information Technology, Velagapudi Ramakrishna Siddhartha Engineering College, Vijayawada, Andhra Pradesh, India

Snehal Bhosale Department of E&TC, Symbiosis Institute of Technology (SIT), Symbiosis International (Deemed University) (SIU), Lavale, Pune, Maharashtra, India

Swarup Kumar Bhunia Department of Commerce, Rishi Bankim Chandra Evening College affiliated to West Bengal State University, Naihati, West Bengal, India

Anindya Bose Department of Computational Science, Brainware University, Calcutta, West Bengal, India

Aratrik Bose Computer Science and Engineering, Dr. B.C. Roy Engineering College, Durgapur, West Bengal, India

Rajesh Bose Department of Computational Science, Brainware University, Calcutta, West Bengal, India

Sourav Kr. Burnwal Department of Electronics and Communication Engineering, St. Thomas' College of Engineering and Technology, Kolkata, India

Mebanphira Cajee Department of Computer Science and Engineering, Symbiosis Institute of Technology (SIT), Symbiosis International (Deemed University) (SIU), Lavale, Pune, Maharashtra, India

Kunal Chakate Department of Computer Science and Engineering, Symbiosis Institute of Technology (SIT), Symbiosis International (Deemed University) (SIU), Lavale, Pune, Maharashtra, India

Subhalaxmi Chakraborty University of Engineering and Management, Kolkata, India

V. Chandrasekaran Department of Medical Electronics, Velalar College of Engineering and Technology, Erode, Tamilnadu, India

Kamlesh Chandravanshi School of Computing Science and Engineering, VIT Bhopal University, Sehore, Madhya Pradesh, India

K. S. Chandru Department of CSBS, Bannari Amman Institute of Technology, Sathyamangalam, India

Vinaykumar Chikaraddi Department of BCA, BLDEA's A.S. Patil College of Commerce (Autonomous), Vijayapura, India

Vishwanath Chikkareddi B.L.D.E.A's V.P. Dr. P.G. Halakatti College of Engineering and Technology, Vijayapura, India

Santosh Chinchali B.L.D.E.A's V.P. Dr. P.G. Halakatti College of Engineering and Technology, Vijayapura, India

Ayushya Chitransh DL Unify, DLT Labs, Hyderabad, India

Pravir Chitre Galgotias University, Greater Noida, UP, India;
Bhai Parmanand DSEU Shakarpur Campus II, Galgotias University, Delhi, India

Sahil Chorghe Sardar Patel Institute of Technology, Mumbai, India

Partha Chowdhuri Department of Computer Science, Vidyasagar University, WB, India

Ananya Roy Chowdhury Bangladesh Army International University of Science and Technology, Cumilla, Bangladesh

Prasun Chowdhury Department of Electronics and Communication Engineering, St. Thomas' College of Engineering and Technology, Kolkata, India

T. Daranya Department of Information Technology, Kongu Engineering College, Erode, Tamilnadu, India

Puja Das Department of Computer Science, HMM College for Women, Dakshineswar, Kolkata, India

Rajarshi Das University of Engineering and Management, Kolkata, India

Ritaban Das Department of Electronics and Communication Engineering, St. Thomas' College of Engineering and Technology, Kolkata, India

Asish Debnath Vidyasagar University, Midnapore, West Bengal, India

S. Deepika School of Computer Science and Engineering, Vellore Institute of Technology, Chennai, India

Ashis Dey Department of Computer Science, Silda Chandra Sekhar College, WB, India

Keval Dhanani Sardar Patel Institute of Technology, Mumbai, India

Rajesh Dhumal Symbiosis Institute of Geoinformatics (SIG), Symbiosis International (Deemed University) (SIU), Pune, Maharashtra, India

D. Divya Symbiosis Centre for Management Studies (SCMS), Symbiosis International (Deemed University) (SIU), Bengaluru, Karantaka, India

Saraswati Dutta Department of Mathematics, Midnapore College [Aotonomus], Midnapore, West Bengal, India

Jyotsna Gaikwad Deogiri College, Aurangabad, Maharashtra, India

Sonali Gaikwad Shree Shivaji Science and Arts College, Chikhli, Maharashtra, India

V. Geetha Department of CSE, SCSVMV (Deemed to be University), Tamilnadu, India

Bibek Ranjan Ghosh Department of Computer Science, Ramakrishna Mission Residential College, Narendrapur, Calcutta University, Kolkata, India

Nandana Ghosh Department of Computer Science, Vidyasagar University, Midnapore, West Bengal, India

Ritam Ghosh ACM Student Member, Kolkata, India

Subha Ghosh University of Engineering and Management, Kolkata, India

Debasis Giri Department of Information Technology, Maulana Abul Kalam Azad University of Technology, Kolkata, West Bengal, India

Govinda Giri Department of Computer Science and Engineering, Symbiosis Institute of Technology (SIT), Symbiosis International (Deemed University) (SIU), Lavale, Pune, Maharashtra, India

Anand Godbole Sardar Patel Institute of Technology, Mumbai, India

C. K. Gomathy Department of CSE, SCSVMV (Deemed to be University), Tamilnadu, India

Deepsubhra Guha Roy IEM Centre of Excellence for Cloud Computing and IoT, Department of CSE (AIML), Institute of Engineering and Management, Kolkata, India

Barnali Gupta Banik DL Unify, DLT Labs, Hyderabad, India

Alok Haldar Department of Computer Science, Kharagpur College, Kharagpur, West Bengal, India;
Department of Computer Science, Vidyasagar University, Midnapore, West Bengal, India

Amrita Haldar Department of Computer Science and Business Studies, IEM, Kolkata, India

Alok Halder Department of Computer Science, Khragpur College, West Midnapore, West Bengal, India

Nishat Tasnim Haque Bangladesh Army International University of Science and Technology, Cumilla, Bangladesh

Vemula Harish Jawaharlal Nehru TechnologicalUniversity, Hyderabad (JNTUH), Telangana, India

S. Hashwanth Vellore Institute of Technology, Chennai, India

Abhijeetsinh Jadeja Department of Computer Science, Shri C.J Patel College of Computer Studies (BCA), Sankalchand Patel University, Visnagar, India

Biswapati Jana Department of Computer Science, Vidyasagar University, West Midnapore, West Bengal, India;
Department of Computer Science, Kharagpur College, Kharagpur, West Bengal, India

Manasi Jana Department of Computer Applications, Haldia Institute of Technology, Haldia, West Bengal, India

Sharmistha Jana Department of Mathematics, Midnapore College [Autonomous], Midnapore, West Bengal, India

Shubhankar Joardar Department of Computer Science and Engineering, Haldia Institute of Technology, Haldia, West Bengal, India

Anshuman Kalla Department of Computer Engineering, CGPIT, Uka Tarsadia University, Bardoli, Gujarat, India

M. Kameswari Kalasalingam Academy of Research and Education, Chennai, Tamilnadu, India

Soumen Kanrar Amity University, Jharkhand, India;
Vlenzor Technologies Pvt. Ltd., Kolkata, India

Priyanshu Kapadia Department of Computer Engineering, CGPIT, Uka Tarsadia University, Bardoli, Gujarat, India

Sunil Karforma Dean(Science) Faculty, Department of Computer Science, The University of Burdwan, Burdwan, India

Parag Ravikant Kaveri Symbiosis Institute of Computer Studies and Research, Symbiosis International (Deemed University), Pune, India

Pooja Khanna Amity University Uttar Pradesh, Lucknow, India

S. Kirthica Department of Computer Science and Engineering, College of Engineering Guindy, Anna University, Chennai, India

Sonali Kothari Department of Computer Science and Engineering, Symbiosis Institute of Technology (SIT), Symbiosis International (Deemed University) (SIU), Lavale, Pune, Maharashtra, India

S. Kranthi Department of Information Technology, Velagapudi Ramakrishna Siddhartha Engineering College, Vijayawada, Andhra Pradesh, India

Srinivasan Krishnaswamy Indian Institute of Technology Guwahati, Assam, India

Pankaj Kumar Srinivasa Ramanujan Department of Mathematics, Central University of Himachal Pradesh, Dharamshala, H.P., India

Sachin Kumar Amity University Uttar Pradesh, Lucknow, India

Kousik Kundu Department of Mathematics, Midnapore College [Aotonomus], Midnapore, West Bengal, India

Pritam Kundu University of Engineering and Management, Kolkata, India

Sumana Kundu Computer Science and Engineering, Dr. B.C. Roy Engineering College, Durgapur, West Bengal, India

Nilesh Kunhare School of Computing Science and Engineering, VIT Bhopal University, Sehore, Madhya Pradesh, India

Chidanand Kusur B.L.D.E.A's V.P. Dr. P.G. Halakatti College of Engineering and Technology, Vijayapura, India

Prathamesh Vijay Lahande Symbiosis Institute of Computer Studies and Research, Symbiosis International (Deemed University), Pune, India

G. Lakshman Department of CSE, SCSVMV (Deemed to be University), Tamilnadu, India

Tzu Chuen Lu Department of Information Management, Chaoyang University of Technology, Taichung, Taiwan, ROC

A. Uma Maheswari PG & Research Department of Mathematics, Quaid-E-Millath Government College for Women (Autonomous), Chennai, India

Soumyadip Maity Ramakrishna Mission Vidyamandira, Howrah, West Bengal, India

Anandaprova Majumder Computer Science and Engineering, Dr. B.C. Roy Engineering College, Durgapur, West Bengal, India

Jyotsna Kumar Mandal Department of Computer Science and Engineering, Kalyani University, Kalyani, India

S. Maragathasundari Kalasalingam Academy of Research and Education, Chennai, Tamilnadu, India

Sudipta Meikap Department of Computer Science, Hijli College, Paschim Medinipur, West Bengal, India;
Department of Computer Science, Vidyasagar University, Midnapore, West Bengal, India

Raaj Anand Mishra Dell EMC, Bangalore, India

Pinaki Mitra Indian Institute of Technology Guwahati, Assam, India

Prachi Mohite Department of Computer Science and Engineering, Symbiosis Institute of Technology (SIT), Symbiosis International (Deemed University) (SIU), Lavale, Pune, Maharashtra, India

Uttam Kr. Mondal Vidyasagar University, Midnapore, West Bengal, India

Megh Naik Department of Computer Engineering, CGPIT, Uka Tarsadia University, Bardoli, Gujarat, India

B. Nandhini Department of Information Technology, Kongu Engineering College, Erode, Tamilnadu, India

Subrata Nandi Narula Institute of Technology, Agarpara, West Bengal, India;
Indian Institute of Technology Guwahati, Assam, India

Pabitra Pal Department of Computer Applications, Maulana Abul Kalam Azad University of Technology, Nadia, West Bengal, India

A. Pandiaraj Department of Computing Technologies, SRM Institute of Science and Technology, Chennai, India

Darshanaben Dipakkumar Pandya Department of Computer Science, Shri C.J Patel College of Computer Studies (BCA), Sankalchand Patel University, Visnagar, India

Chanchal Patra Department of Information Technology, Maulana Abul Kalam Azad University of Technology, Kolkata, West Bengal, India

Soumya Paul St. Mary's Technical Campus Kolkata, Barasat, West Bengal, India

Anil Audumbar Pise Department of Data Science and Machine Learning Computer Science, University of the Witwatersrand, Johannesburg, South Africa

Pragya MVD College, Lucknow, India

Asis Prasad Department of Electronics and Communication Engineering, St. Thomas' College of Engineering and Technology, Kolkata, India

Yogesh Rajput Symbiosis Institute of Geoinformatics (SIG), Symbiosis International (Deemed University) (SIU), Pune, Maharashtra, India

Dirun Reddy Department of Computer Science and Engineering, Symbiosis Institute of Technology (SIT), Symbiosis International (Deemed University) (SIU), Lavale, Pune, Maharashtra, India

Saha Reno Bangladesh Army International University of Science and Technology, Cumilla, Bangladesh

Sandip Roy Department of Computational Science, Brainware University, Calcutta, West Bengal, India

Satyabrata Roy Department of Computer Science and Engineering, Manipal University Jaipur, Jaipur, Rajasthan, India

Shovan Roy Department of Mathematics, Midnapore College [Aotonomus], Midnapore, West Bengal, India

Adapaka Sai Kishore Department of Information Technology, Velagapudi Ramakrishna Siddhartha Engineering College, Vijayawada, Andhra Pradesh, India

Nguyen Kim Sao Department of Computer Science, University of Transport and Communication, Hanoi, Vietnam

Piyush Sharma Srinivasa Ramanujan Department of Mathematics, Central University of Himachal Pradesh, Dharamshala, H.P., India

Rohan Shinde Sardar Patel Institute of Technology, Mumbai, India

Afrin Siddiqui Amity University Uttar Pradesh, Lucknow, India

Moutushi Singh Department of Information Technology, IEM, Kolkata, India

Prabhash Kumar Singh Department of Computer Science, Vidyasagar University, Midnapore, West Bengal, India

Debarpito Sinha Ramakrishna Mission Vidyamandira, Howrah, West Bengal, India

Gaurav Soni School of Computing Science and Engineering, VIT Bhopal University, Sehore, Madhya Pradesh, India

R. Sridevi Department of Computer Science and Engineering, JNTUH University College of Engineering, Science and Technology, Hyderabad, Telangana, India; Jawaharlal Nehru TechnologicalUniversity, Hyderabad (JNTUH), Telangana, India

Srinivasan Sriramulu Galgotias University, Greater Noida, UP, India

Vijayan Sugumaran School of Business Administration, Oakland University, Rochester, MI, USA

C. Swedheetha Vaigai College of Engineering, Madurai, Tamilnadu, India

T. C. Swetha Priya Department of Computer Science and Engineering, JNTUH University College of Engineering, Science and Technology, Hyderabad, Telangana, India

Zerin Tasnim Bangladesh Army International University of Science and Technology, Cumilla, Bangladesh

Garima Thakur Srinivasa Ramanujan Department of Mathematics, Central University of Himachal Pradesh, Dharamshala, H.P., India

Lu Tzu-Chuen Department of Information Management, Chaoyang University of Technology, Taichung, Taiwan, ROC

R. Vanalakshmi Kalasalingam Academy of Research and Education, Chennai, Tamilnadu, India

R. Venkatesan Department of IT, Bannari Amman Institute of Technology, Sathyamangalam, India

K. P. Vijayakumar School of Computer Science and Engineering, Vellore Institute of Technology, Chennai, India

G. Vimalsubramanian Department of CSE, Kalasalingam Academy of Research and Education, Srivilliputhur, India

Thanh Nhan Vo Department of Information Management, Chaoyang University of Technology, Taichung, Taiwan, R. O. C.

Security and Privacy

Machine Learning-Based Identification of DDoS Flood Attack in eHealth Cloud Environment

Anindya Bose, Sandip Roy, and Rajesh Bose

1 Introduction

Use of cloud computing in health care allows easy access to patients' health records, thus enabling medical professionals to analyze the patients' health issues and provide diagnosis even from remote locations. As a central repository for medical research, disease control, and epidemic monitoring, the cloud can be used to support national health policies. However, there are certain limitations. Since cloud is a shared platform it poses security and privacy threat.

The availability of data in critical circumstances is an important aspect of the eHealth system that is often overlooked, including the ability to continue operations despite mischief by some authorities and the capability to continue operations despite a possible security gap. Power outages, hardware failures, system upgrades, and distributed denial-of-service (DDoS) attacks should not cause service interruptions. The most extensive of the attacks on the healthcare cloud is the DDoS attack that is capable of crippling the network and avoids the access to any delicate data. During a distributed denial-of-service attack (DDoS), patient data may not always be available as well as data may be unable to flow across the internetworking system. A DDoS attack can alter a portal's functionality by restricting access to information. It is not just about breaching data, but also about restraining users from accessing information.

A. Bose · S. Roy (✉) · R. Bose (✉)
Department of Computational Science, Brainware University, Calcutta, West Bengal 700124, India
e-mail: sandiproy86@gmail.com

R. Bose
e-mail: bose.raj00028@gmail.com

J. K. Mandal et al. (eds.), *Proceedings of International Conference on Network Security and Blockchain Technology*, Lecture Notes in Networks and Systems 738,
https://doi.org/10.1007/978-981-99-4433-0_1

1.1 DDoS Attack in Health Care

When a network is overwhelmed by traffic caused by distributed denial-of-service attacks (DDoS), it becomes unusable. DDoS assaults frequently serve as a diversion, while malicious actors introduce more dangerous software onto the network of their victims. A DDoS assault might prevent access to vital services for health care, including bed capacity, data exchange, and scheduling appointments. Many ransomware groups launch DDoS attacks at the public-facing websites of the organizations. This can pose a risk for healthcare industry where unavailability of service during the emergency can be fatal for the patients. A DDoS assault happens when several machines collaborate to attack a single target. Cyber-attackers typically utilize botnets, a collection of hacked devices linked to the internet, to exploit device weaknesses and seize control. Once in charge, the assailant can direct the botnet to launch a DDoS assault on the target. DDoS assaults increase the attack strength by allowing for a lot more responses to be issued than a typical DoS attack [1]. DDoS may drastically lower the performance of cloud services in a healthcare cloud setting by harming the virtual servers.

In 2014, Boston Children's Hospital suffered a DDoS attack which caused the loss of the internet connection for everyone else on that network, including Harvard University and all of its hospitals. Some patients and medical staff were incapable of using their online accounts to verify appointments, test results, and other case information during the almost week-long network failures. The hospital incurred a loss of $300,000 [2]. Though the DDoS attack originates from outside the cloud environment, it targets the cloud resources and floods them with heavy traffic [3]. According to Cloud Security Alliance, DDoS is one of the major threats to cloud security environment. Different categories of DDoS attack are shown in Fig. 1. In this work, the authors concentrated on the TCP-SYN or TCP-SYN-ACK flood attack on health cloud environment as this kind of attack quickly consumes all the necessary resources and renders the communication useless.

1.2 TCP-SYN FLOOD Attack

DDOS attacks are more detrimental in clouds because they run client services inside virtual machines, unlike traditional networks. The flood attack will affect all virtual machines in a shared environment if one virtual machine becomes overloaded. TCP uses three-way handshake to set up a connection between the client and the server. The clients send a SYN message to the server to start a connection. The server then replies with a SYN-ACK message. So to begin exchanging data, the client will send an ACK message after receiving the SYN-ACK message. Attackers initiate their attacks and stop the server from serving genuine users by taking advantage of a flaw in this three-way handshake procedure.

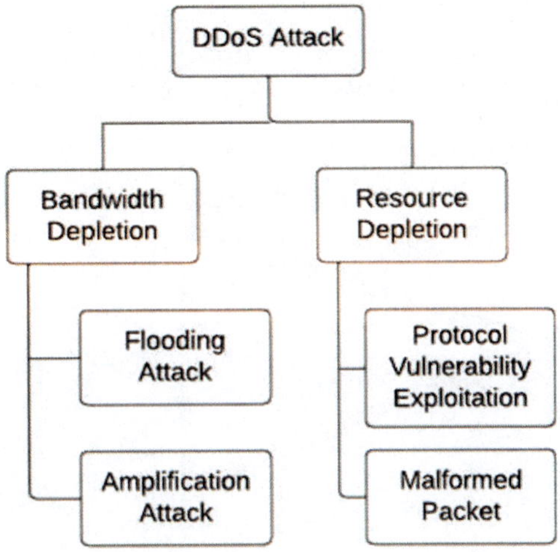

Fig. 1 Categorization of different types of DDoS attack

In a SYN flood attack, the intruder repeatedly transmits SYN packets to all of the server's ports while frequently utilizing a fictitious IP address. The server receives several messages to establish contact that seem to be valid, but it is oblivious of the assault. Each open port sends a SYN-ACK packet in response to each attempt. If the IP address is faked, the malicious client either never gets the SYN-ACK in the first place or fails to transmit the required ACK. In any case, the server that is being attacked will wait a while for the acknowledgment of its SYN-ACK message. A large number of SYN packets keep on arriving while the connection remains half-open waiting for the ACK. These half-open connections remain in the backlog queue which has a finite length. When this length is exceeded the all incoming requests are rejected and the service is blocked. Both the conditions are depicted in Fig. 2a and b.

1.3 Background of the Study

DDoS attacks in the first quarter of 2021 alone were approximately 2.9 million, and it was a 31% increase compared to the previous year. DDoS attacks estimate on quarterly basis and distribution of DDoS attacks by month are shown in Fig. 3a and b. An intrusion detection system is required to identify any abnormal traffic. Detection of the flood attack can be signature-based, behavior-based, or hybrid, i.e., a combination of the two types. The authors in this work opted for signature-based detection since it is easier to implement as well as cost-effective. There are many techniques for flood attack detection, viz.

i. The past understanding of network flow serves as the foundation for the statistical identification techniques. However, fraudulent network flows are evolving as a target in the modern world. Therefore, accurately describing network traffic is a

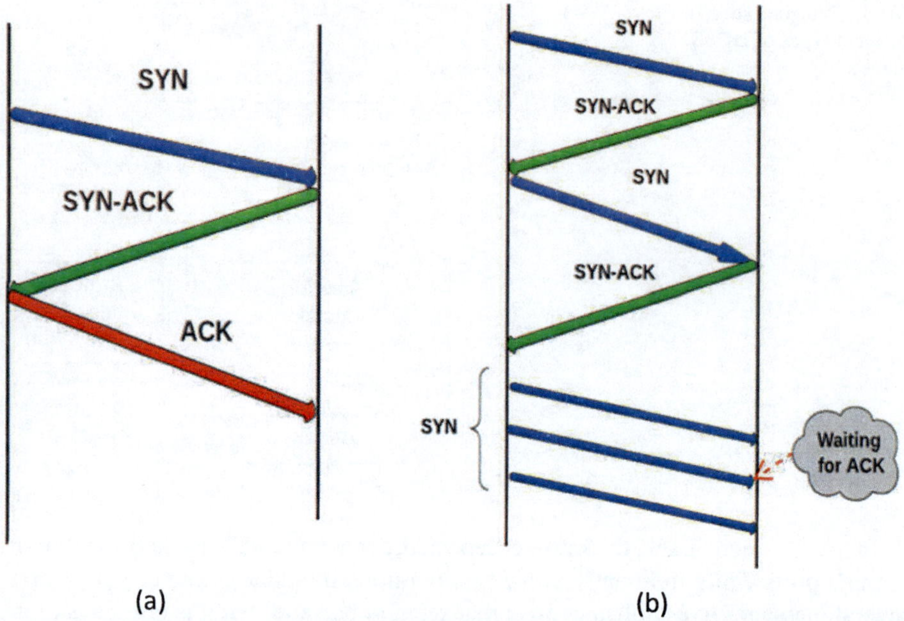

Fig. 2 **a** Normal three-way handshake of the TCP protocol and **b** TCP-SYN FLOOD situation where the server is sapped with numerous SYN packets

difficult undertaking. The majority of statistical DDoS detection techniques rely heavily on different consumer criteria. To keep up with changes in a network, such criteria must be changed dynamically. To choose the appropriate statistical features for the entropy measure of statistical approaches, substantial network expertise and experiments are required. The majority of statistical techniques used to identify SYN flood attacks, use entropy, correlation, etc. which have a lot of computation complexity. Consequently, they cannot be carried out in real time [4].

ii. By applying the rules to a little quantity of data, machine learning functions well. The ML first assesses the statistical properties before classifying or valuing them. Additionally, the model must be updated often to reflect changes in assaults. The SML techniques address the issue by decomposing it into manageable sub-problems, addressing those sub-problems, and providing the solution as a whole.

Only a handful of studies have suggested high-rate online detection at the onset of the attack. To create detection model that can detect the SYN flood attack within the first minute, this study made use of several machine learning-based classification techniques. These proposed models were assessed by the authors using several statistical measures.

The rest of the paper is arranged as follows: Sect. 2 describes some of the other research works in the related field; Sect. 3 describes the methodology of the proposed

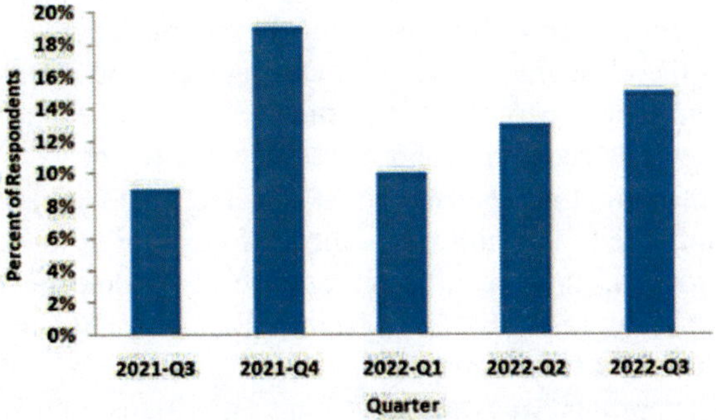

(a)

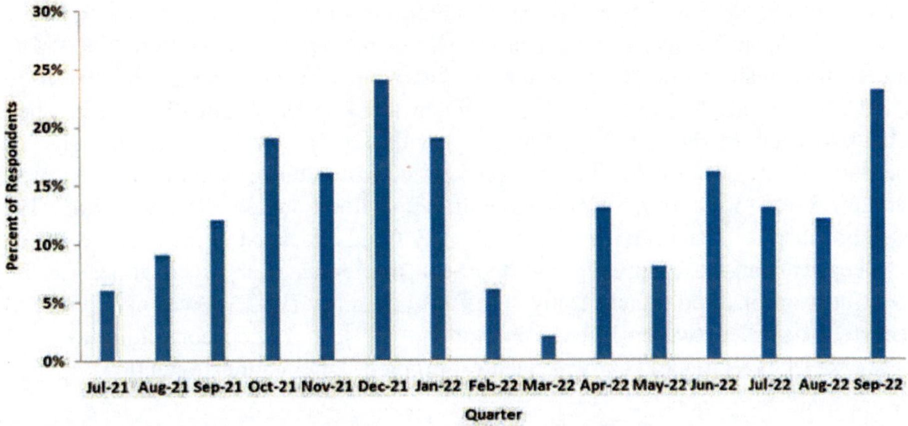

(b)

Fig. 3 **a** Quarter-wise distribution of DDoS attack July 2021–September 2022 and **b** month-wise DDoS attack for the same period [21]

work; result and discussion are given in Sect. 4; and Sect. 5 concludes the current study with some future work prospect.

2 Related Work

Anomaly intrusion detection was proposed by Rawashdeh et al. [5] as a method of reducing DDoS performances among virtual machines in the hypervisor layer. An evolutionary neural network is used to develop the proposed detection method. Evolutionary neural networks incorporate particle swarm optimization (PSO) with neural networks for analysis of traffic data for DDoS attacks. This model detected both UDP and TCP-SYN FLOOD attack with high accuracy.

In a public clouds environment, Sahi et al. [6], to stop DDoS TCP flood attacks, a new classification-based detection system was created. DDoS detection was improved to safeguard stored records by classifying the incoming packets and making a choice according to the classification results. During the flood attack, Wireshark was used to analyze the network traffic. While the prevention phase is in place, the proposed detection methods recognize and establish which packets are regular or created by an attacker. Wani et al. [7] used SVM to detect flood attack in the cloud environment. They compared the result with random forest-based model and found that SVM produced a more accurate result. Kanimozhi et al. [8] detected DDoS attack in the hypervisor layer using radial basis function (RBF) with PSO. They achieved a high detection and classification accuracy.

A machine learning approach was used by Kemp et al. [9] to deploy the proposed model. In total, eight classification algorithms for prediction models were selected: random forest, decision trees, K-nearest neighbor, multilayer perceptron, RIPPER (JRip), SVM, and Naïve Bayes. The models were evaluated based on AUC-ROC curves, true positive rate (TPR), and false positive rate (FPR). Singh et al. [10] used MLP with genetic algorithm to detect DDoS attack in the application layer. They took into account the number of requests within a time window and the size of the data packet because fixed-length packets indicate attack. Lima et al. [11] used random forest to identify various types of DDoS flood attacks. Sreeram et al. [12] used bio-inspired machine learning to identify flood attacks at an early stage.

Deep learning-based models too had been implemented in recent times for the identification of flood attack. Long short-term memory (LSTM) was used to build recurrent neural network (RNN)-based model [13], and deep convolution neural network (DCNN) model [14] was used for efficient DDoS attack detection.

3 Methodology

In this work the authors used Vellore Institute of Technology (VIT) cloud security research dataset. These DDoS data were generated using virtual instances running in a private cloud security testbed. Creators of the dataset deployed Open Stack private cloud, and the virtual instances that run on it are included in the cloud security testbed. The virtual instances of the Open Stack private cloud are behaving like zombies from various public and private networks. The DDoS flood attack in the experiment was annotated as TCP-SYN FLOOD and TCP-SYN NORMAL [15]. The dataset contains 327,000 instances with 20 attributes. The DDoS flood attacks were simulated on two different dates, each date with two timestamps with 1 min difference. The protocols used were TCP, SSH, ICMP, IPv4, and ARP. Figure 4a depicts the cumulative bytes—the blue area is the TCP-SYN FLOOD attack in 1 min window, and the red area marks the normal traffic, and Fig. 4b shows the number of packets generated both under normal circumstances and the flood attack.

The features of the data are: *Date, UTC Time, Time, Relative Time, Absolute Time, Delta Time, Source, Destination, Protocol, Length, Source Port, Destination Port,*

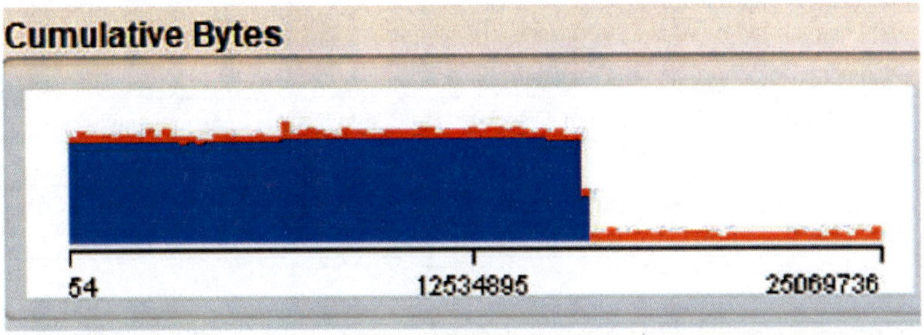

(a)

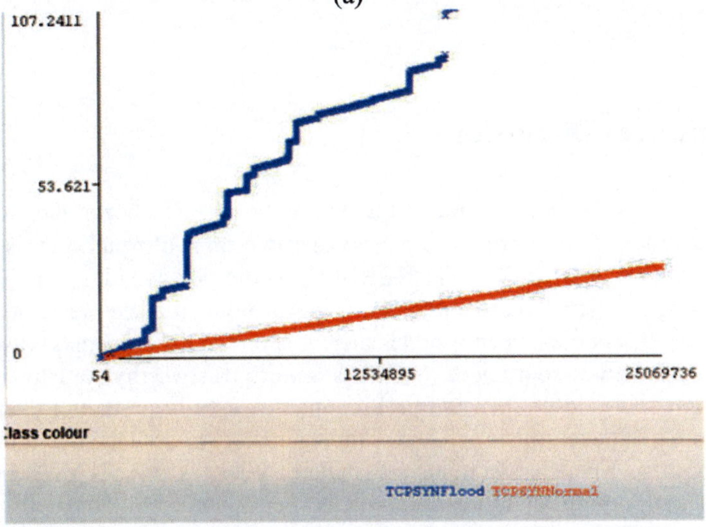

(b)

Fig. 4 **a** Cumulative bytes in 1 min window and **b** the number of packets generated in 1 min window. Blue is the TCP-SYN FLOOD attack, and the red marks the normal traffic

Cumulative Bytes, Hardware destination address, Hardware source address, Unresolved destination port, Unresolved source port, Network source address, Network destination address, Expert info, and Class.

3.1 Classification Using Machine Learning

The authors used multilayer perceptron (MLP), random forest (RF), and support vector machine (SVM) to identify the data traffic as flood attack or normal. Classifiers were trained using tenfold cross-validation. The training dataset is split into ten subgroups of equal size. One subgroup was used for the testing of the model that is trained with the rest of the nine subgroups. This process was repeated ten times, thus allowing each subgroup to be used once as test dataset. The ultimate performance of

Table 1 Hyper-parameters of the classifiers

Machine learning algorithm	Hyper-parameters
Multilayer perceptron (MLP)	Batch size: 100, Hidden layers: 2, Learning rate: 0.4
Random forest (RF)	Batch size: 100, No. of iterations 100, seed: 01
Support vector machine (SVM)	Batch size: 100, Cache size: 40, $\epsilon = 0.001$, $\gamma = 0$, Kernel type = linear

the model is the average performance of each validation set. Hyper-parameters are listed in Table 1.

4 Results and Discussion

Accuracy is the usual metric for the evaluation of the performance of the model if the dataset is symmetric. However, in the class distribution is unequal. In this work the authors have considered TCP-SYN NORMAL as the positive class and TCP-SYN FLOOD as the negative class. Precision is about how sure we are about the true positive (TP). F-score has been used because it is best for uneven class distribution. Matthew's correlation coefficient (MCC) is a more trustworthy statistical rate that yields a high score only when the prediction successfully predicts in each of the four categories of the confusion matrix, proportionately to both the magnitude of the dataset's positive and negative items [16]. The confusion matrix is given in Table 2. Performance of the four classification models is evaluated using the metrics such as accuracy, true positive (TP), false positive (FP), specificity, precision, and F-score. This is shown in Table 3, and a comparison graph is shown in Fig. 5.

Though all the classifiers produced high accuracy, this value does not reflect the true performance measure of the classification model since the dataset is not symmetric. Statistical measures, in case of such bias, may produce dangerously inflated result. Hence, the authors included MCC which indicates that the MLP-based classification model performed better than the SVM or RF in distinguishing between normal traffic from the flood attack.

Table 2 Confusion matrix

Machine learning algorithm		TCP-SYN FLOOD	TCP-SYN NORMAL
MLP	TCP-SYN FLOOD	288,933	1067
	TCP-SYN NORMAL	1158	35,842
RF	TCP-SYN FLOOD	276,820	13,180
	TCP-SYN NORMAL	3673	33,327
SVM	TCP-SYN FLOOD	266,758	23,242
	TCP-SYN NORMAL	3821	33,179

Table 3 Statistical evaluation metrics

Machine learning algorithm	Accuracy (%)	Precision	False positive (FP)	False negative (FN)	True positive (TP)	F-score	MCC
MLP	99.32	0.9963	0.0289	0.0040	0.996	0.9962	0.9659
RF	94.85	0.9546	0.2834	0.0131	0.9869	0.9705	0.7757
SVM	91.73	0.9199	0.4109	0.0141	0.9859	0.9517	0.6846

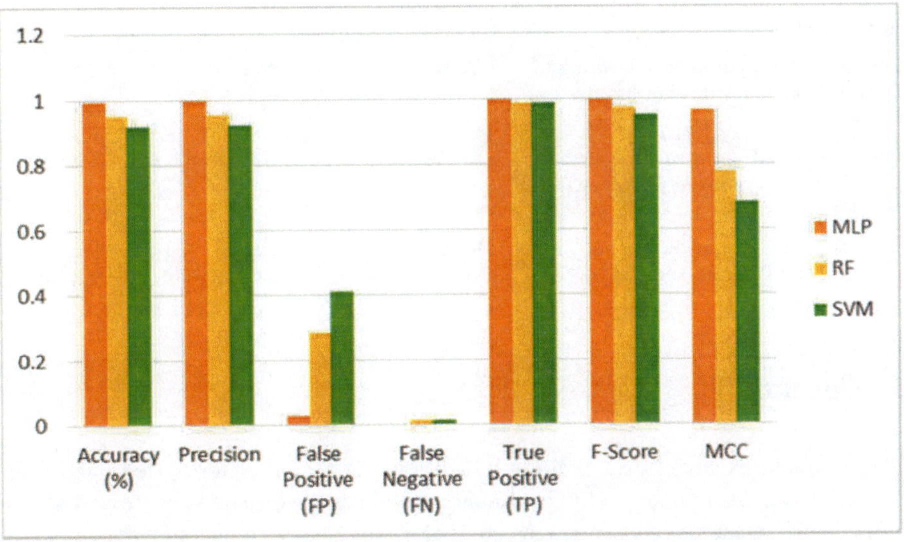

Fig. 5 Graphical representation of the statistical metrics of the performance of the different classifiers

4.1 *Comparison with Other Works*

Comparison with some of the previous works is given in Table 4. Compared to the other traditional machine learning-based models the proposed system produced a better accuracy, though the datasets used in each of these models have different set of features. Only the incremental learning model produced a result comparable to the proposed study.

Table 4 Comparison of the proposed method with other research works

Related work	Classification model	Dataset used	Accuracy (%)
Nawir et al. [17] detected network anomaly	MLP RBFN Naïve Bayes Averaged one-dependence estimator (AODE)	UNSW-NB15	89.76 84.41 76.12 97.26
Sahi et al. [6] detected DDoS flood attack in public cloud	LS-SVM	NA	97
Hwang et al. [18] used deep learning to detect DDoS attack on high-profile websites	CNN	NA	98.8
Novaes et al. [19] proposed a system to characterize and mitigate DDoS attack in the cloud environment	LSTM-fuzzy	NSL-KDD	95.77
Bamasag et al. [20] worked to mitigate real-time DDoS attack in cloud environment	Incremental learning	DDoS-2020 NSL-KDD	99.39 99.3

5 Conclusion

DDoS attack, especially SYN flood attack, causes system interruptions and non-availability of data which in the healthcare cloud can give rise to severe and sometimes life-threatening consequences. It is thus necessary to identify the attack at the onset. The authors in this work concentrated on distinguishing a SYN flood attack from normal packet flow using machine learning-based model. Three machine learning MLPs were found to give the highest accuracy of 99.32%. The contribution of this work is twofold—the attack was identified within the first 1 min of onset and distinguishing normal traffic from attack using machine learning algorithm.

The work can be further improved in many ways. The model can be used to distinguish between different types of DDoS attacks. The model can also be tested on mobile cloud computing platform. In the future, the authors plan to build a deep learning-based model using recurrent neural network (RNN) to distinguish between various types of attacks.

References

1. Xtelligent Healthcare Media (2021) Xtelligent Healthcare. Xtelligent Healthcare website: https://healthitsecurity.com/. Accessed 12 Nov 2022
2. Center for Internet Security. https://www.cisecurity.org/insights/blog/ddos-attacks-in-the-healthcare-sector. https://www.cisecurity.org/, https://www.cisecurity.org/. Accessed 12 Nov 2022

3. Deshmukh RV, Devadkar KK (2015) Understanding DDoS attack & its effect in cloud environment. Proc Comput Sci 49:202–210
4. Hoque N, Kashyap H, Bhattacharyya DK (2017) Real-time DDoS attack detection using FPGA. Comput Commun 110:48–58
5. Rawashdeh A, Alkasassbeh M, Al-Hawawreh M (2018) An anomaly-based approach for DDoS attack detection in cloud environment. Int J Comput Appl Technol 57(4):312–324
6. Sahi A et al (2017) n efficient DDoS TCP flood attack detection and prevention system in a cloud environment. IEEE Access 5:6036–6048
7. Wani A et al (2019) Analysis and detection of DDoS attacks on cloud computing environment using machine learning technique. In: 2019 amity international conference on artificial intelligence (AICAI). IEEE, pp 870–875
8. Kanimozhi S, Radhika D (2022) Detection of DDos attack using machine learning algorithms in cloud computing. Turk Online J Qual Inquiry 13(1):2079–2088
9. Kemp C, Calvert C, Khoshgoftaar T (2018) Utilizing netflow data to detect slow read attacks. In: 2018 IEEE international conference on information reuse and integration (IRI). IEEE, Piscataway, New Jersey, pp 108–116
10. Singh S, Jeong YS, Park JH (2016) A survey on cloud computing security: issues, threats, and solutions. J Netw Comput Appl 75(7):200–222
11. Lima Filho FS et al (2019) Smart detection: an online approach for DoS/DDoS attack detection using machine learning. Sec Commun Netw 2019:1–15
12. Sreeram I, Vuppala VK (2019) HTTP flood attack detection in application layer using machine learning metrics and bio inspired bat algorithm. Appl Comput Inform 15(1):59–66
13. Lin P, Ye K, Xu CZ (2019) Dynamic network anomaly detection system by using deep learning techniques. In: International conference on cloud computing. Springer, Berlin, pp 161–176
14. Haider S et al (2020) A deep CNN ensemble framework for efficient DDoS attack detection in software defined networks. IEEE Access 8:53972–53983
15. Subbulakshmi T et al (2022) https://github.com/VITCCSCSEInformationSecurityGroup/VITCCSCSE-DDoS-Attack-Datasets. Accessed 7 Nov 2022
16. Chicco D, Jurman G (2020) The advantages of the Matthews correlation coefficient (MCC) over F1 score and accuracy in binary classification evaluation. BMC Genom 21(6):1–13
17. Nawir M et al (2019) Effective and efficient network anomaly detection system using machine learning algorithm. Bull Electr Eng Inform 8(1):46–51
18. Hwang RH et al (2020) An unsupervised deep learning model for early network traffic anomaly detection. IEEE Access 8:30387–30399
19. Novaes MP et al (2020) Long short-term memory and fuzzy logic for anomaly detection and mitigation in software-defined network environment. IEEE Access 8(1):1–17
20. Bamasag O et al (2022) Real-time DDoS flood attack monitoring and detection (RT-AMD) model for cloud computing. PeerJ Comput Sci 7:1–21
21. Yoachimik O (2022) Cloudflare DDoS threat report 2022 Q3. Cloudflare. August 20, 2023. https://blog.cloudflare.com/cloudflare-ddos-threat-report-2022-q3/

Machine Learning-Based Phishing E-mail Detection Using Persuasion Principle and NLP Techniques

Chanchal Patra and Debasis Giri

1 Introduction

Now email (electronic mail) plays a crucial role in everyone's life. It is one of the simplest and most efficient ways to transport files and messages.

Phishing attacks, which combine social engineering and technological fraud to obtain user private information, are now one of the cyberattacks that the fastest growing rates. This type of cybercrime involves contacting a target or targets via email and persuading them to provide confidential details including passwords, banking information, credit card details, and others information. The use of these details to access other people's crucial accounts might lead to identity theft and financial damage [1].

Phishing attacks come in a variety of forms. The sort of mass-mail phishing that is most prevalent is generic phishing. The term "spear-phishing" refers to phishing attempts that target certain groups. "Whaling" is a method of targeting a specific person, typically the boss of an organization. For performing this attack require considerable preparation. "Vishing" or voice phishing, is a different term for a phishing attempt that took place over the phone call, while phishing through SMS is known as "smishing" or SMS Phishing [1].

The annual report for 2021 has been issued by the Internet Crime Complaint Center (IC3) of the FBI [8]. They provide details in their 2021 Internet Crime Report on the 847,376 reports of suspected cybercrime they received over the course of the year, with claimed damages totaling $6.9 billion. The research adds that assaults have increased in frequency during the past few years.

Anti-Phishing Working Group (APWG) [2] also highlight that phishing attacks increases in recent years. Figure 1 illustrates that the APWG Phishing Activity Trends Report from third Quarter 2021 to second Quarter 2022. APWG recorded 1,025,968

C. Patra (✉) · D. Giri
Department of Information Technology, Maulana Abul Kalam Azad University of Technology, Kolkata, West Bengal, India
e-mail: chanchalpatra89@gmail.com

J. K. Mandal et al. (eds.), *Proceedings of International Conference on Network Security and Blockchain Technology*, Lecture Notes in Networks and Systems 738,
https://doi.org/10.1007/978-981-99-4433-0_2

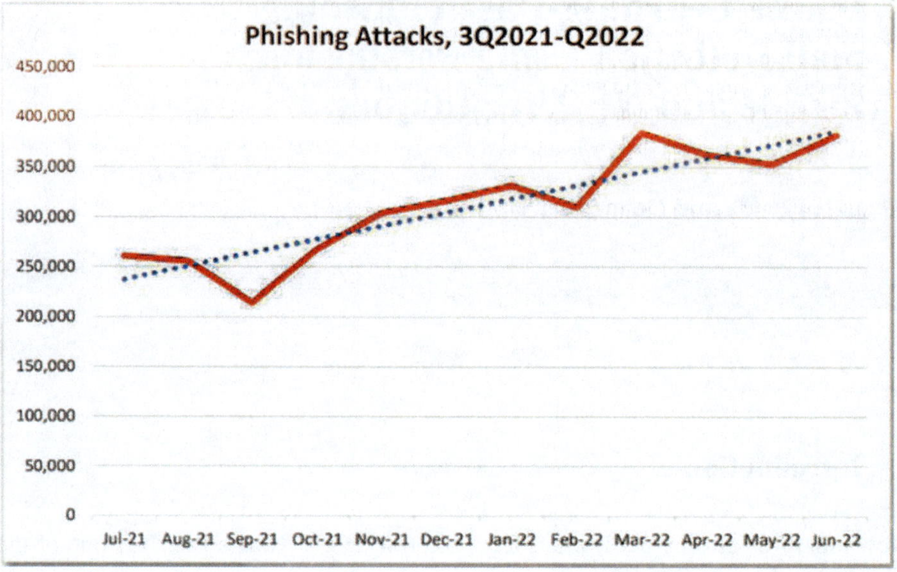

Fig. 1 Phishing activity trends report from the third quarter of 2021 to the second quarter of 2022 by the Anti-Phishing Working Group (APWG)

phishing attempts in total during the first quarter of 2022. A new record and the worst phishing quarter ever was set by the APWG in the second quarter of 2022, when 1,097,811 total phishing assaults were recorded. Since early 2020, there have been four times as many phishing attacks reported to APWG. Increase in smishing and vishing, which together have increased by about 70%, is a major trend in mobile phone-based fraud.

Compare the predicted accuracy of seven different classifiers using collected dataset. This is the key contribution of this work. The classifiers include logistic regression, support vector machines, random forests, decision trees, k-nearest neighbors, naive Bayes, and AdaBoost. A dataset is constructed with 1988 legitimate emails and 2578 phishing emails. For training and testing the classifiers, two categories of features are used here, one is based on persuasion principle and the other is natural language processing-based features.

The rest of this paper organized as follows: In Sect. 2, we talk about related works. Demonstrate in Sect. 3 how the study's various categorization techniques leverage machine learning. In Sect. 4, give a description of the dataset. The feature extraction and selection process is discussed in Sect. 5. In Sect. 6 we explain the experimental methodology. We explain about different evaluation metrics in Sect. 7. The results analysis was covered in Sect. 8. Finally, we described the conclusion and discuss the future work.

2 Related Works

There are three types of phishing: one is web-based phishing, in which a fake website is created to look like a reputable one and deceives people into providing critical information. The next type of phishing is email-based, when a hacker sends emails to several people pretending there is a problem with their account in the hopes that some of them would fall for it. The third one is malware-based phishing, in which a genuine website is compromised with harmful code, and when a user accesses that site, the malicious software is placed on the user's computer [7].

The email-based phishing detection algorithms analyze the emails to identify text-based elements before deploying machine learning (ML) techniques to distinguish phishing emails from legitimate emails. NLP is a technology that is primarily used to extract useful information from emails' content, while machine learning is a classification approach that is used to create prediction models that are used to identify phishing emails [15].

Recently, systems for detecting phishing emails that exploit the semantics of the emails' text have benefited from the development of Natural Language Processing (NLP) tools [6].

Recent research approaches phishing emails identification is a text classification issue, taking merely the text of emails and applying NLP techniques to handle the textual aspects [3].

Bountakas et al. [4] proposes a strategy that uses Natural Language Processing and machine learning techniques for comparison-based phishing email detection.

Gualberto et al. [9] retrieved unique features from the content of phishing email. The authors address a number of issues, including the features extraction, sparsity, and "the curse of dimensionality". They built a model using 10 characteristics and used the XGBoost method to get a fair level of accuracy. After that, they devised a multistage technique for detection of phishing emails, which was reported in [10] as a continuation of their earlier work.

3 Machine Learning Methods

Machine learning approach one of the efficient methods for data analysis. We can apply machine learning models for categorizing as phishing email identification is a classification problem. For comparative experiment we used supervised machine learning model. In following discussed some useful machine learning classification methods that mainly we used.

A. Naive Bayes: The categorization problem is handled by a probabilistic machine learning model. Assuming that each input variable is independent is what gives naïve theory its name. The performance of this classifier is best in two situations that go against each other: functionally dependent features, which is unexpected, and entirely independent features [22].

B. Logistic Regression: It forecasts a discrete or categorical value's output. It may be applied to predict the probability that a binary event will happen. Since the outcome is a probability, the dependent variable can only take values between 0 and 1 [14].
C. Decision Tree: It is a non-parametric supervised learning technique used for both classification and regression applications. Its hierarchical tree structure consists of a root node, branches, internal nodes, and leaf nodes (also known as terminal nodes) [20].
D. Random Forest: An ensemble learning technique for classification. It includes the number of distinct decision trees. Each decision tree provides a forecast, and the highest anticipated class among all classes is the outcome [13].
E. K-Nearest Neighbors: It is a straightforward machine learning approach built on the supervised learning methodology, and it may be used to address classification and regression issues. For the purpose of predicting new data points, this method employs feature similarity. A value will be given to the new data point in this case depending on how closely it resembles the points in the training set [18].
F. Support Vector Machine: It is a well-known and effective supervised learning algorithms, and it is also utilized for classification issues. In order to create the hyperplane, SVM selects the extreme points. Support vectors are used in these severe situations [11].
G. AdaBoost: The AdaBoost classification also a supervised learning algorithm. It is similar to random forest classification model, i.e., the combination groups of weak classifiers which form a strong classification model. Its performance may be enhanced by combining it with a variety of different learning methods. A single machine learning model may occasionally classify objects very poorly. However, combining a number of classifiers for the purpose of overall classification can be beneficial [17].

4 Dataset Description

Dataset preparation is one of the importan and challenging task for phishing detection. Even though there have been a lot of scholarly articles concerning phishing detection published, none of them have detailed their datasets.

The dataset contains both a training and testing data. Total collected emails are 4566 out of 1988 are ham emails and 2578 are phishing emails. The ham emails came from various emails domains as well as SpamAssassin [12] while the phishing emails came from well-known Enron corpus [16] Nazario phishing corpora and some manual generated phishing emails.

5 Feature Extraction

TF-IDF: The full form of TF-IDF is Term Frequency and Inverse Document Frequency. TF-IDF encoding was utilized to express data in numerical format [21]. Find the most distinctive terms in the corpus's whole document using this vector representation. Among all the documents, IDF discovers the singular words. While TF is determined by dividing the total number of terms in the document by the frequency with which each word appears in the document.

$$\text{TF}(t, d) = \frac{x}{y} \tag{1}$$

where x is the count of t in document d, and y is the number of words in document d.

IDF calculates a the uniqueness of the word across the corpus.

$$\text{IDF}(t, d) = \log \frac{N}{n} \tag{2}$$

where N is the total number of documents present in the corpus, and n is the number of documents where the term t appears.

$$\text{TF} - \text{IDF} = \text{TF} \times \text{IDF} \tag{3}$$

Cialdini's Principles:

In 1984, Robert Cialdini released his book "Influence: The Psychology of Persuasion" [5]. Cialdini established six guiding principles that might influence the decision-making process.

1. Reciprocity: It's a little something to get a little something in return, an equal exchange of either rewards or penalties.
2. Commitment: The feeling of responsibility that a person has toward the goals. It is desirable for people's views and ideals to be in harmony.
3. Consensus: The individual people's tendency are to follow the lead of the group. Nothing compares to feeling affirmed by what other people are doing.
4. Authority: It's involves referencing experts and expertise. It is the power or right to give orders, make decisions, and enforce obedience.
5. Liking: Liking is the feeling that you like someone or something. It's means that you are happy being with that person, while loving someone.
6. Scarcity: A situation in which something is not easy to find. When something is short supply we want it more!

In order to identify phishing emails, we conducted an analysis of the dataset using Cialdini's ideas from "The science of persuasion". As illustrated in Table 1, we discover that the three persuasive concepts employed in phishing emails are reciprocity, scarcity, and authority.

Table 1 Three persuasion principle and corresponding words

Persuasion principle	Corresponding words
Reciprocation	Bank, Customers, Accounts, Updates, Benefits
Scarcity	Suspension, Suspended, Terminated, Limited, Services
Authority	Identity, Verify, Fraud, Management, Paypal

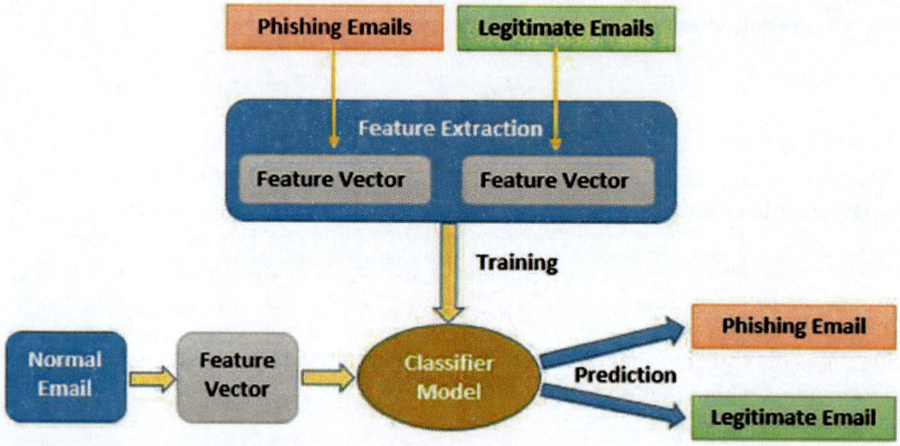

Fig. 2 Phishing email detection framework

6 Experimental Methodology

Our experiment's core three components are used to detect phishing emails. These are features extraction, training the machine learning-based classification model and then prediction new email as phishing or legitimate. The specific framework is shown in Fig. 2 where two types of emails (phishing or legitimate) are used for training purpose.

In machine learning classification technique mainly divided into two stages: First one is training stage, where we utilized the training dataset including the samples and their labels so that the classification algorithm could learn from this dataset. The second stage is the testing phase, when we employed the testing dataset, which comprises the sample data but not their labels. So these steps provides the classification algorithm training and then label prediction of each sample. Finally this classification model will be use for prediction purpose, which will predict the new email as phishing or legitimate.

7 Evaluation Metrics

We use a variety of assessment criteria to assess the effectiveness of the machine learning classification models for evaluation purposes. The assessment measures include things like F1-scores, recall, accuracy, and precision [19]. Based on true positive, false positive, true negative, and false negative scores, each measure is computed. Descriptions of these four basic characteristics (numbers) are: (a) True Positive (TP): Represents the predicted values correctly predicted as actual positive. (b) True Negative (TN): Represents the predicted values correctly predicted as an actual negative. (c) False Positive (FP): Represents the predicted values incorrectly predicted an actual positive, i.e., negative values predicted as positive. (d) False Negative (FN): Represents the Positive values predicted as negative.

Accuracy (Acc): Accuracy tells how many times the ML model was correct overall. It indicates how close a collection of measurements is to being accurate.

$$\text{Acc} = \frac{(\text{TP} + \text{TN})}{(\text{TP} + \text{FP} + \text{TN} + \text{FN})} \tag{4}$$

where (TP + TN) is the number of emails that were correctly identified as phishing and ham, where as (TP + FP + TN + FN) is the total number of emails.

Precision (P): Precision measures how near the calculated results are to one another. It is determined by dividing the actual positives by any positive predictions.

$$\text{P} = \frac{\text{TP}}{(\text{TP} + \text{FP})} \tag{5}$$

where (TP + FP) is the total numbers of emails classified as phishing, and TP is the number of phishing emails classified as phishing.

Recall or sensitivity is known as true positive rate (TPR): The recall measures how well the model can identify positive samples. It is computed as a ratio of Positive samples that have been accurately identified as positive to all positive samples.

$$\text{TPR} = \frac{\text{TP}}{(\text{TP} + \text{FN})} \tag{6}$$

where (TP + FN) = Total No. of Phishing Emails and TP = Number of Phishing Emails Classified as Phishing.

Specificity as true negative rate (TNR):

$$\text{TNR} = \frac{\text{TN}}{(\text{TN} + \text{FP})} \tag{7}$$

where TN = The number of emails classified as ham, and (TN + FP) = The total number of emails classified as ham.

F1 scored: F-Measure provides a single score that balances both the concerns of precision and recall in one number.

$$\text{F1-Score} = \frac{2 * (\text{Precision} * \text{Recall})}{(\text{Precision} + \text{Recall})} \tag{8}$$

AUC-ROC Curve: The AUC-ROC curve offers a performance indicator for classification problems at various threshold settings. AUC is the level or measure of separability, whereas ROC is a probability curve. It shows how effectively the model can distinguish between classes. We employed AUC-ROC curve with regard to each classifier in our trials.

8 Result and Discussion

This section discuss the information on precision, recall, accuracy, and F1-score respect to different classifiers. Performance of each classical machine learning approach for the specified binary classification issue to determine whether an email is legitimate or phishing is shown in Table 2. We have used train-test split to split the training data into training and validation.

In our experiments, for performance evaluation we also use ROC curve with AUC. The ROC curve is one of the useful measurement technique for illustrating the relationship between the true positive rate and the false positive rate. The ideal situation is when the curve is closest to the top left corner of the graph. The performance of a classifier may be measured by the AUC. In Fig. 3 shows that the ROC comparison curve with AUC value. We see the AdaBoost, SVC, Naive Bayes, Random Forest classifiers are almost similar and achieves the maximum AUC value 1.0 whereas k-nearest neighbors (KNN) gives 0.86 AUC value.

Table 2 Comparative results respect to all seven classifiers

Method	TP	TN	FP	FN	Accuracy	Precision	Recall	F1-Score
Logistic regression	836	617	1	53	0.96	0.97	0.96	0.96
K-nearest neighbors	836	405	1	265	0.82	0.87	0.82	0.81
Naive Bayes	758	661	79	9	0.94	0.94	0.95	0.94
Decision tree	812	638	25	32	0.96	0.96	0.96	0.96
Random forest	834	633	3	37	0.97	0.98	0.97	0.97
Support vector machine	836	631	1	39	0.97	0.98	0.97	0.97
AdaBoost	829	656	8	14	0.99	0.99	0.98	0.99

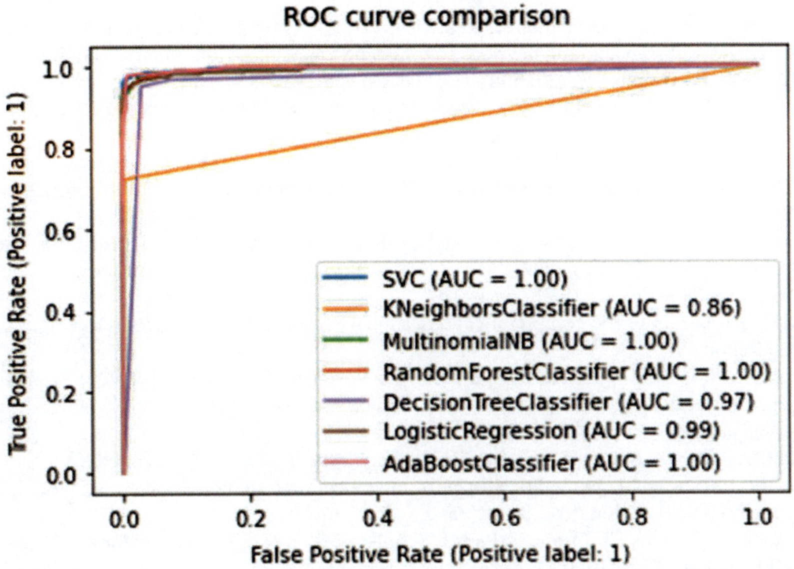

Fig. 3 ROC curve and AUC for different classifiers (SVC, KNeighbors, Naibe Base, Random forest, Decision tree, Logistic regression, and AdaBoost)

9 Conclusion and Future Works

We have developed a phishing detection system that uses various classifiers on the 2578 phishing emails in addition to the 1988 legitimate emails. The examined classifiers are Logistic Regression, Naive Bayes, Decision Tree, Random Forest, Support Vector Machine, AdaBoost. In Table 2, we get very good (highest) performance in ensemble classifier name as AdaBoost respect to accuracy, precision, recall, and F1-scored.

Although the obtained results are quite good. The performance of these models may be enhanced in future by adding additional features to our collected dataset. Another improvement is email attachments analysis which is not implemented in this work. For this method we needs to extract text from attachment and then we can use this for text-based test for phishing detection.

References

1. Phishing-what is phishing? phishing.org (2018). https://www.phishing.org/what-is-phishing
2. APWG: Phishing activity trends report (2022). https://apwg.org/trendsreports/
3. Basit A, Zafar M, Liu X, Javed AR, Jalil Z, Kifayat K (2021) A comprehensive survey of AI-enabled phishing attacks detection techniques. Telecommun Syst 76(1):139–154

4. Bountakas P, Koutroumpouchos K, Xenakis C (2021) A comparison of natural language processing and machine learning methods for phishing email detection. In: The 16th international conference on availability, reliability and security, pp 1–12
5. Cialdini RB, Cialdini RB (2007) Influence: the psychology of persuasion, vol 55. Collins, New York
6. Das A, Baki S, El Aassal A, Verma R, Dunbar A (2019) Sok: a comprehensive reexamination of phishing research from the security perspective. IEEE Commun Surveys Tutorials 22(1):671–708
7. Dong Z, Kapadia A, Blythe J, Camp LJ (2015) Beyond the lock icon: real-time detection of phishing websites using public key certificates. In: 2015 APWG symposium on electronic crime research (eCrime). IEEE, pp 1–12
8. FBI: Ic3 (internet crime complaint center) annual report (2021). https://www.ic3.gov/home/annualreports
9. Gualberto ES, De Sousa RT, Thiago PDB, Da Costa JPC, Duque CG (2020) From feature engineering and topics models to enhanced prediction rates in phishing detection. IEEE Access 8:76368–76385
10. Gualberto ES, De Sousa RT, Vieira TPDB, Da Costa JPCL, Duque CG (2020) The answer is in the text: multi-stage methods for phishing detection based on feature engineering. IEEE Access 8:223529–223547
11. Hearst MA, Dumais ST, Osuna E, Platt J, Scholkopf B (1998) Support vector machines. IEEE Intell Syst Appl 13(4):18–28
12. Henning JL (2006) Spec cpu2006 benchmark descriptions. ACM SIGARCH Comput Archit News 34(4):1–17
13. Ho TK (1995) Random decision forests. In: Proceedings of 3rd international conference on document analysis and recognition, vol 1. IEEE, pp 278–282
14. Hosmer DW Jr, Lemeshow S, Sturdivant RX (2013) Applied logistic regression, vol 398. Wiley
15. Kannoorpatti K, Karim A, Azam S, Sanmugam B (2015) On a comprehensive survey for intelligent spam email detection. IEEE J Comput Intell
16. Klimt B, Yang Y (2004) The enron corpus: a new dataset for email classification research. In: European conference on machine learning. Springer, pp 217–226
17. Korada NK, Kuma N, Deekshitulu Y (2012) Implementation of naïve bayesian classifier and ada-boost algorithm using maize expert system. Int J Inform Sci Tech (IJIST) 2
18. Peterson LE (2009) K-nearest neighbor. Scholarpedia 4(2):1883
19. Powers DM (2020) Evaluation: from precision, recall and f-measure to roc, informedness, markedness and correlation. arXiv preprint arXiv:2010.16061
20. Quinlan JR (1986) Induction of decision trees. Mach Learn 1(1):81–106
21. Ramos J et al (2003) Using tf-idf to determine word relevance in document queries. In: Proceedings of the first instructional conference on machine learning, vol 242. New Jersey, USA, pp 29–48
22. Rish I et al (2001) An empirical study of the naive bayes classifier. In: IJCAI 2001 workshop on empirical methods in artificial intelligence, vol 3, pp 41–46

Cryptanalysis and Improvement of a Mutual Authentication Scheme for Smart Grid Communications

Piyush Sharma, Garima Thakur, and Pankaj Kumar

1 Introduction

An advanced intelligent network that maximizes energy efficiency is the smart grid. A smart grid is an enhanced electrical power grid that aids in controlling power distribution and facilitating communication between customers and service providers. A bidirectional communication link is the main means by which suppliers and consumers can dynamically change the distribution of power in real-time. By doing this, it is feasible to reliably and effectively transmit electricity while preventing the development of excess electricity. The latter will help in upsurging the power operators profit. As a result, the smart grid mitigates the loopholes of our traditional power grid by utilizing bidirectional communication instead of one-way communication. Bidirectional connectivity, self-control, remote verifier, distributed management, and additional consumer options are just a few of the impressive features of the smart grid. Industry and academic researchers have both shown interest in it.

To deliver proficient, reliable, cost-effective, and sustainable power, the smart grid incorporates controls, automation of the framework, computerization, and new technologies. However, communication between legal organizations is vulnerable to cyberattacks since there are no reliable security measures in place. The complex nature of SG and its several security requirements pose challenges to its widespread use. Two imperative issues are the preservation of user privacy and sender authentication. A suitable authentication mechanism should be implemented to ensure that the sender's identity can be verified because the data transmitted by individual appliances impacts the measure of electricity a generator must produce. Keeping in view the aforementioned security prerequisites, we have proposed an authentication scheme in this paper.

P. Sharma · G. Thakur · P. Kumar (✉)
Srinivasa Ramanujan Department of Mathematics, Central University of Himachal Pradesh, Dharamshala, H.P. 176215, India

J. K. Mandal et al. (eds.), *Proceedings of International Conference on Network Security and Blockchain Technology*, Lecture Notes in Networks and Systems 738,
https://doi.org/10.1007/978-981-99-4433-0_3

1.1 Related Work

Ma et al. [1] presented work on smart grid communication in which they discussed the various challenges and opportunities in smart grid communication. Further, in Yan et al. [2] paper, we studied how smart grid communication helps balance power supply and demand. This paper also allows us to know why we need new infrastructure today, the shortcomings in our old infrastructure, and how we can overcome them through smart grid communication. Kabalci et al. [3] also presented a survey paper that examined the technology, applications, and difficulties associated with smart metering and smart grid communication methods. Faheem et al. [4] provide an overview of several smart grid applications, including their advantages, traits, and prerequisites. This study researches and examines several wired and wireless communication technologies, as well as a number of significant difficulties, unresolved problems, and potential future research topics. Chen et al. [5] presented a paper that describes smart attacks and their defenses in a communication network for the smart grid. Thakur et al. [6] highlighted that the scheme proposed by Son et al. is vulnerable to various security assaults including impersonation attacks, offline password-guessing attacks, etc., and proposed an enhanced authentication scheme that can withstands all security attacks. Maitra et al. [7] also proposed an enhanced authentication scheme employing Chebyshev Chaotic Map and demonstrated the robustness of their scheme utilizing the widely accepted random oracle model and AVISPA simulation tool. Numerous other authentication schemes for smart grid environment have been proposed by Ferrag et al. [8], Fouda et al. [9], Liu et al. [10], and Mahmood et al. [11]. Further, an authentication protocol for home and building area networks was put forth by Li et al. [12]; however, the protocol has substantially higher computational cost. Sule et al. [13] also proposed a variable length message authentication code scheme to ensure a safe interaction among AMI devices and collector nodes but could not provide user anonymity, session key agreement, and message authentication as mentioned by [14].

1.2 Motivation and Contribution

Technology today binds us to convenience and ease of living. We formerly lit our homes with fire, but today, we talk of replacing our outdated grid system with the smart grid. In doing so, there are numerous threats, namely man-in-the-middle attacks, password-guessing attacks, replay attacks, insider attacks, smart card loss attacks, impersonation attacks, etc., and security relevance, namely data confidentiality, non-traceability, message authentication, user anonymity, etc. As stated in the preceding section, despite the introduction of numerous authentication schemes [9, 11–14], none of them are entirely proficient of delivering the needed security attributes for the reference of a smart grid. Therefore, this paper discusses the design flaws and cryptanalysis of Khan et al.'s protocol [14]. We discovered that their scheme needs an authenticated key agreement, and also, we have mentioned how this protocol is susceptible to offline password-guessing attacks, user impersonation attacks, and replay attacks. Finally, we proposed an enhanced scheme utilizing the fuzzy extractor function [15, 16] that has no security risks.

2 Review of Khan et al.'s Scheme

This section reviews the Khan et al. [14] scheme. All the symbols used in this paper are given in Table 1.

2.1 Initialization Phase

1. S_g chooses q, E_q (a, b): $y^2 = x^3 + ax + b \bmod q$, where $a, b \in G$ with $4a^3 + 27b^2 \bmod q \neq 0$.
2. S_g selects a base point $P \in G$ and chooses their $h\,(\cdot)$.
3. S_g selects its private key as $s \in Z_q^*$ and public key as $\text{PKs} = s \cdot P$.
4. $\{\text{Eq}(a, b), q, p, P, \text{PKs}, h(\cdot)\}$ are public parameters, and s is kept confidential.

Table 1 Symbols and their description

Symbol	Description
ECC	Elliptic curve cryptography
Q	Large prime number
G	Additive group
P	Generator of G
U	User
s_g	Smart grid server
ID_u	Identity of U
PW_u	Password of U
B_u	Biometric of U
$Z_q{}^*$	Multiplicative group of order $q - 1$
SK_{us}	Session key between u and s
Gen(.), Rep(.)	Fuzzy extractor, reproduction function
A	Adversary
δt	Time stamp
$h\,(\cdot)$	Hash function
$\Vert$, $\oplus$	Concatenation, bitwise XOR operators

2.2 Registration Phase

1. The user U chooses an identity ID_u and a password PW_u, imprints their B_u, and computes $(\sigma_u, \theta_u) = \text{Gen}(B_u)$. A random nonce a is generated by U, thereafter computes $B_1 = h(\text{PW}_u \| \sigma_u) \oplus$ a and transfers the registration request $\{ID_u, B_1, t_{BG1}\}$ to S_g through a secure channel.
2. On receiving registration request from U, S_g verifies $t_{BG2} - t_{BG1} \leq \delta t$. Thereafter, computes $B_2 = \text{h}(\text{ID}_u \|s\| z)$ where s is the private key of S_g and z denotes the counter. Then, S_g computes $B_3 = B_2 \oplus B_1$ and stores $\{B_3, z, p, h(\cdot)\}$ in the database and forwards $\{B_3, z, p, h(\cdot)\}$ to U.
3. After receiving $\{B_3, z, p, h(\cdot)\}$, U computes $B_4 = B_3 \oplus \sigma_u$, $B_5 = h(\text{ID}_u \| \text{PW}_u \| B_4)$ and stores $\{B_3, B_4, B_5\}$ in the database.

2.3 Login and Authentication Phase

The subsequent steps are performed by U and S_g to accomplish mutual authentication:

1. Firstly, user U enters his/her ID'_u, PW'_u, *imprints* B'_u, computes $\sigma'_u = \text{Rep}(B'_u, \theta'_u)$, $B'_4 = B'_3 \oplus \sigma'_u$, $B'_5 = h(\text{ID}'_u \| \text{PW}'_u \| B'_4)$ and verifies $B'_5 \stackrel{?}{=} B_5$. If the validation holds, then U generates a random nonce $r \in Z_q^*$, thereafter computes $M_1 = h(\text{ID}_u \| B_1 \| t_1)$, $\text{ID}_{U1} = \text{ID}_u \oplus (B_1 \oplus t_1)$ and sends $\{M_1, \text{ID}_{U1}, r \cdot p, t_1\}$ to S_g through a public channel.
2. On receiving $\{M_1, \text{ID}_{U1}, r \cdot p, t_1\}$ from user, S_g verifies $t_2 - t_1 \leq \delta t$. Thereafter, S_g computes $\text{ID}_u^* = \text{ID}_{U1} \oplus (B_1 \oplus t_1)$, $M_1^* = \text{h}(\text{ID}_u^* \| B_1 \| t_1)$ and verifies $M_1^* \stackrel{?}{=} M_1$. If the validation holds, then S_g generates a random nonce $b \in Z_q^*$, computes $M_2 = h(\text{ID}_s \| B_3 \| t_2)$ and session key as $\text{SK}_{su} = h(\text{ID}_u^* \| \text{ID}_s \| M_1^* \| M_2 \| B_3 \cdot p \| r \cdot b \cdot p \| s \cdot p \| t_3)$, $\text{ID}_{S1} = \text{ID}_s \oplus (B_3 \oplus t_3)$. Finally, the message $\{M_2, \text{ID}_{S1}, b \cdot p, t_3\}$ is sent to the user U.
3. On receiving data from S_g, U verifies $t_4 - t_3 \leq \delta t$. If yes then, the user computes $\text{ID}_s^* = \text{ID}_{S1} \oplus (B_3 \oplus t_3)$, $M_2^* = h(\text{ID}_s^* \| B_3 \| t_3)$ and verifies $M_2^* \stackrel{?}{=} M_2$. If the validation holds, then the session key is computed by $\text{SK}_{us} = h(\text{ID}_u \| \text{ID}_s^* \| M_1 \| M_2^* \| B_3 \cdot p \| r \cdot b \cdot p \| \text{PK}_s \| t_3)$.

3 Cryptanalysis of the Khan et al.'s Scheme

This segment shows the cryptanalysis of Khan et al.'s scheme [14].

3.1 Offline Password-Guessing Attack

Step 1. Suppose adversary A is a privileged—insider of the server S_g. Then, he/she can access the information of the registration phase, i.e., adversary A has the information $\{ID_u, B_1, t_{BG1}\}$, where $B_1 = h\ (PW_u \parallel \sigma_u) \oplus a$.

Step 2. If adversary A steals user's device, the database of U can be accessed. Therefore, the information $\{B_3, B_4, B_5\}$ is known to A.

Step 3. Now, the adversary guesses a password PW_u^*, computes $B_5^* = h(ID_u \parallel PW_u^* \parallel B_4)$, and verifies $B_5^* \stackrel{?}{=} B_5$. If B_5^* equals B_5, then the adversary successfully guessed the user's password.

3.2 User Impersonation Attack

A can produce a new forged login message in this attack, which is then sent to S_g. If S_g acknowledges this message, the attacker will be successful in user impersonation attack. In Khan et al.'s scheme, this attack is possible if the privileged user performs the task of A. The data $\{ID_u, B_1, t_{BG1}\}$ is accessible to the privileged user. This attack then takes place as follows:

Step 1. A generates its own random nonce a ϵZ_q^*, thereafter A computes $M_1 = h\ (ID_u \parallel B_1 \parallel t_1)$, $ID_{U1} = ID_u \oplus (B_1 \oplus t_1)$, where t_1 is the current time stamp. Then, A sends $\{M_1, ID_{U1}, a \cdot p, t_1\}$ to S_g through a public channel.

Step 2. On receiving $\{M_1, ID_{U1}, a \cdot p, t_1\}$ from user, S_g verifies $t_2 - t_1 \leq \delta t$. After verification S_g computes $ID_u^* = ID_{U1} \oplus (B_1 \oplus t_1)$, $M_1^* = h(ID_u^* \parallel B_1 \parallel t_1)$ and verifies $M_1^* \stackrel{?}{=} M_1$. This verification would be successful because of using correct identification factors. Thus, A is successful in performing a user impersonation attack.

3.3 Replay Attack

The message $\{M_1, ID_{U1}, r \cdot p, t_1\}$ transferred over the public channel is captured by A. Assume A is a privileged—insider of the server S_g. As previously mentioned, A can compose its message $\{M_1, ID_{U1}, r \cdot p, t_1\}$ and transmit it to S_g again. As a result, the attacker's replay attack is successful.

4 Design Flaws of Khan et al.'s Scheme

Khan et al. [14] have the following potential design problems:

- During the login and authentication phase of [14], in step-2 when the S_g generates a random nonce $b \in Z_q^*$ and computes $M_2 = h\ (\mathrm{ID}_s \parallel B_3 \parallel t_2)$, the server used its identity ID_s, B_3 (stored in the database in the registration phase), and used a timestamp t_2. After that S_g compute session key as $\mathrm{SK}_{su} = \mathrm{h}\ (\mathrm{ID}_u{}^* \parallel \mathrm{ID}_S \parallel M_1{}^* \parallel M_2 \parallel B_3 \cdot p \parallel r \cdot b \cdot p \parallel s \cdot p \parallel t_3)$. But in step-3 of login and authentication phase, the user first compute $\mathrm{ID}_S{}^* = \mathrm{ID}_{S1} \oplus (B_3 \oplus t_3)$, then computes $M_2{}^* = h\ (\mathrm{ID}_s{}^* \parallel B_3 \parallel t_3)$ and verifies $M_2{}^* \stackrel{?}{=} M_2$. Here, U uses time stamp t_3 instead of t_2. So, by property of the hash function, $M_2{}^*$ does not equal M_2. Hence, mutual authentication does not hold. Also, U computes session key as $\mathrm{SK}_{us} = h\ (\mathrm{ID}_u \parallel \mathrm{ID}_s{}^* \parallel M_1 \parallel M_2{}^* \parallel B_3 \cdot p \parallel r \cdot b \cdot p \parallel \mathrm{PK}_S \parallel t_3)$.

Since $M_2{}^* \neq M_2$, session keys SK_{su} and SK_{us} are not equal. Therefore, there is no mutual authentication and session key agreement in [14].

- For a moment, if we assume S_g computes M_2 as $M_2 = h\ (\mathrm{ID}_s \parallel B_3 \parallel t_3)$, that is, S_g uses time stamp t_3 in M_2 so that $M_2{}^* = M_2$. But still, there is no session key agreement because the S_g computes the session key as $\mathrm{SK}_{su} = h\ (ID_u{}^* \parallel \mathrm{ID}_s \parallel M_1{}^* \parallel M_2 \parallel B_3 \cdot p \parallel r \cdot b \cdot p \parallel s \cdot p \parallel t_3)$ whereas U computes session key as $\mathrm{SK}_{us} = h\ (\mathrm{ID}_u \parallel \mathrm{ID}_s{}^* \parallel M_1 \parallel M_2{}^* \parallel B_3 \cdot p \parallel r \cdot b \cdot p \parallel \mathrm{PK}_S \parallel t_3)$. Then, U uses the server's public key PK_S instead of $s \cdot p$, but PK_S is computed as $\mathrm{PK}_S = s \cdot P$, where P is a base point belonging to elliptic curve group G, and p is the generator of G. So, if p and P are different, then PK_S does not equal $s \cdot p$. As a result, the hash function's attribute prevents the server-generated session key SK_{su} from being similar to the user-generated session key SK_{us}. Because of this, there is no session key agreement.
- The recommended scheme must include a password change phase if the user wants to change their password for whatever reason. However, the common authentication approach suggested in [14] does not include a step for changing passwords. The user must periodically change his password for security reasons. They will thus be protected from numerous attacks.

5 Discussion and Improvements

This section proposes an enhanced scheme that overcomes the security threats of Khan et al.'s scheme [14]. The proposed scheme has the following four phases:

5.1 Initialization Phase

In this phase, server S_g chooses q, E_q (a, b): $y^2 = x^3 + ax + b \bmod q$, where $a, b \in Z_q$ with $4a^3 + 27b^2 \bmod q \neq 0$ and chooses his/her h $(\cdot)$. S_g generates its private key $s \in Z_q^*$ and computes public key as $PK_S = s \cdot p$, where p is the generator of G. {Eq (a, b), q, p, PK_s, h $(\cdot)$} are public parameters and server keeps its private key s secretly.

5.2 Registration Phase

The following are the steps for user registration:

Step 1. The U selects an identity ID_u and a password PW_u, and he/she imprints his/her B_u and computes $(\sigma_u, \theta_u) = \text{Gen}$ (B_u). Then, a random nonce a is generated by user U and computes $B_1 = h$ $(PW_u \parallel \sigma_u) \oplus a$, $HID_u = h$ $(ID_u \parallel \sigma_u) \oplus a$ and transfers data $\{HID_u, B_1, \}$ toward S_g through a secure channel.

Step 2. On receiving data from U, S_g computes $B_2 = h$ $(HID_u \parallel s \parallel z)$ where s is the private key of S_g and z is the counter. Then, S_g computes $B_3 = B_2 \oplus B_1$. Server S_g stores $\{B_3, z, p, h(\cdot)\}$ in the database and forwards it to U.

Step 3. After receiving $\{B_3, z, p, h(\cdot)\}$, U computes $B_4 = B_3 \oplus \sigma_u$, $B_5 = \text{h}$ $(ID_u \parallel PW_u \parallel B_4) \oplus a$ and stores $\{B_3, B_4, B_5\}$ in database of U.

5.3 Login and Authentication Phase

The subsequent steps are performed by U and S_g to accomplish mutual authentication:

Step 1. Firstly, U enters their ID'_u, PW'_u and imprints B'_u then computes $\sigma'_u = \text{Rep}(B'_u, \theta'_u)$, $B'_4 = B_3 \oplus \sigma'_u$, $a' = HID_u \oplus h(ID'_u \parallel \sigma'_u)$, $B'_5 = h(ID'_u \parallel PW'_u \parallel B'_4) \oplus a'$ and checks $B'_5 \stackrel{?}{=} B_5$. If the verification holds, then U computes $M_1 = h(ID_u \parallel B_1 \parallel a \cdot p)$ and encrypts the message $M_{U1} = E_{K_U}(M_1, ID_u, t_1)$ with the help of key $K_U = h(HID_u \parallel a \cdot s \cdot p)$. Finally, U sends $\{M_{U1}, a \cdot p, t_1\}$ to S_g through a public channel.

Step 2. On receiving $\{M_{U1}, a \cdot p, t_1\}$ from U, S_g verifies $t_2 - t_1 \leq \delta t$. After verification S_g decrypts $(M_1, ID_u, t_1) = D_{K_S}(M_{U1})$ with the help of key $K_S = h(HID_u \parallel a \cdot s \cdot p)$, computes $M_1^* = h(ID_u \parallel B_1 \parallel a \cdot p)$ and verifies $M_1^* \stackrel{?}{=} M_1$. Thereafter, S_g generates a random number $b \in Z_q^*$, computes $M_2 = h$ $(ID_s \parallel B_3 \parallel b \cdot p)$, calculates the session key as $SK_{su} = h(ID_u \parallel ID_s \parallel M_1^* \parallel M_2 \parallel B_3 \cdot p \parallel a \cdot b \cdot p \parallel s \cdot p \parallel t_3)$, and encrypts the message $M_{S1} = E_{K_{S1}}(M_2, ID_S, t_3)$with the help of key $K_{S1} = h(HID_u \parallel a \cdot b \cdot p)$. Then, the message $\{M_{S1}, b \cdot p, t_3\}$ t is sent to U.

Step 3. On receiving message from S_g, the user verifies $t_4 - t_3 \le \delta t$, if yes then, user decrypts $(M_2, \text{ID}_s, t_3) = D_{K_{U1}}(M_{S1})$with the help of key $K_{U1} = h(\text{HID}_u \| a \cdot b \cdot p)$ and computes $M_2^* = h(\text{ID}_s \| B_3 \| t_3)$, verifies $M_2^* \stackrel{?}{=} M_2$. If the verification holds, the user computes the session key as $\text{SK}_{us} = h(\text{ID}_u \| \text{ID}_s \| M_1 \| M_2^* \| B_3 \cdot p \| a \cdot b \cdot p \| \text{PK}_s \| t_3)$.

5.4 *Change Password and Biometric Phase*

If U wants to change their biometric and password, then the following steps are to be followed by U:

Step 1. U inputs $\text{ID}'_u, \text{PW}'_u, B'_u$ and computes $\sigma'_u = \text{Rep}(B'_u, \theta'_u)$, $B'_4 = B_3 \oplus \sigma'_u$, $a' = \text{HID}_u \oplus h(\text{ID}'_u \| \sigma'_u)$, $B'_5 = h(\text{ID}'_u \| \text{PW}'_u \| B'_4) \oplus a'$ and verifies $B'_5 \stackrel{?}{=} B_5$. If verification does not hold, then the session is terminated. Otherwise, U selects a new biometric B_u^* and password PW_u^*. The user computes $(\sigma_u^{\text{new}}, \theta_u^{\text{new}}) = \text{Gen}(B_u^*)$ and $B_1^{\text{new}} = h(\text{PW}_u^* \| \sigma_u^{\text{new}}) \oplus$ a. User sends $\{\text{HID}_u, B_1^{\text{new}}, t_{BG1}\}$ toward S_g.

Step 2. Firstly, S_g verifies $t_{\text{BG1}} - t_{\text{BG2}} \le \delta t$ then computes $B_3^{\text{new}} = B_2 \oplus B_1^{\text{new}}$. Thereafter, S_g replaces B_3 *with* B_3^{new} in the database and sends B_3^{new} to U.

Step 3. After receiving B_3^{new}, the user U computes $B_4^{\text{new}} = B_3^{\text{new}} \oplus \sigma_u^{\text{new}}$, $B_5^{\text{new}} = h(\text{ID}_u \| \text{PW}_u^* \| B_4^{\text{new}}) \oplus a$. Further, U replaces PW_u by PW_u^*, B_u by B_u^*, σ_u by σ_u^{new} and θ_u *by* θ_u^{new}. Finally, user U stores $\{B_1^{\text{new}}, B_3^{\text{new}}, B_4^{\text{new}}, B_5^{\text{new}}\}$ in database replacing $\{B_1, B_3, B_4, B_5\}$, respectively.

6 Informal Security Analysis

In this section, an informal security analysis of the improved scheme has been discussed.

6.1 *Replay Attack*

The most frequent defenses against this attack are the timestamp and random numbers. The user and server generate random integers (a and b) and a time stamp condition $t_i - t_j \le \delta t$ at each stage of the proposed scheme. They are used to ensure the message's freshness. By checking the timestamp of received messages, the user and the server can determine the nature of the assault. As a result, the proposed scheme maintains a replay attack.

6.2 Man-in-the-Middle Attack

Any attacker A may attempt to log in to the server using the previous message. A replay $\{M_{U1}, a \cdot p, t_1\}$, where $M_1 = h(\text{ID}_u \| B_1 \| a \cdot p)$, $K_U = h(\text{HID}_u \| a \cdot s \cdot p) and$ a $\in Z_q^*$ and t_1 is the time stamp that prohibits the replay attack. S_g checks the two verifying conditions $M_1^* \stackrel{?}{=} M_1$ and $t_2 - t_1 \leq \delta t$ after receiving the message. Similarly, when the user receives the message $\{M_{S1}, b \cdot p, t_3\}$, the user verifies $t_4 - t_3 \leq \delta t$, and $M_2^* \stackrel{?}{=} M_2$. A cannot access the user's or server's private keys and therefore cannot determine a real verifier. As a result, A cannot modify a parameter since the verifiers need to be suitably modified. In light of this cryptography attack, the proposed framework is secure.

6.3 Mutual Authentication

Here is a description of message authentication:

- S_g confirms the time stamp conditions $t_2 - t_1 \leq \delta t$ after receiving the message $\{M_{U1}, a \cdot p, t_1\}$. Then, verifies $M_1^* \stackrel{?}{=} M_1$.
- U confirms the time stamp conditions $t_4 - t_3 \leq \delta t$ after receiving the message $\{M_{S1}, b \cdot p, t_3\}$. Then, verifies $M_2^* \stackrel{?}{=} M_2$.

Message security is ensured by checking parameters, and hash values are difficult for an attacker to guess. Therefore, the recommended framework allows for mutual authentication.

6.4 Impersonation Attack

A can get $\{M_{U1}, a \cdot p, t_1\}$ and try to compute M_{U1}, it is very difficult to compute for any attacker because M_{U1} is encrypted with symmetric key $K_U = h(\text{HID}_u \| a \cdot s \cdot p)$. Similarly, M_{S1} is encrypted with key $K_{S1} = h(\text{HID}_u \| a \cdot b \cdot p)$ and protected with secret parameters using the elliptic curve computational Diffie Hellman problem. As a result, A cannot impersonate anyone in correspondence between these forwarded messages. Therefore, our protocol is protected from user impersonation assault.

6.5 Key Freshness

Every step in the suggested scheme uses a new key, such as a random number or a time stamp, so the key freshness criterion holds true throughout each session.

6.6 User Anonymity

The suggested protocol is free from the problem of anonymity, since the hash function and biometrics protect user identity, and random number a, as $\mathrm{HID}_u = h\,(\mathrm{ID}_u \,\|\sigma_u)\oplus a$.

6.7 Offline Password-Guessing Attack

During the registration phase, the user selects their password PW_u, which is then used to compute $B_1 = h\,(\mathrm{PW}_u \parallel \sigma_u) \oplus a$, which is secured by a secure hash value and protected by private parameters like a biometric and a random value. Also, assume that A gets B_5 from the user database, but they can't guess the password from here because the random value a is used in B_5. In a secure medium, it isn't easy to guess the user's password in this way. As a result, the proposed scheme defends against the offline password-guessing attack.

6.8 Session Key Agreement

According to the proposed scheme, the session keys for the user and server are computed as $\mathrm{SK}_{\mathrm{su}} = h(\mathrm{ID}_u\|\mathrm{ID}_s\|M_1^* \,\|M_2\|B_3 \cdot p\|a \cdot b \cdot p\|s \cdot p\|t_3)$ and $\mathrm{SK}_{us} = h(\mathrm{ID}_u\|\mathrm{ID}_s\|M_1 \,\|M_2^*\|B_3 \cdot p\|a \cdot b \cdot p\|\mathrm{PK}_s\|t_3)$. Clearly, $\mathrm{SK}_{su} = \mathrm{SK}_{us}$. As a result, a session key agreement exists.

6.9 Data Confidentiality

- The user encrypts the message $M_{U1} = E_{K_U}(M_1, \mathrm{ID}_u, t_1)$ with the help of key $K_U = h(\mathrm{HID}_u\|a \cdot s \cdot p)$. Thereafter, S_g decrypts $(M_1,\ \mathrm{ID}_u,\ t_1) = D_{K_S}(M_{U1})$ with the help of key $K_S = h(\mathrm{HID}_u\|a \cdot s \cdot p)$ and verifies $M_1^* \stackrel{?}{=} M_1$.
- S_g encrypts the message $M_{S1} = E_{K_{S1}}(M_2, \mathrm{ID}_S, t_3)$ with the help of key $K_{S1} = h(\mathrm{HID}_u\|a \cdot b \cdot p)$. Thereafter, U decrypts $(M_2, \mathrm{ID}_s, t_3) = D_{K_{U1}}(M_{S1})$ with the help of key $K_{U1} = h(\mathrm{HID}_u\|a \cdot b \cdot p)$ and verifies $M_2^* \stackrel{?}{=} M_2$. Thus, the proposed scheme secures data privacy.

Table 2 Security features

Security features	Khan et al. [14]	Fouda et al. [9]	Li et al. [12]	Sule et al. [13]	Proposed
MM	Yes	Yes	Yes	Yes	Yes
RP	No	Yes	Yes	Yes	Yes
UA	No	No	Yes	No	Yes
KF	Yes	Yes	Yes	Yes	Yes
MA	No	No	Yes	No	Yes
IM	No	Yes	Yes	Yes	Yes
SK	No	Yes	Yes	No	Yes

Yes: Prevent the attack
No: Does not prevent the attack

7 Performance Analysis

7.1 *Security Features*

In this section, we compare the security features of proposed scheme with related previous schemes such as Khan et al. [14], Fouda et al. [9], Li et al. [12], and Sule et al. [13]. Table 2 shows that the proposed scheme resists all malicious attacks, namely man-in-the-middle attack (MM), key freshness (KF), replay attack (RP), message authentication (MA), session key agreement (SK), user anonymity (UA), etc. As a result, compared to the other existing schemes, the proposed scheme offers a wider range of security features.

7.2 *Computational Cost*

This section compares the computational cost of various authentication protocols [9, 12–14] for the login and authentication phase. Relying on [14], the execution time for point addition (PA), point multiplication (PM), symmetric encryption/decryption (ESED), modular exponentiation (ME), public key encryption/decryption (PKED), hash-based message authentication (HMAC), and hash operation (HO) is 0.0288, 2.226, 0.0046, 3.85, 3.85, 0.0046, and 0.0023 ms. Table 3 demonstrates that the computational cost of our scheme is higher than [14]; however, the cost is lesser than that of [9, 12, 13]. Therefore, the proposed protocol offers higher security and efficiency.

Table 3 Comparison of the computational costs

Scheme	Operations	Computational cost (ms)
Khan et al. [14]	$4T_{PM} + 7T_{HO}$	$\cong 8.9201$
Fouda et al. [9]	$4T_{ME} + 4T_{PKED} + 2T_{HO}$	$\cong 30.8046$
Li et al. [12]	$7T_{ME} + 6T_{HO}$	$\cong 26.9638$
Sule et al. [13]	$4T_{ME} + 4T_{PKED} + 2T_{HMAC}$	$\cong 30.8092$
Proposed	$11T_{PM} + 12T_{HO}$	$\cong 24.5136$

8 Conclusion

Smart grid technology is gaining popularity and is becoming a new area of interest. Sensitive data stored in the smart grid has upsurged the requirement of security of bidirectional communication. In this study, we have examined the various design flaws and vulnerability of scheme suggested by [14] in opposition of numerous cryptographic attacks like user impersonation attacks, replay attacks, and offline password-guessing attacks. We have also proposed an enhanced authentication framework for smart grid environment. The informal security analysis of the proposed scheme shows the efficiency and security against the various attacks. Further, the proposed scheme is compared to the related protocols in terms of computational efficiency and security features. The results demonstrate the elevated security of the proposed protocol. Therefore, the proposed work is suitable for smart grid environment.

References

1. Ma R, Chen HH, Huang YR, Meng W (2013) Smart grid communication: its challenges and opportunities. IEEE Trans Smart Grid 4(1):36–46
2. Yan Y, Qian Y, Sharif H, Tipper D (2012) A survey on smart grid communication infrastructures: motivations, requirements and challenges. IEEE Commun Surv Tutorials 15(1):5–20
3. Kabalci Y (2016) A survey on smart metering and smart grid communication. Renew Sustain Energy Rev 57:302–318
4. Faheem M, Shah SBH, Butt RA, Raza B, Anwar M, Ashraf MW, Gungor VC (2018) Smart grid communication and information technologies in the perspective of Industry 4.0: opportunities and challenges. Comput Sci Rev 30:1–30
5. Chen PY, Cheng SM, Chen KC (2012) Smart attacks in smart grid communication networks. IEEE Commun Mag 50(8):24–29
6. Thakur G, Kumar P, Jangirala S, Das AK, Park Y (2023) An effective privacy-preserving blockchain-assisted security protocol for cloud-based digital twin environment. IEEE Access 11:26877–26892
7. Maitra T, Singh S, Saurabh R, Giri D (2021) Analysis and enhancement of secure three-factor user authentication using Chebyshev Chaotic Map. J Inform Sec Appl 61:102915
8. Ferrag MA, Maglaras LA, Janicke H, Jiang J, Shu L (2018) A systematic review of data protection and privacy preservation schemes for smart grid communications. Sustain Cities Soc 38:8–835
9. Fouda MM, Fadlullah ZM, Kato N, Lu R, Shen XS (2011) A lightweight message authentication scheme for smart grid communications. IEEE Trans Smart grid 2(4):675–685

10. Liu Y, Cheng C, Gu T, Jiang T, Li X (2015) A lightweight authenticated communication scheme for smart grid. IEEE Sens J 16(3):836–842
11. Mahmood K, Chaudhry SA, Naqvi H, Shon T, Ahmad HF (2016) A lightweight message authentication scheme for smart grid communications in power sector. Comput Electr Eng 52:114–124
12. Li X, Wu F, Kumari S, Xu L, Sangaiah AK, Choo KKR (2019) A provably secure and anonymous message authentication scheme for smart grids. J Parallel Distrib Comput 132:242–249
13. Sule R, Katti RS, Kavasseri RG (2012) A variable length fast message authentication code for secure communication in smart grids. In: 2012 IEEE power and energy society general meeting. IEEE, pp 1–6
14. Khan AA, Kumar V, Ahmad M (2022) An elliptic curve cryptography based mutual authentication scheme for smart grid communications using biometric approach. J King Saud Univ Comput Inform Sci 34(3):698–705
15. Maurya AK, Das AK, Jamal SS, Giri D (2021) Secure user authentication mechanism for IoT-enabled Wireless Sensor Networks based on multiple Bloom filters. J Syst Architect 120:102296
16. Dodis Y, Reyzin L, Smith A (2004) Fuzzy extractors: How to generate strong keys from biometrics and other noisy data. In: Advances in cryptology-EUROCRYPT 2004: international conference on the theory and applications of cryptographic techniques, Interlaken, Switzerland, May 2–6, 2004. Proceedings 23. Springer Berlin Heidelberg, pp 523–540

Collision Avoidance and Drowsiness Detection System for Drivers

Fatima Mohammad Amin

1 Introduction

Traffic accidents are one of the top causes of death worldwide. A total of 1.35 million persons lose their lives in automobile accidents each year. The year 2019 had a total of 437,396 road accidents in India alone, resulting in 154,732 fatalities and 439,262 injuries. Most of the deaths occur due to late intimation of the person to hospital or in police station. Even if the victim is sent to hospital on time, it takes a lot of time to complete all the formalities. In fact, in some cases, doctors get shortage of blood or any commodities that are required from the relatives of the victim. Even relatives are unaware of the mishap and didn't know the wellness of their loved once when they go out. Another problem that people faces is of heavy traffic forcing them to reach late to their destination. This happens because people are unaware of the routes which are having heavy traffic and which routes are clear. Even this applies for the railway system where people miss their trains and have to face difficulties [1, 2].

From the problem statement, it is quite clear that we need to take serious initiative to curb down the number of accidents as well as number of causalities. Accident detection technology and alerting system will help in saving a lot of lives of the people of a country [3].

With the help of this technology, we will be able to get an instant information of any accident at any instant of time, whether at dawn or at dusk. Another important aspect that makes this technology is that it will also provide the location where the accidents had occurred. This will help in providing quick support to the victim and also will inform the local police station. This will help in saving a lot of lives. Another unique advantage of using this technology will help in reducing traffic in a more densely populated area. Since we are using GPS in the technology, it will help

F. M. Amin (✉)
Vellore Institute of Technology, Vellore, India
e-mail: fatimamohammad.amin2020@vitstudent.ac.in

J. K. Mandal et al. (eds.), *Proceedings of International Conference on Network Security and Blockchain Technology*, Lecture Notes in Networks and Systems 738,
https://doi.org/10.1007/978-981-99-4433-0_4

the traffic police access the information of the areas which are densely packed with vehicles as well as the routes that are having less traffic.

After getting the information, the respective officials can curb down the traffic by changing the routes from high traffic area to lower ones. This will save a lot of time from wasting in traffics. Thus, this technology can bring revolution in the road transport saving both the precious lives and the time where people don't lose their loved ones and saving their time from frustrating traffic.

2 Tools

The system has 2 main components-

(a) A collision/accident avoidance subsystem
(b) A drowsiness detection and awareness subsystem.

A collision/accident avoidance subsystem. To avoid collision, the user is fixed with a subsystem which involves an ultrasonic sensor along with an Arduino Uno with buzzer and a neopixel ring. So the user is alerted through a series of lights and buzzer if they are in a close proximity to another vehicle which may cause a collision.

This part of the project is implemented Tinkercad.

A drowsiness detection and awareness subsystem. The second part of my project is drowsiness detection, in this the system detects if the driver is drowsy by calculating the Euclidean distance between their eyes and alerts the user to be awake, to prevent any mishaps from occurring. The drowsiness detection model is run using Python and its libraries on Spyder.

Software Description. The softwares that have been employed in this project are described below.

Tinkercad. Free 3D modelling software called Tinkercad is renowned for being user-friendly. Since it is entirely Web-based, anyone with an Internet connection can use it. It is used by kids, teachers, and enthusiasts to design anything they can think of. Tinkercad creations can be realised via 3D printing, laser cutting, or building blocks. It is a popular teaching tool in schools for projects involving 3D design, electronics, and visual code blocks. I have used Tinkercad to construct the collision prevention system.

Python Idle. Python is a well-known computer programming language used to build programmes, Websites, automate corporate processes, and analyse data. Python was developed to be a general-purpose language that could be used to construct a variety of applications rather than to tackle any specific difficulties. Because of its adaptability and accessibility to novice programmers, it has developed into the most extensively used programming language at the time. The code is written in Python programming language, using the Idle. Its run using Anaconda Prompt (Anaconda3).

Anaconda Prompt (Anaconda3). The Python programming language has an implementation called anaconda. It's convenient, especially for Windows users, as many scientific libraries are already pre-compiled for Windows, saving you from many of the hassles associated with self-compiling them. However, this also means that Anaconda may not (i.e. pretty certainly won't) function properly in a standard command-line window. By default, Anaconda does not alter system variables or add itself to the system path, so it doesn't interfere with other versions of Python already installed on your system. An Anaconda Prompt shortcut is created when Anaconda is installed. This offers you a command-line window, but one in which the environment variables relevant to Anaconda have been set up so that Anaconda will function properly in it.

Figure 1 denotes the flowchart of the entire system. It has been inspired from [4].

3 Methodology and Implementation

The simulation of this subsystem is carried out on TinkerCad (Fig. 2).

The circuit shown in Fig. 3 is constructed on TinkerCad. It consists of an ultrasonic sensor, an Arduino microcontroller, a buzzer, and a motor. Ultrasonic sensors will sense the distance between obstacle and the vehicle. The Arduino is programmed to activate buzzer if the distance between vehicle and obstacle is reduced below a certain threshold. If the sensor senses the object which is too close to vehicle, then it will control the motor to implement braking and slowing down of vehicle.

Drowsiness Detection and Awareness Model. The second part of the constructed system aims at drowsiness detection, for this, we have used computer vision and CNN as a deep learning algorithm. To search for faces in a stream, we set up a camera. Faces can be located via facial landmark detection, in which case the trained model's eye regions are acquired. We may determine whether the eyes are open or closed by computing the eye aspect ratio whilst being aware of where the eyes are. When the eye aspect ratio indicates that the eyes have been closed for some time, the message "DROWSY—Don't Sleep" displays. Once the eye region is extracted, we calculate the aspect ratio of the eye landmarks by using the Euclidian method. If the distance gets lower than a particular threshold, which means the eyes are closed. The alert sounds if the eyes are shown to be closed for a longer amount of time than is typical.

Calculating the distance between landmark points. Starting from the left corner of the eye as seen from a human's perspective, each eye is given six (x, y)-coordinates. The remainder of the eye area is then encircled by a circle that is drawn clockwise using these coordinates (Figs. 4, 5 and 6) [5].

Thus, the aspect ratio can be calculated as

$$\text{EYE} = \frac{||p_2 - p_6|| + ||p_3 - p_5||}{2||p_1 - p_4||}$$

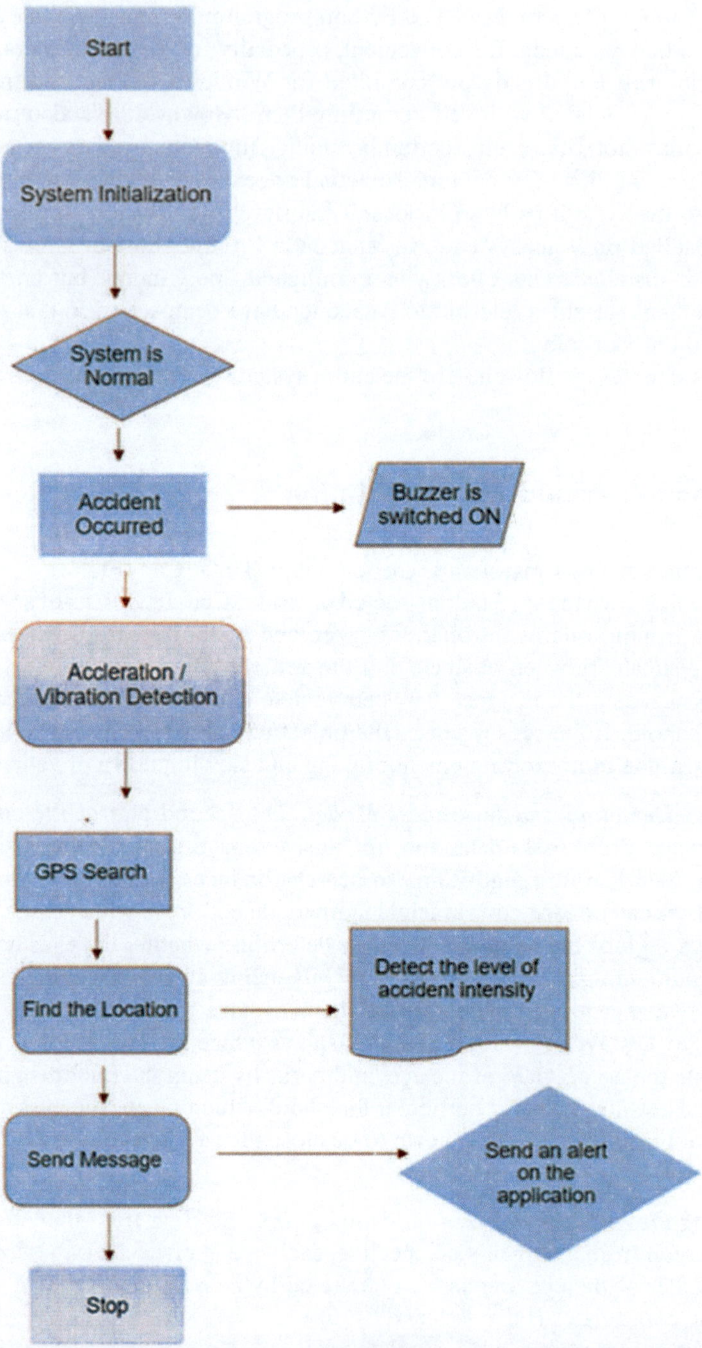

Fig. 1 Flowchart of the constructed system

Accident Alert System

Component List

Name	Quantity	Component
U1	1	Arduino Uno R3
RING1	1	NeoPixel Ring 12
PIEZO1	1	Piezo
DIST1	1	Ultrasonic Distance Sensor

Fig. 2 Components used in TinkerCad

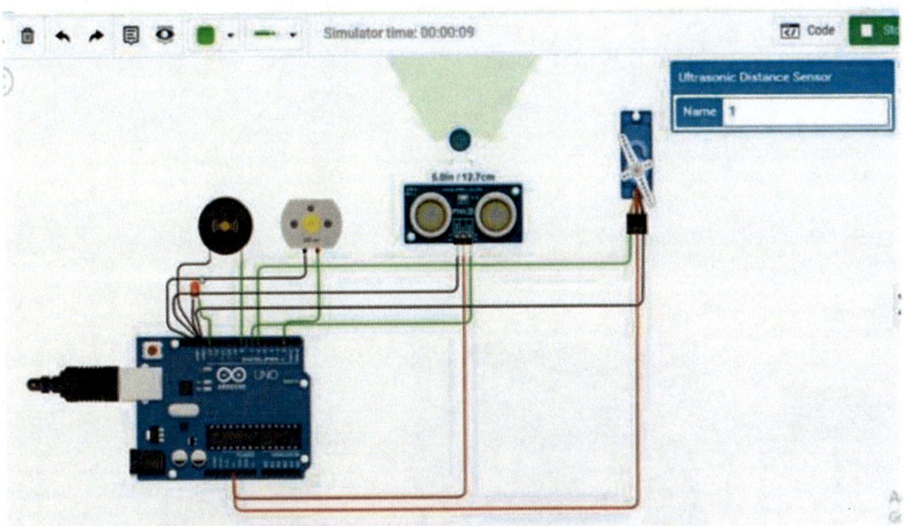

Fig. 3 Constructed circuit diagram

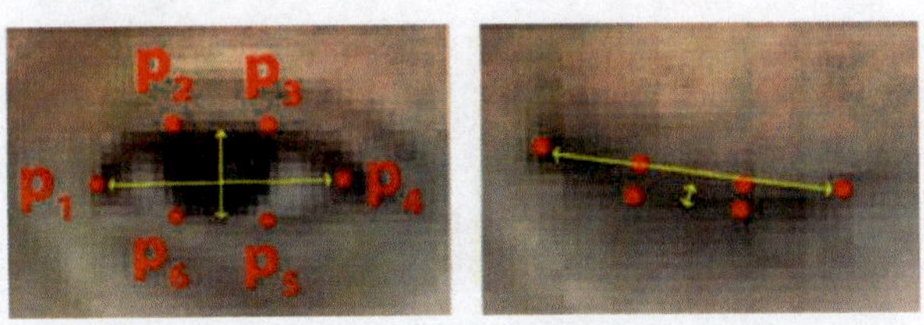

Fig. 4 Calculating the distance between the landmark points

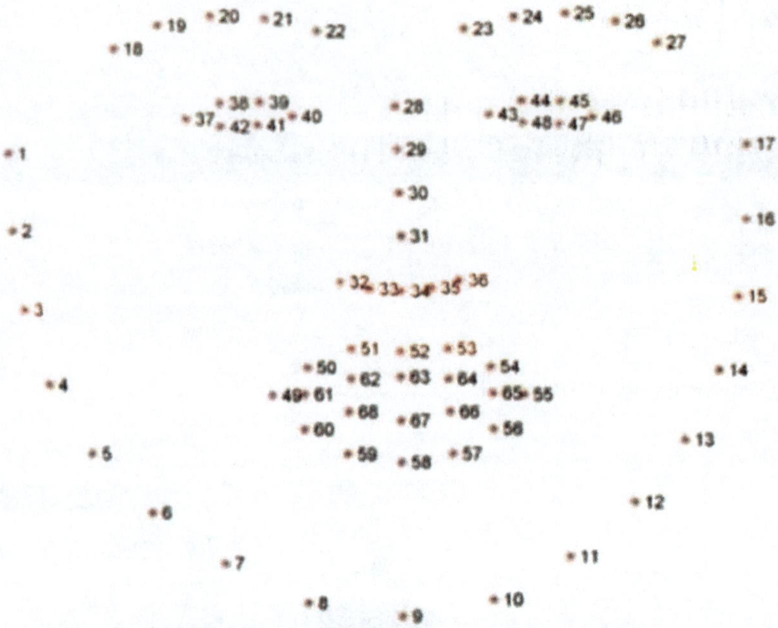

Fig. 5 Default facial landmarks

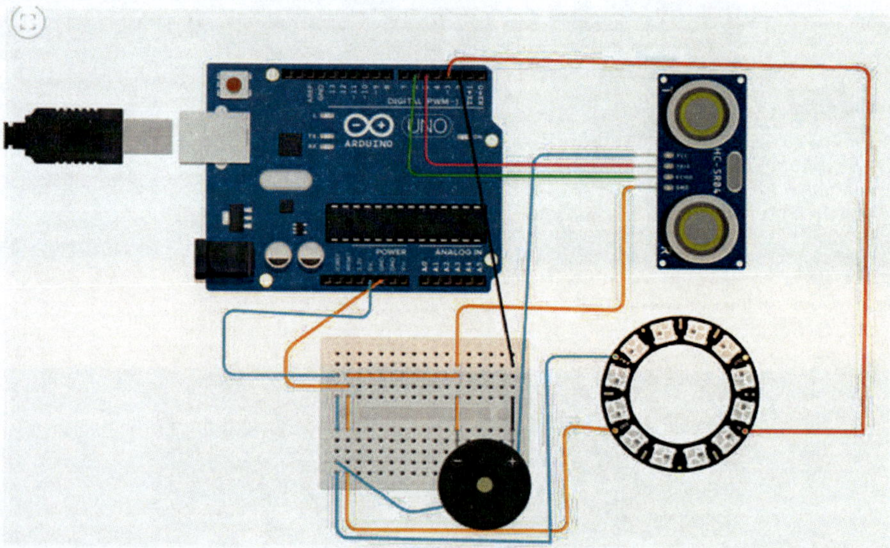

Fig. 6 Collision prevention subsystem

4 Results

Figures 7, 8, and 9 depict the results of the collision prevention subsystem.

Drowsiness Detection and Awareness Model

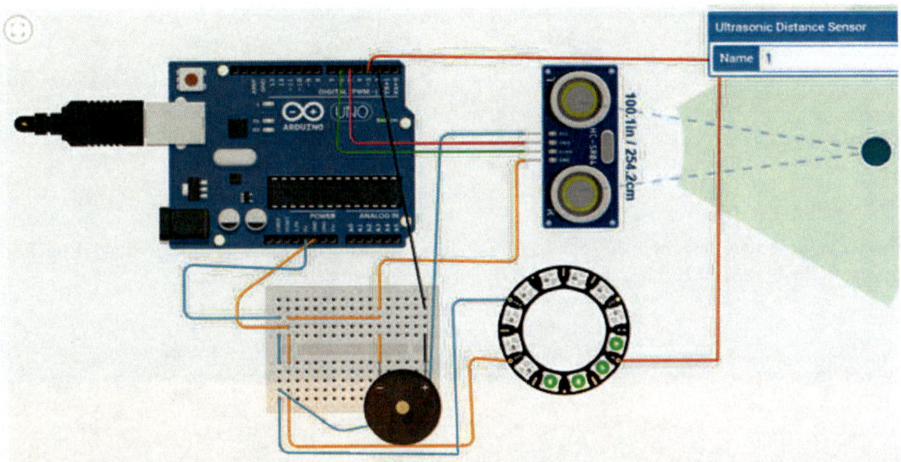

Fig. 7 Vehicle is at a safe distance from the user, which is represented by a green light

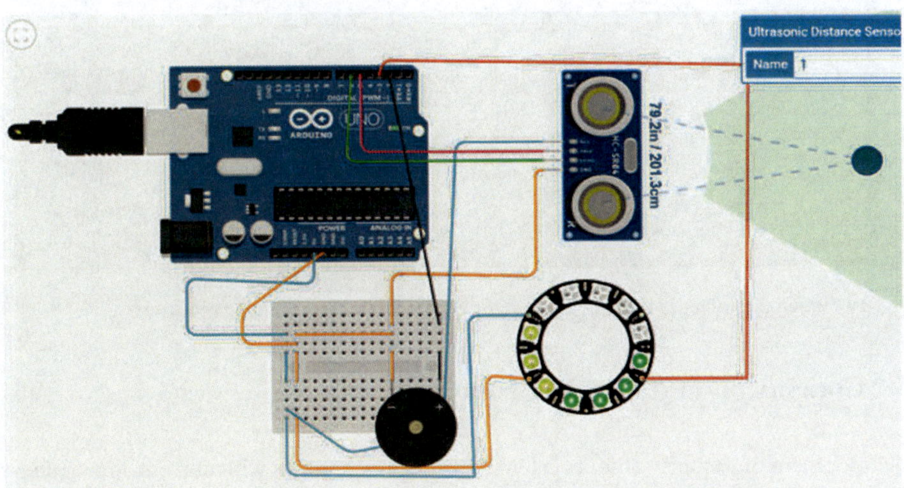

Fig. 8 Vehicle has not breached the set threshold distance from the user but has come closer. This is indicated by Yellow Lights on the Neopixel Ring

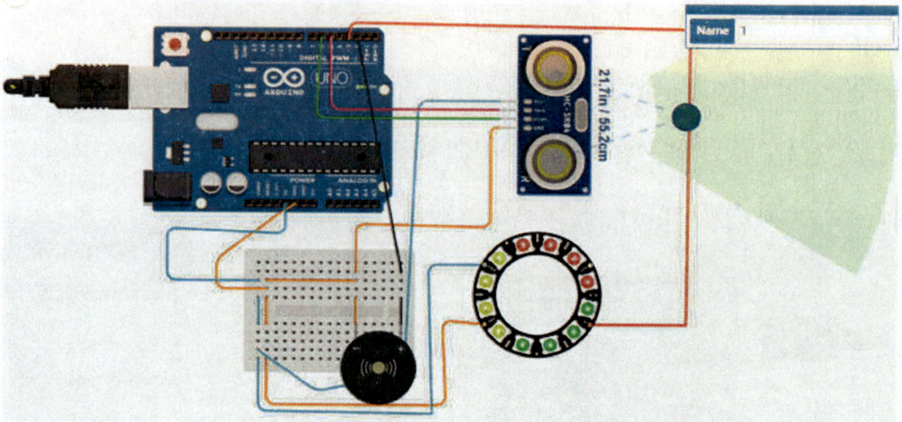

Fig. 9 Vehicle has breached the set threshold distance from the user. This is indicated by Red Lights on the Neopixel Ring, accompanied by an alarm produced by the buzzer

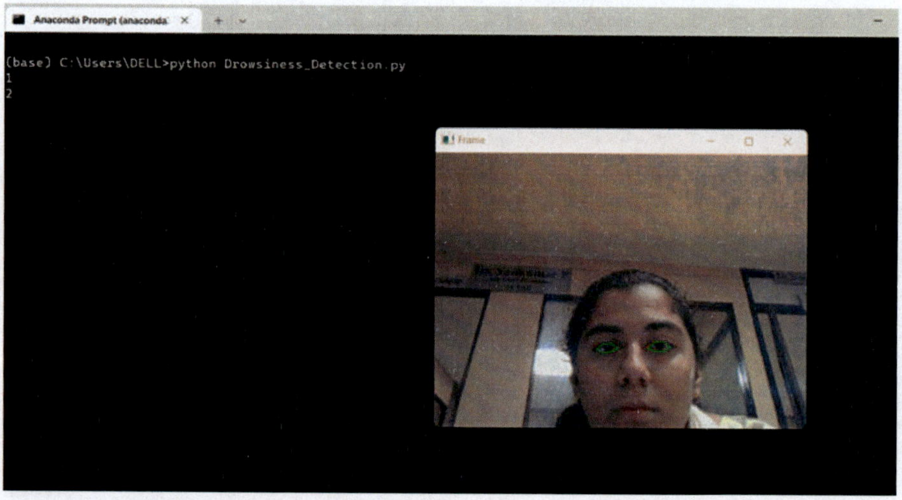

Fig. 10 When the driver is fully awake and the eyes are open, no alert is produced

5 Conclusion and Future Scope

Development of a smart vehicle system that addresses issues with current approaches, enhances vehicle and human security, and reduces the likelihood of accidents was successfully done. When an accident occurs, the GSM/GPRS system will rapidly tell friends in the vicinity of the location of the automobile using the speed and other features of the smart vehicle system. The device also includes a fire sensor and an eye blink sensor. In the drowsiness awareness model, we were able to detect the drowsiness on the face of a person. This model along with incorporation of sound

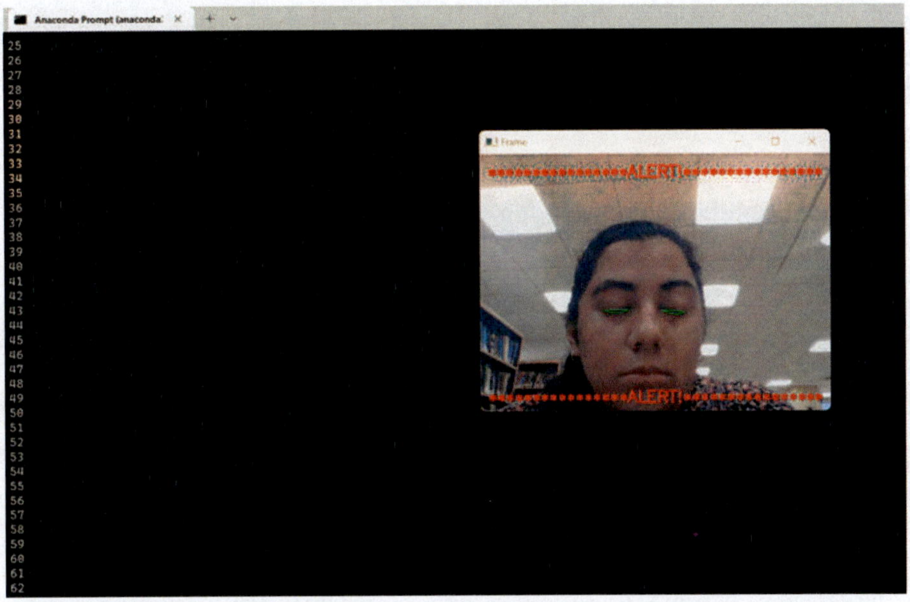

Fig. 11 When the driver is sleepy and closes the eyes, an alert is produced

(for alarm) is used to effectively detect if the driver is asleep in the car whilst driving and thus prevent many accidents.

This system can be extended by using ARM processors and Arduino controllers instead of microcontrollers for very fast operation of processors. In addition, the cameras can be interfaced with this system to see the exact scene of an accident. Moreover, the system can be designed in such a way that it automatically shuts off the vehicle engine whilst accident occurs [6].

LiDAR sensors have proven to be a valuable tool for creating 3D models, particularly for mapping purposes. By utilising LiDAR mapping, it's possible to generate elevation models that can aid in identifying areas prone to flooding, thereby enabling effective allocation of resources. The technology behind LiDAR, or light detection and ranging allows for the creation of detailed 3D elevation maps. Additionally, the long-term objective is to use LiDAR sensors to reduce the frequency of accidents caused by drunk driving, by monitoring incidents and providing valuable information about the vehicles involved. This technology update enhances the safety features of vehicles, making it a significant advancement in the automotive industry. [7]

To offer precise accident detection, a hardware solution can pair physical sensors like accelerometers and gyroscopes with a GPS module and an on-board unit (OBU). The OBU increases the accuracy of accident detection and aids in accident localization. The quantity of accident victims within a car may also be counted using facial recognition technology. The received message's picture may be used to gather facial detection data using the open-source computer vision library (OpenCV). This advanced technology has the potential to improve incident reaction times and passenger safety. We are also planning on coming up with a machine learning model that would give predictions based on the weather in a location and the probability of the occurrence of an accident.

References

1. World Health Organization Road Traffic Injuries Fact Sheet No 358, March 2013. http://www.who.int/mediacentre/factsheets/fs358/en/. Accessed 16 Dec 2017
2. National statistics of road traffic accidents in India, September 2013. http://www.jotr.in/article.asp?issn=0975-7341;year=2013;volume=6;issue=1;spage=1;epage=6;aulast=Ruikar. Accessed 16 Dec 2017
3. Dalai T (2013) Emergency alert and service for automotives for India. Int J Adv Trends Comput Sci Eng Mysore India 2(5):08–12
4. Wakure AR, Patkar AR (2014) Vehicle accident detection and reporting system using Gps and Gsm. IJERGS
5. Malla A, Davidson P, Bones P, Green R, Jones R (2010) Automated video-based measurement of eye closure for detecting behavioral microsleep. In: 32nd annual international conference of the IEEE, Buenos Aires, Argentina
6. Biswal AK, Singh D, Pattanayak B, Samanta D, Yang M-H (2021) IoT-based smart alert system for drowsy driver detection. Wirel Commun Mob Comput 2021:1–13. https://doi.org/10.1155/2021/662721
7. Ray A, Das A, Kundu A, Ghosh A, Rana TK (2017) Prevention of driving under influence using microcontroller. In: 2017 1st international conference on electronics, materials engineering and nano-technology (IEMENTech). Kolkata, India, pp 1–2. https://doi.org/10.1109/IEMENTECH.2017.8077023
8. Meena A, Iyer S, Nimje M, JogJekar S, Jagtap S, Rahman M (2014) Automatic accident detection and reporting framework for two wheelers. In: IEEE international conference on advanced communication control and computing technologies (ICACCCT), pp 962–967

A New Algorithm for Encryption and Decryption Using AUM Block Sum Labeling

A. Uma Maheswari and C. Ambika

1 Introduction

The theory of cryptography is developing rapidly in recent years. The appropriate integration of the cryptographic technique with the graph labeling [1] helps to ensure safe communication by preventing the intrusion of any secret messages during conversion. In 1949, Shannon [2] made a proposal for modern cryptography. In [3], Gallian gave a review of graph labeling. Uma Maheswari and Azhagarasi [4] established the concept of AUM block labeling. AUM block sum labeling is the new block labeling technique developed in the scope for applications to heterogeneous field. In [5–11], the discussion of AUM block sum labeling for various graph families is given. New encoding and decoding methods involving AUM block sum labeling are presented in [12, 13]. Here, we present a new technique using AUM block sum labeling on any block graph by relating the numbers into the perfect square number. A key is required to ensure confidentiality during the encryption and decryption operations. Here, we have given a key as a matrix form, ensuring more secure transmission.

2 Preliminaries

We present the basic graph theory and cryptography concepts relevant for the proposed technique in this part.

A. U. Maheswari
PG & Research Department of Mathematics, Quaid-E-Millath Government College for Women (Autonomous), Chennai, India

C. Ambika (✉)
Department of Mathematics, Ethiraj College for Women, Chennai, India
e-mail: ambika_c@ethirajcollege.edu.in

J. K. Mandal et al. (eds.), *Proceedings of International Conference on Network Security and Blockchain Technology*, Lecture Notes in Networks and Systems 738,
https://doi.org/10.1007/978-981-99-4433-0_5

2.1 Definition: Block Graph [14, 15]

A maximal non-separable subgraph of the graph is the block graph of G.

2.2 Definition: Triangular Snake Graph [16]

The triangular snake T_n is a graph that is obtained by replacing each edge of the path graph P_n with a triangle C_3.

2.3 Definition: Plain Text [17]

The original message that the sender desires to communicate to the receiver is in plain text.

2.4 Definition: Cipher Text [17]

The encrypted message that contains the plain text in an unreadable format is called cipher text.

2.5 Definition: Encryption [17]

Encryption is the process of converting plain text into cipher text. A key and an encryption algorithm is needed for the encryption process.

2.6 Definition: Decryption [17]

Decryption is the process of reversing encryption. It is the transformation of cipher text into plain text. A decryption algorithm and a key are needed for the decryption process.

2.7 *Definition: Key [17]*

A key is a particular symbol or a text with numbers or letters. The key is used to encrypt plain text and decrypt cipher text, respectively.

2.8 *Definition: AUM Block Sum Labeling [4]*

Consider a graph G with p vertices and a vertex set $V(G)$, q edges and an edge set $E(G)$ and block set with b blocks, $p, q, b \geq 1$.

We say that the graph G admits AUM block sum labeling if there exists a bijection.

$f : V(G) \to \{1, 2, 3, \ldots p\}$ and $f^* : E(G) \to Z^+$ induced from f by $f^*(uv) = f(u) + f(v)$ and $f^{**} : B(G) \to Z^+$ defined as follows:

Let B_j be incident with the vertices $v_{j_1}, v_{j_2}, \ldots..v_{j_k}$, $1 \leq j_k \leq p$ and edges e_{j_1}, $e_{j_2}, \ldots e_{j_m}$, $1 \leq j_m \leq q$.

Then, $f^{**}(B_j) = \sum_{i=1}^{k} f(v_{j_i}) + \sum_{i=1}^{m} f^*(e_{j_i})$ and $f^{**}(B_j) \neq f^{**}(B_i)$ for $1 \leq i, j \leq b$ and $i \neq j$.

2.9 *AUM Block Sum Labeling - Triangular Snake Graph [4]*

Consider the triangular snake graph, T_n, $n \geq 2$.

Define $f^{**} : B(T_n) \to Z^+$ by $f^{**}(B_i) = 18i$, $1 \leq i \leq n - 1$.

For $i \neq j$, $f^{**}(B_j) \neq f^{**}(B_i)$ as $18j \neq 18i$ implying the block labels $f^{**}(B_i)$ are distinct.

3 Main Results

3.1 *New Encryption and Decryption Algorithm with Illustration*

In [13], we have presented various coding algorithms based on the concept of relating numbers to geometric mean to encode and decode a message. In this section, we devise a new encryption and decryption algorithm by taking any block graph and assign AUM block sum labeling in the beginning to encode the message. For the illustration, we have considered the triangular snake graph T_{n+1}.

3.2 Encryption Algorithm

- Consider any block graph with n blocks and assign AUM block sum labeling where n indicates message length. The block labels are served as a key in the first row of the matrix.
- Assign numbers from 0 to 25 to the alphabets in the order of 0, 1, 2, ... 12 to N, $O \ldots Z$ and 13, 14, ... 25 to $A, B, C, \ldots M$.
- Find the corresponding number for each alphabet using the encoding table and take it as x_i for $i = 1, 2, 3 \ldots n$.
- Find any positive integer y_i to make $x_i + y_i = z_i$ for $i = 1, 2, 3 \ldots n$ a perfect square number. The numbers y_i are the second row of the key matrix.
- Find the difference between the block label B_i and z_i, i.e., $w_i = (B_i - z_i)(\text{mod}26)$ for $i = 1, 2, 3 \ldots n$.
- Find the alphabet corresponding to each w_i for $i = 1, 2, 3 \ldots n$.
- Send the sequence of the text w_i as cipher text for decryption.

Key

The key is the matrix with 2 rows and n (message length) columns containing block labels in the first row and values of $y_i's$ in the second row.

Decryption Algorithm

- From the cipher text, using the encoding table, determine the number that represents each character, and take it as w_i for $i = 1, 2, 3 \ldots n$.
- Find $z_i = (B_i - w_i)(\text{mod}26)$ for $i = 1, 2, 3 \ldots n$.
- Using key numbers y_i, find $x_i = z_i - y_i$ for $i = 1, 2, 3 \ldots n$.
- Find the alphabet that corresponds to each x_i to get the plain text through the encoding table.

Illustration I

Encryption

Consider the plain text "**RAINBOW**", which is to be converted.

The message length is 7.

Therefore, consider the triangular snake graph T_8 with 7 blocks.

AUM block sum labeling in triangular snake graph T_8 is given in Fig. 1.

Find the corresponding number for each alphabet using the encoding table.

R	A	I	N	B	O	W
4	13	21	0	14	1	9

Denote $x_1 = 4,\ x_2 = 13,\ x_3 = 21,\ x_4 = 0,\ x_5 = 14,\ x_6 = 1,\ x_7 = 9$.

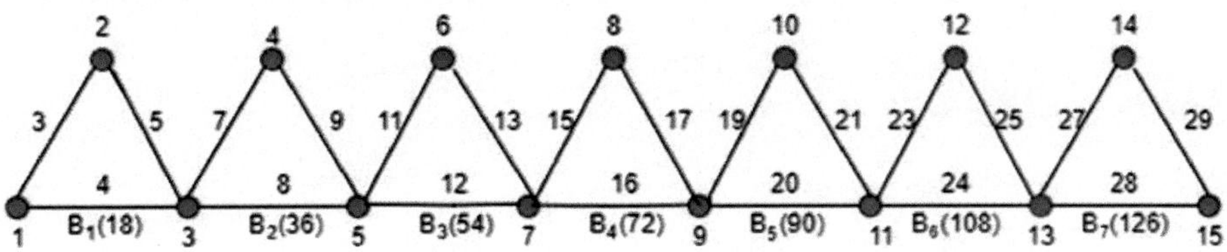

Fig. 1 T_8 AUM block sum labeling

Table 1 Encoding table

A	B	C	D	E	F	G	H	I	J	K	L	M
13	14	15	16	17	18	19	20	21	22	23	24	25
N	O	P	Q	R	S	T	U	V	W	X	Y	Z
0	1	2	3	4	5	6	7	8	9	10	11	12

Find any positive integer y_i to make $x_i + y_i = z_i$ for $i = 1, 2, 3 \ldots 7$ a perfect square number.

Therefore,

$$x_1 + y_1 = 4 + 5 = 9 = z_1$$

$$x_2 + y_2 = 13 + 3 = 16 = z_2$$

$$x_3 + y_3 = 21 + 4 = 25 = z_3$$

$$x_4 + y_4 = 0 + 1 = 1 = z_4$$

$$x_5 + y_5 = 14 + 2 = 16 = z_5$$

$$x_6 + y_6 = 1 + 3 = 4 = z_6$$

$$x_7 + y_7 = 9 + 7 = 16 = z_7$$

The values of y_i are 5, 3, 4, 1, 2, 3, 7.

Find the difference between the block label B_i and z_i.

$$\text{i.e., } w_i = (B_i - z_i)(\text{mod}26) \ \text{for } i = 1, 2, 3 \ldots 7.$$

$$w_1 = (B_1 - z_1)(\text{mod}26) = (18 - 9)(\text{mod } 26) = 9$$

$$w_2 = (B_2 - z_2)(\text{mod}26) = (36 - 16)(\text{mod}26) = 20$$

$$w_3 = (B_3 - z_3)(\text{mod } 26) = (54 - 25)(\text{mod}26) = 3$$

$$w_4 = (B_4 - z_4)(\text{mod } 26) = (72 - 1)(\text{mod } 26) = 19$$

$$w_5 = (B_5 - z_5)(\text{mod } 26) = (90 - 16)(\text{mod } 26) = 22$$

$$w_6 = (B_6 - z_6)(\text{mod } 26) = (108 - 4)(\text{mod } 26) = 0$$

$$w_7 = (B_7 - z_7)(\text{mod}26) = (126 - 16)(\text{mod}26) = 6$$

From the encoding table, the letters that correspond to the above numbers are WHQGJNT.

Therefore, the cipher text for the given message is WHQGJNT.

$$\textbf{Key}: \begin{bmatrix} 18 & 36 & 54 & 72 & 90 & 108 & 126 \\ 5 & 3 & 4 & 1 & 2 & 3 & 7 \end{bmatrix}$$

Decryption

Apply the above decryption procedure to the encrypted text you received to retrieve the plain text.

From the cipher text, the message length is 7.

Here, the cipher text is WHQGJNT.

Using the encoding table, identify the number that each character corresponds to and take it as

$$y_i \text{ for } i = 1, 2, 3 \ldots 7.$$

W	H	Q	G	J	N	T
9	20	3	19	22	0	6

Take $w_1 = 9,\ w_2 = 20,\ w_3 = 3,\ w_4 = 19,\ w_5 = 22,\ w_6 = 0,\ w_7 = 6$.
Find $z_i = (B_i - w_i)(\text{mod}26)$ for $i = 1, 2, 3 \ldots 7$.

$$z_1 = (B_1 - w_1)(\text{mod}26) = (18 - 9)(\text{mod}26) = 9$$

$$z_2 = (B_2 - w_2)(\text{mod}26) = (36 - 20)(\text{mod } 26) = 16$$

$$z_3 = (B_3 - w_3)(\text{mod}26) = (54 - 3)(\text{mod}26) = 25$$

$$z_4 = (B_4 - w_4)(\text{mod}26) = (72 - 19)(\text{mod}26) = 1$$

$$z_5 = (B_5 - w_5)(\text{mod } 26) = (90 - 22)(\text{mod}26) = 16$$

$$z_6 = (B_6 - w_6)(\text{mod}26) = (108 - 0)(\text{mod}26) = 4$$

$$z_7 = (B_7 - w_7)(\text{mod}26) = (126 - 6)(\text{mod}26) = 16$$

Consider the second row of the key matrix, 5, 3, 4, 1, 2, 3, 7, as the values of y_i. Find $x_i = z_i - y_i$ for $i = 1, 2, 3 \ldots 7$ where y_i's are key numbers.

$$x_1 = z_1 - y_1 = 9 - 5 = 4$$

$$x_2 = z_2 - y_2 = 16 - 3 = 13$$

$$x_3 = z_3 - y_3 = 25 - 4 = 21$$

$$x_4 = z_4 - y_4 = 1 - 1 = 0$$

$$x_5 = z_5 - y_5 = 16 - 2 = 14$$

$$x_6 = z_6 - y_6 = 4 - 3 = 1$$

$$x_7 = z_7 - y_7 = 16 - 7 = 9$$

Using the encoding table, we can infer that the plain text represented by the above numbers is "**RAINBOW**".

4 Conclusion

We devised a unique algorithm for the transmission of plain text into cipher text and vice versa. AUM block sum labeling on any block graph and a perfect square number are integrated in this coding for transmission. A triangular snake graph is considered for the illustration. The theory developed can be further extended to other standard graphs by employing new coding techniques.

References

1. Ni B, Qazi R, Rehman SU, Farid G (2021) Some graph-based encryption schemes. J Math 2021:1–8. Article ID 6614172
2. Shannon CE (1949) Communication theory of secrecy systems. Bell Syst Tech J 28(4):656–715
3. Gallian JA (2021) A dynamic survey of graph labeling. Electron J Comb
4. Uma Maheswari A, Azhagarasi S (2022) New labeling for graphs-AUM block sum labeling. Int J Curr Sci 12(1):574–584. ISSN: 2250-1770
5. Uma Maheswari A, Azhagarasi S (2022) AUM block labeling for cycle cactus block graphs. Compl Eng J 13(4):84–96
6. Uma Maheswari A, Azhagarasi S (2002) AUM block sum labeling for some special graphs. Int J Mech Eng 7(Special Issue 5):102–110. ISSN: 0974-5823
7. Uma Maheswari A, Bala Samuvel J (2022) New coloring for blocks—AUM block coloring for standard graphs. J Compl Eng J 13(5):68–77. ISSN: 0898-3577
8. Uma Maheswari A, Azhagarasi S (2022) AUM block sum labelling & AUM block labelling for perfect binary tree $T_{(3,1)}$ $T_{(4,1)}$ & $T_{(5,1)}$. In: Advances in graph labelling, colouring and power domination theory, vol 1. NFED Publications, pp 1–17. E-ISBN: 978-81-95499-8-5
9. Uma Maheswari A, Bala Samuvel J (2022) AUM block coloring for triangular snake graph family. In: Advances in graph labelling, colouring and power domination theory, vol 1. NFED Publications, pp 72–91. E-ISBN: 978-81-95499-8-5
10. Uma Maheswari A, Purnalakshimi AS (2022) Aum block labelling for friendship, tadpole and cactus graphs. Neuro Quantol 20(6):7876–7884. eISSN 1303-5150
11. Uma Maheswari A, Purnalakshimi AS (2022a) AUM block labelling for snake graphs and Dutch windmill graph. Neuro Quantol 20(9):414–421
12. Uma Maheswari A, Azhagarasi S (2022) A new algorithm for encoding and decoding using Aum block labelling. Compl Eng J 13(4):264–274. ISSN No: 0898-3577
13. Uma Maheswari A, Ambika C (2022) New coding algorithms using AUM block SUM labeling. Neuro Quantolgy 20(9):377–385. eISSN 1303-5150
14. Bondy JA, Murty USR (1976) Graph theory with applications. Elsevier Science Publishing Co., Inc., pp 1–264. ISBN: 0-444-19451-7
15. Harary F (1972) Graph theory. Addison-Wesley, Reading, Mass
16. Ponraj R, Sathish Narayanan S (2013) Difference cordiality of some graphs obtained from double alternate snake graphs. Global J Math Sci Theory Pract 5(2013):167–175
17. Agrawal M, Mishra P (2012) A comparative survey on symmetric key encryption techniques. Int J Comput Sci Eng 4(05):877–882. ISSN: 0975-3397

Secured Reversible Data Hiding Scheme with NMI Interpolation and Arnold Transformation

Manasi Jana, Biswapati Jana, Shubhankar Joardar, Sharmistha Jana, and Tzu Chuen Lu

1 Introduction

Data hiding [1] is an approach to hide secret data into digital media for public transmission. A key factor in ensuring the security of digital transmission is the reversible data hiding (RDH) approach [2]. Reversibility is desired with great accuracy in military, satellite communication, medical imaging, and legal investigations. Over the years, interpolated-based RDH approaches [3] have become increasingly prominent. Jung and Yoo [4] first suggested a reversible data hiding (RDH) scheme which is known as neighbor mean interpolation (NMI). Lee and Huang [5] improved the Jung and Yoo's scheme [4] in terms of PSNR and embedding capacity. Tang et al. [6] suggested a high capacity reversible data hiding (RDH) method with multi-layer embedding. But it suffers from low image quality of stego image and low time complexity. The suggested method results in an appreciable increase in data hiding capacity and higher stego image visual quality. The contributions of the proposed method are described below.

M. Jana
Department of Computer Applications, Haldia Institute of Technology, Haldia, West Bengal, India

B. Jana (✉)
Department of Computer Science, Vidyasagar University, West Midnapore, West Bengal, India
e-mail: biswapatijana@gmail.com

S. Joardar
Department of Computer Science and Engineering, Haldia Institute of Technology, Haldia, West Bengal, India

S. Jana
Department of Mathematics, Midnapore College [Autonomous], Midnapore, West Bengal, India

T. C. Lu
Department of Information Management, Chaoyang University of Technology, Taichung, Taiwan, ROC

J. K. Mandal et al. (eds.), *Proceedings of International Conference on Network Security and Blockchain Technology*, Lecture Notes in Networks and Systems 738,
https://doi.org/10.1007/978-981-99-4433-0_6

(i.) A novel RDH (Reversible Data Hiding) method has been suggested for better imperceptibility and payload.
(ii.) At embedding phase, the input image is taken as reference for getting good quality stego image.
(iii.) Secret message has been scrambled by Arnold transformation before embedding to implement the 1st level of security of the suggested method.
(iv.) Instead of difference value between two adjacent pixels, here upper bound (ub) has been taken to achieve high embedding capacity.
(v.) Before hiding, the scrambled secret message has been modified by a range of pixels for implementing 2nd level of security.

The remainder of the article is structured as follows. The theoretical background is presented at Sect. 2. In Sect. 3, the proposed scheme has been illustrated in detail. Section 4 gives some experimental analysis. Finally, the Sect. 5 gives some conclusions.

2 Theoretical Background

In this section, image interpolation technique and Arnold transformation are briefly discussed which are applied in the proposed scheme.

2.1 NMI Interpolation

In 2009, Jung and Yoo [4] first suggested an image interpolated technique (NMI) to hide the secret data into the cover image. The $M \times N$ sized input image I^{im} is downsized to original image O^{im} of size $\left(\frac{M}{2}+1\right) \times \left(\frac{N}{2}+1\right)$. Then the original image O^{im} is scaled up to cover image C^{im} of size $M \times N$ using NMI interpolation technique. For a 3×3 block, the interpolated pixels are calculated using Eq. (1).

$$\begin{aligned}
C^{im}(1,1) &= I^{im}(1,1) \\
C^{im}(1,2) &= (I^{im}(1,1) + I^{im}(1,3))/2 \\
C^{im}(2,1) &= (I^{im}(1,1) + I^{im}(3,1))/2 \\
C^{im}(2,2) &= (I^{im}(1,1) + C^{im}(1,2) + C^{im}(2,1))/3
\end{aligned} \tag{1}$$

2.2 Arnold Transformation

Arnold transform [7] is applied to scramble digital images because of its periodicity. Vladimir Igorevich Arnold first proposed this transformation which was applied on cat images. Therefore, it is also called cat mapping where the positions of pixels are changed to new positions resulting in a distorted image. After a number of cycles, the original image can be recovered in accordance with the periodicity of the Arnold scrambling. Using Arnold transformation, a two-dimensional array has been scrambled using Eq. (2).

$$\begin{bmatrix} x' \\ y' \end{bmatrix} = \begin{bmatrix} 1 & 1 \\ 1 & 2 \end{bmatrix} \begin{bmatrix} x \\ y \end{bmatrix} (\text{mod } N_1) \tag{2}$$

where $x, y \in 0, 2, \ldots, N_1 - 1$; N_1 is the order of a square matrix.

3 The Proposed Scheme

Here, we have proposed a novel data hiding and data extracting algorithm based on NMI interpolation and Arnold transformation. A block diagram of the proposed method has been sketched in Fig. 1. The proposed scheme has been divided into two phases (a) Data hiding phase and (b) Data extracting phase. At data hiding phase, the scrambled secret image is embedded into the cover image producing stego image where as at data extracting phase the original image and the secret image have been retrieved from stego image without any distortion.

3.1 Data Hiding Phase

The process of data hiding is described below.

Step 1: An input image (I^{im}) of sized $M \times N$ is downsized to original image (O^{im}) of size $\left(\frac{M}{2} + 1\right) \times \left(\frac{N}{2} + 1\right)$.

Step 2: Then the NMI interpolation technique has been applied to produce cover image (C^{im}) of $M \times N$ using Eq. (1).

Step 3: The gray values of pixels that is 0 to 255 are grouped into seven ranges where lower bound (lb) starts from 0 and rest are calculated using Eq. (3). The upper bound (ub) is calculated using Eq. (4).

$$lb_i = 2^i, \quad i = 2 \text{ to } 7 \tag{3}$$

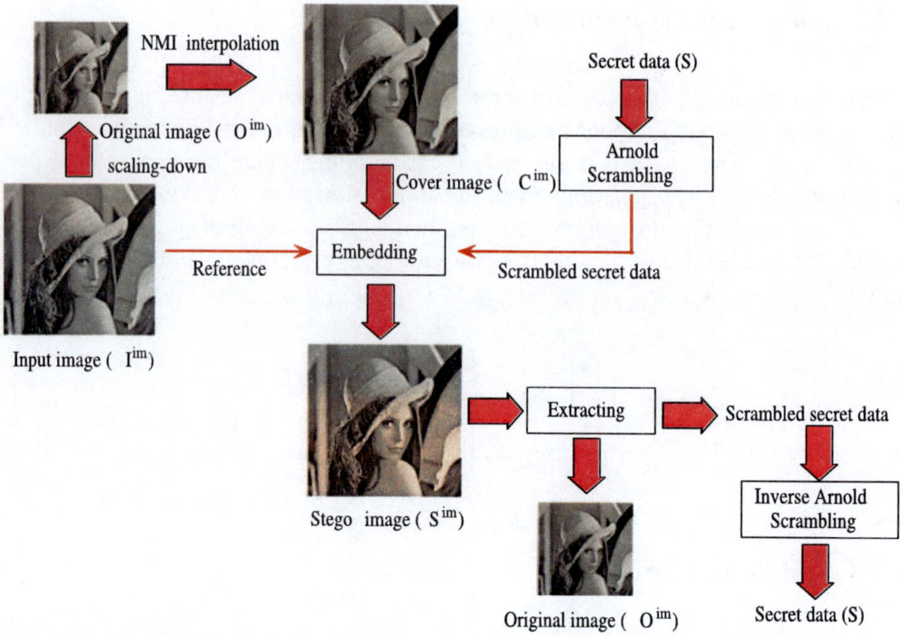

Fig. 1 Block diagram of proposed method

$$ub_{i-1} = 2^i - 1, \quad i = 2 \text{ to } 8 \tag{4}$$

Step 4: The cover image is broken up into overlapping blocks B_i (3×3), $i = 1$ to $\lfloor \frac{M}{3} \rfloor \times \lfloor \frac{N}{3} \rfloor$.

Step 5: The differences (δ_i), $i = 1$ to 3 between adjacent pixels of each block B_i are calculated using Eq. (5) and then find out the upper bound $(ub_i, i = 1 \text{ to } 3)$ and lower bound $(lb_i, i = 1 \text{ to } 3)$ of the range to which the difference value (δ_i) belongs to.

$$\begin{aligned}
\delta_1 &= |C^{im}(e, f) - C^{im}(e, f+2)| \\
\delta_2 &= |C^{im}(e, f) - C^{im}(e+2, f)| \\
\delta_3 &= \max \quad (|C^{im}(e, f) - C^{im}(e, f+2)|, |C^{im}(e, f) - C^{im}(e+2, f)|, \\
&\quad |C^{im}(e, f) - C^{im}(e+2, f+2)|)
\end{aligned} \tag{5}$$

where $e = 1, 3, 5, \ldots, M-2$ and $f = 1, 3, 5, \ldots, N-2$.

Step 6: Instead of using difference value which was used in previous schemes, here the upper bound $(ub_i, i = 1 \text{ to } 3)$ has been used to calculate the number of bits n_k, $k = 1$ to 3, to be embedded in a pixel of a block B_i to enhance the embedding capacity. The difference may be 0 or 1 at smooth region where no data can be embedded in schemes proposed in past. Here upper bound starts from 3, so we can embed at least one bit at smooth region which enhances the embedding capacity of the suggested method. The value of n_k is calculated using Eq. (6).

$$n_k = \lfloor \log_2 ub_k \rfloor, \quad k = 1 \text{ to } 3 \tag{6}$$

Step 7: A secret data (S) has been taken and scrambled using Arnold transformation to implement the 1st level of security of the proposed method using Eq. (2). Let, the cycle of scrambling is τ. If we know the cycle of scrambling, we can restore the original secret message by applying inverse Arnold scrambling method.

Step 8: Then the sub-secret messages (b_k) with n_k bits are taken from scrambled secret message (β) and are adjusted to modified sub-secret messages (b'_k) using Eq. (7) to implement the 2nd level of security of the suggested method.

$$b'_k = |b_k - lb_k|, \quad k = 1 \text{ to } 3 \tag{7}$$

Step 9: Then the modified sub-secret messages (b'_k) are embedded into cover image (C^{im}) to produce a stego image (S^{im}) using Eqs. (8), (9) and (10). Here, The input image I^{im} is taken as reference to enhance the image quality of the stego image S^{im}.

$$S^{im}(e, f+1) = \begin{cases} C^{im}(e, f+1) - b'_1, & \text{if} \quad C^{im}(e, f+1) > I^{im}(e, f+1) \\ C^{im}(e, f+1) + b'_1, & \text{otherwise} \end{cases} \tag{8}$$

$$S^{im}(e+1, f) = \begin{cases} C^{im}(e+1, f) - b'_2, & \text{if} \quad C^{im}(e+1, f) > I^{im}(e+1, f) \\ C^{im}(e+1, f) + b'_2, & \text{otherwise} \end{cases} \tag{9}$$

$$S^{im}(e+1, f+1) = \begin{cases} C^{im}(e+1, f+1) - b'_3, & \text{if} \quad C^{im}(e+1, f+1) > I^{im}(e+1, f+1) \\ C^{im}(e+1, f+1) + b'_3, & \text{otherwise} \end{cases} \tag{10}$$

where $e = 1, 3, 5, \ldots, M-2$ and $f = 1, 3, 5, \ldots, N-2$.

Step 10: **Step 5** to **Step 9** should be repeated up until all the secret messages are hidden into the cover image.

3.2 Data Extracting Phase

The process of extracting data from stego image (S^{im}) is as follows:

Step 1: From stego image (S^{im}), the original image (O^{im}) has been extracted from non-interpolated pixels showing reversibility of the proposed method.

Step 2: The cover image (C^{im}) has been produced from original image (O^{im}) by applying NMI interpolation using Eq. (1).

Step 3: Then the difference δ_k and number of bits n_k has been calculated using Eqs. (5) and (6) respectively.

Step 4: The modified secret sub-messages b'_k have been calculated using Eq. (11).

$$\begin{aligned} b'_1 &= |S^{im}(e, f+1) - C^{im}(e, f+1)| \\ b'_2 &= |S^{im}(e+1, f) - C^{im}(e+1, f)| \\ b'_3 &= |S^{im}(e+1, f+1) - C^{im}(e+1, f+1)| \end{aligned} \tag{11}$$

where $e = 1, 3, 5, \ldots, M-2$ and $f = 1, 3, 5, \ldots, N-2$.

Step 5: The scrambled sub-secret messages b_k have been calculated using Eq. (12) and the inverse Arnold scrambling has been applied on scrambled secret image to get the original secret image.

$$b_k = |lb_k - b'_k|, \quad k = 1 \text{ to } 3 \tag{12}$$

3.3 Numerical Analysis

A numerical analysis of data hiding is explained in Fig. 2. Here, a 3×3 sized input image (I^{im}) is taken and scaled-down to original image (O^{im}) on which NMI interpolation is applied to produce a cover image (C^{im}) of sized 3×3. The gray values of pixels are grouped into seven ranges as 0 to 3, 4 to 7, 8 to 15, 16 to 31, 32 to 63, 64 to 127, and 128 to 255, where 0, 4, 8, ... are lower bounds (lb_i, $i = 1$ to 7) and 3, 7, 15, ... are upper bounds (ub_i, $i = 1$ to 7) of the seven ranges. The Arnold scrambling is applied on secret data (S) using Eq. (2) to produce scrambled secret data $\beta = (100010011\ldots)_2$ which is embedded into cover image to produce a stego image (S^{im}). Here, the cycle of scrambling (τ) = 1. The differences $\delta_1 = 18$, $\delta_2 = 2$ and $\delta_3 = 18$ are calculated using Eq. (5). According to the differences δ_i, the lower bounds and upper bounds $lb_1 = 16$, $ub_1 = 31$, $lb_2 = 0$, $ub_2 = 3$, $lb_3 = 16$, and $ub_3 = 31$ are found out from the set of ranges. The number of bits $n_1 = 4$, $n_2 = 1$, and $n_3 = 4$ are calculated using Eq. (6). So, the total $4 + 1 + 4 = 9$ bits can be hidden into a 3×3 overlapping block B_i. Then the scrambled sub-secret messages

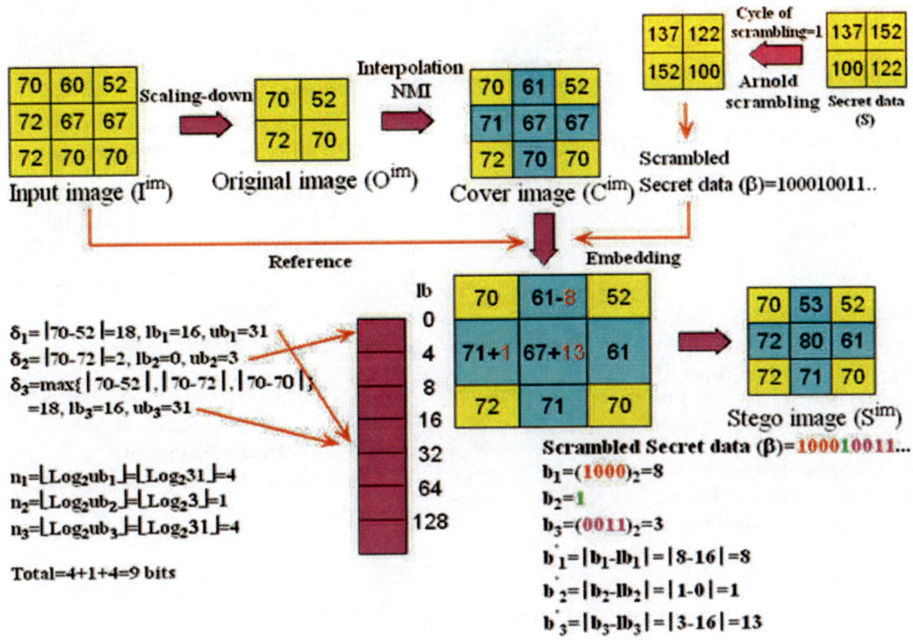

Fig. 2 Numerical analysis of data embedding

b_k are modified to b'_k and embedded to cover image (C^{im}) using Eqs. (8), (9) and (10) which results in a stego image (S^{im}). A numerical example of data extraction is shown in Fig. 3 which is the reverse process of data embedding.

4 Experimental Analysis and Comparisons

In our experiments, six input images "Lena", "Pepper", "Tiffany", "Tank", "Pirate", and "Cameraman" are used, each with size 512 × 512, as shown in Fig. 4. The secret image has been scrambled four times and hidden into the cover image to produce stego image as depicted in Fig. 4. Peak-Signal-to-Noise-Ratio (PSNR), Mean Squared Error (MSE), Pure Payload (bits), Payload (bpp), Q-index and Bit Error Rate (BER) are calculated to analyze the proposed scheme when a secret image of sized 100 × 100 is taken as shown in Table 1. It shows satisfactorily visual quality of stego image and high embedding capacity of the proposed scheme. The proposed scheme is compared with Jung and Yoo [4], Lee and Huang [5], Malik et al. [8], Wahed and Nyeem [9], Tang et al. [6], Hu and Li [10], Zhang et al. [11], and Shaik et al. [12] in terms of average PSNR (dB) value between input and stego images as depicted in Table 2. It depicts that the proposed technique provide high PSNR (dB) value than other methods. This method also has higher embedding capacity than Jung and Yoo [4], Lee and Huang [5], Zhang et al. [11], and Lee et al. [13] as shown in Table 3.

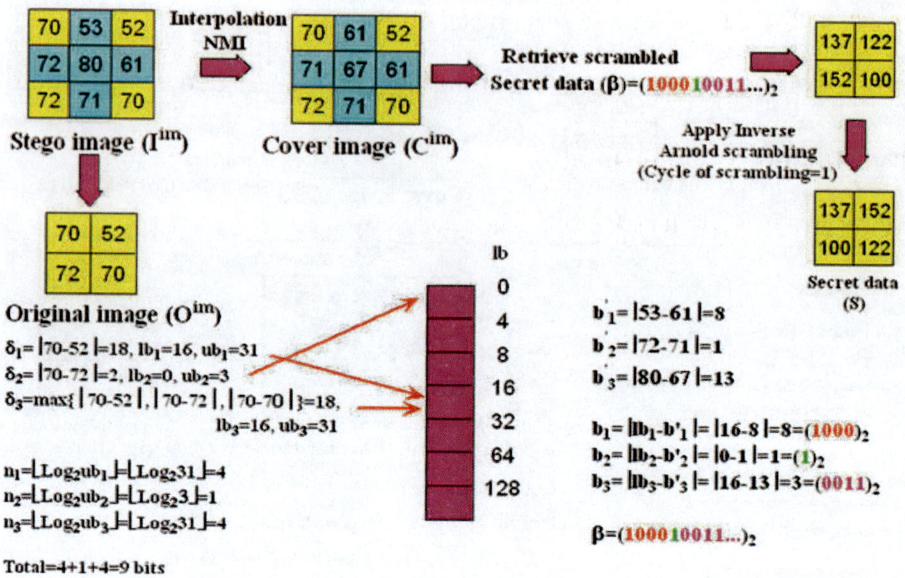

Fig. 3 Numerical analysis of data extraction

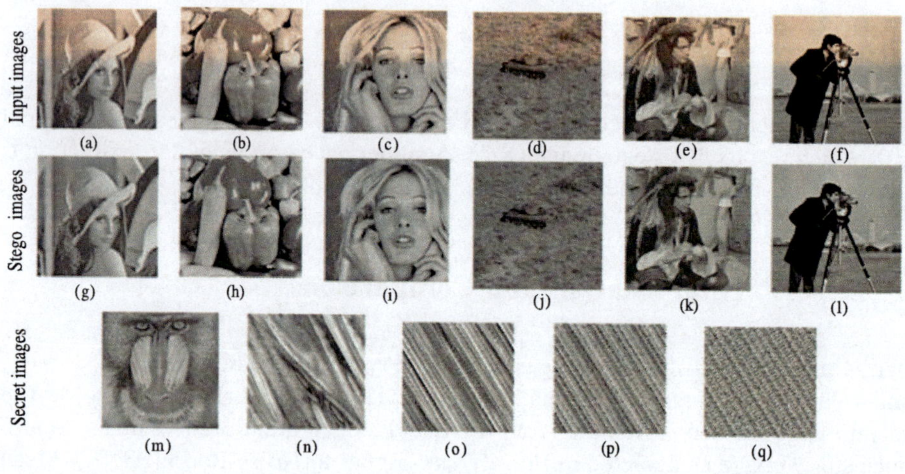

Fig. 4 Input images (512 × 512): **a** Lena **b** Pepper **c** Tiffany **d** Tank **e** Pirate **f** Cameraman. Stego images (512 × 512): **g** Lena **h** Pepper **i** Tiffany **j** Tank **k** Pirate **l** Cameraman. Secret images (100 × 100): **m** Original image **n** Scrambling 1 time of original image **o** Scrambling 2 times of original image **p** Scrambling 3 times of original image **q** Scrambling 4 times of original image

Table 1 PSNR (dB), MSE, pure payload (bits), payload (bpp), Q-index and BER of the sample text images

Images (512 × 512)	PSNR (dB)	MSE	Pure payload (bits)	Payload (bpp)	Q-index	BER
Lena	29.33	75.80	449,974	1.71	0.99	0.21
Pepper	29.57	71.78	450,714	1.71	0.99	0.22
Tiffany	28.87	84.21	493,933	1.88	0.99	0.22
Tank	29.46	73.56	583,011	2.22	0.99	0.25
Pirate	28.26	97.03	506,815	1.93	0.99	0.22
Cameraman	27.68	110.82	378,601	1.44	0.98	0.17
Average	28.86	85.53	477,170	1.81	0.99	0.21

Table 2 Comparison of average PSNR (dB) with other methods

Schemes	[4]	[5]	[8]	[9]	[6]	[10]	[11]	[12]	Proposed
Average PSNR (dB)	28.67	26.91	28.70	27.92	17.06	26.75	27.41	27.66	28.86

Table 3 Comparison of average payload (bpp) with other methods

Schemes	[4]	[5]	[11]	[13]	Proposed
Average payload (bpp)	0.89	1.45	1.63	0.33	1.81

4.1 Overflow and Underflow Analysis

Data embedding involves adding or subtracting of secret data from interpolated pixels, which could cause an overflow or underflow issue. Suppose, we have a modified scrambled secret data $b' = 8$, pixel of input image $I^{im}(e, f) = 254$ and interpolated pixel of cover image $C^{im}(e, f) = 253$ at $(e, f)^{th}$ position. Here, $C^{im}(e, f) < I^{im}(e, f)$. So, addition will be executed. That is, $S^{im}(e, f) = C^{im}(e, f) + b' = 253 + 8 = 261$ which results in overflow problem because the range of pixel value is 0 to 255. To overcome this problem, we have subtracted b' from $C^{im}(e, f)$ instead of addition. The inverse process has been applied to solve the underflow problem of the proposed scheme.

5 Conclusions

In this paper, we have proposed a 2-level secure reversible data hiding (RDH) method using interpolation and Arnold transformation. The proposed scheme keeps a relative balance between image quality and data hiding capacity. The upper bound of the range interval is used to increase the embedding capacity whereas the input image is used as reference to enhance the image quality of the stego image. Before embedding, the secret message is scrambled using Arnold transform. The Arnold transform and the range interval have been used to implement two-level security of the proposed scheme. The experimental results and analyses show the superiority of the proposed method over other methods. In future, the suggested method can be carried out in the transform domain and can be tested its robustness against different statistical attacks.

References

1. Sharma VK, Sharma PC, Goud H, Singh A (2022) Hilbert quantum image scrambling and graph signal processing-based image steganography. Multimedia Tools Appl 81(13):17817–17830
2. Jana M, Jana B (2020) An improved data hiding scheme through image interpolation. In: Computational intelligence in pattern recognition. Springer, pp 157–169
3. Jana M, Joardar S, Jana B (2022) A new reversible data hiding scheme by altering interpolated pixels exploiting neighbor mean interpolation (NMI). In: International conference on computational intelligence in pattern recognition. Springer, pp 393–402
4. Ki-Hyun J, Kee-Young Y (2009) Data hiding method using image interpolation. Comput Standards Interfaces 31(2):465–470
5. Chin-Feng L, Yu-Lin H (2012) An efficient image interpolation increasing payload in reversible data hiding. Exp Syst Appl 39(8):6712–6719
6. Mingwei T, Hu J, Wen S (2014) A high capacity image steganography using multi-layer embedding. Optik-Int J Light Electron Opt 125(15):3972–3976
7. Panchikkil S, Manikandan VM, Zhang Y-D (2022) A convolutional neural network model based reversible data hiding scheme in encrypted images with block-wise Arnold transform. Optik 250:168137
8. Malik A, Sikka G, Verma HK (2017) Image interpolation based high capacity reversible data hiding scheme. Multimedia Tools Appl 76(22):24107–24123
9. Wahed MdA, Nyeem H (2019) Reversible data hiding with interpolation and adaptive embedding. Multimedia Tools Appl 78(8):10795–10819
10. Hu J, Tianrui L (2015) Reversible steganography using extended image interpolation technique. Comput Electr Eng 46:447–455
11. Xianquan Z, Zerui S, Zhenjun T, Yu C, Xiaoyun W (2017) High capacity data hiding based on interpolated image. Multimedia Tools Appl 76(7):9195–9218
12. Ahmad S et al (2019) High capacity reversible data hiding using 2d parabolic interpolation. Multimedia Tools Appl 78(8):9717–9735
13. Chin-Feng L, Chi-Yao W, Cheng-Yu K (2019) Reversible data hiding using Lagrange interpolation for prediction-error expansion embedding. Soft Comput 23(19):9719–9731

Face Mask Detection Exploiting CNN and MobileNetV2

Nandana Ghosh, Biswapati Jana, Sharmistha Jana, and Nguyen Kim Sao

1 Introduction

Since November 2019, the whole world was suffering from pandemic situation. As days were going, the more new-variants were coming to be seen. All viruses, including SARS-CoV-2, the virus that causes COVID-19, was shown changes over time. Most changes have little to no impact on the virus' properties.

But unfortunately, some people are knowingly avoiding to wear mask. The result of which is maximum number of people are getting affected by this virus. Many steps are taken by the government to fight against COVID-19. Some of these are night curfews from around 10 pm, several lockdown in country, vaccination camp in school and colleges were set up. Doctors were first to be vaccinated as they have to look after the patients.

In the year 2021, the month of August, September, November, and December has shown the reduction of cases of COVID-19. But at the end of December and Start of January 2022 has shown high peak raise cases of coronavirus. This is due to the carelessness of people toward COVID-19. As the Christmas Eve and New Year celebration has led this happen. People just avoided wearing mask.

As many persons were seen without mask in their face had come to visit Church, Zoo, Parks, Temples, Fair, Museums, and the tourist places. Many people in the markets, railway stations, bus, trains, Airports were seen without mask. When they

N. Ghosh (✉) · B. Jana
Department of Computer Science, Vidyasagar University, Midnapore, West Bengal, India
e-mail: nandanaghosh18@gmail.com

S. Jana
Department of Mathematics, Midnapore College [Autonomous], Midnapore, West Bengal, India

N. K. Sao
Department of Computer Science, University of Transport and Communication, Đ. Cầu Giấy, Láng Thượng, Đống Đa, Hà Nội, Vietnam

J. K. Mandal et al. (eds.), *Proceedings of International Conference on Network Security and Blockchain Technology*, Lecture Notes in Networks and Systems 738,
https://doi.org/10.1007/978-981-99-4433-0_7

were asked why you are not wearing mask, then just people gave several excuses that they have forgotten or was wearing just open few second ago or they have handkerchief or have towel or muffler, etc. Some of the people start arguing also if they were asked why you have not wear a mask.

So up to when people will not be careful for wearing a mask in their face up to then it would be difficult for everyone to get control over this pandemic situation. If train was closed, then poor people will be affected economically. If lockdowns are done then several shopkeepers, private companies will be affected economically. So the only way to fight with this pandemic situation is to make the mask our companion wherever we would go it must be taken with us and must be worn properly. This paper addresses the situation whether the people is wearing mask which is secure or not. If the people is wearing the mask properly, then it is secure otherwise it is insecure. This model can be used in train, bus, airports, hospitals, market, offices, banks and various working places, etc. As it works like this the place where we have to monitor whether the person is wearing a mask or not. Only one operator is needed it can be the security person also. As we have seen nowadays CCTV camera has become very common in our real life. It is used everywhere starting from malls to our local markets. And in malls, there is one central room from where all the activities of customers are checked. So now the main work starts, the CCTV which is used in various places will be used for capturing the images of people and will be given to the convolutional neural networks (CNNs) which is a class of deep neural network (DNN) commonly used in image recognition and classification. Then, pre-processing will take place and will recognise the faces which have wear mask or not, and it is secure or not towards COVID-19 viruses. If somebody has wear a mask improperly or have not wear a mask then security who is present at that place would ask them to wear a mask or charge a fine against them as for violating COVID protocols.

Also this model can be used in various transports as bus where a bus conductor can see whether the passengers who are travelling in the bus are wearing a mask in proper way or not. As one CCTV camera is needed which will be inside the bus and will capture the images of the passengers travelling in the bus. And the captured images of the passengers will act as the input for this proposed model, and the conductor can observe the passengers by sitting at one place without going near the passengers to ask them for wearing the mask properly. And can warn the passengers by calling their sit numbers or can charge a fine against them. Same way it can be used by the train authorities in trains, by the air hostage inside the aeroplane and by the airport authorities in the airports. This proposed model can also be used in various tourist places where tourists will be advised not to avoid mask otherwise they will be charged a penalty for this. Also it can be used in Temples, Mosques, Church, Gurudwaras.

2 Literature Survey

Due to the increasing cases of COVID-19 along with its variants and Omicron all over the world, it has become very necessary to wear a face mask. As we must make our habit to wear a face mask. So to make it happen various researches had been done [1]. Research work has proposed an optimistic convolution network that helps to ensure whether in public the people are wearing a mask or not by monitoring automatically [2]. Deep learning models struggle to learn in presence of a limited number of samples. So in this research, this problem had been solved by over-sampling where the proposed methodology is split into two phases. The first phase is dealing with over-sampling with image augmentation of the training data, whereas the second phase deals with the detection of face mask using transfer learning of Inception V3. In this research, training images are augmented with eight distinct operations, namely shearing, contrasting, flipping horizontally, rotating, zooming, and blurring. The generated datasets are then rescaled to 224 $\times$ 224 pixels and converted to a single channel greyscale representation [3]. Two stage convolution neural network (CNN) has been used to detect both masked face and unmasked faces using CCTV cameras [4]. Both face mask detection and people are maintaining social distance or not among themselves is done thorough a camera integrated with a Raspberry Pi4. The MobileNetV2 architecture had been used to train the model. For high quality image classification, the SSD multibox detector is used to make it robust model which is a neural network architecture [5]. A real-time face mask detection with accuracy of 98.2% which is a robust model. This model uses ResNet50, AlexNet, and MobileNet for training and testing purposes termed as baseline models [6]. A automated system for a real-time face mask detection uses a technique of transfer learning with faster-RCNN and has achieved an average precision of 81% and average recall of 84%. This model detects masked faces and unmasked faces. It has face mask dataset (FMD) consisting of 853 images and their corresponding XML annotation files. To get more images, an augmentation is applied which includes random horizontal flip.

Song et al. [7] paper focusses on four types of detections in real life which are mask detection, mask type classification, mask position classification, and identity recognition. CNN architecture is used for mask detection, improved AlexNet model architecture is used for mask type detection, improved VGG16 model architecture is used for mask position detection, and a face mask facial recognition pipeline (FMFRP) is used to predict user ID with name. For FMFRP, faces extracted from the input image using multi-task CNN. Features are extracted by using a tool call FaceNet. Then, these extracted features are given to the SVM classifier for the identification of user ID with name. Three datasets are used. First dataset is CASIA WebFace dataset for face detection problems which contains 453,453 images over 10,575 identities, second dataset is MaskedFace Net which contains two datasets containing correctly masked face dataset and incorrectly masked face dataset containing 67,193 images and 66,900 images, respectively. Again incorrect images are divided into three categories uncovered chin, uncovered nose, and uncovered nose and mouth. Mask type detection detects for types of mask surgical, homemade, n95, and no mask. The accuracy achieved by this model is 97%.

Mahmoud and Mengash [8] this proposed model at first determines the existence of human beings which is done by detecting head and shoulders. If head and shoulders are detected, then it goes for the detection of faces. Then, faces are clustered to cluster patches. Then, determination of presence or absent of human skin is determined. If the determination shows that it is a human skin, then the model concludes that the human is not wearing a face mask otherwise concludes that the mask is worn by human being. For this model, two datasets had been used. First dataset contains 650 images of skin patches, and the second datasets consist of 800 images of face. The accuracy given by this model is 97.51%. This paper shown a hybrid approach that combines both normalised RGB and YCbCr space colour in the proposed model.

Loey et al. [9] the proposed model uses ResNet50 for feature extraction, and classification is done using support vector machine (SVM), decision trees, and ensemble algorithm. Three datasets have been used. First dataset consists of 5000 masked faces and 90,000 unmasked faces known as real-world masked face dataset (RMFD), second dataset consists of 1570 images among which 785 images are simulated masked faces and 785 unmasked faces known as simulated masked face dataset (SMFD), third dataset consists of 13,000 images of celebrities masked faces around the world known as labelled faces in the wild (LFW). The accuracy achieved by SVM classifier is 99.64% in RMFD, in SMFD 99.49%, and in LFW 100% [10]. Object detection techniques are used. Object detection is divided into two broad categories. One is traditional approach, and other is deep learning based approaches. Two stages are used for facemask detection face identification and feature extraction.

Jiang et al. [11] this model uses OpenCV and TensorFlow with DNN module. DNN module consists of single shot multibox detector. Two datasets are used one containing masked images and other unmasked images taken from Kaggle' Medical Mask Dataset. For classification architecture, ResNet10 is used as a backbone for the model and for image classification MobileNetV2 had been used [12]. The proposed model uses RetinaFaceMask architecture. For feature extraction, ResNet50 is used as a backbone network, and feature maps are generated by with distinct receptive fields, allowing for the detection of objects of various sizes [13]. Datasets used are masked and unmasked faces. Face detection is done using Viola-Jones face detector, and histogram oriented gradient (HOG) is used for feature extraction. SVM classifier is used. For classification process, a comparison is done between the binary-SVM-model against the queried features vector of face image.

Koklu et al. [14] long short-term memory and bi-directional long-short term memory architectures are used instead of classification layers. Transfer learning is done through two convolutional neural networks are AlexNet and VGG16. Datasets consist of four different images masked, unmasked, masked but nose open and masked but under the chin [15]. Convolutional network is used to build the model and MobileNetV2 to train the deep learning model. Two datasets have been used masked containing 800 images and unmasked 750 images. This model gives 80% accuracy rate. Data augmentation is also done for data expansion.

Loey et al. [16] this is a real-time face mask detection which is used to detect whether a person is wearing a medical face mask or not. For this proposed model, two datasets are used. First dataset contains medical masks dataset which consists of 682 images, and second dataset contains public masked face dataset known as face mask dataset consisting of 853 images. Two components are used in this model first component is ResNet-50, and a deep learning model used for features extraction and the other component is YOLO v2 used for the detection of medical face masks. Adam optimizer is used which achieved the precision of 81% [17]. CNN is used, and ReLU is used for activation function. After features extraction, supervised learning is applied. 80% of data is used for training, and 20% is used for testing. ADAM optimizer is used with learning rate of 0.0001 which is set for optimization [18]. Face mask detection is done by combining image resolution and classification networks (SRCNet). Dataset used here public dataset medical masks. Dataset is divided into three categories 671 images of no facemask-wearing, 134 images of facemask-wearing, and 3030 images of facemask-wearing. 98.70% accuracy is achieved by this model [19]. The model uses multi-task cascaded convolutional neural network (MTCNN). Deep convolutional network comprises of three stages fully convolutional network, refine network, O-Net. Activation function used here is ReLU6. MobileNetV2 is used for feature extraction.

Rahman et al. [20] the proposed model uses convolutional neural network as an architecture for learning. Two datasets are used one for masked images and other for unmasked images. Accuracy achieved by this model is 98.7% [21]. A three stage model is used which includes people detection, tracking, inter-distance estimation, CSPDarkNNet53 is applied along with an SPP/PAN and SAM neck, YOLO head, Mish activation function. Complete IoU loss function and a Mosaic data augmentation on multi-viewpoint MS COCO had been applied. To enrich the training phase, Google Open Image datasets are used [22]. For face mask detection, MobileNet with Global pooling block is used. Softmax function is used for classification which connected to a fully connected layer. Transfer learning is applied to the datasets. Fully connected dense layer contains 64 neurons. Two datasets are used, named as DS1 and DS2. Both are used to evaluate the proposed model. DS1 consists of 1918 images of unmasked people and 1915 images of masked people. DS2 consists of 824 images of unmasked people and 826 images of masked people. Accuracy achieved by this proposed model over DS1 is 99%, and over DS2 is 100% [23]. FMD-YOLO framework is used to monitor the masked people. To design the deep network, three major components are used Im-Res2Net-101, En-PAN, and Classification + Regression. Im-Res2Net is responsible for extracting the features from the inputs which is the modified form of Res2Net which acts as a backbone for FMD-YOLO [24]. The proposed face mask detection system which is used in the classroom to monitor the students' wearing a face mask or not. Those who are not wearing the names are added to the list of students not wearing a mask. The CNN architecture ResNet50 is used for this proposed model.

Oumina et al. [25] for feature extraction VGG19, Xception, and MobileNetV2 are for deep neural networks. For classification purposes, KNN and SVM classifiers are used. Cropping, padding, and horizontal flipping are the augmented techniques that are applied. Two datasets are used one for masked images and other for unmasked images. The accuracy achieved by MobileNetV2–SVM is 97.11%, VGG19-KNN is 96.65%, MobileNetV2-KNN is 94.92%, and Xception-SVM is 94.57%.

3 Motivation and Objective

- The main motivation and objective of this paper are to detect those people who are not wearing their face mask in a proper way and also to make people aware of how to wear face mask.
- As wearing a face mask in a proper way are seen very less among the people during this COVID time. As we can see many people wears their face mask in a wrong way.
- Some people wears their face mask below their nose which make them place in a dangerous position towards COVID 19. Some people wear their mask in such a way where their nose and mouth are left uncovered that is the mask is wear below their chin.
- Some people wear mask in such a way that their chin is left uncovered. So this paper objective is to detect such type of people who are not wearing their mask in a proper way in the road side, malls or market, etc.
- This model can be used on road side by traffic police to keep track on the people passing by.
- Also this model can be used in shopping malls where a security person can keep a track on the customers whether they are following the rules to wear a face mask in a proper way or not.
- This model can also be used in schools, colleges, universities to keep track on their students whether they are wearing a face mask or not.

4 Contribution of This Investigation

- This paper will help us to control the spread of the disease so called COVID 19. As, it has become very necessary to alert people to wear their face mask in a proper way and help the world to overcome this disease.
- In this paper, we have used two types of proposed model, one is CNN and other is MobileNetV2.
- In CNN, ReLU activation function is applied to the output of each layer, and final layer uses the softmax activation function to get the probabilities of input being in a particular class (Mask: Secure, Improper Mask: Insecure, No mask: Danger).
- MobileNetV2 acts as a backbone for feature extraction. There is pretrained model of ImageNet which assigns the weights.

- In this model, MobileNetV2 acts as the base model, and ReLU activation function is applied to the other output layers, and softmax activation function is applied to get the particular class (Mask: Secure, Improper Mask: Insecure, No mask: Danger).
- Here, a particular class is detected in this way. If the person is wearing a mask in a proper way, then it will detect by surrounding the face with a box of green colour showing a message as "Mask: Secure 100%". Here, the percentage may vary also.
- If it detects that the person is wearing a mask but not in a proper way, then it will detect by surrounding the face with a blue colour box showing a message as "Improper mask 100%". Here also, the percentage may vary.
- Now, if it detects that the person is not wearing a mask, then it will detect by surrounding the face with a red colour box showing a message as "No mask: Danger 100%". Here also, the percentage may vary as it take the input of the image continuously.

5 Methodology

CNN and MobileNetV2 are used for the proposed model. We have developed this proposed model in two ways one using CNN and other using CNN with MobileNetV2 and have compared their accuracy. Here, we have used three datasets named as Incorrect_mask, (Fig. 1) with_mask (Fig. 2) and without_mask (Fig. 3). Incorrect mask dataset contains 703 images of people who have worn the mask in a wrong way, with_mask dataset contains 690 images of people who have worn the mask in a right way, and without_maskdataset contains 686 images of people who have not worn the mask. This dataset has been taken from the Kaggle datasets shown in Table 1.

This model captures the images of people where capturing of images is done by CCTV camera which acts as an input and then is fed to the CNN architecture where

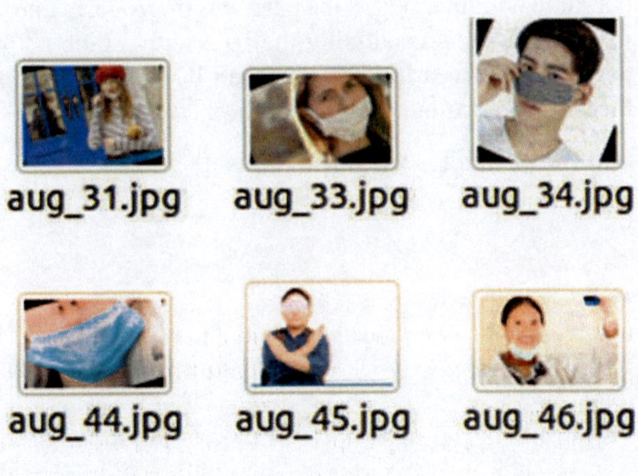

Fig. 1 Improperly masked people

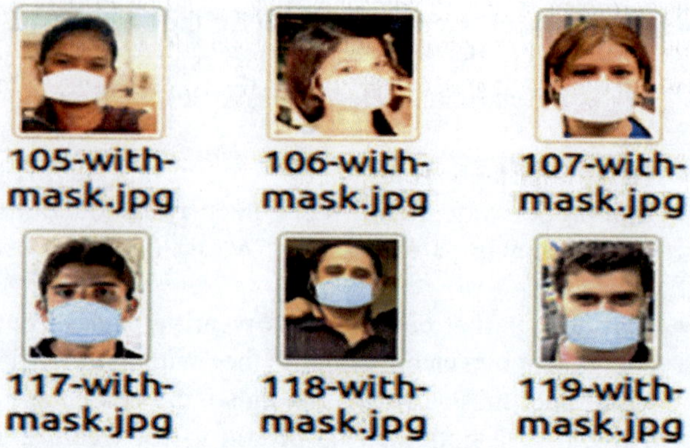

Fig. 2 Properly masked people

Fig. 3 No masked people

Table 1 Dataset image combination

	Incorrect_mask	With_mask	Without_mask
No. of images	703	690	686

high-level features extraction takes place. At last, fully connected layer with Softmax activation function classification of the images takes place. If the input image was of masked person, then it will show Mask: Secure surrounding the face with green colour rectangular box, it the person is improperly masked, then it will show Improper mask: Insecure surrounding the face with a blue colour rectangular box, and if the person is not wearing a mask, then it will show no mask: danger surrounding the face with red colour rectangular box.

Sample of images from the dataset

6 Proposed Method

6.1 Convolutional Neural Network (CNN)

A convolutional neural network which is shortly known as ConvNet/CNN which is a deep learning algorithm. It takes an input image assigns learnable weights and biases in the image, and then, each layer generates several activation functions that are passed to the next layer. Each layer extracts different features of the image. ConvNet reduces the images into a form which is easier to process, without losing features. The output from each layer is computed by matrix multiplication of output of the previous layer with learnable weights of that layer and then by the addition of learnable biases followed by activation function which makes the network nonlinear. The pooling layers which are used max pooling which prevents the model from overfitting. The application of convolutional neural network in the proposed model is shown in Fig. 4.

Steps Followed by Convolutional Neural Network

Input layer: This layer takes the image as an input. The image can be of masked, improper masked, or unmasked people.

Convolution layer: Images have to pass through three convolutional layer where different features of image are extracted. First convolution layer is responsible for capturing the low-level features of image such as size of edges, colour, and gradient orientation. Second and third layer extract the more high-level features giving us a network which has the wholesome understanding of images in the dataset. After each CNN, there is pooling layer. Here, max pooling is applied.

ReLU activation function: ReLU activation function is applied to the output of each convolution layer. The activation function will be applied element wise.

Max pooling: Pooling is the process of merging. So it helps to reduce the size of data. It returns the maximum value from the portion of the image covered by the kernel. It also removes the noise along with dimensionality reduction.

Flattening: This layer converts the data into a 1-dimensional array for inputting it to the next layer. We flatten the output of the convolutional layers to create a long

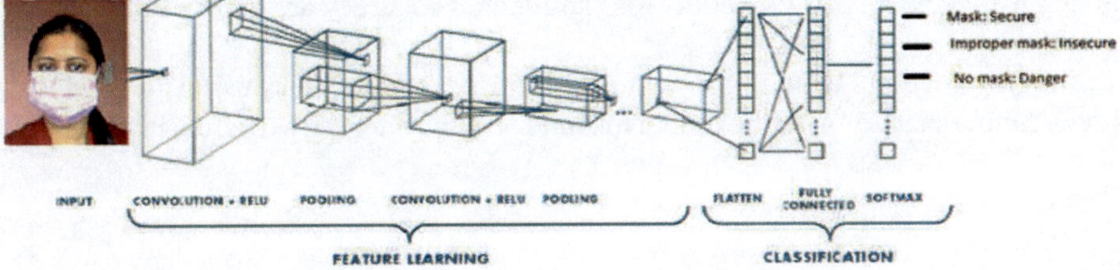

Fig. 4 Application of convolutional neural network in the proposed model

feature vector. After that it is connected to the final classification layer known as a fully connected layer.

Fully connected layer: This layer is a regular neural network layer that takes the input from the previous layer and computes the class scores and outputs the 1D array of size equal to the number of classes.

Softmax activation function: After passing through the fully connected layers, the final layer uses the softmax activation function which is used to get probabilities of the input being in a particular class (Mask: Secure, Improper mask: Insecure, No mask: Danger).

6.2 MobileNetV2

MobileNetV2 (Fig. 5) is an updated version of MobileNetV1. In MobileNetV2, a better module is introduced with inverted residual structure. Non-linearities in narrow layers are removed this time. MobileNetV2 acts as a backbone for feature extraction. In proposed MobileNetV2 (Fig. 6), there are two types of blocks. One is residual block with stride of 1. Another one is block with stride of 2 for downsizing. There are three layers for both type of blocks. This time, the first layer is 1×1 convolution with ReLU6. The second layer is the depth wise convolution. The third layer is another 1×1 convolution but without any non-linearity. There is pretrained model of image known as ImageNet which assigns weights to the images. The application of MobileNetV2 in the proposed model is shown in Fig. 7.

7 Experimental Result

Accuracy achieved by this proposed model using convolutional neural network is 97.83%. Number of epochs applied for this model is 25. Output shown during training phase is shown below, whereas accuracy achieved by this proposed model using MobileNetV2 is 99.76%. Number of epochs applied for this model 25. Output shown during training phase is shown below.

The graph in Figs. 8 and 9 showing training loss and accuracy for the proposed model using CNN. Also the model loss and model accuracy are shown below in the graph.

The graph in Figs. 10 and 11 showing training loss and accuracy using the proposed model MobileNetV2. Also the model loss and accuracy are shown below in the graph.

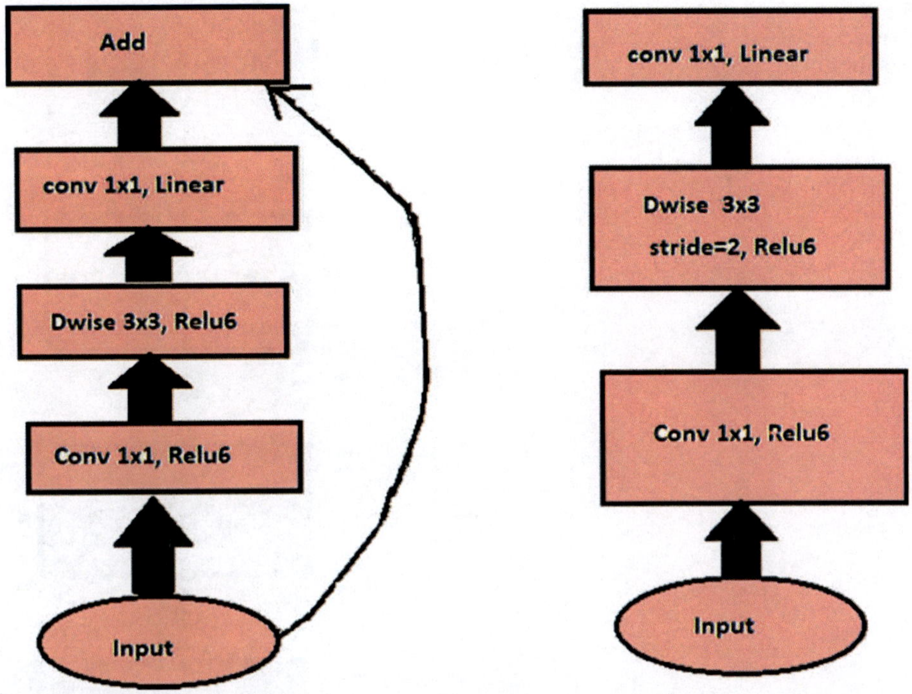

Fig. 5 MobileNetV2

8 Comparison Between the Proposed Models Done with CNN and MobileNetV2

The accuracy achieved by the model done with CNN is 97.83%, and accuracy achieved by the model done with MobileNetV2 is 99.76% shown in Table 2. Time taken for training the model done by CNN is more than the time taken by the model MobileNetV2. So, the proposed model done with the MobileNetV2 is the robust model.

There are three types of face mask detection which is done through this proposed model. First is masked face detection shown in Fig. 12, second is improper masked face detection shown in Fig. 13, and third is unmasked face detection shown in Fig. 14. There are three types of improper mask; first type is mouth and chin is covered with mask, second type is only chin is covered with mask, and the third type is only nose and mouth is covered with mask. Face mask detection done by this proposed model is shown below.

Fig. 6 Proposed model architecture using MobileNetV2

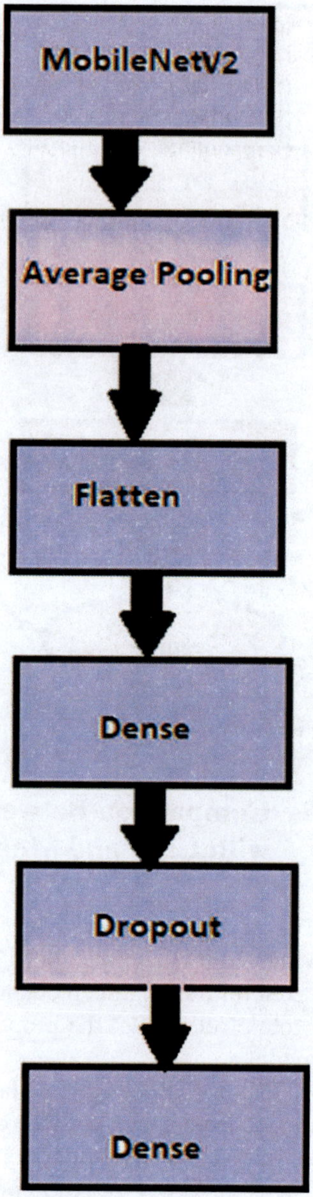

9 Output of from Experiments

We have compared the proposed model with existing models MobileNetV2-KNN, MobileNetV2-SVM, VGG19-KNN, Xception-SVM, and CNN shown in Table 3. It has been observed that our proposed model achieves 97.83 using CNN and 99.76 using MobileNetV2.

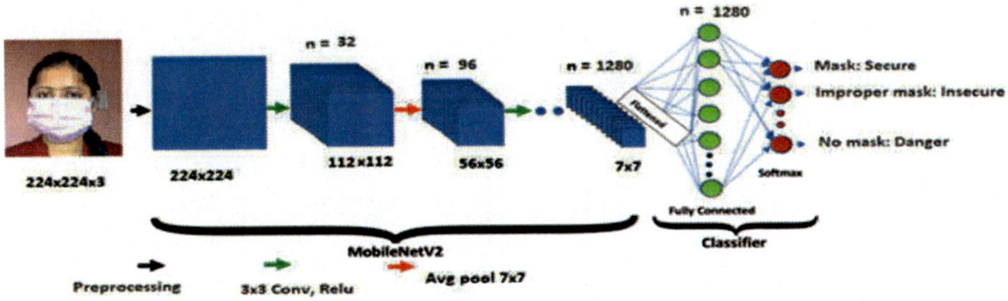

Fig. 7 Application of MobileNetV2 in the proposed model

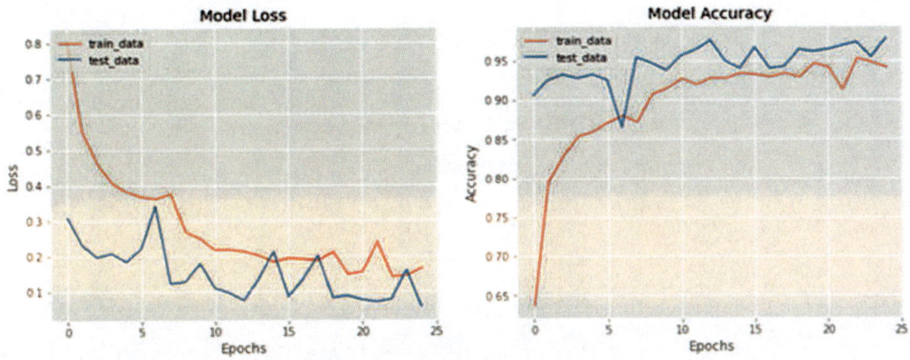

Fig. 8 Model loss and model accuracy using CNN

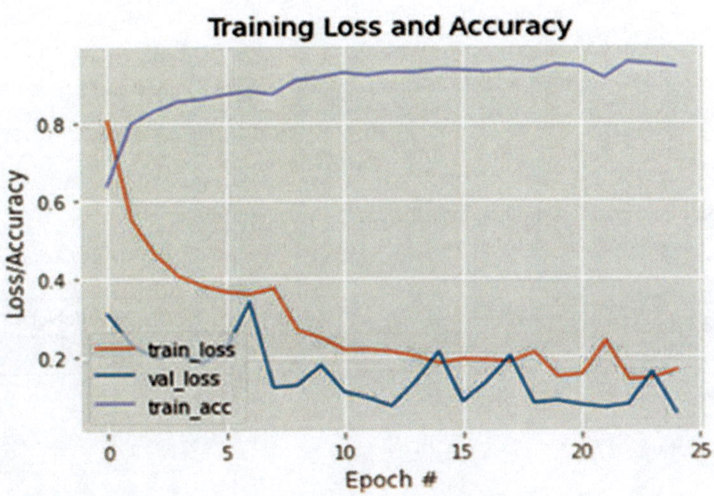

Fig. 9 Combined result of training loss and accuracy using CNN

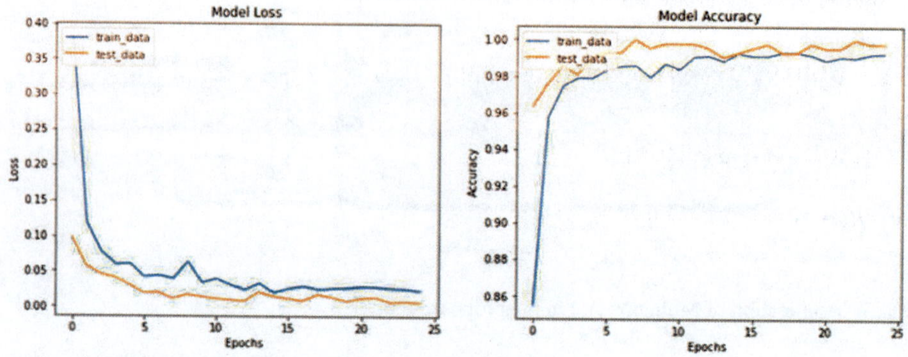

Fig. 10 Model loss and model accuracy using MobileNetV2

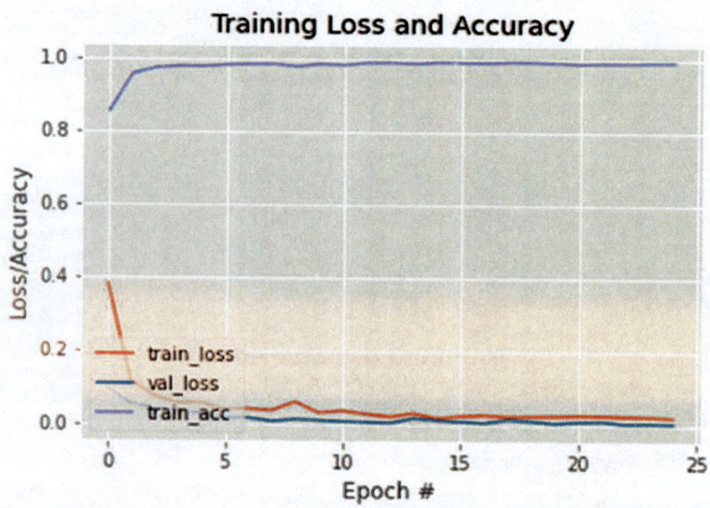

Fig. 11 Combined result of training loss and accuracy using MobileNetV2

Table 2 Accuracy obtained by the proposed model over CNN and MobileNetV2

	CNN	MobileNetV2
Accuracy	97.83%	99.76%

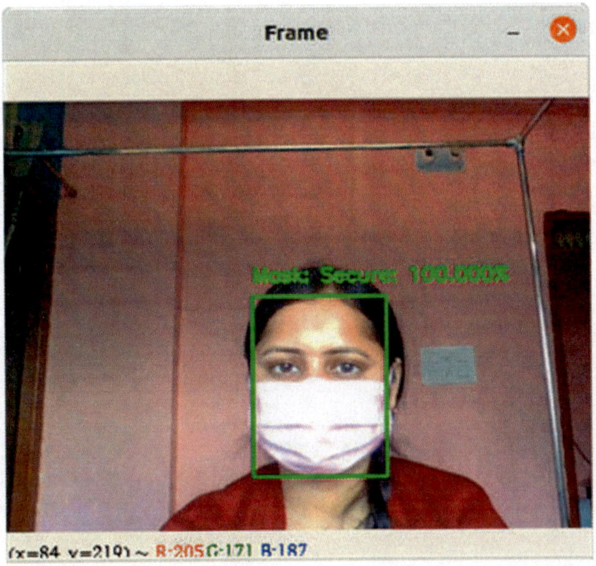

Fig. 12 Secure mask face detection

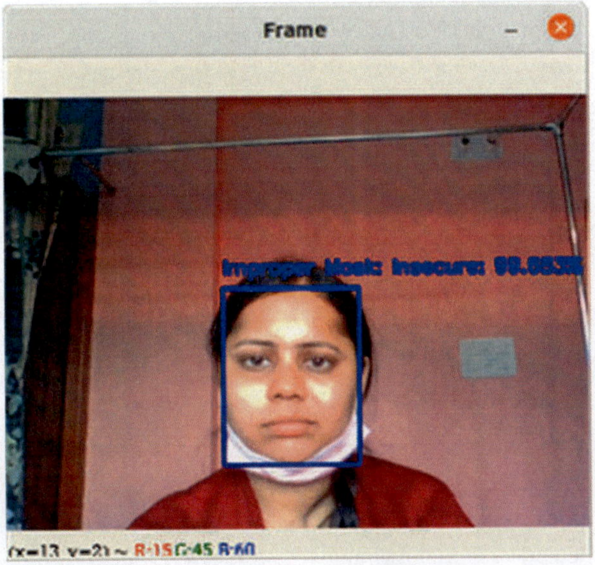

Fig. 13 Insecure mask face detection

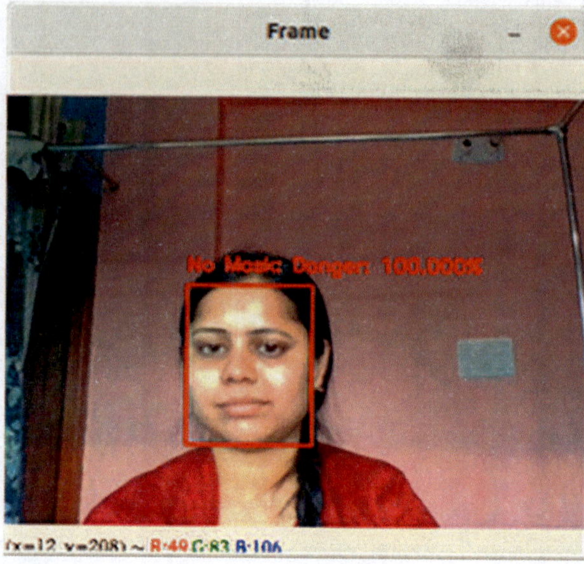

Fig. 14 No mask face detection

Table 3 Comparison with the state-of-the-art methods

Method	Accuracy (%)
MobileNetV2-KNN	94.92
MobileNetV2-SVM	97.11
VGG19-KNN	96.65
Xception-SVM	94.57
CNN	98.7
Proposed method CNN	97.83
Proposed method MobileNetV2	99.76

10 Conclusion

When the world was fighting from COVID-19 situation. It has become necessary to wear a face mask. Only this is the path that everyone should follow to get rid of COVID-19 situation. So, we have proposed a secure and unsecure face mask detection model through which it can be checked the person who is wearing a mask and who is not wearing a mask. And also the person is wearing the face mask in a proper way or not. This model can be used in Hospitals, Schools, Colleges, Universities, Offices, mall, etc. This proposed model is shown by using CNN and also by using MobileNetV2. The accuracy given by using CNN is 97.83% and by using MobileNetV2 is 99.76%. Both these models detect masked person, improperly masked person and unmasked person. Masked person is detected by surrounding a green colour box with the written text "Mask: Secure", improperly masked person is detected by surrounding a blue colour box with the written text "Improper Mask:

Insecure" and Unmasked person is detected by the surrounding a red colour box with the written text "No Mask: Danger" with percentage showing the accuracy.

References

1. Suresh K, Palangappa MB, Bhuvan S (2021) Face mask detection by using optimistic convolutional neural network. In: 2021 6th international conference on inventive computation technologies (ICICT). IEEE, pp 1084–1089
2. Jignesh Chowdary G, Singh Punn N, Sonbhadra SK, Agarwal S (2009) Face mask detection using transfer learning of InceptionV3. arXiv e-prints. Published online 2020: arXiv-2009
3. Chavda A, Dsouza J, Badgujar S, Damani A (2021) Multi-stage CNN architecture for face mask detection. In: 2021 6th international conference for convergence in technology (I2CT). IEEE, pp 1–8
4. Yadav S (2020) Deep learning based safe social distancing and face mask detection in public areas for COVID-19 safety guidelines adherence. Int J Res Appl Sci Eng Technol 8(7):1368–1375. https://doi.org/10.22214/ijraset.2020.30560
5. Sethi S, Kathuria M, Kaushik T (2021) Face mask detection using deep learning: an approach to reduce risk of Coronavirus spread. J Biomed Inform 120:103848
6. Sabir MFS, Mehmood I, Alsaggaf WA et al (2022) An automated real-time face mask detection system using transfer learning with faster-RCNN in the era of the covid-19 pandemic. Comput Mater Contin 4151–4166. Published online 2022
7. Song Z, Nguyen K, Nguyen T, Cho C, Gao J (2022) Spartan face mask detection and facial recognition system. In: Healthcare, vol 10. Multidisciplinary Digital Publishing Institute, p 87
8. Mahmoud HAH, Mengash HA (2021) A novel technique for automated concealed face detection in surveillance videos. Pers Ubiquitous Comput 25(1):129–140
9. Loey M, Manogaran G, Taha MHN, Khalifa NEM (2021) A hybrid deep transfer learning model with machine learning methods for face mask detection in the era of the COVID-19 pandemic. Measurement 167:108288
10. Nowrin A, Afroz S, Rahman MS, Mahmud I, Cho Y-Z (2021) Comprehensive review on facemask detection techniques in the context of covid-19. IEEE Access. Published online 2021
11. Jiang M, Fan X, Yan H (2020) Retinamask: a face mask detector. arXiv Prepr arXiv200503950. Published online 2020
12. Nagrath P, Jain R, Madan A, Arora R, Kataria P, Hemanth J (2021) SSDMNV2: a real time DNN-based face mask detection system using single shot multibox detector and MobileNetV2. Sustain Cities Soc 66:102692
13. Abdulmajeed AA, Tawfeeq TM, Al-jawaherry MA (2022) Constructing a software tool for detecting face mask-wearing by machine learning. Baghdad Sci J 19(3):642
14. Koklu M, Cinar I, Taspinar YS (2022) CNN-based bi-directional and directional long-short term memory network for determination of face mask. Biomed Signal Process Control 71:103216
15. Bhadani AK, Sinha A (2020) A facemask detector using machine learning and image processing. Eng Sci Technol Int J 1–8
16. Loey M, Manogaran G, Taha MHN, Khalifa NEM (2021) Fighting against COVID-19: a novel deep learning model based on YOLO-v2 with ResNet-50 for medical face mask detection. Sustain Cities Soc 65:102600
17. Militante SV, Dionisio NV (2020) Real-time facemask recognition with alarm system using deep learning. In: Proceedings of 2020 11th IEEE control and system graduate research colloquium, ICSGRC 2020, August, pp 106–110. https://doi.org/10.1109/ICSGRC49013.2020.9232610
18. Qin B, Li D (2020) Identifying facemask-wearing condition using image super-resolution with classification network to prevent COVID-19. Sensors (Switzerland) 20(18):1–23. https://doi.org/10.3390/s20185236

19. Joshi AS, Joshi SS, Kanahasabai G, Kapil R, Gupta S (2020) Deep learning framework to detect face masks from video footage. In: Proceedings of 2020 12th international conference on computational intelligence and communication networks, CICN 2020, pp 435–440. Published online 2020. https://doi.org/10.1109/CICN49253.2020.9242625
20. Rahman MM, Manik MMH, Islam MM, Mahmud S, Kim JH (2020) An automated system to limit COVID-19 using facial mask detection in smart city network. In: IEMTRONICS 2020—International IOT, electronics and mechatronics conference, proceedings. https://doi.org/10.1109/IEMTRONICS51293.2020.9216386
21. Rezaei M, Azarmi M (2020) Deepsocial: social distancing monitoring and infection risk assessment in covid-19 pandemic. Appl Sci 10(21):1–29. https://doi.org/10.3390/app10217514
22. Venkateswarlu IB, Kakarla J, Prakash S (2020) Face mask detection using MobileNet and global pooling block. In: 4th IEEE 4th conference on information and communication technology, CICT 2020, pp 0–4. Published online 2020. https://doi.org/10.1109/CICT51604.2020.9312083
23. Wu P, Li H, Zeng N, Li F (2022) FMD-Yolo: an efficient face mask detection method for COVID-19 prevention and control in public. Image Vis Comput 117:104341. https://doi.org/10.1016/j.imavis.2021.104341
24. Nithiyasree K, Kavitha T (2021) Face mask detection in classroom using deep convolutional neural network. Turkish J Comput Math Educ 12(10):1462–1466
25. Oumina A, El Makhfi N, Hamdi M (2020) Control the COVID-19 pandemic: face mask detection using transfer learning. In: 2020 IEEE 2nd international conference on electronics, control, optimization and computer science, ICECOCS 2020. Published online 2020. https://doi.org/10.1109/ICECOCS50124.2020.9314511

Malicious Transaction URL Detection Using Logistic Regression

Aratrik Bose, Anandaprova Majumder, and Sumana Kundu

1 Introduction

Machine learning has helped in cyber security domain by letting systems analyze patterns and learn from past security attacks and prevent such attacks by responding to changing behavior. The world has seen significant frauds in the field of E-Commerce as more and more people have started shifting toward cash-free transactions. Since COVID-19 quarantine, people have shifted to online purchase more to stay safe or because the products they need have been unavailable in local shops. However, as we look into the worst part cybercrimes have increased at an alarming rate. With increasing online businesses, Website impersonation can be regarded as one of the easiest forms of cyber-attacks [1]. A common type of Website impersonation attack is typo squatting, where an attacker impersonates a popular Website using a closer variant of the domain name which is to be impersonated (For example, www.@mazon.com instead of www.amazon.com). Hackers usually make a copy of our intended destination by creating a similar interface as that of the original Website so that we do not understand that we are on a different Website. Sometimes, such sites are created to sell products that are in direct competition to the products we are actually looking for but most often such practices are used by hackers to get hold of personal highly sensitive data such as credit card details, CVV numbers or maybe passwords. Such sites are also quite risky as malicious software could be downloaded to our devices simply by visiting these links. So even accepting a download or clicking any link is not necessary for installing a dangerous code on any system.

A. Bose (✉) · A. Majumder · S. Kundu
Computer Science and Engineering, Dr. B.C. Roy Engineering College, Durgapur, West Bengal, India
e-mail: aratrikstudy@gmail.com

A. Majumder
e-mail: anandaprova.majumder@bcrec.ac.in

J. K. Mandal et al. (eds.), *Proceedings of International Conference on Network Security and Blockchain Technology*, Lecture Notes in Networks and Systems 738,
https://doi.org/10.1007/978-981-99-4433-0_8

This is called drive-by download, and many typo squatters use this method to spread malicious software in order to steal our personal information. Environment of online businesses and growth of E-Commerce platforms have also increased occurrences of virtual transactions. As more monetary transactions are being conducted, more monetary frauds are also being conducted side by side. According to RBI data, a total of 4071 fraud cases were reported by Indian lenders between April and September 2021 [2]. To prevent such illegal activities, this proposed work deals with certain links that can be considered for transaction purposes and are checked using a machine learning approach to determine whether it is safe for the same purpose or not.

2 Prior Researches

A malicious URL resembles normal at a simple glance. The biggest threats to present digital world are these malicious links which can cause heavy damage to our digital systems. Such attacks can also occur in our system by opening any file that might contain viruses or any link received through an e-mail. On the other hand, phishing Websites [3] are often cloned Websites of certain famous Websites. Mostly, online purchase Websites are cloned in order to conduct transaction-related attacks. E.g.: Amazon, Flipkart, Ekart, eBay, etc. Users think that they are on safe sites seeing the interface similar to original sites but they overlook the slits through which attackers get their work done. When it comes to user infections, malicious sites can be detected by identifying certain noticeable patterns. Many researchers already developed models to overcome these problems. In [4], Lakshmanarao et al. proposed a machine learning-based method for identifying malicious Websites. A Kaggle dataset with more than 5,000,000 URLs is used for the experiments. Then, they built a phishing Website detection model using four machine learning classifiers—logistic regression, KNN, decision tree, and random forest—and used three techniques for text feature extraction: the count vectorizer, the hashing vectorizer, and the IDF vectorizer. With a hash vectorizer and random forest, the machine learning model was accurate to 97.5%. Using Flask, they also developed a Web application for determining whether a URL entered is malicious or not. The main focus in [5] was on using super learner ensemble to implement a machine learning classifier model for classifying malicious URLs. To support offline and real-time detection, the static feature set is extracted from the URL information with less latency and computational complexity. The proposed multi-class classifier model divides URLs into multiple categories of attacks (phishing, malware, spam, and defacement), while the proposed binary classifier model is used to distinguish between benign and malicious URLs. A dataset of approximately 750,000 URLs is used to test these classifiers. The empirical results demonstrate that the proposed model is effective at detecting malicious URLs. The multi-class classifier has a precision of 96.234% and accuracy of 94.69%, while the binary classifier has a precision of 95.145%. In [6], Manyumwa et al. compared the following ensemble learners' performance: extreme gradient boosting (XGBoost), adaptive gradient boosting (AdaBoost), light gradient boosting (LightGBM), and categorical gradient boosting (CatBoost). The authors looked at how well a few URL features like the Kullback–Leibler Divergence (KL divergence), bag of words segmentation, and additional word-based features were performed. The outcomes demonstrated that the experiments with and without their chosen features performed better. These algorithms were trained on 126,983 URLs from benchmark datasets, and each of the four learners produced results with an overall accuracy greater than 0.95. In [7], Peng et al. demonstrated that adversarial samples are sensitive to the vulnerabilities of the existing DL-based malicious URL detection models. URL adversarial samples are constructed based on perturbations at the character and component levels. Then, these adversarial samples were used to attack conventional DL-based detection models, resulting in decreases in detection accuracies. In the meantime, the perturbations were constrained by the fact that each adversarial sample URL can be easily distinguished from the original URL. In addition, adversarial samples constructed by altering 14 different character types and all other components (with the exception of the scheme component) resulted in the greatest increase in missed blocking of malicious URLs, i.e., a greater decrease in accuracy than other constructed methods. The adversarial samples continued to function and exhibit oblique decreased in accuracy despite the adversarial training being applied to them. The use of a multilayer convolutional neural network (CNN) for malicious URL detection was suggested in [8]. The proposed model first looked at one CNN layer. After that, a two-layer CNN will be utilized to enhance accuracy. The result showed that when the model uses two layers of CNN, the accuracy of detecting malicious Websites increases from 89 to 91%. In [9], Ren et al. proposed an attentional-based BiLSTM model called AB-BiLSTM for detecting malicious URLs. Pre-trained Word2Vec was used to preprocess the URLs and turn them into word vectors. Next, BiLSTM and an attention mechanism were trained to extract and classify the features of URL sequences. The model was tested on a dataset that was collected. The experimental results show that the proposed model can achieve an F1-score of 95.92%, an accuracy rate of 98.06%, a precision rate of 96.05%, and a recall rate of 95.79%.

3 Proposed Methodology

This proposed model is being carried out due to increased cybercrimes in recent days. As the world is shifting toward the digital world since the occurrence of pandemic more and more people have found online platforms reliable for money transaction as they find it easier to transfer money wherever and whenever they want. But handling money on a network is also dangerous as hackers or frauds are present to steal money, passwords, card details, personal information, identity, and much more by unethical means. To prevent such fraudulent attacks, we came with an approach to determine whether a particular URL is safe to make a transaction. A model is trained using machine learning algorithm to recognize the pattern of malicious URL and safe URL and in turn the trained model predicts if any site is safe for transaction or not

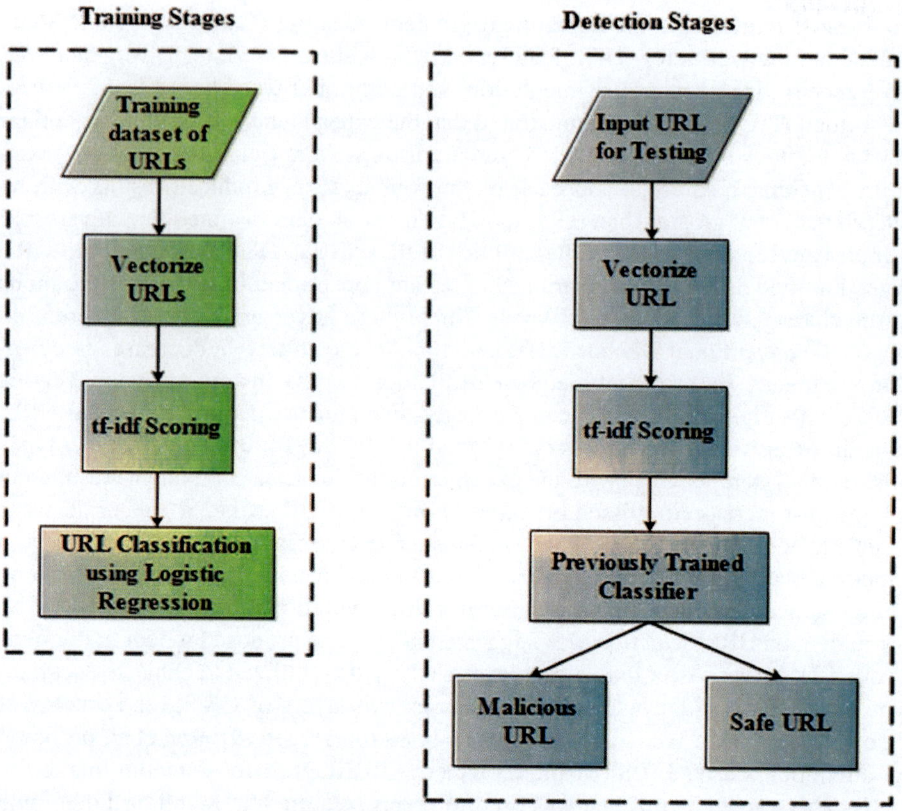

Fig. 1 Block diagram of the proposed model

(Refer to Fig. 1). The data is initially classified into safe and unsafe URLs depending on its encryption and occurrence of secure socket layer certification.

3.1 Identification Learning and Testing

1. We have a dataset of the different URLs where transactions are made, which we use for processing and extract the features.
2. So initially, the dataset of URLs was classified into two parts whether they were safe or unsafe.
3. Furthermore, we used a vectorizer function on the URLs to vectorize the words in URLs in order to find out words that have positive or negative impact. For e.g: words like 'virus', 'dat', 'exe' being in a URL needs more attention for considering whether safe or not.
4. Here, we used TF-IDF scoring. TF-IDF was basically used to find out the importance of specific words whether negative or positive impact in terms of URL

safety. The URLs were broken down into small tokens, and the tokenized words were further vectorized and fed into the model.

5. Next, the dataset was split for training and testing data in 80–20 ratio, and train_test_split was used.
6. Then, the logistic regression classifier was used to train the model. Logistic regression is one of the most commonly known machine learning algorithms that come under the category of supervised learning technique. A given set of independent variables are used to detect the categorical dependent variable. The result must be a categorical or discrete value in logistic regression. Logistic regression is represented using an equation much similar to linear regression. The independent values (*X*) are linearly combined with certain valued weights or coefficients for finding dependent variable (*Y*). A noticeable difference of logistic regression from linear regression is that the output value of the model is a binary value (0 or 1) instead of a numeric value. Below is the equation of logistic regression:

$$Y = \frac{e^{(b_0 + b_1 \times x)}}{(1 + e^{(b_0 + b_1 \times x)})}$$

Finally, the model was tested by using the test dataset's URL and got expected results.

4 Experimental Results and Comparative Analysis

More than 11,500 URLs related to malware Websites were obtained from DNS-BH which is a project that maintain list of malware sites [10]. The data is put into a data frame using Python package pandas.

```
In [2]: #Load URL data
        proj_data=pd.read_csv("Malware_dataset.csv")
```

The dataset has been further categorized on the basis of their encryption layer. This is being done by the separate function that has been created named Url_classifier.

```
In [4]: k=Url_classifier()
        print(k)

0        UnSafe
1        UnSafe
2        UnSafe
3        UnSafe
4        UnSafe
          ...
11561    UnSafe
11562    UnSafe
11563    UnSafe
11564    UnSafe
11565    UnSafe
Length: 11566, dtype: object
```

The URLs are further broken into small tokens to make it suitable to be vectorized using another separate function named make_Tokens to find out each feature that could be depicted through the URL.

```
In [7]: def makeTokens(f):
            tkns_BySlash = str(f.encode('utf-8')).split('/')    # make tokens after splitting by slash
            total_Tokens = []
            for i in tkns_BySlash:
                tokens = str(i).split('-') # make tokens after splitting by dash
                tkns_ByDot = []
                for j in range(0,len(tokens)):
                    temp_Tokens = str(tokens[j]).split('.')# make tokens after splitting by dot
                    tkns_ByDot = tkns_ByDot + temp_Tokens
                total_Tokens = total_Tokens + tokens + tkns_ByDot
            total_Tokens = list(set(total_Tokens)) #remove redundant tokens
            if 'com' in total_Tokens:
                total_Tokens.remove('com') #removing .com since it occurs a lot of times and it should not be included in our features
                print(total_Tokens)
            return total_Tokens
```

The usage of TF-IDF vectorizer shows the importance given to each word in the URL such as spam words found in the URL shall be marked as unsafe, and the words are further converted to vectors used to fit into the model.

```
In [10]: # Using Custom Tokenizer
         vectorizer = TfidfVectorizer(tokenizer=makeTokens)

In [11]: X = vectorizer.fit_transform(url_list)
```

```
['', 'gzzax.livechatvalue.com', 'chatclient', "chatbox.jsp?companyid=11806&enterurl=file%3a%2f%2f%2fc%3a%2fusers%2famelia%2fd
esktop%2f%25e7%25bd%2591%25e4%25b8%258a%25e8%25b5%258c%25e5%259c%25ba.html&pagetitle=&pagereferrer=&firstenterurl=file%3a%2f%
2f%2fc%3a%2fusers%2famelia%2fdesktop%2f%25e7%25bd%2591%25e4%25b8%258a%25e8%25b5%258c%25e5%259c%25ba.html&lan=zh&tm=1420533810
210'", 'jsp?companyid=11806&enterurl=file%3a%2f%2f%2fc%3a%2fusers%2famelia%2fdesktop%2f%25e7%25bd%2591%25e4%25b8%258a%25e8%25
b5%258c%25e5%259c%25ba', 'chatbox', 'gzzax', "html&lan=zh&tm=1420533810210'", 'html&pagetitle=&pagereferrer=&firstenterurl=fi
le%3a%2f%2f%2fc%3a%2fusers%2famelia%2fdesktop%2f%25e7%25bd%2591%25e4%25b8%258a%25e8%25b5%258c%25e5%259c%25ba', 'livechatvalu
e', "b'http:", 'chat']
['', 'gzzax.livechatvalue.com', 'chatclient', "chatbox.jsp?companyid=11806&enterurl=file%3a%2f%2f%2fc%3a%2fusers%2famelia%2fd
esktop%2f%25e7%25bd%2591%25e4%25b8%258a%25e8%25b5%258c%25e5%259c%25ba.html&pagetitle=&pagereferrer=&firstenterurl=file%3a%2f%
2f%2fc%3a%2fusers%2famelia%2fdesktop%2f%25e7%25bd%2591%25e4%25b8%258a%25e8%25b5%258c%25e5%259c%25ba.html&lan=zh&tm=1420533810
210'", 'jsp?companyid=11806&enterurl=file%3a%2f%2f%2fc%3a%2fusers%2famelia%2fdesktop%2f%25e7%25bd%2591%25e4%25b8%258a%25e8%25
b5%258c%25e5%259c%25ba', 'chatbox', 'gzzax', "html&lan=zh&tm=1420533810210'", 'html&pagetitle=&pagereferrer=&firstenterurl=fi
le%3a%2f%2f%2fc%3a%2fusers%2famelia%2fdesktop%2f%25e7%25bd%2591%25e4%25b8%258a%25e8%25b5%258c%25e5%259c%25ba', 'livechatvalu
e', "b'http:", 'chat']
['', 'gzzax.livechatvalue.com', 'chatclient', "chatbox.jsp?companyid=11806&enterurl=file%3a%2f%2f%2fc%3a%2fusers%2famelia%2fd
esktop%2f%25e7%25bd%2591%25e4%25b8%258a%25e8%25b5%258c%25e5%259c%25ba.html&pagetitle=&pagereferrer=&firstenterurl=file%3a%2f%
2f%2fc%3a%2fusers%2famelia%2fdesktop%2f%25e7%25bd%2591%25e4%25b8%258a%25e8%25b5%258c%25e5%259c%25ba.html&lan=zh&tm=1420533810
210'", 'jsp?companyid=11806&enterurl=file%3a%2f%2f%2fc%3a%2fusers%2famelia%2fdesktop%2f%25e7%25bd%2591%25e4%25b8%258a%25e8%25
b5%258c%25e5%259c%25ba', 'chatbox', 'gzzax', "html&lan=zh&tm=1420533810210'", 'html&pagetitle=&pagereferrer=&firstenterurl=fi
```

The model further predicts an accuracy of 99%.

```
lgs = LogisticRegression(max_iter=500)
lgs.fit(X_train,y_train)
```

```
LogisticRegression(max_iter=500)
```

```
print("Accuracy:",lgs.score(X_test,y_test)*100)
```

```
Accuracy: 99.61106309420916
```

The model is tested with different URLs, and it gives desired output though it is giving too much accuracy value which might be a sign of overfitting. Also best results are achieved when popular links are being used which can be regarded as a drawback. As for example, we tested it on one of the most popular URLs 'https://google.com/' and got expected results, which can be obviously termed as success.

```
In [32]: X_predict=["https://google.com"]
         X_predict= vectorizer.transform(X_predict)
         Y_predict = lgs.predict(X_predict)

In [33]: print(Y_predict)

         ['safe']
```

4.1 Comparative Study

This paper works on binary classification of the dataset which means a probabilistic outcome. The dataset has limited data labels and with no noise in it. In such a situation, choosing logistic regression as the classifier is more beneficial as it is easy to be implemented, does not give discrete output and gives output in probabilistic manner which is our goal. On the other hand, though other classifiers give better accuracies but they are complex in structure, and for linearly separable balanced data, linear regression is the best classifier. Our proposed work is compared with other existing works, and its performance can be summed up as shown in the following Table 1. Also from Table 2, it can be shown that the time complexity of our proposed model is less than other conventional models. On the other hand, Table 3 and Fig. 2 show that our proposed model gives higher accuracy than other traditional models.

Table 1 Comparison with other existing works

Proposed work	Previous works
Our model can derive the significance of features quite fast	Other works could not derive the significance of features or even if could was quite slow
In our model, the algorithm has convex loss function, so it will not hang in local minima	In other works, the algorithms used complex non-linear mathematical functions at different times and most often the loss function is non-convex, thus it is quite possible to get stuck in local minima
Our model has an extra feature of checking the secured layer of the links and hence can be better used for checking transaction URLs	The previous works did not have any feature of checking the encryption
Our model can derive confidence level (about its prediction)	Other models could only output the labels
Our model is based on statistical approaches (that is dealing probabilistic data here)	Other models were based on geometrical properties of the data more and less focus on probabilistic data

Table 2 Comparison based on time complexity

Comparison perspective	Time taken by proposed model	Time taken by other existing models		
	Using logistic regression	Using decision tree	Using SVM	Using K-nearest neighbors
Train time complexity	**O(n*m)**	O(n*log(n)*m)	O(n^2)	O(k*n*m)
Test time complexity	**O(m)**	O(m)	O(n'*m)	O(n*m)

Table 3 Comparison with respect to accuracy

Models	Accuracy (%)
Ref. [4]	97.5
Ref. [5]	94.69
Ref. [6]	95
Ref. [8]	91
Ref. [9]	98.06
Proposed model	99

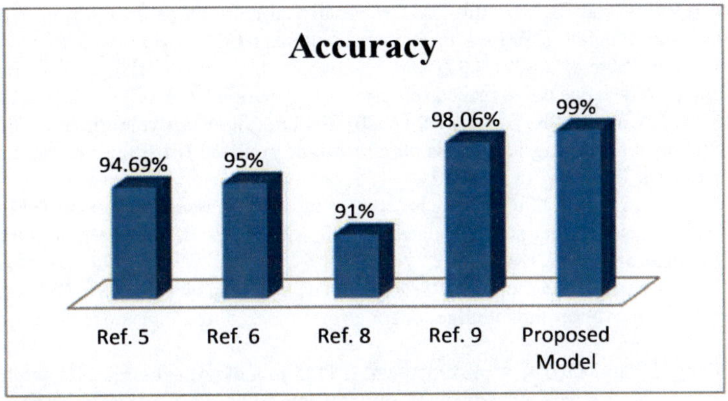

Fig. 2 Graphical representation of comparative analysis with respect to accuracy

5 Conclusion and Future Scope

We can conclude that with the help of certain methods and processes of hacking which is known to be social attack, a hacker can steal money, obtain card details, and much more. This can either be private or can also be public data, where they can make alterations using user's profile for their selfish needs. By getting into the account, the user might have unknowingly given access with just a single click on the respective malicious Website. Therefore, to make a safer transaction and protect data, we have implemented a machine learning algorithm to detect and help users in tracking malicious Websites before performing any transaction on that particular site.

In future, our target will be to implement these algorithms such that they can categorize the URLs more precisely. We shall try to work on the URLs so that they predict the output in terms of rating and percentage of the safeness of the URL. We shall also implement it's working in the browser in which user can copy a transaction URL and paste to check whether it is safe. The machine learning algorithm shall automatically help the user by showing a message whether to continue if the site is malicious or else proceed if it is safe.

References

1. Bendovschi A (2015) Cyber-attacks—trends, patterns and security countermeasures. Procedia Econ Finan 28:24–31
2. Digital commerce transactions. Report https://www.moneycontrol.com/news/business/banks/digital-commerce-transactions-increased-30-in-2021-but-so-are-digital-frauds-says-rbi-report-8256031.html
3. Karabatak M, Mustafa T (2018) Performance comparison of classifiers on reduced phishing website dataset. In: 2018 6th international symposium on digital forensic and security (ISDFS), pp 1–5
4. Lakshmanarao A, Babu MR, Bala Krishna MM (2021) Malicious URL detection using NLP, machine learning and FLASK. In: 2021 international conference on innovative computing, intelligent communication and smart electrical systems (ICSES), pp 1–4
5. Hevapathige A, Rathnayake K (2022) Super learner for malicious URL detection. In: 2022 2nd international conference on advanced research in computing (ICARC), pp 114–119
6. Manyumwa T, Chapita PF, Wu H, Ji S (2020) Towards fighting cybercrime: malicious URL attack type detection using multiclass classification. In: 2020 IEEE international conference on big data (Big Data), pp 1813–1822
7. Peng Z, He Y, Sun Z, Ni J, Niu B, Deng X (2022) Crafting text adversarial examples to attack the deep-learning-based malicious URL detection. In: ICC 2022—IEEE international conference on communications, pp 3118–3123
8. Singh A, Roy PK (2021) Malicious URL detection using multilayer CNN. In: 2021 international conference on innovation and intelligence for informatics, computing, and technologies (3ICT), pp 340–345
9. Ren F, Jiang Z, Liu J (2019) A bi-directional LSTM model with attention for malicious URL detection. In: 2019 IEEE 4th advanced information technology, electronic and automation control conference (IAEAC), pp 300–305
10. Malware dataset. https://www.unb.ca/cic/datasets/url-2016.html

Secured Information Communication Exploiting Fuzzy Weight Strategy

Alok Haldar, Biswapati Jana, Sharmistha Jana, Nguyen Kim Sao, and Thanh Nhan Vo

1 Introduction

Today, everyone in the world uses the internet and finds it to be quite useful. Digital data protection is both important and difficult. Using steganography and digital watermarking techniques is one way to protect the security of digital data. This is especially important for different types of multimedia data, such text, images, audio files, and video files. Information concealment using established and emerging scientific techniques such as steganography and digital watermark embedding. In order to secure and safeguard digital materials from any illegal access, this is the primary criterion. In order to defend the legitimate rights, it resolves copyright disputes, offers proof to counter illegal copying, and authenticates, ownership identification, detects copy move fraud, uses digital forensics, and detects tampering with multimedia content.

In steganography, extra information that is embedded in a digital cover entity is protected. Steganographic techniques' primary objective is to hide embedded data from an attacker. Because only the sender and recipient will be aware that the secret

A. Haldar · B. Jana (✉)
Department of Computer Science, Kharagpur College, Kharagpur, West Bengal, India
e-mail: biswapatijana@gmail.com

Department of Computer Science, Vidyasagar University, Midnapore, West Bengal, India

S. Jana
Department of Mathematics, Midnapore College [Autonomous], Midnapore, West Bengal, India

N. K. Sao
Department of Computer Science, University of Transport and Communication, Hanoi, Vietnam
e-mail: saonkoliver@utc.edu.vn

T. N. Vo
Department of Information Management, Chaoyang University of Technology, Taichung 41349, Taiwan, ROC
e-mail: vtnhan@tdmu.edu.vn

J. K. Mandal et al. (eds.), *Proceedings of International Conference on Network Security and Blockchain Technology*, Lecture Notes in Networks and Systems 738,
https://doi.org/10.1007/978-981-99-4433-0_9

information is included inside the transmitted cover object, this assertion allows for the safe transmission of sensitive information over an open communication channel [1].

The digital cover items themselves are protected via digital watermarking techniques. A digital object's integrity and authenticity are attested to by a digital watermark. Digital watermarks are used to control data integrity and to authenticate the origins of data in order to safeguard the authorship of multimedia files [2].

Secret communication, authentication, copyright protection, ownership identification, copy move forgery, digital forensics, and tamper detection of multimedia documents are just a few of the uses for data hiding. Digital watermarking, cryptography, steganography, data encryption, and other methods fall under the category of data concealing and are used to protect digital media and information. Data concealing is a technique of incorporating information in the form of signal that can withstand noise, such as an audio, video, or image file. The invisibility of embedding is typically the primary need for steganographic methods to be effective. These days, robust embedding algorithms that can find embedded data in a digital item are developed as data hiding algorithms. However, a number of authors are creating algorithms with additional quality markers.

Digital watermarking [4] and steganography [3] are two techniques to achieve this goal. Based on their level of robustness, we can classify digital watermarking techniques into three groups: fragile, semi-fragile, and robust. A minimal alteration to the watermarked signal would cause the detection of the message at the receiver end to fail in a fragile watermarking. This type of watermarking is helpful for integrity proof. Moderate alterations are resistant to semi-fragile watermarking. Therefore, we can use this kind of watermarking to protect against a certain type of attack. When using robust watermarking, the watermark message can withstand numerous attacks. For the protection of copyright, this is helpful.

Spatial and transform domain algorithm are two categories of data concealment or steganography [5]. Regarding hiding ability, visual quality, security, storage space, robustness, and execution time, each of these categories has advantages and disadvantages of their own. The cover image's pixels immediately encode the secret message in the spatial domain [7], However, in the transform domain [8], the cover media must first undergo a transformation to acquire frequency coefficients and modified coefficients for data embedding, such as discrete fourier transform (DFT) [8], discrete wavelet transform (DWT) [11], or discrete cosine transform (DCT) [10]. The robustness is then achieved by embedding the secret message bits into significant coefficients. Despite the higher computational complexity of transform domain techniques, it is believed that watermarks inserted in transform domains withstand attacks better than those embedded in spatial domains.

A special focus has been placed on designing protocols that take advantage of the features included in the cover image in order to boost resilience.

The following are some important key features of data hiding schemes:

(i) **Reversibility**: One approach to authentication is data concealment. Watermarking, however, has the potential to deteriorate the cover data that was present in the original cover media after embedding. The receiver end makes it challenging to obtain a precise cover medium. However, in recent times, it has become crucial to retrieve original cover media in a variety of human-centric application fields like the military and medicine. Such programs use reversible data hiding as opposed to traditional data concealing.

(ii) **Security**: Security of a method is assessed by how well it defends against potential assaults. The data concealment technique utilised in real applications has been determined to have some security flaws based on the work that has already been done. Regarding unauthorised detection, the stego image should not provide any indications as to the existence of the hidden message.

(iii) **Payload**: Payload is the quantity of information inserted within the cover image. It is a key component of data hiding schemes, therefore it was evident from the literature review that many researchers are attempting to boost embedding capacities while preserving image quality. The payload should be as high as possible while maintaining a discernible level of image quality.

(iv) **Integrity/Tamper detection**: Tampering is the planned alteration of documents so that the consumer would be harmed. Therefore, it is crucial for the authorised user to expose both the secret information and the cover image throughout the extraction process.

(v) **Imperceptibility/visual quality**: Any technique for hiding data must meet the imperceptibility criteria first and foremost. There shouldn't be any visual deterioration due to the embedding data that is included in the original image. For the hidden message to remain imperceptible and harmful, it must not be seen by humans, cannot be detected by their eyes, and must not cause any visual distortion in the stego image. Therefore, maintaining acceptable visual quality after embedding the hidden data is a crucial quality of any invisible data concealment strategy.

(vi) **Robustness**: A media action, such as filtering, lossy compression, or alteration, is considered robust if the hidden message can still be discovered. There are times when delicate confidential data may be required. Fragile data means that it should not resist manipulation or would only resist to a limited degree. It is a critical prerequisite for data concealing.

2 Literature Survey

Wenyin and Shih [21] demonstrated a semi-fragile watermarking technique utilising Local Binary Pattern (LBP) operators, pixel contrast, and a multilevel picture watermarking system. A straightforward texture descriptor called the LBP operator which was developed by Ojala et al. [22]. With the help of suggested methodology, the author has demonstrated that the watermarking techniques can withstand standard image processing operations including JPEG compression, brightness alteration, contrast modification, and additive noise. The results, however, were not thoroughly compared with any other LBP schemes and were not robust against significant geometrical attacks.

Fan et al. [23] developed a data concealing method that can only conceal four secret data bits within a (3 × 3) block using a weighted matrix for a grayscale image. Tseng et al. [24] suggested an effective binary image data embedding technique that makes use a weighted matrix W with key matrix K. There is only one modular sum of entry-wise multiplication that can be performed using a weighted matrix W and a (3 × 3) pixel block for both of these matrix-based data hiding methods. A significant research question is still how to achieve large capacity with reversibility in watermarking while preserving decent visual quality. Reversible data hiding (RDH) becomes a crucial and difficult issue in hidden data communication for authentication, copy right protection and ownership identification in medical and military applications. Through the use of picture interpretation, Jana et al. [25] created a weighted matrix-based reversible data hiding strategy that can conceal 2.97 bpp. However, Chowdhuri et al. [26] created a weighted matrix-based reversible data concealing method that uses the colour image and provides image authentication and tamper detection. To repeatedly embed the secret data in each block, they divided the original image into (3 × 3) pixel blocks and multiplied the sum of the entries using a modified weighted matrix. Additionally, the proposed data hiding strategy increased visual quality PSNR 50.03 and data embedding capacity up to 8.03 (bpp) (dB). Their technique was successful in detecting and authenticating tampered images, however it may only partially recover the original cover picture from a manipulated stego image. In their innovative method for image watermarking in the spatial domain [27], Kumar and Dutta combine the well-known LSB substation technique with the idea of information theory. The watermark is positioned in the block with the highest entropy after dividing the cover image into a number of blocks. None of geometric attack difficulties were supported by the experimental findings. Cao et al. [28] developed hamming code-based data hiding techniques with high payload and embedding rates as high as 3 (bpp) and PSNR as low as 51(dB). Bai and Chang [29] have created a high payload steganographic technique for compressed images. They have a 2 (bpp) payload, however their PSNR is under 30(dB). Jana et al. [5, 30] recently developed a partial and dual image based reversible data hiding technique using the (7, 4) hamming code, which may improve data concealing capacity, visual quality, and accomplish reversibility. Su and Chen [31] presented an algorithm that combines frequency and spatial domain to protect copyright.

Watermarked images with good visual quality and good resistance to common image processing attacks, this technique ensures an appropriate compromise between the computational complexity and the authentication level of host images. Due to the watermark being embedded in multiple blocks, the technique's key value is giving the ability to retrieve the watermark either it cropping or rotation attacks. The goal of Hassan and Gutub [34] was to enhance the reversible data concealment strategy with interpolation. The proposed method scales up the original image utilising the existing enhanced neighbour mean interpolation (ENMI) and modified neighbour mean interpolation (MNMI) approaches before inserting the secret data.

Using a dual watermark approach and blind removal, Rangel-Espinoza et al. [39] suggested a removable visible data concealment system. The brightness and texture properties of the watermark and original images are taken into consideration when placing a visible watermark pattern in the DCT domain to create a visible watermarked image. Without other information, such as the original watermark or host image, the watermark could only be removed using the keys of the right user. The experimental findings demonstrate that the suggested system surpasses earlier comparable efforts with the help of blind removal, retention of the quality of the undamaged restored images, and higher visual degradation of the recovered the content of images in the event of an illegal removal attempt. In the value of the interpolation pixel, Jana et al. [40] presented an effective data hiding strategy that takes advantage of the Centre Folding Strategy (CFS) and Fuzzy Logic System (FLS).

3 Proposed Method

This work uses the Fuzzy weight strategy to construct a novel image interpolation scheme. The interpolated pixel values are produced by taking into consideration each pair of pixels in a particular block's fuzzy weight value. Each input pixel pair's fuzzy membership values have been taken to represent the range between the block's minimum and maximum value. The input membership value is fed into the fuzzy output function, which calculates the fuzzy rule's strength using the Max–Min composite principle. Then, through a defuzzification process, interpolated pixel values are calculated from the fuzzy output function dependent on the fuzzy rule's strength. In actuality, fuzzy weight based interpolation algorithms create virtual pixels, which are superior to the interpolation techniques now in use.

FWS based interpolated method

To provide fuzzy interpolated pixel values, a fuzzy rule-based controller with four features—fuzzification, fuzzy rule base, fuzzy inference mechanism, and defuzzification—is developed. As Demirci (Demirci, 2006) demonstrated in his study for similarity calculation, the visual evaluation of a fuzzy interpolated pixel value may vary from one skillful person to another skillful person. Because fuzzy interpolated pixel values between the pixels are imprecise, it is advisable to employ fuzzy rule-based logic to handle them. Two variables, pixel P_i and pixel P_j, are taken into consideration as input for fuzzification in order to determine the fuzzy interpolated pixel value between the pixels. Each component is shown with the fuzzy set for the following fuzzy linguistic variables: Very Low (VL), Low (L), Medium (M), High (H), and Extremely High (VH). These language words can be modelled using a variety of membership functions, including trapezoidal, gaussian, triangular, etc. Nonetheless, due to its simplicity and symmetric representation, the selected representation of the provided linguistic words is a triangular fuzzy number (TFN). The range of possible values is 0–255, with 255 being the maximum. Figure 1 depicts the

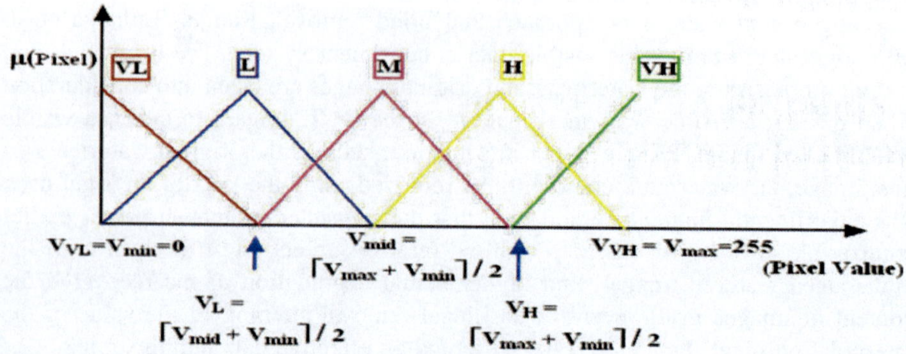

Fig. 1 Input pixel pair of a block's membership functions

modelling for both of these factors. A fuzzy inference engine is set up after fuzzification and operates in accordance with predefined rules. A degree of association between two pixels is produced by combining the fuzzy inputs created during fuzzification with the rule basis. The degree to which a fuzzy interpolated pixel value exists is measured by this connection. The *i*th fuzzy rule according to Eq. (18) is thus expressed as: With n inputs such that x_1X_1, x_2X_2, ..., x_nX_n and one output y Y R_i: If x_1 is 1, x_2 is 2, etc., then x_n must be n. The set of inputs in the rule base is also referred to as the antecedent, and the output is referred to as the subsequent, if y is B. Table 1 displays the most effective fuzzy rules that were successful for the recommended job. The linguistic words VVL, VL, L, ML, M, MH, H, VH, and VVH define the form in which the consequent is created. To create a fuzzy output set as the finished product, each rule's actions are superimposed. The fuzzy output is produced by combining the results of the rules that have been shot. Further defuzzification techniques produce a crisp value on the fuzzy output. Two common defuzzification approaches are area and maxima based methods. The centroid approach is a well-liked area-based defuzzification method. The crisp value, as its name suggests, marks the location at which the output membership function divides the region in half.

The following is a list of the main steps in image interpolation:

Step-1: Imagine a colour cover image that is (M x N) pixels in size, divided into the three basic colours of Red (R), Green (G) and Blue (B), and then divided into ($n \times n$) pixels for each colour block, where $n = 3, 5, 7, 9\ldots$

Step-2: Choose $V_{\min} = 0$ as the block's minimum and maximum pixel values, respectively, and use those values to calculate the fuzzy membership of each pair of pixels using $V_{\min}$ as the lower value and $V_{\max}$ as the higher value with the medium value is $V_{\text{mid}} = \left\lceil \frac{V_{\min}+V_{\max}}{2} \right\rceil$. Assume five linguistic inputs, such as VL for very low, low (L), medium (M), high (H) and very high (VL).

Consider of Fig. 1 as being very highly illustrated.

Table 1 Proposed fuzzy rules

Rule	Antecedent	Consequent
R^1:	If μ_{pi} is VL and μ_{pj} is VL	then μ_z is VVL
R^2:	If μ_{pi} is VL and μ_{pj} is L	then μ_z is VL
R^3:	If μ_{pi} is VL and μ_{pj} is M	then μ_z is L
R^4:	If μ_{pi} is VL and μ_{pj} is H	then μ_z is ML
R^5:	If μ_{pi} is VL and μ_{pj} is VH	then μ_z is M
R^6:	If μ_{pi} is L and μ_{pj} is VL	then μ_z is VL
R^7:	If μ_{pi} is L and μ_{pj} is L	then μ_z is L
R^8:	If μ_{pi} is L and μ_{pj} is M	then μ_z is ML
R^9:	If μ_{pi} is L and μ_{pj} is H	then μ_z is M
R^{10}:	If μ_{pi} is L and μ_{pj} is VH	then μ_z is MH
R^{11}:	If μ_{pi} is M and μ_{pj} is VL	then μ_z is L
R^{12}:	If μ_{pi} is M and μ_{pj} is L	then μ_z is ML
R^{13}:	If μ_{pi} is M and μ_{pj} is M	then μ_z is M
R^{14}:	If μ_{pi} is M and μ_{pj} is H	then μ_z is MH
R^{15}:	If μ_{pi} is M and μ_{pj} is VH	then μ_z is H
R^{16}:	If μ_{pi} is H and μ_{pj} is VL	then μ_z is ML
R^{17}:	If μ_{pi} is H and μ_{pj} is L	then μ_z is M
R^{18}:	If μ_{pi} is H and μ_{pj} is M	then μ_z is MH
R^{19}:	If μ_{pi} is H and μ_{pj} is H	then μ_z is H
R^{20}:	If μ_{pi} is H and μ_{pj} is VH	then μ_z is VH
R^{21}:	If μ_{pi} is VH and μ_{pj} is VL	then μ_z is M
R^{22}:	If μ_{pi} is VH and μ_{pj} is L	then μ_z is MH
R^{23}:	If μ_{pi} is VH and μ_{pj} is M	then μ_z is H
R^{24}:	If μ_{pi} is VH and μ_{pj} is H	then μ_z is VH
R^{25}:	If μ_{pi} is VH and μ_{pj} is VH	then μ_z is VVH

Step 3: The fuzzy output membership function has been developed using a certain range of pixel values and nine linguistic terms, including VVL-very very low, VL-very low, L-low, ML-medium low, M-medium, MH-medium high, H-high, VH-very high, and VVH-very very high. This function is shown in Fig. 2. The VVL will belong to the scale's lower valuc, P_i, while the VVH will belong to the scale's higher value, P_j.

Step 4: The fuzzy weighted strategy rule has been built using the nine output parameters VVL, VL, L, ML, M, MH, H, and VH displayed in Table 1 of the rule together with the five input parameters VL, L, M, H, and VH.

Step 5: The strength of the rule has been calculated using the Max–Min composite rule and the actual input values of each pair of pixels from the selected specified block.

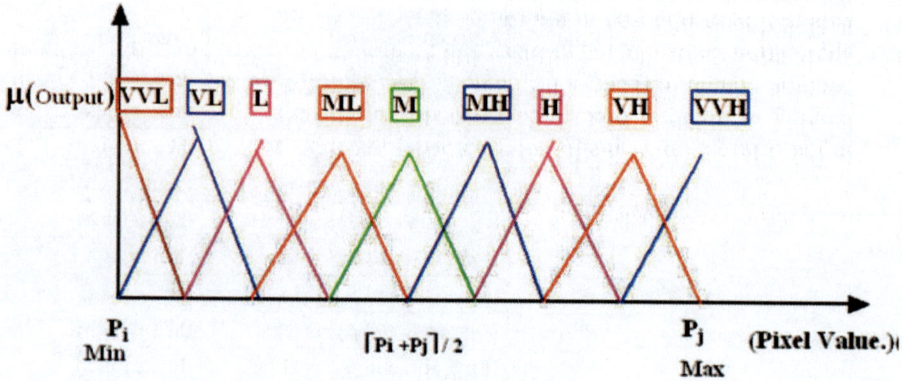

Fig. 2 Membership functions for block virtual pixel calculation

Step 6: The next step is to select the appropriate membership value from the fuzzy output function based on the rule's strength. The weighted fuzzy function output is then used to construct the interpolated virtual pixel, and in cases where there are several values, the interpolated pixel value of that block for the selected pixel pair will be the average value.

Numerical Example

Think about the image block in Fig. 3b. As seen in Fig. 3c, We reduce the size of the block by removing one row and one column of pixels. Figure 1 shows the anticipated pixel block produced by the Fuzzy Weight Strategy (FWS). Now think about the pixel pair with the values $P_i = 70$ and $P_j = 120$ and $V_{\min} = 0$ and $V_{\max} = 255$. VL $= 0$, $L = 64$, $M = 128$, $H = 192$, and VH $= 255$ are the results.

The following formula is used to determine fuzzy membership values:

Numerical Example

Table 2

120	68	70
181	177	124
200	165	180

(a): Original block

Table 3

120	70
200	180

(b): Reduced block

Table 4

120	89	70
170	151	139
200	196	180

(c): The Predicted block

Fig. 3 Numerical example of fuzzy weight strategy

$$\mu_{\mathrm{VL}}(x) = \frac{64 - x}{64 - 0} = \frac{64 - x}{64}, \quad 0 \le x \le 64,$$

$$\frac{x - 64}{64 - 0} = \frac{x - 64}{64}, \quad 0 \le x \le 64,$$

$$\mu_L(x) = \frac{128 - x}{128 - 64} = \frac{x - 128}{64}, \quad 64 \le x \le 128$$

$$\frac{x - 64}{128 - 64} = \frac{x - 64}{64}, \quad 64 \le x \le 128$$

$$\mu_M(x) = \frac{192 - x}{192 - 128} = \frac{192 - x}{64}, \quad 128 \le x < 192$$

$$\frac{x - 128}{192 - 128} = \frac{x - 128}{64}, \quad 128 \le x < 192$$

$$\mu_H(x) = \frac{255 - x}{255 - 192} = \frac{255 - x}{63}, \quad 192 \le x < 255$$

$$\mu_{\mathrm{VH}}(x) = \frac{x - 192}{255 - 192} = \frac{x - 192}{63}, \quad 192 \le x < 255,$$

Similar to that, each colour component of a particular block's membership values for the output fuzzy function were defined, taking into account the lower pixel value of 70 and the higher pixel value of 120, as shown below.

$$\mu_{\mathrm{VVL}}(z) = \frac{77 - z}{77 - 70} = \frac{77 - z}{7}, \quad 70 \le z < 77,$$

$$\frac{z - 70}{77 - 70} = \frac{z - 70}{7}, \quad 70 \le x < 77$$

$$\mu_{\mathrm{VL}}(z) = \frac{83 - z}{83 - 77} = \frac{83 - z}{6}, \quad 77 \le z < 83$$

$$\frac{z - 77}{83 - 77} = \frac{z - 77}{6}, \quad 77 \le z < 83$$

$$\mu_L(z) = \frac{89 - z}{89 - 83} = \frac{89 - z}{6}, \quad 83 \le z < 89$$

$$\frac{z - 77}{89 - 83} = \frac{z - 83}{6}, \quad 83 \le z < 89$$

$$\mu_{\mathrm{ML}}(z) = \frac{95 - z}{95 - 89} = \frac{95 - z}{6}, \quad 89 \le z < 95$$

$$\frac{z - 89}{95 - 89} = \frac{z - 89}{6}, \quad 89 \le z < 95$$

$$\mu_M(z) = \frac{102 - z}{102 - 95} = \frac{102 - z}{7}, \quad 95 \le z < 102$$

$$\frac{z-95}{102-95}=\frac{z-95}{7},\quad 95\le x<102$$

$$\mu_{\mathrm{MH}}(z)=\frac{108-z}{108-102}=\frac{108-z}{6},\quad 102\le z<108$$
$$\frac{z-102}{108-102}=\frac{z-102}{6},\quad 102\le z<108$$

$$\mu_H(z)=\frac{114-z}{114-108}=\frac{114-z}{6},\quad 108\le z<114$$
$$\frac{z-108}{114-108}=\frac{z-108}{6},\quad 108\le z<114$$

$$\mu_{\mathrm{VH}}(z)=\frac{120-z}{120-114}=\frac{120-z}{6},\quad 114\le z<120$$

$$\mu_{\mathrm{VVH}}(z)=\frac{120-z}{120-114}=\frac{120-z}{6},\quad 114\le z<120$$

Then, assess the input for each pair of pixels you collected from the block's input fuzzy membership value. In this case, $P_i = 70$, which is between 64 and 128. Thus, it is necessary to calculate,

$$\mu_L(x)=\frac{128-x}{64}=\frac{128-70}{64}=\frac{58}{64}$$

$$\mu_M(x)=\frac{x-128}{64}=\frac{70-64}{64}=\frac{6}{64}$$

In this case, $P_j = 120$ which is between $64 \le x < 128$. Thus, it is necessary to calculate,

$$\mu_L(x)=\frac{128-x}{64}=\frac{128-120}{64}=\frac{8}{64}$$

$$\mu_M(x)=\frac{x-64}{64}=\frac{120-64}{64}=\frac{56}{64}$$

The following formula has been used to determine the strength of the fuzzy rule formed by the first four values:

S_1: P_i is (VL) Very Low and P_j is (L) Low, So,

$$S_1=\min(\mu_L(70),\mu_M(120))=\min\left(\frac{58}{64},\frac{8}{64}\right)=\frac{8}{64}$$

S_2: P_i is (VL) Very Low and P_j is (M) Medium, So,

$$S_1 = \min(\mu_L(70), \mu_M(120)) = \min\left(\frac{58}{64}, \frac{56}{64}\right) = \frac{56}{64}$$

S_3: P_i is (VL) Very Low and P_j is (M) Medium, So,

$$S_1 = \min(\mu_L(70), \mu_M(120)) = \min\left(\frac{6}{64}, \frac{8}{64}\right) = \frac{6}{64}$$

S_4: P_i is (VM) Very Medium and P_j is (M) Medium, So,

$$S_1 = \min(\mu_L(70), \mu_M(120)) = \min\left(\frac{6}{64}, \frac{56}{64}\right) = \frac{6}{64}$$

The first four numbers together create a fuzzy rule and the following formula has been used to determine the strength of this rule:

MAX $\{S_1, S_2, S_3, S_4\}$, that is

$$\text{MAX}\left(\frac{8}{64}, \frac{56}{64}, \frac{6}{64}, \frac{6}{64}\right) = \frac{56}{64} \quad \text{i.e } S_2$$

As a result, S_2, which correlates to P_i being low and P_j being medium, is the rule with the highest strength out of the four. Now, using the defuzzification technique and analyze the fuzzy function that results to determine the virtual pixel:

$$\frac{z - 83}{6} = \frac{56}{64} \quad \text{i.e } z_1 = 88.25$$

$$\frac{95 - z}{6} = \frac{56}{64} \quad \text{i.e } z_2 = 89.75$$

Therefore $z = \left\lceil \frac{z_1+z_2}{2} \right\rceil = \left\lceil \frac{88.25+89.75}{2} \right\rceil = 89$.
We have determined the projected values 158 and 144 for the diagonal input sets 120,180 and 70,200 respectively, for the central value 151 of the Fig. 1c.

```
if ( interpol > origin )
   m =interpol – origin
   s = ceil(m/2)
   new_interpol =interpol –s
else
   m = origin – interpol
   s = ceil(m/2)
  new-interpol = interpol + s.
```

Lastly, the average between 158 and 144, or $\lceil \frac{158+144}{2} \rceil = 151$, is used to get the centre predicted value. In Fig. 3c displays the final, identical-sized predicted pixel block that was produced using FWS.

4 Experimental Results with Comparative Analysis

We conducted the following investigation [41] using the six common benchmark photos from the tests: (a) aeroplane, (b) sailboat, (c) Lena, (d) baboon, (e) peppers, and (f) Tiffany in Fig. 4. These 512 × 512-pixel images are fed into the suggested system to evaluate the effectiveness of the suggested research. Several comparisons are used in this part to demonstrate the viability of the created strategy. Secondly, five previously developed strategies are compared to the suggested interpolation technique.

To evaluate the advantages and disadvantages of the suggested innovative interpolation method, NMI, INP, ENMI, CRS, and MNMI were used. The suggested embedding strategy is then evaluated using the outcomes of the experiments, and various comparisons are used to investigate the scheme.

The PSNR is computed as follows:

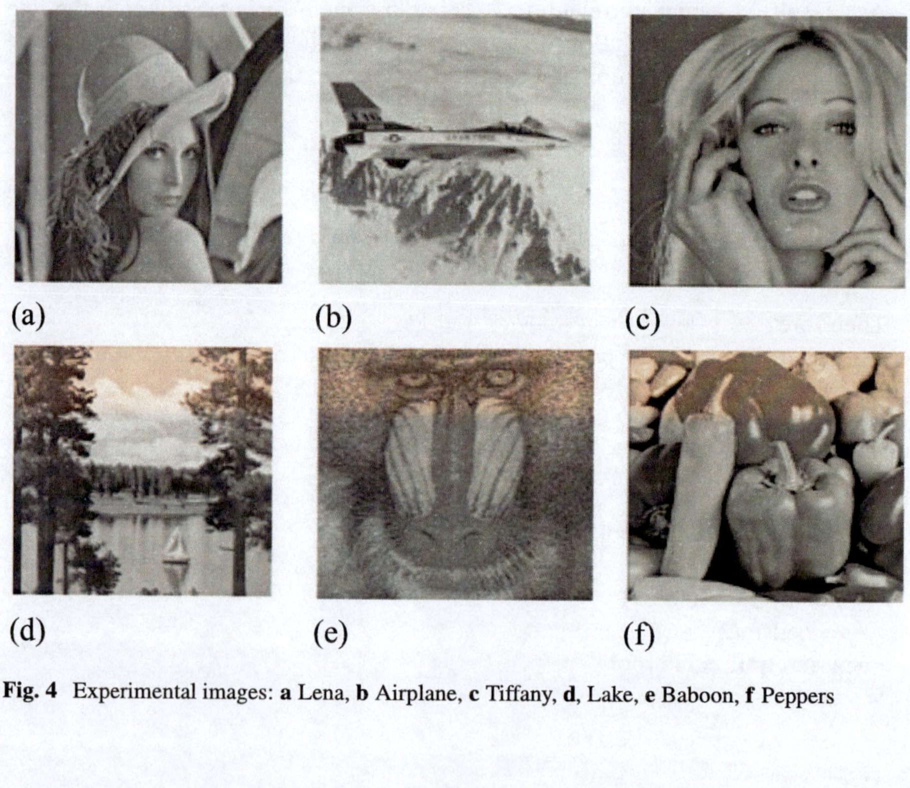

Fig. 4 Experimental images: **a** Lena, **b** Airplane, **c** Tiffany, **d**, Lake, **e** Baboon, **f** Peppers

Table 5 Six picture interpolation methods have been compared in terms of the visual attributes assessed by PSNR (dB)

Cover image	PSNR of both original and interpolated image					
	NMI	INP	ENMI	CRS	MNMI	FWS
Lena	28.06	26.41	25.64	26.29	22.64	33.95
Airplane	27.30	27.16	25.48	25.24	25.11	34.15
Tiffany	21.82	26.29	28.18	26.64	25.70	35.05
Lake	26.56	22.81	27.54	27.12	25.49	31.13
Mandrill	25.30	25.25	22.07	26.39	24.81	32.88
Pepper	22.09	21.44	23.84	21.51	24.45	33.34
Average	26.06	25.18	26.24	25.27	29.18	33.41

$$\text{MSE} = \frac{1}{h \times w} \sum_{i=1}^{h} \sum_{j=1}^{w} C(i, j) - S(i, j)^2$$

$$\text{PSNR} = 10 \log_{10}\left(255^2/\text{MSE}\right)(\text{dB})$$

Where $C(i, j)$ is the cover image's pixel value, $S(i, j)$ is the stego image's pixel value at position, and MSE is the mean square error. To assess the effectiveness of the interpolation approaches, the PSNR is calculated. The NMI, INP, ENMI, CRS, and MNMI interpolation techniques are compared to the FWS methodology. Table 5 displays the outcomes. The NMI, INP, ENMI, CRS, and MNMI recommended interpolation approaches are all inferior than the FWS based image interpolation scheme. The suggested method showed that, in nearly all of the tested input images, the FWS methodology achieved the greatest PSNR, according to the results reported in Table 5. The PSNR average is roughly estimated to be 33 dB. As a result, the FWS was employed in the suggested technique to produce the enhanced cover image.

5 Conclusion

In this study, after performing picture interpolation with fuzzy weight, an innovative, efficient, and reversible data hiding strategy was created using the centre folding approach. The fuzzy weight of each pair of pixels in a particular image block is taken into account to produce the interpolated pixel values. Subsequently, data hiding operations were carried out in each interpolated pixel by centre folding and comparison with the signal of the secret data. The suggested technique reverses the folded value to match the secret data if the signal differs from that of the interpolated pixel in order to lessen the difference in image distortion between the interpolated picture and the stego-image. You can figure out the hidden data using the discrepancy between the outputs of these interpolated values and the receiver's centre folding scheme.

Experimental results and their analysis from many angles show that the suggested method is superior than state-of-the-art approaches. Additionally, security research of RS assaults reveals that the stego image has less noise, which makes it difficult to use it to detect the existence of secret information. The proposed technique may be utilised for authentication or secret data exchange in military, medical, and academic institutions. This method can be applied in the future for image authentication, copyright protection, and ownership identification using cellular automata, local binary patterns, and other methods.

References

1. Fridrich J (2009) Steganography in digital media: principles, algorithms, and applications. Cambridge Univ. Press, Cambridge, UK
2. Panah AS, Van Schyndel R, Sellis T, Bertino E (2016) On the properties of non-media digital watermarking: a review of state of the art techniques. IEEE Access 4:2670–2704. https://doi.org/10.1109/ACCESS.2016.2570812
3. Singh S, Singh AK, Ghrera SP (2017) A recent survey on data hiding techniques. In: 2017 international conference on I-SMAC (IoT in Social, Mobile, Analytics and Cloud) (I-SMAC). IEEE, pp 88–886
4. Chowdhury FS, Dhar PK, Deb K, Koshiba T (2020) Blind image watermarking in canonical and cepstrum domains based on 4-connected t-o'clock scrambling. Symmetry 12(2):266
5. Jana B, Giri D, Mondal SK (2017) Partial reversible data hiding scheme using (7, 4) hamming code. Multimedia Tools Appl 76(20):21691–21706
6. Jana B (2016) High payload reversible data hiding scheme using weighted matrix. Optik 127(6):3347–3358
7. Jana M, Jana B (2020) An improved data hiding scheme through image interpolation. In: Computational intelligence in pattern recognition. Springer, pp 157–169
8. Gunjan R, Pandia P, Mohnot R (2017) Secure extraction of image data based on optimized transform method. In: 2017 IEEE 4th international conference on cyber security and cloud computing (CSCloud). IEEE, pp 217–222
9. Gonge SS, Ghatol AA (2017) An enhancement in security and copyright protection technique used for digital still image. In: 2017 international conference on nascent technologies in engineering (ICNTE). IEEE, pp 1–9
10. Pushpad A, Potnis AA (2017) Improved image security scheme using combination of image encryption and reversible watermarking. In: 2017 4th international conference on signal processing and integrated networks (SPIN). IEEE, pp 293–297
11. Rasti P, Anbarjafari G, Demirel H (2017) Colour image watermarking based on wavelet and qr decomposition. In: 2017 25th signal processing and communications applications conference (SIU). IEEE, pp 1–4
12. Kelkar V, Mehta JH, Tuckley K (2018) A novel robust reversible watermarking technique based on prediction error expansion for medical images. In: Proceedings of 2nd international conference on computer vision and image processing. Advances in intelligent systems and computing, vol 703, pp 131–143
13. Shehab A, Elhoseny M, Muhammad K, Sangaiah AK, Yang P, Huang H, Hou G (2018) Secure and robust fragile watermarking scheme for medical images. IEEE Access 6:10269–10278
14. Liu J, Li J, Ma J, Sadiq UN, Yang A (2019) A robust multi-watermarking algorithm for medical images based on DTCWT-DCT and Henon map. Appl Sci 9(4):1–23
15. Rehman A, Sultan K, Aldhafferi N, Alqahtiani A, Mahmood M (2018) Reversible and fragile watermarking scheme for medical images. Comput Math Methods Med 18:7

16. Zhang H, Wang C, Zhou X (2017) Fragile watermarking for image authentication using the characteristic of SVD. Algorithms 10(1):1–12
17. Konstantinides K, Natarajan B, Yovanof GS (1997) Noise estimation and filtering using block-based singular value decomposition. IEEE Trans Image Process 6(3):479–483
18. Liu R, Tan T (1997) An SVD-based watermarking scheme for protecting rightful ownership. IEEE Trans Image Process 6(3):479–483
19. Roy S, Pal AK (2017) An indirect watermark hiding in discrete cosine transform singular value decomposition domain for copyright protection. Roy Soc Open Sci 4(6). Art. no. 170326
20. Makbol NM, Khoo BE, Rassem TH (2016) Block-based discrete wavelet transform-singular value decomposition image watermarking scheme using human visual system characteristics. IET Image Process 10(1):34–52
21. Wenyin Z, Shih FY (2011) Semi-fragile spatial watermarking based on local binary pattern operators. Opt Commun 284(16–17):3904–3912
22. Ojala T, Pietikainen M, Maenpaa T (2002) Multiresolution gray-scale and rotation invariant texture classification with local binary patterns. IEEE Trans Pattern Anal Mach Intell 24(7):971–987
23. Fan L, Gao T, Cao Y (2013) Improving the embedding efficiency of weight matrix-based steganography for grayscale images. Comput Electr Eng 39(3):873–881
24. Tseng YC, Chen YY, Pan HK (2002) A secure data hiding scheme for binary images. IEEE Trans Commun 50(8):1227–1231
25. Jana B (2016) High payload reversible data hiding scheme using weighted matrix. Optik-Int J Light Electron Opt 127(6):3347–3358
26. Chowdhuri P, Pal P, Jana B (2019) Improved data hiding capacity through repeated embedding using modified weighted matrix for color image. Int J Comput Appl 41(3):218–232
27. Kumar S, Dutta A (2016, April) A novel spatial domain technique for digital image watermarking using block entropy. In: 2016 international conference on recent trends in information technology (ICRTIT). IEEE, pp 1–4
28. Cao Z, Yin Z, Hu H, Gao X, Wang L (2016) High capacity data hiding scheme based on (7, 4) Hamming code. Springer Plus 5(1):1–13
29. Bai J, Chang CC (2016) A high payload steganographic scheme for compressed images with hamming code. Int J Netw Sec 18(6):1122–1129
30. Jana B, Giri D, Mondal SK (2018) Dual image based reversible data hiding scheme using (7, 4) hamming code. Multimedia Tools Appl 77(1):763–785
31. Su Q, Chen B (2018) Robust color image watermarking technique in the spatial domain. Soft Comput 22(1):91–106
32. Laouamer L (2019, March) A new image watermarking technique in spatial domain using DC coefficients and graph representation. In: International conference on advanced machine learning technologies and applications. Springer, Cham, pp 633–644
33. Peng F, Zhao Y, Zhang X, Long M, Pan WQ (2020) Reversible data hiding based on RSBEMD coding and adaptive multi-segment left and right histogram shifting. Sig Process Image Commun 81:115715
34. Hassan FS, Gutub A (2021) Efficient image reversible data hiding technique based on interpolation optimization. Arab J Sci Eng 46(9):8441–8456
35. Maheswari S, Rameshwaran K, Malarselvi KM (2015) DCT-PCA based watermarking on E-governance documents. Res J Appl Sci Eng Technol 9(7):507–511
36. Abbas NH, Ahmad SMS, Ramli ARB, Parveen S (2016) A multi-purpose watermarking scheme based on hybrid of lifting wavelet transform and Arnold transform. In: International conference on multidisciplinary in IT and communication science and applications
37. Ghazvini M, Hachrood EM, Mirzadi M (2017) An improved image watermarking method in frequency domain. J Appl Sec Res 12(2):260–275
38. Yuan Y, Huang D, Liu D (2006) An integer wavelet based multiple logo-watermarking scheme. In: First international multi-symposiums on computer and computational sciences, Hanzhou, vol 2, pp 175–179

39. Rangel-Espinoza K, Fragoso-Navarro E, Cruz-Ramos C, Reyes-Reyes R, Nakano-Miyatake M, Perez-Meana HM (2018) Adaptive removable visible watermarking technique using dual watermarking for digital color images. Multimedia Tools Appl 77(11):13047–13074
40. Jana S, Jana B, Lu T-C, Vo TN (2022) Reversible data hiding scheme exploiting center folding with fuzzy weight strategy
41. USC-SIPI (2017). University of Southern California. The usc-sipi image database. http://sipi.usc.edu/database/database.php. Accessed 20 Sept 2017

Secure Data Communication Through Improved Multi-level Pixel Value Ordering Using Center-Folding Strategy

Sudipta Meikap, Biswapati Jana, Prabhash Kumar Singh, Debkumar Bera, and Tzu Chuen Lu

1 Introduction

The secret information is hidden under text documents, cover pictures, or any other types of multimedia as a data carrier in the data concealing technique in order for the recipient to effectively decipher the message. Reversible and irreversible information concealment techniques are based on the reconstruction of the original picture. Reversible data concealing techniques, also known as lossless information hiding, should always be used in sectors that are particularly susceptible to picture distortion, such as the medical, military, and scientific ones. Although non-reversible data hiding techniques normally have high embedding capacities, there are several limitations in the application domains since the cover image was irreparably lost after stuffing the secret information. After the extraction procedure is successfully finished, the secret data and the cover picture may be rebuilt. This study fixes secrets into image pixels using the RDH technique as its foundation. Reversible approaches can recreate the true cover picture after data extraction, whereas irreversible methods are insufficient to do so. In order to integrate hidden messages into cover graphics, this research explores reversible data concealing techniques.

In 2003, Tian [17] wrote up a method for hiding information within a pair of pixels called difference expansion (DE). It outlines how the picture was produced in order to get high embedding data with little distortion. A lossless information concealing

S. Meikap
Department of Computer Science, Hijli College, Paschim Medinipur, West Bengal 721306, India

S. Meikap (✉) · B. Jana · P. K. Singh · D. Bera
Department of Computer Science, Vidyasagar University, Midnapore, West Bengal 721102, India
e-mail: sudiptameikap@gmail.com

T. C. Lu
Department of Information Management, Chaoyang University of Technology, Taichung 41349, Taiwan, ROC
e-mail: tclu@cyut.edu.tw

J. K. Mandal et al. (eds.), *Proceedings of International Conference on Network Security and Blockchain Technology*, Lecture Notes in Networks and Systems 738,
https://doi.org/10.1007/978-981-99-4433-0_10

approach that allows distortion to be completely erased from the annotated picture was presented by Lee et al. [3] in 2005. Additionally, the secret data was extracted from the marked picture. Li et al. [4] presented pixel value ordering-based information concealing techniques in 2013 where minimum and maximum pixels are regulated by information embedding. This demonstrates a high-fidelity RDH system for pictures using a novel approach that incorporates the information via prediction error expansion (PEE). Peng et al. [15] improved the PVO that Li et al. had shown by computing additional pixel differences and using new histogram modifications. Jana [1] has described a high payload-based data concealing approach using sub-sample. The remainder of the study is organized as follows: Sect. 2 explores the literature review. Section 3 explores the suggested embedding and extraction procedure. In Sect. 4, the comparisons and observational findings are discussed. The conclusions of the suggested approach are examined in Sect. 5.

2 Literature Review

Several reversible data hiding methods have recently been published by researchers. The center-folding method was used by Lu et al. [5] to approach reversible data concealing. It employs two copies of the original picture to achieve high embedding. The hidden message is folded and after that embedded into dual marked-images in this approach. In 2018, a technique [2] was put out to reduce distortion and preserve the authenticity of forgery alteration in dual stego-images. Meikap et al. [9–11] spoke about PVO-based RDH methods that improved the high data bit capability. In 2021, the [5] was changed from having a single layer to having numerous levels in order to present an improved interpolation-based concealment strategy [7]. Center-folding-based reversible information hiding methods [12, 13] were introduced in the year 2021. In 2022, [14] employed dual-image to build a shift technique that enables secret data to be inserted in those pixels that were previously exclusively altered in PVO-based schemes. The difficulty is to enhance payload by thinking about embedding in more than two pixels using references and neighboring pixels inside an image block after folding the message and inserting it into image pixels. To address these issues, we developed an improved multi-level PVO (IMPVO) with various block size information concealing mechanism based on center-folding. The strength of security is increased by our recommended approach. The q data set is used to translate the secrets into decimal values during the first embedding. Second, during second embedding, the embed procedure depends on the quantity and size of picture blocks. Third, the data is split across two pictures. The size of the image block, the quantity of image blocks, the value of q, and two stego-images are necessary for message retrieval. The aforementioned criteria are required if an unauthorized individual wants to view the message. Unauthorized individuals may find it challenging.

3 Proposed Method

In this part, we suggested the process of improved multi-level PVO (IMPVO) secret embedding and extraction using the center-folding approach. By maintaining picture quality along with our recommended improved multi-level PVO method, we seek to maximize the embedding capacity. The following stages are used to organize the whole procedure.

3.1 Data Embedding Phase

Step 1: By averaging the neighboring pixels, we enlarge these pictures using the interpolation technique presented in Fig. 1. If the picture size is $(w \times w)$, the interpolated image will be $(w + (w - 1)) \times (w + (w - 1))$, where $(w - 1)$ is the interpolated row/column. For making rows and columns even, one is added to each, and the previous row's and column's value is copied.
Step 2: The cover picture is $IC = \{pix_{1,1}, \ldots, pix_{w,h}\}$, where w as well as h stand for width and height of image, respectively. Every q bit is taken as a set by the secret information, which is then modified to a decimal using the hidden symbol d_s. The symbol range is then adjusted from $Q = \{0, 1, \ldots, 2^q - 1\}$ to $R' = \{-2^{q-1}, -2^{q-1} + 1, \ldots, -1, 0, 1, \ldots, 2^{q-1} - 2, 2^{q-1} - 1\}$ using the concealed symbol d_s. The following formula may be used to determine the folded hidden symbol d_s':

$$d_s' = d_s - 2^{q-1} \tag{1}$$

where 2^{q-1} denotes in-between values. The process of inserting concealed data can be done by

$$\begin{cases} d_{s1}' = \left\lfloor \frac{d_s'}{2} \right\rfloor, \\ d_{s2}' = \left\lceil \frac{d_s'}{2} \right\rceil. \end{cases} \tag{2}$$

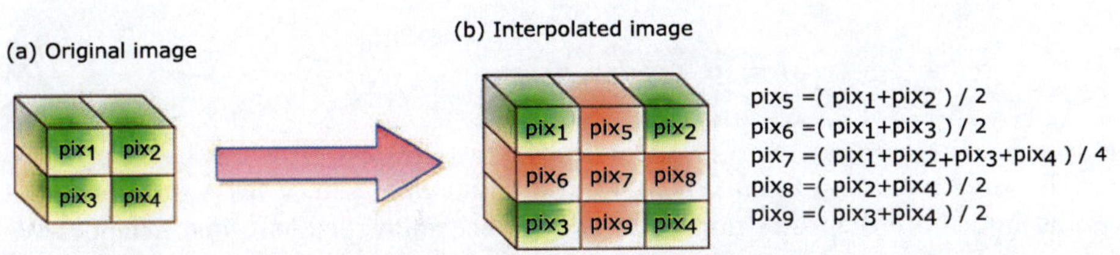

Fig. 1 Creation of interpolated image. Pink color represents interpolate row and column

Here, d'_{s1} as well as d'_{s2} are obtained from d'_s. The following equations create dual marked pixels $pix^1_{i,j}$ and $pix^2_{i,j}$ by inserting these values into the pixel $pix_{i,j}$.

$$\begin{cases} pix^1_{i,j} = pix_{i,j} + d'_{s1}, \\ pix^2_{i,j} = pix_{i,j} - d'_{s2}. \end{cases} \tag{3}$$

The hidden data is concealed in pixels between 2^{q-1} and $256 - 2^{q-1}$.

Step 3: The dual interpolated stego pictures are divided into separate, overlapping K blocks in this stage. All pixels are arranged in ascending order inside each block. The threshold T value between 0 and 255 is set to have the least amount of distortion throughout the data embedding procedure. If pixels are close together, median and neighbor pixel differences are modest. When $CL \leq T$, data may be hidden, and the picture block is viewed as smooth. The picture block will be handled as a rough block if complexity level $CL > T$ since it cannot conceal the contents. The median value is calculated from a block's sorted pixels. The reference pixel M used to represent this median value. The n pixels that surround the current block in the left, bottom, right, and top positions are referred to as its neighbor pixels O_P. The complexity level CL is calculated as the sum of the absolute differences between each neighboring pixel and the reference pixel. This is produced using the formula (4).

$$CL = \sum |O_P - M| \tag{4}$$

It has two median values for even-numbered pixels. For this, the median value with the high positioned is chosen as M. It comprises one median value that is chosen for M for pixels with an odd number. Equations (6) and (5) are used to produce the reference pixel M and its location ml from an image block $(wbl \times hbl)$ comprising n pixels.

$$ml = \begin{cases} \frac{(wbl \times hbl)+1}{2} & \text{if } n = \text{ODD}, \\ \frac{(wbl \times hbl)}{2} + 1 & \text{if } n = \text{EVEN}. \end{cases} \tag{5}$$

$$M = \begin{cases} b_{\sigma \frac{(wbl \times hbl)+1}{2}} & \text{if } n = \text{ODD}, \\ b_{\sigma \frac{(wbl \times hbl)}{2}+1} & \text{if } n = \text{EVEN}. \end{cases} \tag{6}$$

The arrangement of the pixels in a block is hampered since the secrets are integrated into more than one minimum and/or maximum pixel in this instance. We introduce the η value to keep the order. Maximum pixels are increased by eta, while minimum pixels are decreased. The value of η changes after each repetition. Lemma 1 and Lemma 2 affect how much η is worth.

Lemma 1 *If the number of pixels of a block is* $bl_{w\times h}$=*ODD and increasing wise sorted pixels are* $(pix_1, pix_2, \ldots, b_{(w\times h)})$ *where the range of i is* $1 \leq i \leq (w \times h)$. *Then, the altered pixels are* $(pix_1 - \eta_{fix(\frac{w\times h}{2})-1}, pix_2 - \eta_{fix(\frac{w\times h}{2})-1-1}, \ldots, pix_{(w\times h)-1} + \eta_{fix(\frac{w\times h}{2})-1-1}, pix_{(w\times h)} + \eta_{fix(\frac{w\times h}{2})-1})$, *where, the* $\eta_{fix(\frac{w\times h}{2})-1} = fix(\frac{w\times h}{2}) - 1$. *The maximal and minimal values of* η *are* $(fix(\frac{w\times h}{2}) - 1)$ *and 0, respectively.*

Lemma 2 *If the number of pixels of a block is* $bl_{w\times h}$ = *even_number and ascending order sorted pixels are* $(pix_1, pix_2, \ldots, pix_{(w\times h)})$ *where the range of i is* $1 \leq i \leq (w \times h)$. *Then, the altered pixels are* $(pix_1 - \eta_{fix(\frac{w\times h}{2})-1}, pix_2 - \eta_{fix(\frac{w\times h}{2})-1-1}, \ldots, pix_{(w\times h)-1} + \eta_{fix(\frac{w\times h}{2})-2-1}, pix_{(w\times h)} + \eta_{fix(\frac{w\times h}{2})-2})$, *where the* $\eta_{fix(\frac{w\times h}{2})-1} = fix(\frac{w\times h}{2}) - 1$. *The maximal and minimal* η *are* $(fix(\frac{w\times h}{2}) - 1)$ *and 0 respectively for minimal numbered pixels and the maximal and minimal* η *are* $(fix(\frac{w\times h}{2}) - 2)$ *and 0 respectively for maximal numbered pixels.*

The pixels in an image block contain the relevant data. The following is the embedding approach:

Embedding in Minimum-Modification Assume that a sub-block bl has n pixels in it. All pixels in the sub-block are sorted in a rising-up order to produce $(pix_{\sigma(1)}, \ldots, pix_{\sigma(n)})$. We compute

$$d_{\min_r} = pix_E - pix_F, \text{ where } \begin{cases} E = \max(\sigma((1)+r), \sigma(ml)), \\ F = \min(\sigma((1)+r), \sigma(ml)), \\ r = (0, 1, \ldots, \text{fix}(n/2) - 1). \end{cases} \tag{7}$$

Rounding the output data toward 0 is done using the fix(). The smallest pixels are now converted to $pixel'$. Each action involves a modification to η. The changed minimum pixels are obtained by

$$pixel' = \begin{cases} (pix_{\sigma((1)+r)} - \eta) - D, & \text{if } d_{\min_r} = 0, \\ (pix_{\sigma((1)+r)} - \eta) - 1, & \text{if } d_{\min_r} > 0, \\ (pix_{\sigma((1)+r)} - \eta) - D, & \text{if } d_{\min_r} = -1, \\ (pix_{\sigma((1)+r)} - \eta) - 1, & \text{if } d_{\min_r} < -1. \end{cases} \tag{8}$$

where pixels with the value $D \in \{0, 1\}$ are added.

Embedding in Maximum-Modification The following is a discussion of the data that is inserted into pixels for maximum-modification: Determine,

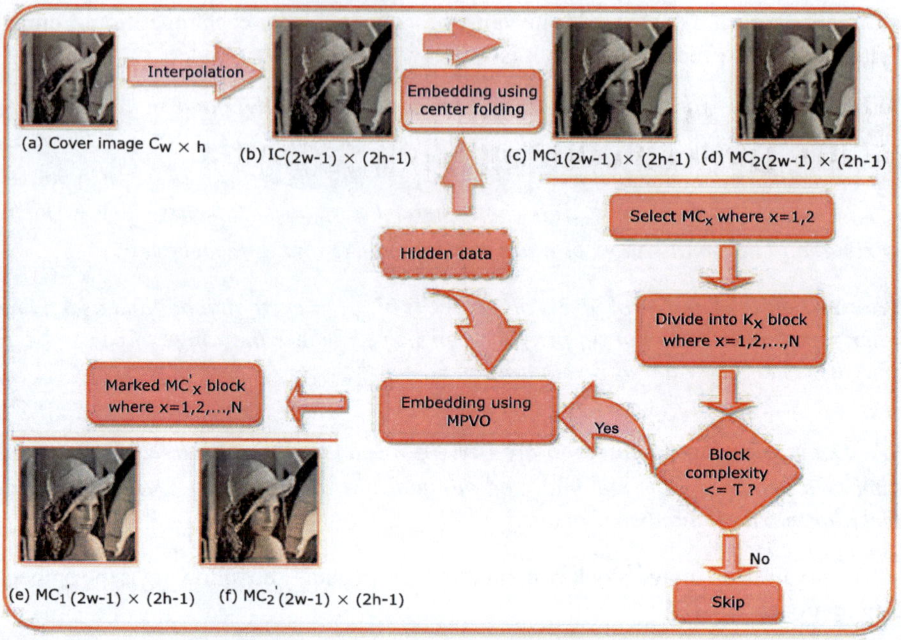

Fig. 2 Overall data embedding process in proposed IMPVO scheme

$$d_{\max_r} = pix_G - pix_H, \text{ where } \begin{cases} G = \max(\sigma((n) - r), \sigma(ml)), \\ H = \min(\sigma((n) - r), \sigma(ml)), \\ r = (0, 1, \ldots, \text{fix}(n/2) - 1) \text{ if } n = \text{ODD}, \\ \quad = (0, 1, \ldots, \text{fix}(n/2) - 2) \text{ if } n = \text{EVEN}. \end{cases} \tag{9}$$

The maximum pixels are now converted to $pixel'$. Each action involves a modification to η. The changed maximum pixels are obtained by

$$pixel' = \begin{cases} (pix_{\sigma((n)-r)} + \eta) + D, & \text{if } d_{\max_r} = 0, \\ (pix_{\sigma((n)-r)} + \eta) + 1, & \text{if } d_{\max_r} > 0, \\ (pix_{\sigma((n)-r)} + \eta) + D, & \text{if } d_{\max_r} = -1, \\ (pix_{\sigma((n)-r)} + \eta) + 1, & \text{if } d_{\max_r} < -1. \end{cases} \tag{10}$$

where pixels with the value $D \in \{0, 1\}$ are added. Figure 2 shows the whole embedding procedure.

3.2 Extraction Phase

Step 1: The cover image and secret data are retrieved in the reverse order at the receiver end using the same process. The stego pictures are transformed into the overlapping K blocks. The arrangement of the pixels in each block remains the same after sorting. Equations (6) and (4) reused to create the M and CL, respectively. The same threshold value is used in the data extraction. The data in this case are taken from many minimum and/or maximum values. Pixel order may thus vary. The value of η maintains the order of the pixels. Lemma 3 and Lemma 4 both have an impact on the value of η.

Lemma 3 *If the number of pixels of a block is $bl_{w\times h}$=ODD and increasing wise sorted pixels are $(pix_1, pix_2, \ldots, pix_{(w\times h)})$ where the range of i is $1 \le i \le (w \times h)$. Then, the altered pixels are $(pix_1 + \eta_{fix(\frac{w\times h}{2})-1}, pix_2 + \eta_{fix(\frac{w\times h}{2})-1-1}, \ldots, pix_{(w\times h)-1} - \eta_{fix(\frac{w\times h}{2})-1-1}, pix_{(w\times h)} - \eta_{fix(\frac{w\times h}{2})-1})$, where the $\eta_{fix(\frac{w\times h}{2})-1} = fix(\frac{w\times h}{2}) - 1$. The maximal and minimal values of η are $(fix(\frac{w\times h}{2}) - 1)$ and 0, respectively.*

Lemma 4 *If the number of pixels of a block is $bl_{w\times h}$ = EVEN and increasing wise sorted pixels are $(pix_1, pix_2, \ldots, pix_{(w\times h)})$ where the range of i is $1 \le i \le (w \times h)$. Then, the altered pixels are $(pix_1 + \eta_{fix(\frac{w\times h}{2})-1}, pix_2 + \eta_{fix(\frac{w\times h}{2})-1-1}, \ldots, pix_{(w\times h)-1} - \eta_{fix(\frac{w\times h}{2})-2-1}, pix_{(w\times h)} - \eta_{fix(\frac{w\times h}{2})-2})$, where the $\eta_{fix(\frac{w\times h}{2})-1} = fix(\frac{w\times h}{2}) - 1$. The maximal and minimal β are $(fix(\frac{w\times h}{2}) - 1)$ and 0, respectively, for minimal numbered pixels and the maximal and minimal η are $(fix(\frac{w\times h}{2}) - 2)$ and 0, respectively, for maximal numbered pixels.*

Extraction in Minimum-Modification The extraction of message as well as reconstruction of the picture are done with the minimum-pixel alteration. Assume that the value has changed to $(cpixel_1, cpixel_2, \ldots, cpixel_n)$. The mapping σ is constant. We derive the equation $d'_{\min_r} = cpixel_E - cpixel_F$, where (E, F, r) are described in Eq. (7).

- When $d'_{\min_r} > -1$, then $cpixel_E \ge cpixel_F$. Here, $E = \sigma(ml)$, $F = \sigma((1) + r)$ and $\sigma((1) + r) < \sigma(ml)$:
  - When $d'_{\min_r} \in \{0, 1\}$, the data is $D = d'_{\min_r}$. The smallest pixel is $cpixel_{\sigma((1)+r)} = (cpixel_F + \eta) + D$;
  - When $d'_{\min_r} > 1$, their is no data. The smallest pixel is $cpixel_{\sigma((1)+r)} = (cpixel_F + \eta) + 1$.
- When $d'_{\min_r} \le -1$, then $cpixel_E < cpixel_F$. Here, $E = \sigma((1) + r)$, $F = \sigma(ml)$ and $\sigma((1) + r) > \sigma(ml)$:
  - When $d'_{\min_r} \in \{-1, -2\}$, the data is $D = -d'_{\min_r} - 1$. The smallest pixel is $cpixel_{\sigma((1)+r)} = (cpixel_E + \eta) + D$;

- When $d'_{\min_r} < -2$, their is no data. The smallest pixel is $cpixel_{\sigma((1)+r)} = (cpixel_E + \eta) + 1$.

Extraction in Maximum-Modification We find that $d'_{\max_r} = cpixel_G - cpixel_H$. Where (G, H, r) is a list in the Eq. (9).

- When $d'_{\max_r} > -1$, then $cpixel_G \geq cpixel_H$. Here, $G = \sigma((n) - r)$, $H = \sigma(ml)$ and $\sigma(ml) < \sigma((n) - r)$:
  - When $d'_{\max_r} \in \{0, 1\}$, the data is $D = d'_{\max_r}$. The largest pixel is $cpixel_{\sigma((n)-r)} = (cpixel_G - \eta) - D$;
  - When $d'_{\max_r} > 1$, their is no data. The largest pixel is $cpixel_{\sigma((n)-r)} = (cpixel_G - \eta) - 1$.
- When $d'_{\max_r} \leq -1$, then $cpixel_G < cpixel_H$. Here, $G = \sigma(ml)$, $H = \sigma((n) - r)$ and $\sigma(ml) > \sigma((n) - r)$:
  - When $d'_{\max_r} \in \{-1, -2\}$, the data $D = -d'_{\max_r} - 1$. The largest pixel is $cpixel_{\sigma((n)-r)} = (cpixel_H - \eta) - D$;
  - When $d'_{\max_r} < -2$, their is no data. The largest pixel is $cpixel_{\sigma((n)-r)} = (cpixel_H - \eta) - 1$.

Now, recovered the secret message as well as images of $MC1_{(2w-1)\times(2h-1)}$ and $MC2_{(2w-1)\times(2h-1)}$.

Step 2: In this stage, using the following formulas, we extract the message from the stego pixels $pix^1_{i,j}$ and $pix^2_{i,j}$ of the interpolated marked pictures $MC1_{(2w-1)\times(2h-1)}$ and $MC2_{(2w-1)\times(2h-1)}$, respectively.

$$d'_s = pix^1_{i,j} - pixel^2_{i,j} \tag{11}$$

$$d_s = d'_s + 2^{q-1} \tag{12}$$

where q represents how many concealed bits make up a group. The following is the reconstruction of the actual pixel, $pix_{i,j}$.

$$pix_{i,j} = \left\lceil \frac{pix^1_{i,j} + pix^2_{i,j}}{2} \right\rceil \tag{13}$$

Now construct the interpolated cover image $IC_{(2w-1)\times(2h-1)}$.

Step 3: Now, rebuild the cover image $C_{w \times h}$ from interpolated cover image $IC_{(2w-1)\times(2h-1)}$ by removing any interpolated columns and rows.

4 Experimental Results and Comparisons

This paragraph compares the proposed scheme's execution to those of the previous dual-image-based secrets concealment systems developed by Authors [2, 5, 6, 16]. Peppers, Lena, Baboon, Fishing boat, Airplane F 16, and Barbara were selected from [18] database, while CXR1000_IM-0003-1001 and CXR1025_IM-0020-1001 were obtained from the [8] database as test inputs for study. All gray-scale image size is (256 × 256). Only USC-SIPI photos are shown in Fig. 3.

The embedding of secrets depends on q and block size. The picture quality is found to be higher in small values of q with large blocks but not much larger block

Fig. 3 For our experiments, the six standard photos are used as input

Table 1 The secrets embedding (EC) in bits with separate p of two separate database pictures with average PSNR (dB)

Database	Cover image (C)	q	$PSNR_1$ (Avg.)	$PSNR_2$ (Avg.)	Average PSNR	EC_1 (IMI_1')	EC_2 (IMI_2')	$EC = EC_1 + EC_2$
USC-SIPI	Lena	2	51.87	53.27	52.57	320,829	280,512,983	601,341
		3	50.21	49.30	49.75	398,613	396,112	794,725
		4	46.39	43.55	44.97	482,731	529,805	1,012,536
	Airplane F16	2	52.38	50.72	51.55	289,437	305,515	594,952
		3	49.13	47.06	48.09	387,659	409,034	796,693
		4	45.89	44.63	45.26	465,120	501,269	966,389
	Fishing boat	2	51.89	51.56	51.77	276,813	309,320	586,133
		3	46.65	46.32	46.48	399,295	396,892	796,187
		4	44.52	43.89	44.20	481,715	505,216	986,931
National library of medicine	CXR1025_IM-0020-1001	2	51.12	52.89	52.00	308,224	290,617	598,841
		3	49.66	48.06	48.86	425,320	398,612	833,932
		4	45.18	45.62	45.40	510,291	495,697	1,005,988
	CXR1000_IM-0003-1001	2	51.40	52.16	51.78	298,032	326,514	624,546
		3	48.37	46.62	47.49	402,910	411,853	814,763
		4	46.53	45.19	45.86	503,841	525,472	1,029,313

Table 2 Comparison among the other schemes and the proposed scheme with image quality in PSNR (dB) as well as embedding capacity (EC) in bits

Schemes	Measure	Lena	Baboon	Peppers	Barbara	Fishing boat
Qin et al. [16]	$PSNR_1$	52.11	52.04	51.25	52.12	52.11
	$PSNR_2$	41.58	41.56	41.52	41.58	41.57
	Avg. PSNR	46.85	46.80	46.39	46.85	46.84
	EC	557,052	557,096	557,245	557,339	557,194
Lu et al. [6]	$PSNR_1$	49.20	49.21	49.19	49.22	49.20
	$PSNR_2$	49.21	49.20	49.21	49.20	49.21
	Avg. PSNR	49.21	49.21	49.20	49.21	49.21
	EC	524,288	524,204	524,192	524,288	524,284
Lu et al. ($k = 2$) [5]	$PSNR_1$	49.89	49.89	49.89	49.89	49.89
	$PSNR_2$	52.90	52.87	52.92	52.90	52.90
	Avg. PSNR	51.40	51.38	51.41	51.40	51.40
	EC	524,288	524,172	523,780	524,288	524,286
Jung [2]	$PSNR_1$	48.18	48.17	48.18	–	48.18
	$PSNR_2$	48.18	48.16	48.18	–	48.18
	Avg. PSNR	48.18	48.16	48.18	–	48.18
	EC	519,180	519,180	519,180	–	519,180
Proposed method ($q = 2$)	$PSNR_1$	51.87	52.30	51.89	52.66	51.89
	$PSNR_2$	53.27	51.38	52.21	51.18	51.66
	Avg. PSNR	52.57	51.84	52.05	51.92	51.77
	EC	601,341	591,322	594,751	604,643	586,133

size than in large values of q, but the message embedding rate is low. For instance, in the Lena image, 601,341 bits are added with an average peak signal-to-noise ratio (PSNR) of 52.57 dB when $q = 2$. In the picture Lena presented in Table 1, when $q =$ 3,794,725 bits are inserted with an average PSNR of 49.75, whereas when $q =$ 41,012,536 bits are placed with an average PSNR of 44.97.

Table 2 shows the PSNR (dB) comparisons between the suggested and previous techniques. The new approach improves the secrets capacity (EC) calculated using other PVO-based approaches. While the value of q is 2 as stated in Table 2, the hidden message capacity is 442,89, 77,053, 77,053, and 82,161 bits larger than that of Qin et al.'s [16], Lu et al.'s [6], Lu et al.'s [5] and Jung [2], respectively, for the picture Lena. The capacity of information concealment in input images is increased by the proposed approach. It has been determined that our approach outperforms existing PVO techniques in terms of payload while maintaining the same level of picture quality. Additionally, it is obvious that the picture quality is superior to other current schemes, as illustrated in Fig. 4.

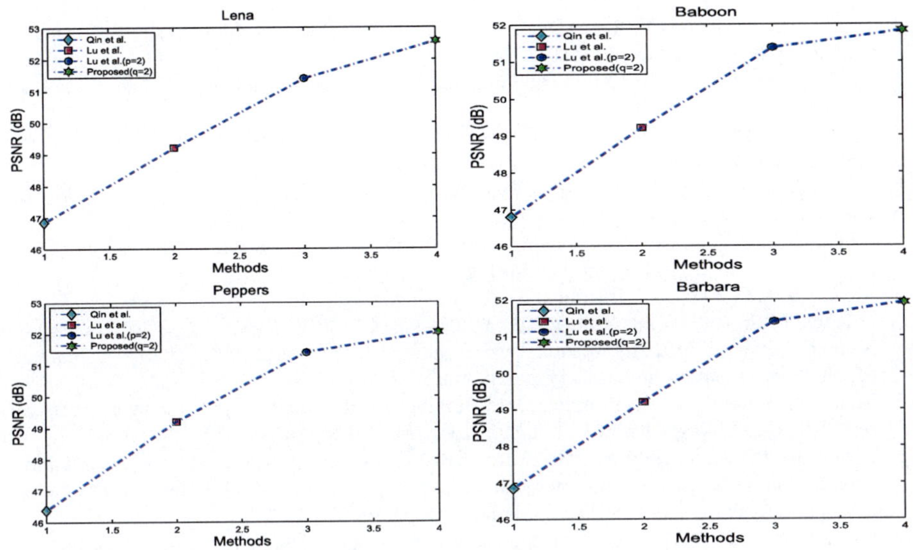

Fig. 4 Image quality comparisons between approaches in terms of PSNR (dB) of [5, 6, 16] and proposed technique with $2 = 2$

5 Conclusion

With varied sizes for the q (set of data) and picture block, this work suggested an improved multi-level PVO (IMPVO) utilizing the center-folding technique. Huge messages are put into and taken from image pixels using the algorithms for inserting and retrieving, respectively. The proposed approach enables safe message transmission by inserting secrets between two pictures. When q is 2, our technique produces a PSNR value over 51 dB and an embedding capacity over 585,000 bits. The suggested approach yields successful outcomes. It is found that the recommended system looks to perform better than other PVO efforts.

References

1. Jana B (2017) Reversible data hiding scheme using sub-sampled image exploiting Lagrange's interpolating polynomial. Multimedia Tools Appl 1–17
2. Jung KH (2018) Authenticable reversible data hiding scheme with less distortion in dual stego-images. Multimedia Tools Appl 77:6225–6241
3. Lee SK, Suh YH, Ho YS (2004, November) Lossless data hiding based on histogram modification of difference images. Pacific-Rim conference on multimedia. Springer, Berlin, Heidelberg, pp 340–347
4. Li X, Li J, Li B, Yang B (2013) High-fidelity reversible data hiding scheme based on pixel-value-ordering and prediction-error expansion. Sig Process 93(1):198–205

5. Lu TC, Wu JH, Huang CC (2015) Dual-image-based reversible data hiding method using center folding strategy. Sig Process 115:195–213
6. Lu TC, Tseng CY, Wu JH (2015) Dual imaging-based reversible hiding technique using LSB matching. Sig Process 108:77–89
7. Lu TC, Huang SR, Huang SW (2021) Reversible hiding method for interpolation images featuring a multilayer center folding strategy. Soft Comput 25:161–180
8. The National Library of Medicine presents MedPix Superscript Registered MedPix®. https://openi.nlm.nih.gov/gridquery.php?q=&it=x
9. Meikap S, Jana B (2017) Extended directional IPVO for reversible data hiding scheme. Communication, devices, and computing. Springer, Singapore, pp 47–58
10. Meikap S, Jana B (2018) Directional PVO for reversible data hiding scheme with image interpolation. Multimedia Tools Appl 77(23):31281–31311
11. Meikap S, Jana B (2019) Directional pixel value ordering based secret sharing using subsampled image exploiting Lagrange polynomial. SN Appl Sci 1(6):645
12. Meikap S, Jana B (2021) Improved center-folding based directional pixel value ordering for reversible data hiding scheme. Multimedia Tools Appl 80(4):5617–5652
13. Sudipta M, Biswapati J, Prasenjit B, Kumar SP (2021) High payload RDH through directional PVO exploiting center-folding strategy. In: Proceedings of international conference on frontiers in computing and systems. Springer, Singapore, pp 659–670
14. Niu Y, Shen S (2022) A novel pixel value ordering reversible data hiding based on dual-image. Multimedia Tools Appl 81(10):13751–13771
15. Peng F, Li X, Yang B (2014) Improved PVO-based reversible data hiding. Digital Sig Process 25:255–265
16. Qin C, Chang CC, Hsu TJ (2015) Reversible data hiding scheme based on exploiting modification direction with two steganographic images. Multimedia Tools Appl 74(15):5861–5872
17. Tian J (2003) Reversible data embedding using a difference expansion. IEEE Trans Circuits Syst Video Technol 13(8):890–896
18. University of Southern California. The USC-SIPI Image Database. http://sipi.usc.edu/database/database.php?volume=misc

Perseverance of the Audio Data using RNN Implied Matrix Segmentation based Lossless Encoder

Asish Debnath and Uttam Kr. Mondal

1 Introduction

Lossless encoding [4] is the compression technique used when it's required to preserve the quality of the original and the decompressed material. Furthermore, it reduces the file's size without altering the data's original content. In essence, lossless audio puts preservation of detail above file size reduction [6]. Today, a variety of classical [9, 11] and neural network-based [10] techniques are utilised to handle a variety of problems. RNN is a special type of neural network which is being applied particularly to these kinds of audio encoding, privacy as well as perseverance problems. Each layer's output is recorded and sent back into the system's input via RNN [12], which uses this principle to forecast each layer's output.

The proposed lossless audio encoder is designed utilising 2D matrix segmentation based on recurrent neural networks, i.e. RNN. The encoder technique consists of the following 3 phases.

In the first stage, audio samples are divided into three components.

(a) The signed integer parts.
(b) The first two digits following the decimal point, that is, the tenths and hundredths.
(c) Next two digits, i.e., digits of the thousandths and ten thousandths positions.

Construct a 2D matrix to represent these data. This statistical technique was developed utilising a regression model built on RNN. Each of these integers is transformed into a 7-bit binary stream and represented in the latent space in the second step.

A. Debnath (✉) · U. Kr. Mondal
Vidyasagar University, Midnapore, West Bengal, India

U. Kr. Mondal
e-mail: Uttam_ku_82@yahoo.co.in

J. K. Mandal et al. (eds.), *Proceedings of International Conference on Network Security and Blockchain Technology*, Lecture Notes in Networks and Systems 738,
https://doi.org/10.1007/978-981-99-4433-0_11

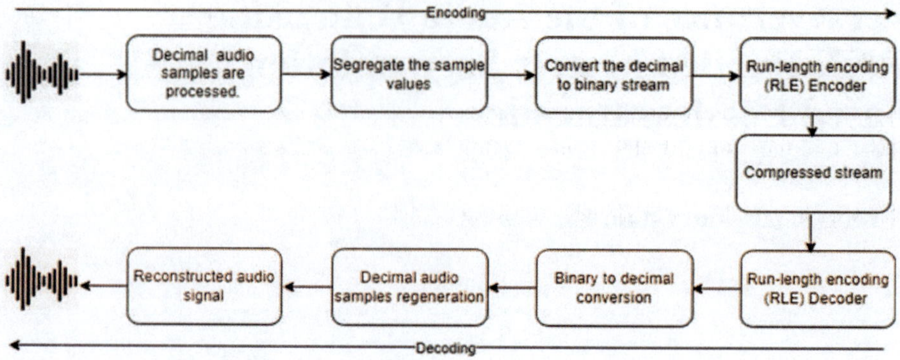

Fig. 1 Flow diagram of the proposed technique

Finally, the repetition eliminating standard Run-length encoding (RLE) [5] technique is used to further compress the binary encoded streams.

The proposed encoding and decoding flow diagram is depicted in Fig. 1.

The remaining sections of this document are structured as follows: The method is described in Sect. 2. In Sects. 2.1 and 2.2, respectively, the encoding and decoding techniques are covered. Network architecture and training are displayed in Sect. 3. Section 4 presents the findings and analyses. Conclusions are drawn in Sect. 5 that follows the references.

2 The Technique

The present method produces compressed audio using a 2D matrix segmentation algorithm which is implied by the rules of RNN. Dynamic 2D matrix-based segmentation is used to divide decimal signed audio samples during the encoding stage (considered a 4-point post fraction for the system constraint). A signed integer component and two sets of double digits after the decimal point constitute each of the three sections of the audio samples. Figure 2 shows how the encoding process works.

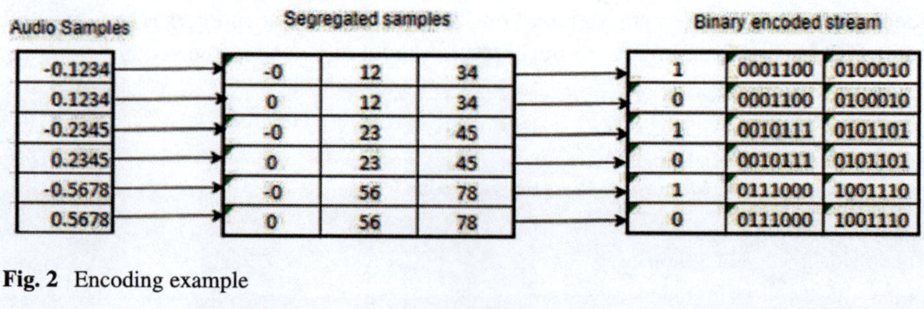

Fig. 2 Encoding example

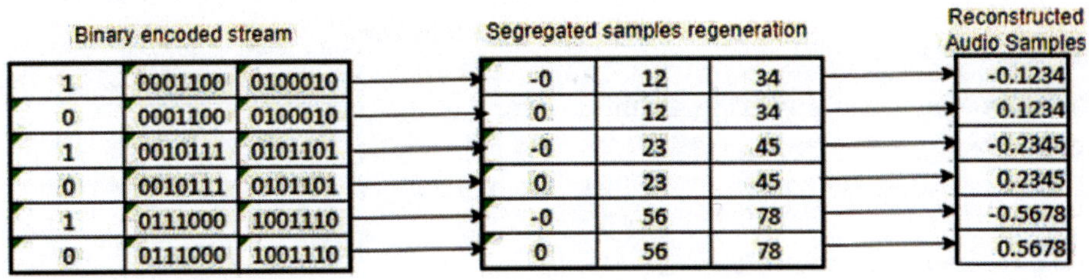

Fig. 3 Decoding example

Figure 2 explains the overview of the encoding process with 6 audio sampled values.

Encoding and decoding are done in the opposite order. The decoding processes for the previously encoded data are shown in Fig. 3.

The encoding and decoding procedures are depicted in Sects. 2.1 and 2.2 respectively.

2.1 *Encoding Technique*

The encoding algorithm is described in Algorithm 1. It demonstrates the segmentation of the audio samples and their eventual encoding into binary streams.

Algorithm 1: Matrix Segmentation Method
Input: A fragment of the audio

Output: Encoding audio

Method: The steps of the encoding process are as follows:

Step 1: The input audio samples are checked whether they are positive or negative. If positive go to step 1.1 (positive_module) else step 1.2 (negative_module).

The dynamic rules as listed below are used to segment each audio samples.

Step 1.1 (positive_module):

i. The whole number portion of the passed decimal number, i.e. 0 is captured. It is added to the first column of the matrix.

Let the matrix defined by $M[i][j]$, where i is the number of audio samples and j is number of columns (considering, $j = 3$). That means the matrix is 3 column matrices. Therefore, $M[i][0] = 0$.

ii. Multiply the decimal number by 10,000.
iii. Divide by 100 and take the quotient. The quotient part is moved to the second column of the same row.

$$M[i][1] = \text{Quotient part}$$

iv. Remainder part pushed to the third column of the same row.

$$M[i][2] = \text{remainder part}$$

Step 1.2 (negetive_module):

The dynamic rule listed below is used to segment each audio samples:

i. The whole number portion of the passed decimal number, i.e. −0 is captured. It is added to the first column of the matrix.

$$M[i][0] = -0$$

ii. The decimal number is multiplied by 10,000.
iii. Divide by 100 to get the quotient. The quotient part is moved to the second column of the same row.

$$M[i][1] = \text{Quotient part}$$

iv. Remainder part pushed to the third column of the same row.

$$M[i][2] = \text{remainder part}$$

Step 2: All the audio sampled values are passed to the segregation rule engine mentioned in step 1. After processing samples, the matrix formed of data passed to step 3 for binary conversion.

Step 3: The first column of first row and first column element is checked.

```
For i = 0 to length (matrix row):
    If M[i][0] < 0 then:
        Transform element at position M[i][0] with single
        binary bit '1'
    Else:
        Transform element at position M[i][0] with single
        binary bit '0'
```

Each element of M[i][1] and M[i][2] is converted to a binary stream using a 7-bit binary number.

Therefore, each audio sample is encoded as a 15-bit binary stream.

At the end of this step, a binary encoded binary stream is generated.

Step 4: Binary encoded stream is passed to run-length encoding (RLE) encoder for further compression.

2.2 Decoding Technique

Algorithm 2 describes the decompression technique. The decompression algorithm reconstructs the original audio from the encoded data stream.

Algorithm 2: Decompression technique
Input: Encoded binary stream.

Output: Reconstructed audio signal.

Method: The decoding procedure is described in following steps:

Step 1: Input encoded stream is fed to Run-length encoding (RLE) decoder. At the end of this step the binary stream is reconstructed.

Step 2: Binary stream segregated into blocks of 15 binary bits.

Step 3: Process each of the blocks using the below dynamic rule:

i. First binary bit is decoded to 0 or −0 corresponds to binary bit '0' or '1', respectively.
ii. Append decimal point after 0 or −0.
iii. Next 7 binary bits are converted to decimal values.

capture the double digit as is.

iv. Adjoining, 7 binary bits are also converted to decimal values.

capture the double digit as is.

v. Process all the blocks and regenerate the audio samples.

Therefore, finally the audio signal is reconstructed.

3 Network Architecture and Training

The sequential order of time series data, like audio, must be followed in order to be understood. Traditional feed-forward networks assume that each input is independent of the others, however in a time series setting, each input depends on the inputs that came before it.

One of the most challenging areas is deep neural network (DNN)-based sequential data processing for audio signals since DNN requires inputs with a fixed dimension. Recurrent neural networks (RNN)-based sequence to sequence learning has been proposed in machine translation to learn fixed length representations of variable length sequences [13].

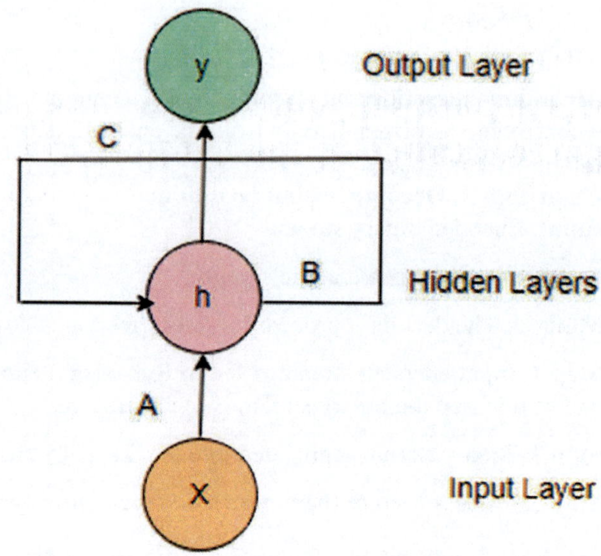

Fig. 4 Fully connected recurrent neural network

Hence, for the encoder and decoder strategy, the RNN-based network architecture is applied.

The proposed recurrent neural network model was trained with encoding and decoding algorithm described in Sects. 2.1 and 2.2 respectively. The mean squared error loss is applied to analyse the network. Single layer selected for the experiment with 16 hidden units and 3 output. All experimentation was performed with Keras using TensorFlow its back end, running under Windows 10 operating system.

A fully connected recurrent neural network is shown in Fig. 4. In this instance, "X" represents the input layer, "h" the hidden layer, and "y" the output layer. The network parameters A, B, and C are used to enhance the model's output. The input at any given time, t, is a mixture of the input at $x(t)$ and x. $(t-1)$. The output at any given time is fetched back to the network in order to improve it.

Figure 5 shows the processing in the recurrent neural network.

$$h(t) = f_c((h(t-1), x(t)) \tag{1}$$

here, $h(t)$ denotes new state, f_c denotes function with parameter c, $h(t-1)$ denotes old state, and $x(t)$ denotes input vector at time step t.

5000 epochs were considered for the experiment to evaluate the performance of the suggested model. The hyperparameters of the experiment were designed by trial and error. The configuration details of the suggested model parameters are described in Table 2. The suggested model accurately predicts and recreates the original audio data with incredibly minor deviations. The suggested model's estimated mean square error is 0.0426. We have carried out a performance evaluation of the proposed model using two additional, pre-existing prediction benchmarks, the Lasso regression, and

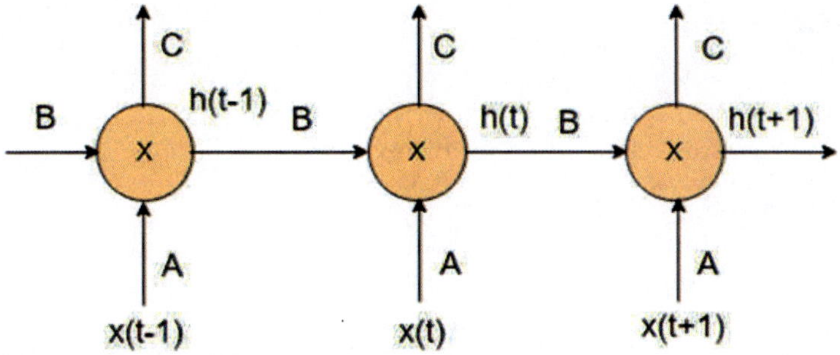

Fig. 5 Processing recurrent neural network

Table 1 Compares the suggested model's performance to those of other deep learning models

Ridge	0.0578
Lasso	0.0516
Proposed model	0.0426

Table 2 Network parameters

Learning rate	0.01
Epoch	5000
Hidden unit	16
Output	3
Timestep	1

the Ridge regression. The suggested model's loss value is contrasted with that of the two other models described in Table 1.

3.1 Dataset

For the model's training, the "audio_model_dataset" custom dataset has been created. 1000 audio files totalling 3 s each make up the dataset. There are five different categories of audio songs: (I) Rabindra Sangeet, (II) Classical, (III) Rock, (IV) Pop, and (V) Sufi. The training dataset has not divided into separate portions for testing. Instead, 25 audio songs are used, each with a standard length of 5 s, that fall into the aforementioned 5 categories.

3.2 *Environment*

The proposed model is programmed in Python 3.6 using the Keras and Tensorflow framework. The infrastructure used for the experiment is described below:

i. 64-bit operating system.
ii. 16 GB of RAM.
iii. Intel Core i7-4790S Processor.
iv. 1 TB Hard drive.

3.3 *Network Configuration Parameters*

During the experiment, various parameters have been used to train the model and measure the loss of the model and accuracy model. Present suggested model produces less loss when compared to other models of a similar nature. The list of parameters utilised in the model network is shown in Table 2.

4 Results and Analysis

Compression ratio [10, 11] is used to judge the compression capacity of a compression technique. Compression ratio is calculated as below:

$$\text{Compression ratio} = \frac{\text{Uncompressed audio file size}}{\text{Compressed audio file size}} \tag{2}$$

Therefore, after applying the compression technique, required space is to be reduced. Equation 3 represents the space saving metric respect to the compressed audio.

$$\text{Space saving}(\%) = 1 - \frac{\text{Compressed audio file size}}{\text{Unompressed audio file size}} * 100 \tag{3}$$

The compression ratio of the selected songs is compared using FLAC [3], WavPack Lossless [2], and Monkey's Audio [1]. The compression quality of the proposed method is compared graphically to that of three different systems as shown in Fig. 6. Comparing the proposed strategy to other referred techniques, Table 3 demonstrates that it has the highest compression ratio. Additionally, it offers each group the highest compression ratio (i.e. Rabindra Sangeet, Pop, Classical, Sufi, and Rock).

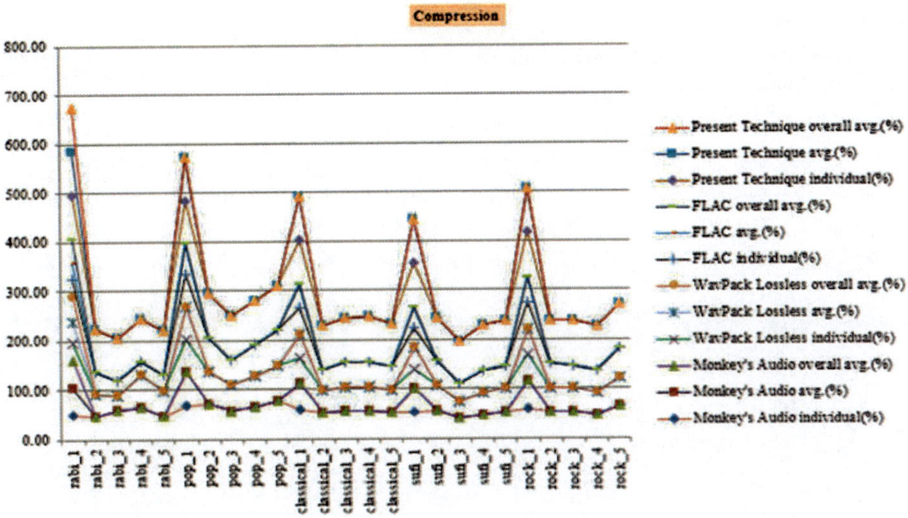

Fig. 6 Graphs showing the compression ratios for the proposed method, FLAC, WavPack lossless, and Monkey's Audio

5 Conclusion

The proposed lossless audio compression technique achieved a higher compression rate in comparison to already accessible tools for lossless audio compression. The reconstructed song is also achieved similar quality as original. The broadened focus of this research is the possibility for further compression ratio enhancement along with increasing privacy and perseverance of the audio data. In the future, it will also intend to further reduce the loss by expanding the dataset and adding more epochs.

Table 3 Compression performance

Song type	Audio files (.wav) (10 s)	Compression ratio (%)							
		Techniques							
		Monkey's Audio [1]		WavPack [2]		FLAC [3]		Present technique	
		Avg.	Over all avg.	Avg.	Over all avg.	Avg.	Over all avg.	Avg.	Over all avg.
Rabindra Tagore	rabi_1	54.89	56.35	48.46	50.14	48.18	48.28	88.02	89.81
	rabi_2								
	rabi_3								
	rabi_4								
	rabi_5								
Pop	pop_1	67.41		64.13		64.34		88.36	
	pop_2								
	pop_3								
	pop_4								
	pop_5								
Classical	classical_1	55.55		50.03		50.78		89.81	
	classical_2								
	classical_3								
	classical_4								
	classical_5								
Sufi	sufi_1	49.68		43.18		43.21		88.96	
	sufi_2								
	sufi_3								
	sufi_4								
	sufi_5								
Rock	rock_1	54.22		44.97		34.9		90.23	
	rock_2								
	rock_3								
	rock_4								
	rock_5								

References

1. https://www.monkeysaudio.com/
2. https://www.wavpack.com/
3. Coalson J (2017) Xiph.org foundation. FLAC: Free lossless audio codec. https://xiph.org/flac/index
4. Ghido F, Tabus I (2012) Sparse modeling for lossless audio compression. IEEE Trans Audio Speech Lang Process 21(1):14–28
5. Arif M, Anand RS (2012) Run length encoding for speech data compression. In: 2012 IEEE international conference on computational intelligence and computing research, pp 1–5. https://doi.org/10.1109/ICCIC.2012.6510185
6. Sharma K, Gupta K (2017) Lossless data compression techniques and their performance. In: 2017 international conference on computing, communication and automation (ICCCA). IEEE, pp 256–261
7. Nowak N, Zabierowski W (2011) Methods of sound data compression–comparison of different standards. (4)
8. Mohdar FJ, Al-Otaibi MS, Aboalsamh HA (2011) Audio compression testing tool for multimedia applications. In: Image processing and communications challenges, vol 3. Springer, pp 409–418
9. Mondal UK, Debnath A (2021) Developing a dynamic cluster quantization based lossless audio compression (DCQLAC). Multimed Tools Appl 80:8257–8280. https://doi.org/10.1007/s11042-020-09886-3
10. Mondal UKr, Debnath A, Mandal JK (2020) Intelligent computing: image processing based applications, vol 1157. ISBN 978-981-15-4287-9
11. Mondal UKr, Debnath A (2022) Multimedia Tools Appl 81(28):40385. https://doi.org/10.1007/s11042-022-12556-1
12. Sherstinsky A (2020) Fundamentals of recurrent neural network (RNN) and long short-term memory (LSTM) network. Physica D Nonlinear Phenomena 404:132306. https://doi.org/10.1016/j.physd.2019.132306
13. Sutskever I, Vinyals O, Le QV (2014) Sequence to sequence learning with neural networks. In: Advances in neural information processing systems, vol 27
14. Pedro HTC, Larson DP, Coimbra CF (2019) A comprehensive dataset for the accelerated development and benchmarking of solar forecasting methods. J Renew Sustain Energ 11:036102

SVD-Based Watermarking Scheme for Medical Image Authentication

Ashis Dey, Partha Chowdhuri, Pabitra Pal, and Lu Tzu-Chuen

1 Introduction

In the current pandemic COVID-19 situation, medical syst. Doctors treat patients through online consultancy and give e-prescription and telemedicine. Medical information is badly at risk due to misuse of medical data. So, security of medical information is a challenge. Nowadays, due to the advancement of Internet, medical information is distributed among various medical personnel for diagnosis, among students and researcher for research, etc. Any minor modification in the content of medical image can be a harmful for patients actual diagnosis. Hence, it is necessary to protect medical information from any type of attacks. In health management systems handle huge amount of patients data every day which are generated by various type of hospitals, clinics, etc. This huge amount of patients data contains text and images which necessary to process, store, and share this data for e-diagnosis and e-treatment. These factors draw attention to security concerns including integrity, authenticity, and privacy.

In medical science digital watermark gives copyright protection of patients documents from misuse and also help to identify the actual legal owner. Robustness is

A. Dey (✉)
Department of Computer Science, Silda Chandra Sekhar College, WB 721515, India
e-mail: ashismou14@gmail.com

P. Chowdhuri
Department of Computer Science, Vidyasagar University, WB 721102, India

P. Pal
Department of Computer Applications, Maulana Abul Kalam Azad University of Technology, Nadia, WB 741249, India

L. Tzu-Chuen
Department of Information Management, Chaoyang University of Technology, Taichung, Taiwan, ROC
e-mail: tclu@cyut.edu.tw

J. K. Mandal et al. (eds.), *Proceedings of International Conference on Network Security and Blockchain Technology*, Lecture Notes in Networks and Systems 738,
https://doi.org/10.1007/978-981-99-4433-0_12

one of the essential aspects for watermarking technique. It must be robust to illegal detection and decoding. The major challenges of watermarking technique are to resist modification, scaling, cropping, etc.

Some essential requirements of digital image watermarking are imperceptibility, robustness, tamper detection, embedding capacity , reversibility, etc. Watermarking technology can resolve the authenticity and security problem for digital medical image.

2 Literature Review

In this section, few related digital image watermarking schemes in medial image are discussed.

Alhaj et al. [2] invented a new watermarking technique for transmitted medical images for various telemedicine applications. They used joint watermarking techniques for security services. Their scheme achieved reversibility when extracting the embedding information from watermarked image and recovered original unaffected image . In 2020, Anand et al. [4] developed a DWT and SVD-based medical image watermarking technique. They inserted multiple watermarks in to medical images to provide better robustness. Adnan et al. [1] introduced an innovative watermarking technique using YCbCr and 2-Level DWT technique to achieved the imperceptibility and robustness. They achieved 59.78 dB PSNR. Bamal et al. [6] suggested a hybrid adaptive lossless data hiding technique using SVD with FWT and SLT transformation technique for medical image authentication. They achieved excellent embedding capacity (EC) within 1.8–7.5 bpp with high energy compaction using distinct filtering technique. Kahlessenane et al. [14] suggested another new DWT-based blind watermarking strategy to check authenticity of various medical images. They have applied DWT in their developed scheme to achieve the robustness of the watermarked image. But DWT consumes more computational power when applied for high resolution digital images. They achieved above 70 dB PSNR value but payload capacity is 43,690 which is low comparing some existing scheme. An innovative watermarking technique was introduced by Liu et al. [16]. They used DWT Wavelet transformation as well as DCT both in their proposed work. They tried to improve the quantization tables of encoders for better robustness of the watermarking scheme. In 2018, Majdi Alqdah [3] proposed a innovative hybrid watermarking mechanism for digital medical images. They built their hybrid watermarking scheme using the three important transformation techniques DWT, DCT, and SVD. Shehab et al. [20] invented a fragile digital watermarking technique for self-recovery and authentication of various medical documents based on SVD. They achieved better balanced between imperceptibility, robustness, and data hiding capacity. Verma et al. [26] invented an unique hybrid watermarking scheme depending on SVD and DWT both technique in medical image. They have used SVD value into the LL-sub-band of original cover image to achieve the robustness. Yang et al. [28] described a watermarking scheme that exhibits high embedding capacity without reducing the original medical image visual quality as well as authenticity also. Zermi et al. [29] proposed a novel scheme to enhance medical image security and integrity using blind watermarking technique. They used DWT-SVD both technique to enhance the protection of medical image. Their scheme achieved good imperceptibility with 57.41 dB PSNR value and SSIM value is near about 1. Mohammed [18] presented a review paper describing watermarking technique based on medical image. His paper give a brief description about objective, encoding technique, and flaws of various medical image related scheme. Favorskaya et al. [11] developed a secured watermarking method that can be used for various medical tools for helping doctors to take decision. Fan et al. [8] suggested a new sensitive watermarking scheme for several medical image that provides better robustness and imperceptibility. Falgun et al. [24] invented a innovative blind watermarking scheme using both SVD-DWT technique in medical documents. They achieved good PSNR value after embedding watermark. Singh et al. [21] invented a new secure multiple watermarking method in medical science using both DWT-DCT transformation technique and SVD. They used medical image as the image watermark and patient information as the text watermark. Eze et al. [7] presented a highly secured watermarking and tampered detection technique for medical images using Spread Spectrum technique. They consider patient records as watermarked bit and embedded this bit in each sub-block of original medical image. The experimental findings demonstrate that the suggested method increased accuracy and tamper detection with minimum computational cost. Anand et al. [5] proposed a survey report using different watermarking methods in the medical field. They gives a brief discussion about the various watermarking scheme, important features, current applications, concepts of embedding, and extraction process of watermarking.

Liu et al. [17] invented a reliable zero watermarking technique for medical image protection. They achieved robustness against various attacks. They extracted features from digital medical image using DRZW and CLBP technique to provide resistance to a variety of threats. Gao et al. [12] invented a new watermarking method using both RDWT and MHM technique in medical image. They used MHM algorithm to enhance embedding capability of the several medical images. The results of proposed scheme gives better embedding capacity and robustness compare than any other existing method. Eze et al. [7] suggested a reliable watermarking technique depending on Spread Spectrum method for detecting the tamper area of medical image to enhance the security. They embedded patients information into the original medical image's sub-block as watermarking bit. The experimental findings depicts that suggested technique optimized the tamper detection and accuracy at minimal computational cost. Iskanda et al. [13] invented an innovative watermarking scheme for Android phone using graph coloring and vector quantization technique. At first, they segmented the original medical image into ROI and RONI region. After that they chosen RONI region for embedded the watermark based on various techniques such as graph generation and graph coloring. Fares et al. [9] invented two blind watermarking schemes using Schur decomposition and DCT technique to enhance the security for medical image. They achieve high robustness and imperceptibility in their proposed scheme to safeguard the patient personal information. Su et al. [22] suggested an efficient self-embedding fragile watermarking technique for various

Table 1 Comparisons table for several medical image watermarking scheme

Scheme name	Method used	Performance	Drawback
Fares et al. [10]	LSB, AES, SHA-265 and MAC	Blind, fragile, PSNR = 70 dB	Computational complexity is high
Kelkar et al. [15]	LBPTI and RDH	Reversible and Secure, PSNR = 49, Capacity bit = 49,734	Embedding capacity is low
Bamal et al. [6]	FWT, AES, SLT, and ANN	–	Low imperceptibility
Thakur et al. [25]	NSCT, RDWT and chaotic encryption	Robust and secure	Enhanced the Embedding capacity (EC)
Eze et al. [7]	SVD, DWT, and BPNN	Secure, robust, and imperceptible	Computational complexity is high
Xia et al. [27]	IoRHFMs, FoRHFMs	Robust to geometric attacks, common attacks, and Secure	Not work for color medical images
AlQdah et al. [3]	SVD, DWT	Blind watermark and secure	Not robust against several attacks
Zermi et al. [29]	SVD, DWT	Imperceptible, Robust against some attack noise addition and JPEG compression, PSNR = 57 dB, SSIM = 0.999	Computational complexity is high
Nazari et al. [19]	LSB, Chaotic sequence and IWT	PSNR = 75 dB,	Not robust against several attacks, High computational complexity
Singh et al. [21]	DWT and DCT	PSNR = 35.84 dB, and NC value = 0.9992	Computational complexity of an algorithm is high

medical images. They achieved 42.11 dB PSNR and 99.83% tampered detecting rate which is outperformed than some other current schemes.

Table 1 depicts the comparisons of different watermarking schemes based on their method used, performance, and limitation.

3 Embedding Process

In this proposed scheme, a novel SVD-based reversible watermarking technique has been developed. The sequence of steps for this scheme has been discussed below. The block diagram of the entire embedding process has been shown in Fig. 1.

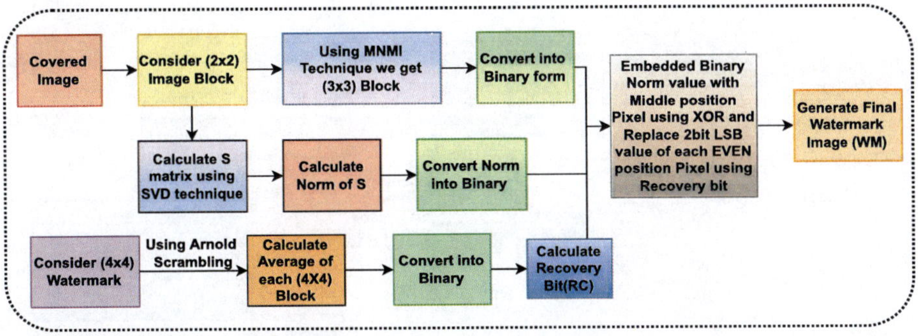

Fig. 1 Block diagram for extraction process

Step I: At first, Cover medical image (CMI) is divided into (2 × 2) image block.

Step II: Using Modified NMI (MNMI) interpolation technique in each (2 × 2) image block to get up-sampled (3 × 3) image block and it convert into binary form.

Step III: The SVD technique is used each (2 × 2) image block to calculate a matrix, say S. Then evaluate the norm of matrix S and convert into binary form.

Step IV: Consider a (4 × 4) watermark image block and Arnold Scrambling technique is applied into each (4 × 4) watermark image block.

Step V: Calculate average from each (2 × 2) watermark image block and store into a (2 × 2) matrix. Taking integer part of each average value it is convert into binary.

Step VI: Taking 2 bit LSB bit from each binary average value consider as a 8 bit recovery bit (RC).

Step VII: Using XOR operation the binary Norm value is embedded into the middle position of binary pixel value each (3 × 3) up-sampled image block and replace 2 bit LSB value of each even position binary pixel value using Recovery bit (RC).

Step VIII: Finally, the watermark image (WM) is created.

3.1 Numerical Example of Embedding Process

The numerical example of proposed scheme is described in Fig. 2.

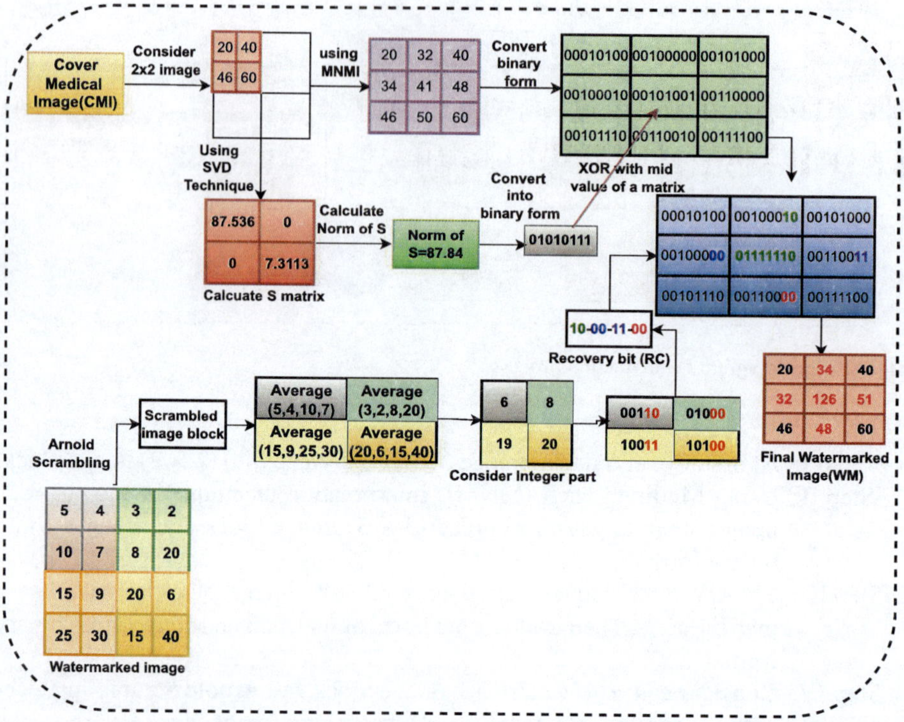

Fig. 2 Numerical example of embedding process

4 Extraction Procedure

The extraction procedure and authentication process has been describe in this section. The extraction operation is just opposite to the embedding operation. The proposed extraction procedure has been explained sequentially in below. The block diagram of the entire extraction procedure has been shown in Fig. 3.

Step I: At first, the watermarked medical image (WM) convert into binary form, say BWMI.

Step II: Watermarked interpolated medical image WM is downed-sampled to recover the (2×2) original cover image.

Step III: Calculate S' matrix from (2×2) original cover image using SVD technique.

Step IV: Again, calculate Norm of S' matrix and convert only integer part of the Norm value into binary form.

Step V: XOR operation is done with binary Norm of S' matrix and Middle position pixel value of BWMI matrix.

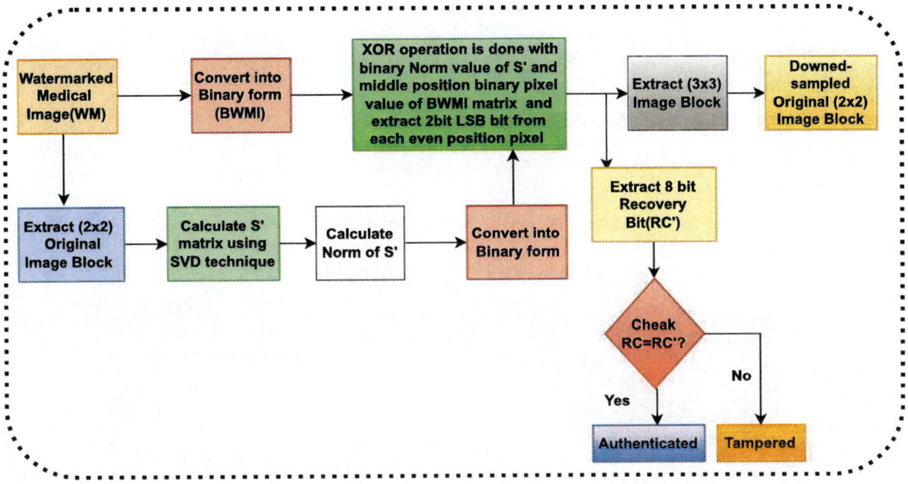

Fig. 3 Block diagram for extraction process

Step VI: After that, extract 2 bit LSB value from each even position binary pixel value the 8 bit Recovery bit (RC') which known as Authentication bit has been recovered.

Step VII: The original Recovery bit RC and extracted Recovery bit RC' is compared to check the authenticity of this proposed scheme. If RC and RC' is equal then proves the authenticity of original medical image.

4.1 Numerical Example of Extraction Process

The numerical example of extraction process describe in Fig. 4.

5 Experimental Results and Discussions

The experimental results has been discussed in this section. The LungCT-Diagnosis CT image database [23] contain lung images with dimension (512 × 512) (Fig. 5) is used for the experimental purpose.

The original medical image and the watermarked image with embedding 64,980 bits watermark information are shown in Fig. 6i, ii, respectively .

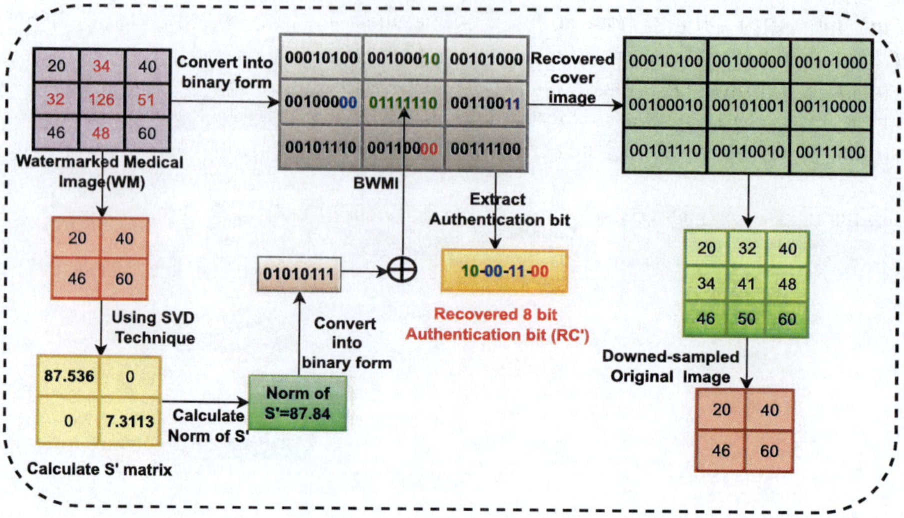

Fig. 4 Numerical example of extraction process

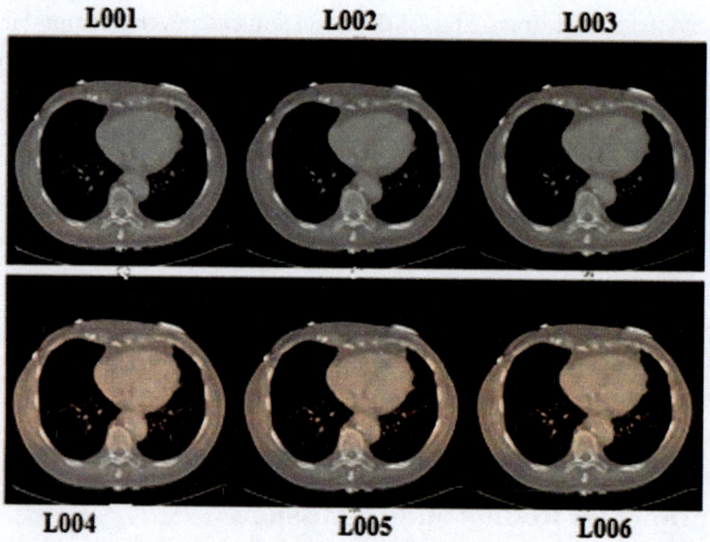

Fig. 5 The lung images used to analysis the proposed method

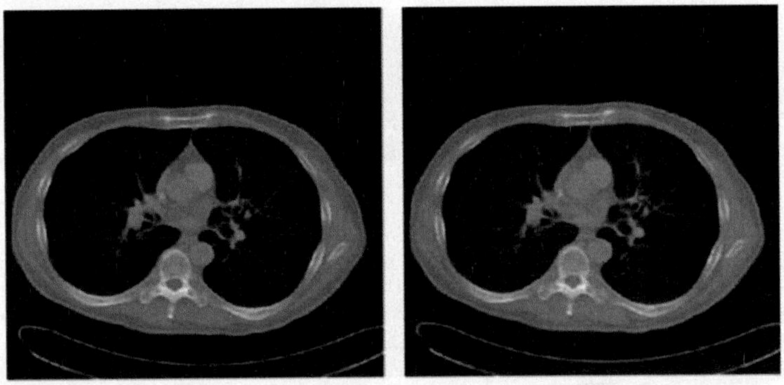

Fig. 6 **i** Original image, **ii** watermarked image

5.1 Imperceptibility Test

In case of digital image, imperceptibility is one of the most important feature. Because doctor diagnose the patients diseases by checking only the scanned digital medical report. The imperceptibility of the proposed scheme is measured using the following parameters.

Peak Signal to Noise Ratio (PSNR) The accuracy of the restored image is calculated by PSNR for $(M \times N)$ images. Usually, image compression quality is measured by using PSNR. The PSNR usually applied to asses the image quality of reconstruction compression images . If the value of PSNR is high, then the quality of the compressed, or reconstructed image is better. The PSNR value of an image can be computed as:

$$\text{PSNR} = 10 \log_{10} \frac{[\max(\text{MSE})]}{\text{MSE}}, \tag{1}$$

Embedding Capacity (EC) Embedding capacity indicate that the amount of external watermark bit can be embedded into original cover image without creating visual distortions. In other words, this determines the quantity of the information that can be carried in a watermark. The Embedding Capacity (EC) of an image is computed as:

$$\text{EC} = \frac{\text{MEB}}{L} \text{bpp}, \tag{2}$$

where, MEB = Maximum number of bits are embedded into original cover image. L = Whole bits in original image.

Structural Similarity Index Measurement (SSIM) The SSIM index is used for assessing the relationship between two digital images. The SSIM estimate the quality

Table 2 Results of PSNR, SSIM and EC

Image	PSNR	SSIM	EC
L001	73.50	0.98	64,980
L002	73.20	0.97	64,980
L003	73.66	0.99	64,980
L004	73.80	0.99	64,980
L005	73.20	0.99	64,980
L006	73.50	0.99	64,980

Table 3 Comparison of imperceptibility with some existing scheme

Scheme	PSNR
Kelkar et al. [15]	49.00
Bamal et al. [6]	34.88
Eze et al. [7]	40.00
Fares et al. [10]	45.45
Thakur et al. [25]	24.64
Adnan et al. [1]	59.78
Su et al. [22]	42.11
Zermi et al. [29]	57.41
Kahlessenane et al. [14]	70
Proposed scheme	73.36

of digital images depending on an initial distortion-free standerd image. The range of SSIM is −1 to +1. The equation of SSIM is shown in bellow:

$$\mathrm{SSIM}(a, b) = \frac{(2\mu_a\mu_b + p_1)(2\sigma_{ab} + p_2)}{(\mu_a^2 + \mu_b^2 + p_1)(\sigma_a^2 + \sigma_b^2 + p_2)} \tag{3}$$

where, μ_a and μ_b are the mean of a and b, the variance of a and b are denotes σ_a and σ_b and the covariance of a and b denote as σ_{ab}; $p_1 = (k_1 L)^2$ and $p_2 = (k_2 L)^2$ denotes are two variables and L represents the range of pixels.

The experiment has been carried out on all lung images of the LungCT-Diagnosis CT image database [23]. Table 2 shows that the first 6 images are consider for experimental results from LungCT-Diagnosis CT image database [23]. From the Table 2, it is shows that the proposed methods provides average 73.36 dB PSNR and near about 0.99 SSIM value after embedding 64,980 watermark bits.

The PSNR comparison results of the proposed scheme with respect to some of the state of the art scheme is given in Table 3. The table clearly shows that the proposed scheme exhibits better PSNR value than other existing schemes.

6 Conclusion

Telemedicine and online consultation is making a very positive contribution to healthcare system during the pandemic situation and is being used in a variety of ways. So that the security of patient digital medical report is important issue because it contains sensitive information of patients. A new SVD based watermarking scheme has been proposed which is robust against some conventional attacks. A recovery bit have been used for the authentication purpose. The experimental findings depicts that the suggested scheme acquire better result compared to some other existing scheme.

References

1. Adnan MM, Mashagba HA, Alfilh RHC, Aljawaheri K, Jaafarz A, Hammood DA, Alkhayyat A, Rahim HA (2021) Watermarking scheme for using ycbcr based on 2-level dwt. J Phys Conf Ser 1962
2. Al-Haj AM, Abdelnabi H (2021) An efficient watermarking algorithm for medical images. Multim Tools Appl 80:26021–26047
3. Al-Qdah M (2018) Secure watermarking technique for medical images with visual evaluation. Sig Image Process Int J 9:01–09
4. Anand A, Singh AK (2020) An improved dwt-svd domain watermarking for medical information security. Comput Commun 152:72–80
5. Anand A, Singh AK (2020) Watermarking techniques for medical data authentication: a survey. Multimedia Tools Appl 1–33
6. Bamal R, Kasana SS (2018) Dual hybrid medical watermarking using walsh-slantlet transform. Multimedia Tools Appl 1–29
7. Eze PU, Udaya P, Evans RJ (2018) Medical image watermark and tamper detection using constant correlation spread spectrum watermarking. Int J Comput Electr Autom Control Inform Eng 12:107–114. World Academy of Science, Engineering and Technology
8. Fan TY, Chao HC, Chieu BC (2019) Lossless medical image watermarking method based on significant difference of cellular automata transform coefficient. Sig Process Image Commun 70:174–183
9. Fares K, Khaldi A, Kafi R, Euschi S (2021) Dct and dwt based watermarking scheme for medical information security. Biomed Sig Process Control 66:102403
10. Fares K, Khaldi A, Kafi R, Euschi S (2021) A dwt based watermarking approach for medical image protection. J Ambient Intell Humanized Comput 12:2931–2938
11. Favorskaya M, Savchina E, Gusev K (2019) Feature-based synchronization correction for multilevel watermarking of medical images. Procedia Comput Sci 159:1267–1276
12. Gao L, Jin Zhang Y, Li G (2020) Reversible watermarking in medical images using sub-sample and multiple histogram modification. J Inform Technol Res 13:75–90
13. Iskandar MW (2019) Adiwijaya: an implementation of text hiding in medical images based on graph coloring for android devices. J Phys Conf Ser
14. Kahlessenane F, Khaldi A, Kafi R, Euschi S (2021) A dwt based watermarking approach for medical image protection. J Ambient Intell Humanized Comput 12(2):2931–2938
15. Kelkar V, Mehta JH, Tuckley KR (2018) A novel robust reversible watermarking technique based on prediction error expansion for medical images. In: CVIP
16. Liu XL, Lin CC, Yuan SM (2016) Blind dual watermarking for color images' authentication and copyright protection. IEEE Trans Circuits Syst Video Technol 28(5):1047–1055
17. Liu X, Lou J, Wang Y, Du J, Zou B, Chen Y (2018) Discriminative and robust zero-watermarking scheme based on completed local binary pattern for authentication and copyright identification of medical images. In: Medical imaging

18. Mohammed RT (2021) Review of medical image authentication techniques and their recent trends. Multim Tools Appl 80:13439–13473
19. Nazari M, Mehrabian M (2021) A novel chaotic iwt-lsb blind watermarking approach with flexible capacity for secure transmission of authenticated medical images. Multim Tools Appl 80:10615–10655
20. Shehab A, Elhoseny M, Muhammad K, Sangaiah AK, Yang P, Huang H, Hou G (2018) Secure and robust fragile watermarking scheme for medical images. IEEE Access 6:10269–10278
21. Singh AK, Dave M, Mohan A (2016) Hybrid technique for robust and imperceptible multiple watermarking using medical images. Multimedia Tools Appl 75:8381–8401
22. Su GD, Chang CC, Lin CC (2020) Effective self-recovery and tampering localization fragile watermarking for medical images. IEEE Access 8:160840–160857. https://doi.org/10.1109/ACCESS.2020.3019832
23. TCIA: The Cancer Imaging Archive (TCIA) collections. http://www.cancerimagingarchive.net/ (2020). Accessed date: 21. 07.2018
24. Thakkar FN, Srivastava VK (2016) A blind medical image watermarking: Dwt-svd based robust and secure approach for telemedicine applications. Multimedia Tools Appl 76:3669–3697
25. Thakur S, Singh AK, Ghrera SP, Mohan A (2018) Chaotic based secure watermarking approach for medical images. Multimedia Tools Appl 79:4263–4276
26. Verma U, Sharma N (2019) Hybrid mode of medical image watermarking to enhance robustness and imperceptibility. Int J Innov Technol Explor Eng 9:351–359
27. Xia Z, Wang X, Wang C, Wang C, Ma B, Li Q, Wang M, Zhao T (2021) A robust zero-watermarking algorithm for lossless copyright protection of medical images. Appl Intell 1–15
28. Yang H, Qi S, Niu P, Wang X (2020) Color image zero-watermarking based on fast quaternion generic polar complex exponential transform. Sig Process Image Commun 82:115747
29. Zermi N, Khaldi A, Kafi R, Kahlessenane F, Euschi S (2021) A dwt-svd based robust digital watermarking for medical image security. Forensic Sci Int 320:110691

Watermark-Based Image Authentication with Coefficient Value Differencing and Histogram Shifting

Bibek Ranjan Ghosh, Siddhartha Banerjee, Jyotsna Kumar Mandal, Arpan Baiagi, and Rahul Deb Bhandari

1 Introduction

The reliability of the huge numbers of digital images generated and distributed over the Internet has become less as they are often compromised. Digital image watermarking has evolved as a tool for image authenticity, where a small image, i.e. "watermark" along with associated authentication data, is imputed to the original image [1]. The originality can be confirmed if the authentication data remains intact after watermark image extraction. The embedding and extraction may involve both spatial and spectral domain techniques [2]. Spectral technique like discrete wavelet transform (DWT) is widely used to solve such problems as DWT has multilevel decomposition and perfect reconstruction capability creating ample embedding space and high robustness [1]. But DWT-based methods produce real coefficients in detail sub-bands leading to complex computation in embedding and extraction phases and are error prone [3]. Integer wavelet transform (IWT) which generate integer coefficients only has become a suitable alternative, which involve integer arithmetic during embedding and extraction [4, 5]. Lifting scheme-based IWT (LWT) involving multiple primal and dual lifting steps along with update and are considered to be very efficient and scalable [6]. Specifically, 3/1 LWT involves 3 high-pass and 1 low-pass filter in analysis and synthesis filter banks [5]. Least significant bit (LSB) embedding is widely used but is easier to detect [1, 2, 5, 7]. However, pixel value differencing (PVD) exhibits higher embedding capacity, and histogram shifting (HS) has

B. R. Ghosh (✉) · S. Banerjee · A. Baiagi · R. D. Bhandari
Department of Computer Science, Ramakrishna Mission Residential College, Narendrapur, Calcutta University, Kolkata, India
e-mail: bibekghosh2019@gmail.com

J. K. Mandal
Department of Computer Science and Engineering, Kalyani University, Kalyani, India

J. K. Mandal et al. (eds.), *Proceedings of International Conference on Network Security and Blockchain Technology*, Lecture Notes in Networks and Systems 738,
https://doi.org/10.1007/978-981-99-4433-0_13

high reversibility [8, 9]. These two essential aspects motivated this work to investigate their performance in image authentication domain with 3/1 LWT. The relevant literature survey is summarized as follows.

Mandal et al. [10] utilized discrete Hartley transform and LSB to achieve 39.02 dB PSNR with 163,840 byte payload in RGB cover. Vaidya et al. [11] proposed adaptive scheme using DWT with Bhattacharya distance and kurtosis to achieve 49.93 dB PSNR with 64 × 64 watermark and 256 × 256 cover. Maheshwari et al. [12] method used DCT and principal component analysis (PCA) to achieve 55.12 dB PSNR with 256 × 256 watermark and 256 × 1024 cover. Agarwal et al. [13] fused DCT and IWT on 8 × 8 blocks and achieved 45.62 dB PSNR with 32 × 32 and 512 × 512 watermark and cover, respectively. Ansari et al. [14] proposed a method involving DWT, SVD, and bee colony optimization to achieve 33 dB average PSNR with 64 × 64 watermark. Joshi et al. [15] used Y plane of YUV planes and DWT to achieve PSNR of 79.193 dB 512 × 512 × 3 cover and watermark 128 × 128. Haribabu et al. [16] method used HSI planes, DWT and entropy computation to achieve 59 dB PSNR. He et al. [17] watermarking scheme used YUV colour space with DWT, DCT, SVD, and Arnold transform and achieved 48.14 dB PSNR with 64 × 64 watermark and 512 × 512 cover. Chen et al. [18] used LWT and randomized location map to achieve PSNR of 46 dB. Abdulrahman et al. [19] used Arnold transform, discrete cosine transform (DCT), DWT in RGB channel and achieved 48 dB PSNR with 1024 × 1024 × 3 cover and 96 × 96 watermark. Ambedekar et al. [20] method used random key encryption, DWT, and Euclidean distance to achieve 54.96 dB PSNR achieved with 90 × 90 watermark. Prasad et al. [21] scheme used encrypted authentication code to exhibit 42.108 dB PSNR, 0.999 SSIM with 393,216 bit payload. Singh et al. [22] used "all phase sine bi orthogonal transform" (APSBT), "dynamic stochastic resonance" (DSR), and SVD to achieve NCC 0.999 with 256 × 256 watermark. Ahmadi et al. [23] employed "human visual system" (HVS), "singular value decomposition" (SVD), and DWT with "particle swarm optimization" (PSO) in the blue channel of RGB image with 52 dB PSNR and 0.9733 SSIM 512 × 512 × 3 cover and 32 × 32 watermark. Naouti et al. [24] scheme used DWT and SVD to report 48 dB PSNR and SSIM 1. Ghosh et al. [7] used random LSB-based method with IWT to achieve 59 dB PSNR for 32,768 bit payload for 512 × 512 cover.

The rest of the sections are organized as follows. Section 2 develops the notion of 3/1 LWT technique, Sect. 3 reveals the detailed proposed methodology adopted, Sect. 4 reveals and analyze the experimental observations, and Sect. 5 makes a conclusion of the work stating future prospects.

2 Formulation of the 3/1 IWT

The 3/1 IWT analysis filter bank has 3 high-pass and one low-pass filter with down sampling by 2. The synthesis filter bank uses the reverse process with up sampling by 2 [4, 25]. Let Org is a digital signal of $2n$ integer elements. The forward transform with Eq. 1 creates average (Avg) and detail (Det) sub-signals each having n

integer elements. Whereas, the inverse transform takes input two one dimensional sub-signals Avg and Det, each of size *n*, to generate original signal Org using Eq. 2.

$$\begin{cases} \text{Det}'[k] = \text{Org}[2k+1] - \text{Org}[2k],\ \text{Avg}[k] = \text{Org}[2k] + \lfloor \text{Det}'[k]/2 \rfloor (\text{Avergae}) \\ \text{Det}[k] = \text{Det}'[k] + \lfloor \text{Avg}[k-1]/4 - \text{Avg}[k+1]/4 + 1/2 \rfloor (\text{Detail}) \end{cases}$$
$$\text{where } k = 0 \ldots (n-1) \qquad (1)$$

$$\begin{cases} \text{Det}'[k] = \text{Det}[k] - \lfloor \text{Avg}[k-1]/4 - \text{Avg}[k+1]/4 + 1/2 \rfloor \\ \text{Org}[2k] = \text{Avg}[k] - \lfloor \text{Det}'[k]/2 \rfloor \text{and}\ \text{Org}[2k+1] = \text{Org}[2k] + \text{Det}'[k] \end{cases} \qquad (2)$$
$$\text{where } k = 0\, to (n-1).$$

To achieve forward transform of grayscale image IM of size $M \times N$, first, Eq. 1 is applied on each row and then on each column of IM individually, generating LL, LH, HL, and HH sub-images. Inverse transform uses Eq. 2 in opposite order.

3 Proposed Method

The proposed system is divided into embedding and authentication phases. In the first phase, after preprocessing, the cover image C of size $M \times N$ is applied 3/1 forward transform to generate three detailed sub-band images HH1, HL1, and LH1 with one average sub-band image LL1, each resulting image with half size of the original. The hash value HW of the watermark image W is computed with SHA256 algorithm. Thereafter, the binarised stream of bits of W is embedded in the HH1 using coefficient value difference method described in Algorithm 3.4 to get WHH1. Further, HW is embedded using histogram shifting method in the HH2 sub-band of the decomposed LH1 sub-band computed earlier using Algorithm 3.2. Finally, the 3/1 inverse transformation is applied level wise to wrap up both the decomposition to generate stego-image S. In the authentication phase, first, the watermark EW is extracted from the difference of coefficients of the HH1 sub-band using Algorithm 3.5. SHA256 hash value H2 is computed from EW. Then, the hash value H1 embedded in first phase in the HH2 sub-band of the LH1 component is extracted from the modified histogram using Algorithm 3.3. Finally, H1 and H2 are compared to confirm the authenticity of the watermark using Algorithm 3.7. The details of each phases are depicted in the Figs. 1 and 2, respectively.

3.1 *Preprocessing*

In order to tackle overflow and underflow (i.e. pixel value greater than 255 or below 0, respectively), which may arise when executing inverse IWT after embedding data in

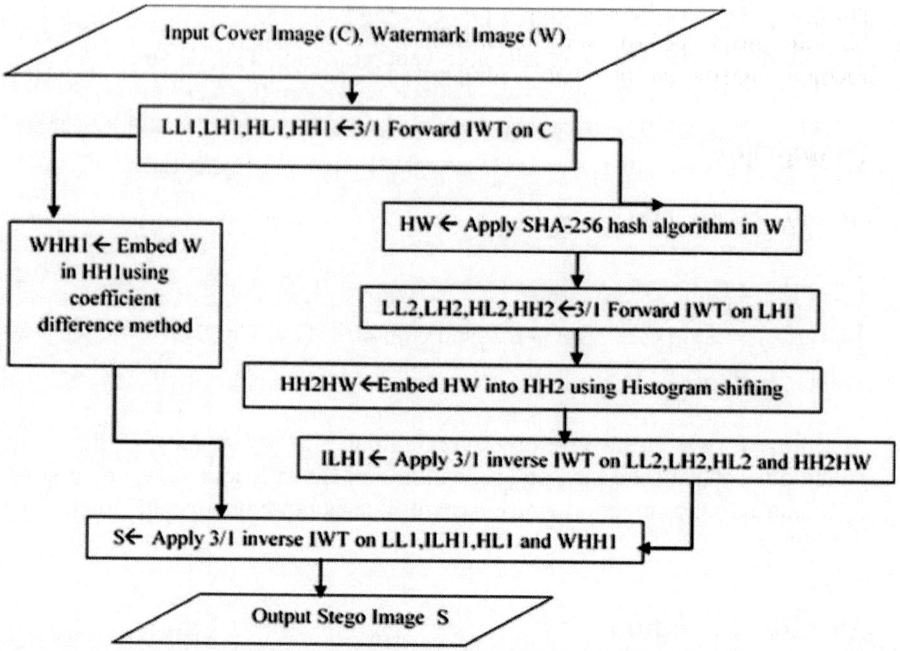

Fig. 1 Workflow of the embedding phase

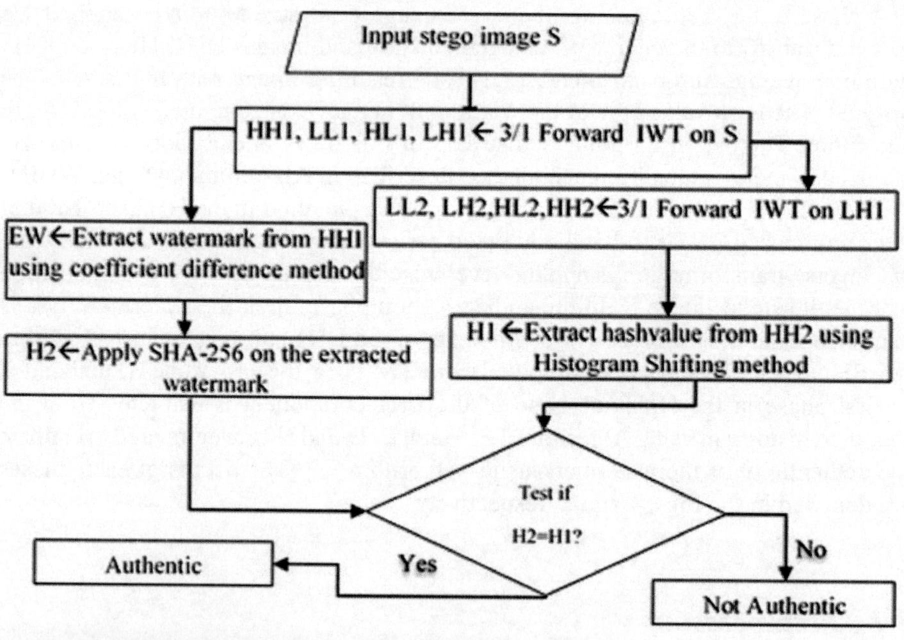

Fig. 2 Workflow of authentication phase

high frequency component of IWT histogram modification is applied to the "original image", i.e. the pixel values are clipped within 8 or 247 [26].

3.2 *Coefficient Histogram Shifting Embedding Algorithm*

Input: Coefficient Matrix IM of size M × N of intensity level L, Message bit stream MSG
Output: Embedded image EM
Step-1: Compute histogram H(k) = f, ∀ k ∈ [0, L − 1] of IM
Step-2: Sequentially probe the histogram H[k] and find a k such that H[k] > = length (MSG) and best fits at k with minimum shifting
Step-3: BP = k, where BP stands for best point intensity in histogram H
Step-4: Shift the histogram to the right by 1 making BP + 1 bin empty
Step-5: Each message bit MB of MSG will be embedded in the empty bin as follows

$$\mathrm{EM}(x, y) = \begin{cases} \mathrm{IM}(x, y) + 1, \ \text{if}, \mathrm{IM}(x, y) = \mathrm{BP} \text{ and } \mathrm{MB} = 1 \\ \mathrm{IM}(x, y), \ \text{if}, \mathrm{IM}(x, y) = \mathrm{BP} \text{ and } \mathrm{MB} = 0 \end{cases}$$
$$\text{for} \forall (x, y) \text{where}, x = 0 \ldots (M-1) \text{ and } y = 0 \ldots (N-1) \quad (3)$$

3.3 *Coefficient Histogram Shifting Extraction Algorithm*

Input: Embedded image EM of size M × N with intensity level L, Best point BP
Output: Bit stream MSG
Step-1: Generate histogram H(k) = f, ∀ k ∈ [0, L-1] of EM.
Step-2: Each bit MB of retrieved message MSG may be retrieved as

$$\mathrm{MB} = \begin{cases} 1, \text{if}, \mathrm{EM}(x, y) = \mathrm{BP} + 1 \\ 0, \text{if}, \mathrm{EM}(x, y) = \mathrm{BP} \end{cases}$$
$$\text{for } \forall (x, y) \text{ where}, x = 0 \ldots (M-1) \text{ and } y = 0 \ldots (N-1) \quad (4)$$

3.4 *Coefficient Value Difference Embedding Algorithm*

Input: Coefficient matrix CM of size M rows and N columns, binary bit stream MSG
Output: Embedded coefficient matrix ECM
Step-1: Co = Convert CM to one dimensional row major form, index = 0

Step-2: Repeat Step-3 to Step-16 for i = 1 to M*N − 1
Step-3: D = Absolute (Co[i] − Co [i + 1])
Step-4: If D < = 255 then
Step-5: x = $\lfloor \log_2^D \rfloor$, LB = 2^x, UB = 2^{x+1}
Step-6: Val = DecimalOf(MSG[index]..MSG[index + x])
Step-7: Dnew = LB + Val, M = Absolute(Dnew-D)
Step-8: Case1: If (Co[i] > = Co[i + 1] and Dnew > D) then
Step-9: Co[i] = Co[i] + Ceiling(M/2),Co[i + 1] = Co[i + 1]-Floor(M/2)
Step-10: Case2: If (Co[i] < Co[i + 1] and Dnew > D) then
Step-11: Co[i] = Co[i] − Floor(M/2), Co[i + 1] = Co[i + 1] + Ceiling(M/2)
Step-12: Case3: If(Co[i] > = Co[i + 1] and Dnew < = D) then
Step-13: Co[i] = Co[i]-Ceiling(M/2), Co[i + 1] = Co[i + 1] + Floor(M/2)
Step-14: Case4:If(Co[i] < Co[i + 1] and Dnew < = D) then
Step-15: Co[i] = Co[i] + Ceiling(M/2), Co[i + 1] = Co[i + 1] − Floor(M/2)
Step-16: index = index + x
Step-17: ECM = Convert Co into 2D row major form
Step-18: return ECM.

3.5 Coefficient Value Difference Extraction Algorithm

Input: An embedded coefficient matrix CM of size M rows and N columns
Output: Extracted message MSG
Step-1: F = Convert CM to one dimensional row major form, MSG = NULL
Step-2: Repeat Step-3 to 7 for i = 1 to M*N − 1
Step-3: D = Absolute (F[i]-F[i + 1])
Step-4: If D < 255 then
Step-5: x = $\lfloor \log_2^D \rfloor$, LB = 2^x, UB = 2^{x+1}.
Step-6: Val = Absolute (D − LB)
Step-7: MSG = Concatenate (MSG, DecimaltoBinary (Val))
Step-8: return MSG.

3.6 Watermark Embedding Algorithm

Input: Preprocessed cover image C of size M × N, Watermark image W
Output: Stego-image IS of size M × N
Step-1: Apply Eq. 1 on C to create HH1, HL1, LH1, and LL1 each of size M/2 × N/2
Step-2: Apply SHA256 hash algorithm to W to produce the hash value HW.
Step-3: Apply Eq. 1 to LH1 to create HH2, HL2, LH2, and LL2 of size M/4xN/4.
Step-4: To get rid of negative values in HH2, convert it as follows:

$$\begin{cases} \text{HH2}(x, y) = \text{MAG}(x, y) * \text{SIGN}(x, y), \text{MAG}(x, y) = \text{Absolute}(HH2(x, y)) \\ \text{SIGN}(x, y) = \begin{cases} +1, \text{if, HH2}(x, y) >= 0 \\ -1, \text{if, HH2}(x, y) < 0 \end{cases} \end{cases}$$

$$\text{for} \forall (x, y) \text{where, } x = 0 \ldots \left(\frac{M}{4} - 1\right) \text{ and } y = 0 \ldots \left(\frac{N}{4} - 1\right) \quad (5)$$

Step-5: Embed HW in MAG using Algorithm 3.2 to generate MAGHW
Step-6: Combine MAGHW and SIGN as follows:

$$\text{HH2HW}(x, y) = \text{MAGHW}(x, y) * \text{SIGN}(x, y)$$
$$\text{for} \forall (x, y) \text{where, } x = 0 \ldots \left(\frac{M}{4} - 1\right) \text{ and } y = 0 \ldots \left(\frac{N}{4} - 1\right) \quad (6)$$

Step-7: Apply inverse IWT to LL2, LH2, HL2, and HH2HW to generate ILH1
Step-8: Bit streams of W are embedded into the HH1 with Algorithm 3.4 if for any successive coefficient values CM(p,q) and CM(u, v) in HH1 having $|CM(p, q) - CM(u, v)| <= 255$ and produce WHH1 coefficient image.
Step-9: Apply inverse 3/1 IWT to ILH1, LL1, HL1, WHH1 to generate Stego-image S.

3.7 Authentication Algorithm

Input: Stego-image S of size M rows and N columns.
Output: Yes or No
Step-1: Apply Eq. 1 to S to create HH1, HL1, LH1, and LL1 of size M/2xN/2.
Step-2: Apply Algorithm 3.5 to extract watermark EW from HH1.
Step-3: Apply SHA256 algorithm to EW to generate 256 bit hash value H2.
Step-4: Apply Eq. 1 to LH1 to create HH2, HL2, LH2, and LL2 of size M/4 × N/4.
Step-5: Use Algorithm 3.3 to extract the hash value H1 from HH2.
Step-6: If H2 = H1, then the image is authentic.

4 Experimental Result and Analysis

This section specifies experimental results which has been obtained by applying the proposed algorithm on 512 × 512 grayscale images as cover from "USC SIPI image database" with resized watermark image [27]. Figure 3 depicts one such result. Sections 4.2 and 4.3 evaluate the proposed method using parameters illustrated in Sect. 4.1.

Fig. 3 **a** Cover image (IC). **b** Watermark image (*W*). **c** Stego-image (IS). **d** Extracted watermark image (EW)

4.1 Parameters

PSNR measured between cover and stego-image reveals degree of imperceptibility of secret data. Higher PSNR value indicates better stego-image. It is computed using Eq. 7, for an 8-bit grayscale cover image (IC) and stego-image (IS) having size (M × N). Again SSIM measures similarity between cover and stego-images given in Eq. 7.

$$\begin{cases} \text{Mean Square Error (MSE)} = \dfrac{1}{\text{MN}} \displaystyle\sum_{x=0}^{M-1} \sum_{y=0}^{N-1} [IC(x, y) - IS(x, y)]^2 \\ \text{PSNR} = 20 \log \dfrac{255}{\sqrt{\text{MSE}}}, \ \text{SSIM} = \dfrac{(2\mu_{IC}\mu_{IS} + C1)(2\sigma_{ICS} + C2)}{\left(\mu_{IC}^2 + \mu_{IS}^2 + C1\right)\left(\sigma_{IC}^2 + \sigma_{IS}^2 + C2\right)} \end{cases}, \quad (7)$$

where μ_{IC} and μ_{IS} are mean of IC and IS, respectively, σ_{IC}^2 and σ_{IS}^2 are variance of IC and IS, respectively, and σ_{ICS} is the covariance of IC and IS. $C1$ and $C2$ are constants to balance the ratio when mean and variance are nearly 0. NCC is a measure to detect level of correlation or similarity between two signals. If the signals are highly correlated the value approaches 1 and becomes 1 when two signals are exactly same. On the other hand, NCC $= -1$ signifies the signals has no similarity at all. For two discrete one dimensional signals S and T each with N elements, the NCC is computed as given in Eq. 8.

$$\text{NCC}(S, T) = \frac{\sum_{i=1}^{N} (S[i] - \overline{S})(T[i] - \overline{T})}{\sqrt{\sum_{i=1}^{N} (S[i] - \overline{S})^2} \sqrt{\sum_{i=1}^{N} (T[i] - \overline{T})^2}}, \quad (8)$$

where $\overline{S}$ and $\overline{T}$ are the mean of signals S and T, respectively.

Table 1 Proposed system performance in imperceptibility test

	Watermark size 32 × 32			Watermark size 64 × 64		
Cover images	MSE	PSNR	SSIM	MSE	PSNR	SSIM
Airplane	0.03246	63.02	0.9998	0.12366	57.21	0.9991
Barbara	0.03532	62.65	0.9998	0.12587	57.13	0.9992
Elaine	0.03897	62.22	0.9997	0.18127	55.55	0.9988
Fishing boat	0.04536	61.56	0.9998	0.16801	55.88	0.9991
Gold hill	0.04515	61.58	0.9999	0.16967	55.83	0.9995
House	0.05839	60.47	0.9998	0.20475	55.02	0.9995
Lena	0.03149	63.15	0.9998	0.14043	56.66	0.9992
Peeper	0.06664	59.90	0.9997	0.17225	55.77	0.9989
Sailboat	0.05936	60.40	0.9998	0.22671	54.58	0.9993
Splash	0.03161	63.13	0.9998	0.11244	57.62	0.9989
Tank	0.04917	61.21	0.9998	0.18532	55.45	0.9992
Average	0.04490	61.75	0.9998	0.16458	56.06	0.9992

4.2 *Imperceptibility Test*

This section illustrates the results of the imperceptibility test conducted on each cover image and corresponding stego-image of size 512 × 512 in terms of PSNR, MSE, and SSIM parameters. Table 1 depicts some of the results obtained with 32 × 32 and 64 × 64 watermark along with 256 bit SHA256 output making payload of 8192 bit and 33024bits, respectively in total.

Table 1 shows that the algorithm performs very well as the MSE is nearly zero, average PSNR is 61.75 dB and 56.06 dB for 32 × 32 and 64 × 64 watermarks, respectively. Average SSIM 0.999 also exhibits high resemblance of stego and cover images in these two cases. In all the cases, the hash value reconstructed form the extracted watermark and initial hash value computed from the watermark perfectly matched, showing hundred percent accuracy in terms of authentication.

4.3 *Robustness Test*

The different types of active attacks on the stego-images were carried out to test the robustness of the system. The attacks under consideration are noising attacks (salt and pepper noise, Gaussian noise), de-noising attacks (blurring, median blur), and image processing operations like histogram equalization, sharpening, cropping, scaling, and JPEG compression [25]. NCC is computed from original and the extracted watermark of the attacked image. The NCC values are listed in Table 2.

Table 2 Robustness test with different attacks on stego-image with NCC values

Cover image	Histogram equalization	Gaussian noise (0.01)	Sharpening	Crop (10%)	Crop (20%)	Blur (3 × 3)	Salt and Peeper Noise (0.01)	Median blur	Scaling (512-256-512)	JPEG compression (10%)
32 × 32 *watermark image*										
Airplane	0.798	0.802	0.797	0.872	0.737	0.679	0.823	0.718	0.651	0.507
Barbara	0.801	0.794	0.787	0.724	0.727	0.703	0.806	0.772	0.682	0.619
Elaine	0.796	0.797	0.782	0.795	0.746	0.694	0.803	0.754	0.628	0.424
Fishing boat	0.819	0.783	0.786	0.805	0.725	0.717	0.810	0.758	0.654	0.513
Gold hill	0.795	0.790	0.792	0.807	0.733	0.747	0.801	0.783	0.669	0.656
House	0.794	0.790	0.810	0.793	0.746	0.732	0.816	0.774	0.669	0.700
Lena	0.792	0.789	0.795	0.850	0.762	0.699	0.793	0.746	0.648	0.410
Peeper	0.784	0.795	0.780	0.776	0.725	0.703	0.798	0.748	0.658	0.572
Sailboat	0.796	0.812	0.800	0.773	0.746	0.727	0.800	0.784	0.668	0.647
Splash	0.770	0.795	0.801	0.802	0.699	0.675	0.794	0.727	0.608	0.391
64 × 64 *watermark image*										
Airplane	0.785	0.791	0.791	0.807	0.805	0.698	0.797	0.731	0.655	0.544
Barbara	0.795	0.799	0.788	0.794	0.759	0.705	0.806	0.751	0.665	0.568
Elaine	0.800	0.794	0.793	0.796	0.762	0.706	0.803	0.755	0.659	0.454
Fishing boat	0.789	0.799	0.792	0.802	0.772	0.726	0.800	0.780	0.667	0.559
Gold hill	0.790	0.794	0.799	0.803	0.759	0.725	0.799	0.772	0.659	0.638
House	0.796	0.790	0.794	0.804	0.777	0.720	0.805	0.769	0.658	0.663

(continued)

Table 2 (continued)

Cover image	Histogram equalization	Gaussian noise (0.01)	Sharpening	Crop (10%)	Crop (20%)	Blur (3 × 3)	Salt and Peeper Noise (0.01)	Median blur	Scaling (512-256-512)	JPEG compression (10%)
Lena	0.789	0.792	0.793	0.911	0.745	0.707	0.797	0.752	0.651	0.518
Peeper	0.791	0.787	0.796	0.797	0.759	0.711	0.804	0.754	0.661	0.542
Sailboat	0.789	0.798	0.795	0.823	0.762	0.733	0.798	0.779	0.682	0.651
Splash	0.786	0.792	0.796	0.802	0.744	0.692	0.800	0.730	0.643	0.391

Histogram equalization, sharpening, and Gaussian noise attack exhibit NCC value nearly 0.79. Cropping by 10% and 20% shows NCC value of 0.80 and 0.75, respectively. Blurring by 3 × 3 kernel shows NCC value nearly 0.7. Salt and pepper noise attack show values around 0.8. Median blurring shows result near to 0.72. Scaling and JPEG compression attack give the lowest result around 0.6.

4.4 Comparison with Other Methods

This section compares the proposed method performance with other state of the art similar existing methods. Table 3 lists down some of the competitive methods available in this category of image authentication and compares them with our proposed method. Abdulrahman et al. [19] exhibit enhanced payload with RGB channel but with lesser imperceptibility, i.e. 48 dB PSNR. Ahmadi et al. [23] also achieve better payload, but the PSNR is lesser, i.e. 52 dB. Proposed method is superior both in terms of carrier size and PSNR in comparison with He et al. [12]. The performance of Ghosh et al. [7] is almost similar. The proposed method outperforms Agarwal et al. [13] and Ansari et al. [14] in terms of PSNR, i.e. 45.62 dB and 33 dB, respectively. These observations suffice to establish the credibility of the proposed method to be a contemporary and strong contender in terms of imperceptibility and robustness.

Table 3 Comparison with other methods

Methods	Method	Cover image size	Payload (bits)	PSNR (dB)
[19]	Arnold transform, DCT, and DWT	1024 × 1024 × 3	73,728	48.00
[23]	SVD, DWT, and PSO	512 × 512 × 3	40,960	52.00
[17]	DWT, DCT, and SVD	512 × 512 × 3	32,768	48.14
[7]	S transform IWT	512 × 512	8192	65.00
			32,768	59.00
[13]	IWT and DCT	512 × 512	8192	45.62
[14]	DWT, SVD, and bee colony optimization	512 × 512	32,768	33.00
[11]	DWT	256 × 256	32,768	49.93
[20]	DWT	228 × 228	64,800	54.96
Proposed method	3/1 IWT	512 × 512	8448	61.75
			33,024	56.06

5 Conclusion

All the sections so far have discussed the legitimacy of the proposed works. Section 1 discussed the need to propose the system in image authentication with watermarking. Experiments are performed to measure imperceptibility parameters of the hidden watermark message to deceive human visual system, and the results are quite impressive, e.g. PSNR nearly 60 dB (well above the accepted threshold value of 30 dB), MSE nearly 0 and SSIM nearly 1. The robustness tests with nine type of standard attacks are also quite satisfactory. Finally, the system performed better in terms of performance in comparison with other contemporary state-of-the-art methods. Many improvements can be adopted to enhance the efficiency of the proposed system. Some of them may be: various IWT-based methods with different filter banks, efficient spatial domain embedding methods, chaotic map for randomization of watermark, error correction codes for tamper recovery, encryption keys, multilevel decomposition, and many more.

Acknowledgements The authors express heartfelt gratitude to the faculty and staff of the Department of Computer Science, Ramakrishna Mission Residential College Narendrapur, Kolkata and University of Kalyani, Kalyani as well as DST PURSE II, University of Kalyani, Kalyani for their overall support for this work.

References

1. Khadim IJ, Premaratne P, Vial PJ, Halloran B (2019) Comprehensive survey of image steganography: techniques, evaluations, and trends in future research. Neurocomputing 335:300–306
2. Li B, He J, Huang J, Shi YQ (2011) A survey on image steganography and steganalysis. J Inf Hiding Mult Sig Proc 2(2):142–172
3. Subburam S, Selvakumar S, Geetha S (2018) High performance reversible data hiding scheme through multilevel histogram modification in lifting integer wavelet transform. Multimedia Tools Appl 77(6):7071–7095
4. Calderbank AR, Daubechies I, Sweldens W, Yeo BL (1998) Wavelet transforms that map integers to integers. Appl Comput Harmonic Anal 5(3):332–369
5. Kalita M, Tuithung T, Majumder S (2019) A new steganography method using integer wavelet transform and least significant bit substitution. Comput J 62(11):1639–1655
6. Reichel J, Menegaz G, Nadenau MJ, Kunt M (2001) Integer wavelet transform for embedded lossy to lossless image compression. IEEE Trans Image Proc 0(3):383–392
7. Ghosh BR, Banerjee S, Mandal JK (2022) Watermark based image authentication using integer wavelet transform. In: 2022 international conference on inventive computation technologies (ICICT 2022). IEEE. Nepal, pp 312–319
8. Wu DC, Tsai WH (2003) A steganographic method for images by pixel-value differencing. Pattern Recogn Lett 24(9–10):1613–1626
9. Hwang J, Kim J, Choi J (2006) A reversible watermarking based on histogram shifting. In: International workshop on digital watermarking. Springer, Berlin, Heidelberg, pp 348–361
10. Mandal JK, Ghosal SK (2012) A fragile watermarking based on separable discrete Hartley transform for color image authentication (FWSDHTCIA). Sig Image Proc: Int J (SIPIJ) 3(6):24–33

11. Vaidya SP, Chandra M (2015) PVSSR: adaptive digital watermarking for copyright protection of digital images in wavelet domain. Procedia Comp Sci 58:233–240
12. Maheswari S, Rameshwaran K, Malarselvi KM (2015) DCT-PCA based watermarking on E-governance documents. Res J Appl Sci Eng Technol 9(7):507–511
13. Agrwal SL, Yadav A, Kumar U, Gupta SK (2016) Improved invisible watermarking technique using IWT-DCT. In: 5th international conference on reliability, infocom technologies and optimization (trends and future directions) (ICRITO). IEEE, Noida, India, pp 283–285
14. Ansari IA, Pant M, Ahn CW (2016) ABC optimized secured image watermarking scheme to find out the rightful ownership. Optik 127(14):5711–5721
15. Joshi AM, Bapna M, Meena M (2016) Blind image watermarking of variable block size for copyright protection. In: Afzalpulkar N, Srivastava V, Singh G, Bhatnagar D (eds) Proceedings of the international conference on recent cognizance in wireless communication & image processing. Springer, New Delhi, pp 853–859
16. Haribabu MI, Bindu CH, Veera Swamy K (2016) A secure and invisible image watermarking scheme based on wavelet transform in HSI color space. Procedia Comput Sci 93:462–468
17. He Y, Hu Y (2018) A proposed digital image watermarking based on DWT-DCT-SVD. In: 2nd IEEE advanced information management, communicates, electronic and automation control conference (IMCEC 2018). IEEE, Xian, China, pp 1214–1218
18. Chen H, Xu W (2018) Secure and robust color image watermarking for copyright protection based on lifting wavelet transform. In: 25th international conference on mechatronics and machine vision in practice (M2VIP). IEEE , Stuttgart Germany, pp 1–5
19. Abdulrahman AK, Ozturk S (2019) A novel hybrid DCT and DWT based robust watermarking algorithm for color images. Multi Tools Appl 78:17027–17049
20. Ambadekar SP, Jain J, Khanapuri J (2019) Digital image watermarking through encryption and DWT for copyright protection. In: Bhattacharyya S, Mukherjee A, Bhaumik H, Das S, Yoshida K (eds) Recent trends in signal and image processing. Advances in intelligent systems and computing, vol 727. Springer, Singapore, pp 187–195
21. Prasad S, Pal AK (2020) A tamper detection suitable fragile watermarking scheme based on novel payload embedding strategy. Multimed Tools Appl 79:1673–1705
22. Singh SP, Bhatnagar G (2020) A robust watermarking scheme for copyright protection. In: Chaudhuri B, Nakagawa M, Khanna P, Kumar S (eds) Proceedings of 3rd international conference on computer vision and image processing. Advances in intelligent systems and computing, vol 1024. Springer, Singapore, pp 431–443
23. Ahmadi SBB, Zhang G, Rabbani M et al (2021) An intelligent and blind dual color image watermarking for authentication and copyright protection. Appl Intell 51:1701–1732
24. Naffouti SE, Kricha A, Sakly A (2022) A sophisticated and provably grayscale image watermarking system using DWT-SVD domain. Visual Comput 1–21
25. Shaik A, Thanikaiselvan V (2021) Comparative analysis of integer wavelet transforms in reversible data hiding using threshold based histogram modification. J King Saud Univer - Comput Inf Sci 33(7):878–889
26. Gulve AK, Joshi MS (2015) An image steganography method hiding secret data into coefficients of integer wavelet transform using pixel value differencing approach. Math Prob Eng, Article ID 684824, 1–11
27. Weber AG (2006) The USC-SIPI image database: version 5. http://sipi.usc.edu/database/

IEMS3: An Image Encryption Scheme Using Modified SNOW 3G Algorithm

Subrata Nandi, Satyabrata Roy, Srinivasan Krishnaswamy, and Pinaki Mitra

1 Introduction

In today's world, digital images play a critical role in multimedia communications that are happening at a breakneck pace because of Internet of Things (IoT) applications. However, if the data is sensitive, it is critical to ensure its protection. Even though the Internet of Things architecture provides numerous advantages to humanity, data transfers pose various security threats, particularly when sensitive images are transferred. These sensors gather data from their surroundings and broadcast it across unsecured public networks. In this context, the adversary can play a crucial role in taking advantage of the system by manipulating the data via various security attacks. As IoT application sensors have some storage and computation constraints, providing security to those applications becomes difficult. Traditional ciphers, as a result, cannot be utilized in IoT devices. SNOW 3G, on the other hand, may be used to provide security to IoT devices in this resource-constrained context because this stream cipher is capable of constructing complicated patterns and pseudorandom sequences efficiently at high speed. It is also simple to put into hardware. Role of

S. Nandi
Narula Institute of Technology, Agarpara, West Bengal, India

S. Nandi (✉) · S. Krishnaswamy · P. Mitra
Indian Institute of Technology Guwahati, Assam, India
e-mail: subrata.nandi@nit.ac.in; subrata.nandi@iitg.ac.in

S. Krishnaswamy
e-mail: srinikris@iitg.ac.in

P. Mitra
e-mail: pinaki@iitg.ac.in

S. Roy
Department of Computer Science and Engineering, Manipal University Jaipur, Jaipur, Rajasthan, India
e-mail: satyabrata.roy@jaipur.manipal.edu

J. K. Mandal et al. (eds.), *Proceedings of International Conference on Network Security and Blockchain Technology*, Lecture Notes in Networks and Systems 738,
https://doi.org/10.1007/978-981-99-4433-0_14

cloud-based IoT infrastructure is vital in the present digital world. In this configuration, small devices use sensors to collect data and send it over the Internet to cloud storage servers. Word-oriented stream cipher [1–4] plays important role in modern days communication. In 4G and 5G communications, SNOW 3G [5] cipher is used to keep the confidentiality and integrity of the data. Moreover, it gives 128-bit security, and it has high throughput. It can be used efficiently in hardware, software, or embedded system devices. In short, the major contributions of this work are listed below.

- A modified SNOW 3G KSG by changing the Linear Feedback Shift Register (LFSR) with 64 input–output, 8 delay blocks σ –LFSR is proposed.
- A feedback configuration matrix [6, 7] with less number of gain matrices has been used instead of the feedback matrix of SNOW 3G to reduce the encryption time of traditional cipher and enhance the randomness of Fig. 1. As the look up table-based implementation of SNOW 3G causes cache timing attacks[8, 9], we use efficient vector-matrix multiplication in the state transition equation of modified SNOW 3G.
- The randomness of the proposed KSG is evaluated by the NIST randomness test suit, and the proposed image encryption scheme is validated by standard tests like information entropy test, histogram analysis, correlation coefficients, NPCR, UACI, etc.

The following is the order in which this paper has been organized. Section 2 comprises earlier related contributions in this field. Section 3 offers some fundamental notions necessary for readers to comprehend proposed scheme, which is presented in Sect. 4. Section 5 contains experimental results, security analysis, and performance analysis of the proposed scheme. Finally, Sect. 6 concludes the work.

2 Related Work

In this section, some of the recent studies have been presented. A good review of various types of the image encryption scheme is presented in [10]. The majority of the techniques are a mixture of more than two different techniques. A novel technique to generate a pseudorandom key for encryption was proposed by Kumar et al. [11]. This pseudorandom key is used to build three secure and efficient methods for satellite image encryption: logistic map (LM) [12], cosine transformed logistic map (CTLM) [13, 14], and cosine transformed logistic-sine map (CTLSM) [15]. Their scheme achieved better security features when compared to other existing schemes. Image encryption scheme using DNA computing was proposed by many researchers in recent years [16–19]. These techniques make use of DNA computing along with the chaotic map. It results in the key space being large enough to resist brute-force attacks. Besides, these techniques provide better correlation coefficient values, MSE and PSNR. Babaei et al. presented an image encryption scheme based on CA and DNA sequences; however, it requires a significant amount of CPU and memory, making it unsuitable for IoT applications. Similarly, meta-heuristic-based image encryption techniques [20, 21] also suffer from slow encryption speed because of their operational complexities. Although there exist many image encryption techniques in the literature that produce high-quality cipher images, most of these are slow and unsuitable for use in resource-constrained environments. These are not a single technique but a mixture of two or more techniques to introduce randomness in the cipher image. Moreover, most of these techniques cannot be implemented easily in hardware or embedded systems. The proposed scheme takes care of the resource constraints, implementation simplicity, and speed limitations. It generates high-quality cipher images as validated in the result and analysis discussed in Sect. 5.

3 Preliminary Concepts

σ−LFSR is one kind of word-based LFSR [22] to generate vector based pseudorandom number.

Definition 1 We define a σ−LFSR [23] as a word-based LFSR which follows the following recurrence relation,

$$\mathbf{D_{n+b}} = \mathbf{C}_{b-1} D_{n+b-1} + \mathbf{C}_{b-2} D_{n+b-2} + \cdots + \mathbf{C}_0 D_n, \tag{1}$$

where each $D_i \in \mathbb{F}_2^m$ and C_i s are in $\mathbb{F}_2^{m\times m}$. Each delay block contains m-inputs and m-outputs D flip flof, and the σ-LFSR produces a series of vectors in $\mathbb{F}_2^m$. The σ-LFSR's gain matrices are denoted by the matrices $C_0, C_1, \cdots, C_{b-1}$, and their configuration matrix is denoted by the matrix shown below.

$$C = \begin{bmatrix} \text{ze} & \text{Id} & \text{ze} & \cdots & \text{ze} \\ \text{ze} & \text{ze} & \text{Id} & \cdots & \text{ze} \\ \vdots & \vdots & \vdots & \dots & \vdots \\ \text{ze} & \text{ze} & \text{ze} & \cdots & \text{Id} \\ C_0 & C_1 & C_2 & \cdots & C_{b-1} \end{bmatrix} \in \mathbb{F}_2^{mb\times mb}, \tag{2}$$

where, ze, Id $\in \mathbb{F}_2^{m\times m}$ are the all-zero and identity matrices, respectively. We will refer to this matrix structure as the m-companion structure or configuration matrix.

To study how to generate M companion matrix, article [24] explains the algorithm with $\mathcal{O}(n^4)$ time complexity.

In this article, every operation is performed over Galois Field(GF(2)), $\oplus$ is used for XOR between two integers, $<<$ sign is used for left shift operator,$||$ is used for bitwise OR operator, $\boxplus$ is used for addition of two integers modulo 2^{64}, F_2^n signifies n dimensional vector over GF(2), $F_2^{n\times n}$ represents a matrix of dimension n over GF(2), and F_{2^8} is a finite fields with 2^8 elements over F_2.

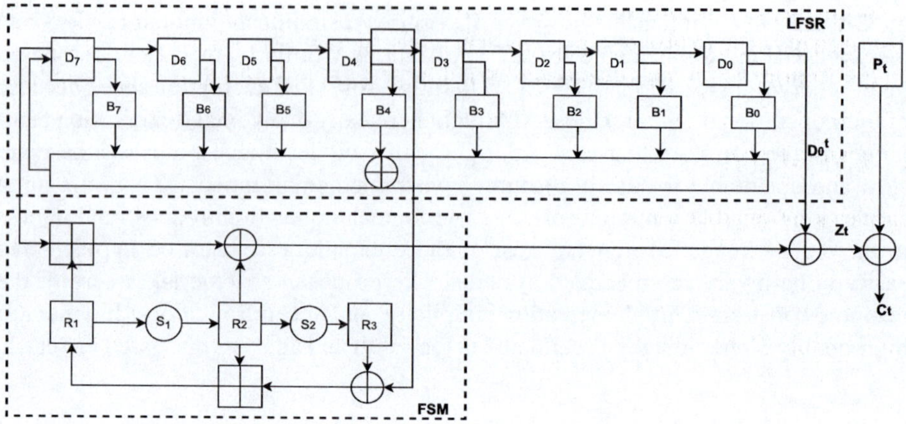

Fig. 1 Modified SNOW 3G

4 Proposed Model

The modified SNOW 3G algorithm comprises a σ—Linear Feedback Shift Register (LFSR) and nonlinear Finite State Machine(FSM) used in SNOW 3G. The σ—LFSR has $b = 8$ delay blocks($D_i | i \in [8]$) of each is of size $m = 64$ bit. Figure 1 describes the sketch of the proposed scheme.

The LFSR part of Fig. 1 has a feedback polynomial which has 8 gain matrices ($B_i \in F_2^{64\times 64}$). The feedback polynomial $x^8 + B_7x^7 + B_6x^6 + \cdots + B_1x + B_0$ over matrix ring $M_{64}(F_2)$is generated from the primitive polynomial $x^{512} + x^{419} + x^{321} + x^{125} + 1$ over F_2. In these contexts, we use the configuration matrix generation algorithm proposed by [24] to find the feedback polynomial of the σ—LFSR. Next, we describe the driving equation of σ—LFSR and Finite State Machine (FSM).

$$D_7^{t+1} = \begin{cases} \sum_{i=0}^{7} B_i D_i^t \oplus F^t & \text{if } t <= 32, \\ \sum_{i=0}^{7} B_i D_i^t, & \text{otherwise.} \end{cases} \quad (3)$$

In the above equation D_i^t is the value ith delay block at tth time stamp and F^t is output of an FSM at tth time stamp which is described below.

The FSM part of SNOW 3G consists of three registers R_1^t, R_2^t, $R_3^t \in F_2^{64}$ and two substitution boxes [5] $S1$ and $S2$. We use 8 parallel S-BOXes (composed of AES Subbyte and Mixcolumn operation) to compute each S-BOX function.

The registers R_1, R_2, R_3 in the KSG are updated as follows:

$$R_3^{t+1} = S2(R_2^t). \tag{4}$$

$$R_2^{t+1} = S1(R_1^t). \tag{5}$$

$$R_1^{t+1} = R_2^t \boxplus (R_3^t \oplus D_5^t), \tag{6}$$

where $\boxplus$ is integer addition modulo 2^{64} and $\oplus$ is vector wise XOR operation.

Proposed Algorithm of Modified SNOW 3G for image encryption and decryption In this section, we explain the various functions of the modified SNOW 3G algorithm and use those algorithms for image encryption and decryption algorithm. The main algorithm can be broken into three parts such as $INITIALIZATION()$, $LFSRUPDATE()$, and $FSMUPDATE()$ function. Each of them is described as follows:

Algorithm 1 Modified SNOW 3G Initialization Process

procedure INITIALIZATION(K=(K_0, K_1, K_2, K_3),IV=(IV_0, IV_1, IV_2, IV_3))
  $D_7^0 \leftarrow (K3 \oplus IV_0) << 32||(K_2)$
  $D_6^0 \leftarrow K_1 << 32||(K_0 \oplus IV_1)$
  $D_5^0 \leftarrow (K_3 \oplus 1) << 32||(K_2 \oplus 1 \oplus IV_2)$
  $D_4^0 \leftarrow (K_1 \oplus 1 \oplus IV_3) << 32||(K_0 \oplus 1)$
  $D_3^0 \leftarrow (K_3 << 32)||(K_2)$
  $D_2^0 \leftarrow (K_1 << 32)||(K_0)$
  $D_1^0 \leftarrow (K_3 \oplus 1) << 32||(K_2 \oplus 1)$
  $D_0^0 \leftarrow (K_1 \oplus 1) << 32||(K_0 \oplus 1)$
  $R_3^0, R_2^0, R_1^0 \leftarrow 0, 0, 0$
  $t \leftarrow 0$
  while $t <= 32$ **do**
    $LFSRUPDATE()$
    $FSMUPDATE()$
    $F^t \leftarrow R_1^{t+1} \oplus Rt2 \boxplus (R_3^t \oplus D_5^t)$
    $D_7^{t+1} \leftarrow \sum_{i=0}^{7} D_i^t B_i \oplus F^t$
  end while
end procedure

Algorithm 1 encompasses the initialization of the KSG. It takes key $K \in F_2^{128}$ which is broken into 4 sub-keys K_0, K_1, K_2, K_3, each $K_i \in F_2^{32}$. It also takes an initialization vector IV $\in F_2^{128}$, which must be changed after each communication. The IV is also divided into 4 sub-parts ($\text{IV}_0, \text{IV}_1, \text{IV}_2, \text{IV}_3$) of size 32 bit each. Here, we initialize the delay blocks of the σ—LFSR by K/IV combination. 1 represents the value $2^{32} - 1$. Keystream runs for 32 clock cycles and is not accessible to the adversary. This procedure adds more entropy to the key than the initial.

Algorithm 2 Modified SNOW 3G LFSR Update Function

1: **procedure** LFSRUPDATE( )
2: $State = (D_7^0 << 448||D_6^0 << 384||D_5^0 << 320||D_4^0 << 256||D_3^0 << 192||D_2^0 << 128||D_1^0 << 64||D_0^0)$
3: $temp \leftarrow \sum_{i=0}^{7} B_i D_i^t$
4: $State \leftarrow State << 64||temp$
5: **end procedure**

The Algorithm 2 presents the state updation of σ—LFSR with the help of delay blocks $D_i \in F_2^{64}$ and gain matrices $B_i \in F_2^{64\times 64}$. $||$ and $<<$ operators are used as bitwise OR and AND operators. Line number 2 in the algorithm states that the state of the LFSR (size 512) is formed using the values of delay blocks D_i. Line numbers 3 and 4 explain the 64-bit updation of the state of the LFSR and the 64-bit left shift of the state value of σ—LFSR. The state of the σ—LFSR is updated in each clock pulse if we call the LFSRUPDATE() function of the KSG.

Algorithm 3 Modified SNOW 3G FSM Update Function

Require: Value of delay blocks D_{15}^t, D_5^t.
Ensure: Output of FSM F_t
procedure FSMUPDATE( )
  while $t \neq I_1$ **do**
    $F_t \leftarrow (R_1^t \boxplus D_{15}^t) \oplus R_2^t$
    $R_3^{t+1} \leftarrow S2(R_2^t)$
    $R_2^{t+1} \leftarrow S1(R_1^t)$
    $R_1^{t+1} \leftarrow Rt2 \boxplus (R_3^t \oplus D_5^t)$
    $R_1^t \leftarrow R_1^{t+1}$
    $R_2^t \leftarrow R_2^{t+1}$
    $R_3^t \leftarrow R_3^{t+1}$
  end while
end procedure

Algorithm 3 represents the nonlinear part of the key stream generator. It uses three nonlinear functions $\boxplus$, nonlinear SBOX $S1$ and $S2$(function $F_2^{64} \rightarrow F_2^{64}$) which is used in between three registers R_1, R_2, R_3. How each register $R_i \in F_2^{64}$ for $i \in [3]$ is updated is given in the equations. The Algorithm outputs $F^t \in F_2^{64}$ in each clock pulse.

Algorithm 4 Image Encryption Algorithm

Require: Plaintext block of Image P_t, Key K, Initialization vector IV
Ensure: One block of Ciphertext C_t.

procedure ENCRYPT(P_t, K, IV) ▷ P_t is the image pixel of size 64 bit.
    Input : Key/IV for the KSG, Plaintext P_t.
    Find size of the image file(I_1)=m.
    Initialization(K, IV)
    $t \leftarrow 0$
    while $t \neq m$ **do**
        $Z_t = D_0^t \oplus F^t$
        $C_t = Z_t \oplus P_t$
        LFSRUPDATE()
        FSMUPDATE()
    end while
end procedure

Algorithm 5 Image Decryption Algorithm

Require: Ciphertext block of Encrypted Image C_t, Key K, Initialization vector IV
Ensure: One block of plaintext P_t.

procedure ENCRYPT(C_t, K, IV) ▷ C_t is the Cipher image pixel of size 64.
    Input : Key/IV for the KSG, Ciphertext C_t.
    Find size of the image file(I_1)=m.
    Initialization(K, IV)
    $t \leftarrow 0$
    while $t \neq m$ **do**
        $Z_t = D_0^t \oplus (R_1^t \boxplus D_{15}^t) \oplus R_2^t$
        $P_t = Z_t \oplus C_t$
        LFSRUPDATE()
        FSMUPDATE()
    end while
end procedure

Algorithms 4 and Algorithm 5 are encryption and decryption algorithms that use modified SNOW 3G. To run the Algorithm, we need to break the image file into 64-bit plaintext. Besides, the key K and the initialization vector IV are also used to run the keystream. Here, Z_t is the key generated from the KSG, and it is exored with the plaintext and the ciphertext as in Algorithms 4 and 5.

5 Experimental Result and Analysis

On an Intel (R) Core (TM) i5-3230M 2.60 GHz CPU, 8 GB RAM, and WINDOWS 10 operating system, the proposed scheme was tested. The comparison with known algorithms such as AES [25], DES [26], 3-DES [27], and the scheme proposed by Babaei et al. [17] is carried out using Python language. The original images used were Lena, Baboon, and Cameraman. The original images and the corresponding cipher images are shown in Figs. 2 and 3, respectively.

(a) Lena (b) Cameraman (c) Baboon

Fig. 2 Original images

(a) Enc_Lena (b) Enc_Cam (c) Enc_Baboon

Fig. 3 Encrypted images

5.1 Performance Analysis

In this section, a detailed discussion is presented regarding various performance analysis of the proposed model.

Analysis of Histogram The histogram plots of various plain and cipher images are shown in Fig. 4. The distribution of the cipher pictures is relatively uniform, as seen from the histogram. It adequately justifies the suggested scheme's security.

Correlation Coefficient Analysis Table 1 displays the exact values in each instance for both plain and encrypted images.

NIST Randomness Tests Various NIST-recommended randomness tests [28] were carried out to ensure that the suggested system is random. The test results for Lena's cipher image are provided in Table 2. Table 2 shows that all *p*-values are more significant than the decision threshold value.

Comparative Analysis Here, we present a detailed comparative analysis of the proposed technique with the state-of-the-art techniques. The comparison has been presented against various standard metrics as given in Table 3.

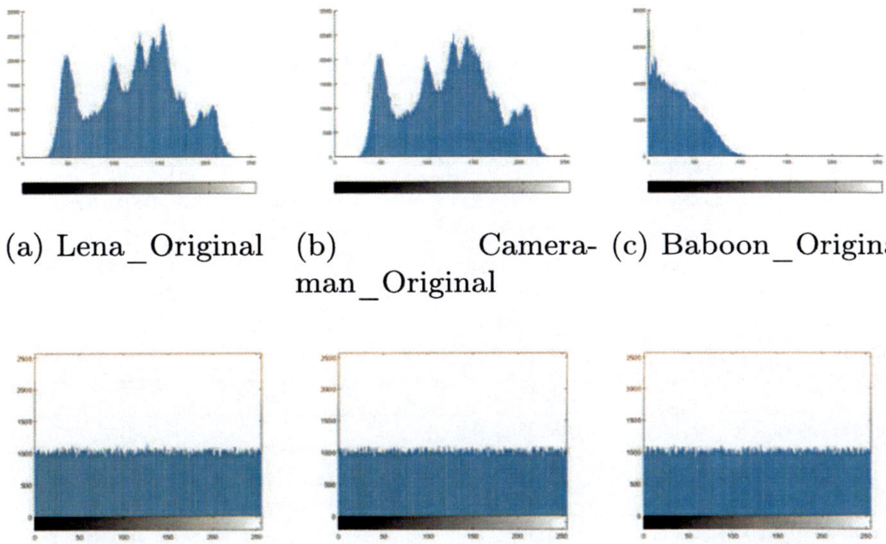

(a) Lena_Original (b) Cameraman_Original (c) Baboon_Original

(d) Enc_Lena (e) Enc_Cameraman (f) Enc_Baboon

Fig. 4 Original and encrypted images' histogram plots

6 Conclusion

This article presents an image encryption based on σ—LFSR oriented stream cipher SNOW 3G with a modified block size of 64. This system generates cipher images as random images, negating the effects of assaults on the Fog nodes' insecure communication route between the perception layer and the network layer. The proposed approach can be used in critical real-world settings in which raw data is crucial, for instance, in health care, military, and biological image communication, to name a few. Besides, efficient vector-matrix multiplication in our scheme Sect. 4 reduces the probability of cache timing attacks. The scheme ensures the delivery of raw data from the sensors to the recipient in all circumstances. The proposed approach also has a low time complexity, which will help to speed up communication. A cache-efficient algorithm with a 128-bit keystream will be more efficient and help extend the sensor's battery life by reducing the number of matrix-vector multiplication.

Table 1 Values of correlation coefficients for different images

Image	Type	Horizontal	Vertical	Diagonal
Lena	Original	0.973321	0.986331	0.966332
	Encrypted	0.012859	0.040757	−0.013456
Cameraman	Original	0.936282	0.919205	0.865661
	Encrypted	0.011704	0.023504	−0.041105
Baboon	Original	0.878303	0.758934	0.732143
	Encrypted	−0.004029	0.015977	0.009952

Table 2 Result of NIST randomness tests

Test name	*P*-value	Result
Longest run of 1's	0.198533	✓
Frequency	0.88342	✓
Block frequency	0.194323	✓
The binary matrix rank	0.182659	✓
Cumulative sum	0.976602	✓
Runs	0.195945	✓
Non-overlapping template matching	0.990263	✓
Discrete fourier (spectral)	0.049341	✓
Maurer's universal statistical	0.992466	✓
Linear complexity	0.120487	✓
Serial	0.130323	✓
Approximate entropy	0.988676	✓

Table 3 Comparison of proposed scheme with state-of-the-art techniques

Technique	Images	Information entropy	NPCR	UACI	MSE	PSNR
Ref [29]	Lena	7.9919	99.6225	33.4293	$1.485\ 10^{-5}$	96.51
	Cameraman	–	–	–	–	–
	Baboon	7.9879	99.7258	32.9826	$1.526\ 10^{-5}$	96.45
Ref [30]	Lena	7.9993	99.6347	33.4653	$1.446\ 10^{-5}$	96.53
	Cameraman	–	–	–	–	–
	Baboon	7.9983	99.7384	33.4105	$1.521\ 10^{-5}$	96.31
Ref [31]	Lena	7.9976	99.6002	33.4592	11143.4939	7.6606
	Cameraman	7.9971	99.6078	33.4026	9460.8759	8.3715
	Baboon	–	–	–	–	–
Proposed scheme	Lena	7.9993	99.78634	33.85143	113.5879	9.2713
	Cameraman	7.9994	99.71278	33.78347	107.6587	8.4415
	Baboon	7.9994	99.91445	33.98965	117.8115	9.7820

References

1. Ekdahl P, Johansson T (2002) A new version of the stream cipher snow. In: International workshop on selected areas in cryptography. Springer, pp 47–61
2. Ekdahl P, Johansson T, Maximov A, Yang J (2019) A new snow stream cipher called snow-v. IACR Trans Symmet Crypto 1–42
3. Kitsos P, Sklavos N, Skodras AN (2011) An FPGA implementation of the ZUC stream cipher. In: 2011 14th Euromicro conference on digital system design. IEEE, pp 814–817
4. Nandi S, Krishnaswamy S, Mitra P (2022) Recent results on some word oriented stream ciphers: snow 1.0, snow 2.0 and snow 3g
5. Orhanou G, El Hajji S, Bentaleb Y (2010) Snow 3g stream cipher operation and complexity study. Contemp Eng Sci-Hikari Ltd 3(3):97–111
6. Krishnaswamy S, Pillai HK (2011) On the number of linear feedback shift registers with a special structure. IEEE Trans Information Theo 58(3):1783–1790
7. Krishnaswamy S, Pillai HK (2012) On multisequences and their extensions. arXiv preprint arXiv:1208.4501
8. Leander G, Zenner E, Hawkes P (2009) Cache timing analysis of IFSR-based stream ciphers. In: IMA international conference on cryptography and coding. Springer, pp 433–445
9. Brumley BB, Hakala RM, Nyberg K, Sovio S (2010) Consecutive s-box lookups: a timing attack on snow 3g. In: International conference on information and communications security. Springer, pp 171–185
10. Kaur M, Kumar V (2020) A comprehensive review on image encryption techniques. Arch Comput Meth Eng 27(1):15–43
11. Kumar A, Dua M (2021) Novel pseudo random key & cosine transformed chaotic maps based satellite image encryption. In: Multimedia tools and applications, pp 1–21
12. Tarasova VV, Tarasov VE (2017) Logistic map with memory from economic model. Chaos, Solitons Fract 95:84–91
13. Phatak S, Rao SS (1995) Logistic map: a possible random-number generator. Phys Rev E 51(4):3670
14. Parah SA, Loan NA, Shah AA, Sheikh JA, Bhat G (2018) A new secure and robust watermarking technique based on logistic map and modification of dc coefficient. Nonlinear Dyn 93(4):1933–1951
15. Hua Z, Zhou Y, Huang H (2019) Cosine-transform-based chaotic system for image encryption. Inf Sci 480:403–419
16. Malik MA, Bashir Z, Iqbal N, Imtiaz MA (2020) Color image encryption algorithm based on hyper-chaos and DNA computing. IEEE Access 8:88093–88107
17. Babaei M (2013) A novel text and image encryption method based on chaos theory and DNA computing. Nat comput 12(1):101–107
18. Zhang Y (2018) The image encryption algorithm based on chaos and DNA computing. Multimedia Tools Appl 77(16):21589–21615
19. El-Shafai W, Khallaf F, El-Rabaie E-SM, El-Samie FEA (2021) Robust medical image encryption based on DNA-chaos cryptosystem for secure telemedicine and healthcare applications. J Ambient Intell Human Comput 12(10):9007–9035
20. Enayatifar R, Abdullah AH, Isnin IF (2014) Chaos-based image encryption using a hybrid genetic algorithm and a DNA sequence. Opt Lasers Eng 56:83–93
21. Abdullah AH, Enayatifar R, Lee M (2012) A hybrid genetic algorithm and chaotic function model for image encryption. AEU-Int J Electron Commun 66(10):806–816
22. Nandi S, Krishnaswamy S, Zolfaghari B, Mitra P (2022) Key-dependent feedback configuration matrix of primitive σ-IFSR and resistance to some known plaintext attacks. IEEE Access
23. Zeng G, Han W, He K (2007) High efficiency feedback shift register: sigma-IFSR. IACR Cryptol ePrint Arch 2007:114
24. Krishnaswamy S, Pillai HK (2014) On the number of special feedback configurations in linear modular systems. Syst Control Lett 66:28–34

25. Zhang X, Parhi KK (2004) High-speed VLSI architectures for the AES algorithm. In: IEEE transactions on very large scale integration (VLSI) systems, vol 12, no 9, pp 957–967
26. Nie T, Zhang T (2009) A study of des and blowfish encryption algorithm. In: Tencon 2009-2009 IEEE region 10 conference. IEEE, pp 1–4
27. Ardiansyah G, Sari CA, Rachmawanto EH, et al (2017) Hybrid method using 3-DES, DWT and ISH for secure image steganography algorithm. In: 2017 2nd international conferences on information technology, information systems and electrical engineering (ICITISEE). IEEE pp 249–254
28. Rukhin A, Soto J, Nechvatal J, Smid M, Barker E (2001) A statistical test suite for random and pseudorandom number generators for cryptographic applications. In: Technical report booz-allen hamilton Inc mclean va
29. Shrivastava M, Roy S, Kumar K, Pandey CV, Grover J (2021) LICCA: a lightweight image cipher using 3-d cellular automata. Nonlinear Dyn 106:2679–2702
30. Roy S, Shrivastava M, Rawat U, Pandey CV, Nayak SK (2021) IESCA: An efficient image encryption scheme using 2-d cellular automata. J Inf Secur Appl 61:102919
31. Wang X, Guan N (2020) A novel chaotic image encryption algorithm based on extended zigzag confusion and RNA operation. Opt Laser Technol 131:106366

Detection of Deepfakes in Financial Transactions Using Algorand Blockchain Consensus Mechanism

S. Anitha, N. Anitha, N. Ashok, T. Daranya, B. Nandhini, and V. Chandrasekaran

1 Introduction

Deepfakes are synthetic media [1] in which a person's likeness is replaced in an existing photograph or video. Deepfakes have received a lot of attention for their involvement in financial fraud, hoaxes, fake news, revenge porn, celebrity pornography, and false news. Algorand [2] is the world's first Pure Proof of Stake (PPoS) blockchain, which offers scalability, decentralization, and security in a sustainable and environmentally friendly manner. In comparison to its predecessor Proof of Stake (PoS), Pure Proof of Stake is based on Byzantine consensus and has a more democratic posture.

1.1 The Need for Consensus in Distributed Networks

The key characteristic of a blockchain is distributed consensus, which guarantees the network's immutability and decentralization. The node, which can be any server computer, is the key element of a public blockchain network. It contains both

S. Anitha (✉) · N. Ashok · T. Daranya · B. Nandhini
Department of Information Technology, Kongu Engineering College, Erode, Tamilnadu, India
e-mail: anitha4ciet@gmail.com

N. Anitha
Head-Talent Development, 1.Ms.N.Anitha, Head-TalPixelExpert Technology & Services Pvt. Ltd., Chennai, India
e-mail: cbmkkec@gmail.com

V. Chandrasekaran
Department of Medical Electronics, Velalar College of Engineering and Technology, Erode, Tamilnadu, India

J. K. Mandal et al. (eds.), *Proceedings of International Conference on Network Security and Blockchain Technology*, Lecture Notes in Networks and Systems 738,
https://doi.org/10.1007/978-981-99-4433-0_15

the chain's software and its continuously updated history of records. In addition to decentralization, each node simultaneously updates the ledger to include new transactions.

A set of established rules that would organize the nodes taking part in block formation is necessary for a distributed system to operate properly while retaining decentralization. This is achieved using consensus protocol. A blockchain's blocks are always linked in chronological sequence to prevent any transaction from ever being changed or canceled. The chain's growth is unidirectional; therefore, adding new blocks is the only way to update a blockchain. Every node on open networks has the ability to add new blocks, and the consensus algorithm determines which nodes act as "miners" or validators. The consensus process is essential because it guards against malevolent nodes that might attempt to conduct fraudulent transactions, blatantly breach the rules, or plan Distributed Denial of Service (DDoS) assaults.

To summarize, the following goals need the use of a consensus mechanism:

- To maintain network decentralization while ensuring everyone has an equal chance to create new blocks.
- To prevent malicious activity and hacker attacks on the network.
- To ensure blockchain contains only accurate records of all transactions.
- To appropriately reward block developers in accordance with the established rules.

1.2 *The Blockchain Challenge*

PoS is more advanced than Bitcoin's Proof of Work (PoW) technology in terms of scalability [3] and transaction speed. This still leaves open the so-called "Blockchain Trilemma" which asserts that only two of a distributed ledger technology's three main purposes may be achieved. Security, decentralization, and scalability are the three essential components, and frequently, one of them is sacrificed in favor of the other two. Each of the three requirements can be satisfied by the solution provided by Algorand. One of the most important developments in the blockchain industry (PPoS) may be Pure Proof of Stake.

1.3 *Proof of Stake*

Based on the market value of the blockchain networks that employ it, PoS is the second most popular consensus algorithm after Proof of Work (PoW). Proof of Work is a consensus algorithm that requires a participant node to prove that the work done and submitted by it is accurate, therefore giving them the right to add more transactions to the blockchain.

PoS posed a challenge to the earliest consensus mechanism, PoW, which was originally introduced with Bitcoin. Based on the market value of the blockchain networks that employ it, PoS is the second most popular consensus algorithm after

Proof of Work (PoW). Additionally, the computer power utilized by miners in PoW systems result in huge energy requirements.

The PoS algorithm was first mentioned on Bitcointalk in 2011. The consensus algorithm has swiftly become widespread across all its forms, even though Peercoin was the first to deploy it in conjunction with PoW. Blockchains running on PoS are maintained by validators, as opposed to PoW networks, which are dependent on miners. In PoS, the validation procedure is known as "forging". A node only has to stake the native token if it wants to take part in the block creation process. Neither money for power nor specialist gear is required. Although PoS networks have different regulations in each instance, the fundamental idea remains the same: nodes wishing to become validators must store a certain minimum number of tokens as collateral.

Each block is limited to 30 million gas (2*target block size = 2*15 million gas). Following that, block sizes will change based on network demands. The network's space and speed requirements will be impacted by huge blocks. More computer power is needed to process larger blocks in time for the next slot.

1.4 Pure Proof of Stake: Silvio Micali's Invention

Even though PoS blockchains had overcome scalability issues, stakers with substantial token holdings can still control them, which has a detrimental effect on decentralization. To overcome the most significant issues with PoS networks, Algorand developed the so-called Pure Proof of Stake (PPoS). The "rich grow richer" issue that plagues most PoS networks is addressed by the PPoS protocol, which offers a more fair solution. Algorand is the first application of the PPoS consensus process.

The system can choose any online user, which has various benefits for the distributed network. First of all, it reinforces the decentralized structure of the blockchain and guarantees system security. Algorand's goal was to create a scalable blockchain system that would meet contemporary needs and address the Blockchain Trilemma. Financial transactions take time and are expensive. Transaction fees of all kinds result in an annual loss of about $5 trillion, with no benefit to end users.

1.5 Comparison of Proof of Stake Variants

In Proof of Stake, based on the number of tokens (stake) a user bet is eligible for participating in the validation process. Combining PoW and PoS in Peercoin also does not meet the security, scalability, and decentralization altogether [4].

Delegates' Proof of Stake (DPoS) has a fixed number of elected nodes known as delegates. They oversee building blocks and are picked on a round-robin basis. Owners of tokens cast votes for the delegates; the number of tokens they own directly determines the number of votes they can cast. The Blockchain Trilemma is not solved, despite the system's strong transaction throughput performance achieved at the cost of decentralization. The PPoS protocol, in contrast, does not have a smaller number of chosen participants; however, regardless of their stake, the algorithm randomly selects validators. Because there isn't a single group of validators for the system, it is difficult for hackers to hack it.

Delegates are absent from BPoS-based blockchains, but users must lock up a specific portion of their tokens in order to stake them and have details of how blocks are built. To become validators, they must lock up their share for a predetermined period of time. Richer participants gain from their stake-related voting power as well. Therefore, wealthier participants profit from both DPoS and BPoS.

The PPoS mechanism, in contrast to the BPoS method, does not need users to lock up any tokens to take part in the block-building process. Users' capacity to use or transfer their stake is unaffected by their participation in the consensus method, ensuring scalability and transaction speed. PPoS systems are more decentralized since validators are chosen at random by the system [5].

1.6 Algorand: PPoS Blockchain

Algorand is a blockchain platform based on a Pure Proof of Stake (PPoS) consensus mechanism providing security, decentralization, and scalability. It is an open-source ledger used as a payment system employing a Byzantine Agreement protocol for message-passing. Users are randomly and secretly selected to both propose blocks and vote on block proposals so that the security of the system is ensured.

Every online user can vote and be chosen to propose. The stake of a user directly affects the likelihood that it will be selected. The amount of tokens each user has in relation to the total number of tokens in the system determines their stake in the system.

Algorand is completely a permissionless blockchain in which anyone can join the network and take part in the protocol without the need for authorization from any reputable authority. Anyone can utilize the blockchain to conduct transactions, take part in the creation of new blocks, view every block, and add their own transactions to upcoming blocks.

Due to the system's reliance on dependable technology, anyone can create decentralized applications (DApps) and issue NFTs on Algorand. The Algorand chain can handle any need for high transaction throughput without having to worry about congestion. Because blocks are completed in a matter of seconds, it may compete with international payment and banking networks in terms of transaction speed. As the first blockchain to provide immediate transaction finality, the PPoS-powered network is used for frictionless banking. Algorand selects the people who will contribute the following.

The evaluation of user-submitted block proposals is done by a committee of voters. A block is deemed legitimate if the vast majority of votes come from individuals who are being completely honest. For insurance, the Internet of Things [6], public safety, financial services, and news media sectors, Attestiv offers tamper-proof media validation platforms and goods. By confirming the legitimacy of digital media and data, Attestiv assists businesses in establishing effective operations, enhancing the customer experience, and offering the highest quality for information transfer. Attestiv uses blockchain and artificial intelligence to guarantee the authenticity of digital images shot by any person or device. Attestiv assists companies of all sizes in establishing their brands and gaining the trust of their customers by offering new services, cost reductions, and fraud prevention.

A revolution in the insurance sector that will enhance automation, client delight, and trust is about, to begin with the merging of Attestiv and Algorand. By initially utilizing the immutability, scale, and sharing attributes of the Algorand distributed ledger, insurance stakeholders like insured, agencies, adjusters, carriers, repair vendors, and public safety can benefit from a single system that validates assets across parties, removing pointless redundancies and lowering the risk of fraud. Smart contracts could someday transform the business by bringing even more automation and efficiency.

Verifiable Random Function (VRF) creates a pseudorandom output with a proof that anyone may use to validate the outcome using a secret key and a value. It is comparable to a weighted lottery where accounts with more Algos have a higher likelihood of proposing and confirming blocks. The use of numerous accounts is not advantageous while using VRF. An anonymous group of users can discreetly choose themselves to take part in Algorand's consensus mechanism via cryptographic sortition (Anonymity).

The Protocol Structure of Algorand is,

- Transactions

  Transfer value are validated by all participating nodes.

- Blocks

  Group of transactions that have been validated by the consensus algorithm.

- Consensus

  Participating users validate the block in accordance with the rules of the Algorand protocol and also compensate the users.

The objectives of the paper are as follows:

- To create a highly scalable, decentralized, and secure blockchain for financial transactions.
- To increase transaction throughput while identifying malicious insurers.
- To detect deepfakes with less resource consumption than other consensus algorithms like Bitcoin, Ethereum, Ripple, and Zcash.

2 Literature Survey

PoW [7] eliminates the problems of double spending and fork creation by allowing a group of peers to agree on a single ledger state. PoW failed to meet the demand for higher throughput. PoW consumes a lot of energy and processing resources.

PoS [8] eliminate the need for the nodes to compete with one another in order to solve a mathematical puzzle, making the process more effective and energy-efficient. Although PoS are scalable, stakeholders own many tokens, which has a detrimental effect on decentralization.

A node that has been elected will be promptly removed and replaced if it misbehaves or performs poorly. Since token holders vote for delegates, the quantity of tokens they own directly affects how many votes they can cast. As a result, only more token holders will receive opportunities [9]. Blockchains built on the BPoS protocol need delegates. To stake tokens and have an impact on the block generation process, users must lock up a minimum number of tokens [10].

Because the lone blockchain project, Vault, which uses the Algorand infrastructure to move assets, is still in the Beta stage, no statistics on the overall situation are now accessible. More research will be done in the future and the reader will have access to more data. Algorand was one of the initial to define and use cryptographic sortition. Using sub-quadratic communication and nearly perfect resilience, it resolves BA WHP [11–14].

The erasure model uses VRFs to sample committees and makes use of a trusted public key infrastructure (PKI). They presented their initial research as a synchronous model, with a solution presuming eventual synchrony. However, they both rely on the temporal presumption to move forward. The majority of financial transactions are based on conventional contractual agreements, which frequently entail paperwork stating the terms between the parties and frequently involve a third party to verify the fulfillment of the intricate provisions to be implemented. As a fundamental blockchain capability that enables contemporary, seamless transaction processes and offers end users more control with fewer middlemen, smart contracts will eventually take the place of these antiquated traditional agreements.

The fact that Algorand's Smart Contracts are executed on a tamper-proof (trustless), secure, decentralized network without forking and provide quicker, error-free, immutable applications that are cost-effective for financial companies makes them extremely reliable [15].

FinTech newcomers like Neo and Challenger Banks, who are revolutionizing how consumers and institutions bank, trade, finance, and provide payments, are debuting with the benefit of cheap overhead that they can pass on to their clients. Consumers, companies, and institutions now have more options, and they also have higher expectations for quick, effective, and dependable services [16]. Therefore, in order to improve customer experiences and maintain market value, incumbent financial institutions must reconsider their business practices and upgrade antiquated technology.

With the proliferation of deepfakes as the technology and skill needed to create fake images and video so good that the human eye can't detect them gets easier to acquire. Attestiv [17] believes its approach can stop "tampered media" from entering the systems of an insurer in the first place. On the open blockchain of Algorand, tamper-proof validation platform Attestiv now intends to develop new tools [18]. Digital media, such as pictures, videos, and documents, can be validated using artificial intelligence using Attestiv's core technology, either at the time of capture or throughout a forensic investigation.

The summary of the existing works are as follows:

- Using PoW, financial transaction takes more power and verification time therefore cost increases.
- Using PoS, malicious insurers increases as it takes into account only stake for validating the transaction.
- Using DPoS, delegates must be trustful, otherwise decentralization misleads.
- Using BPoS, locking stake may lead to bias the validation process and throughput decreases.

3 Proposed System

As long as non-malevolent users have the majority of the stake, Algorand may tolerate malicious users and reach consensus decentralized. Any desire for high transaction throughput can be met by the Algorand chain without concern for congestion. It can compete with international payment and financial networks in terms of transaction speed because blocks are finalized in a matter of seconds.

The aim of the proposed system is to detect deepfakes using Algorand because other consensus algorithms like Bitcoin, Ethereum, Ripple, and Zcash have technical limitations such as resource wastage [19], utilization of high power, vulnerable to attacks, low scalability and ambiguity. The main key idea of Algorand algorithm is that achieving consensus through Byzantine agreement protocol and use gossip protocol for communication. Next, the key assumption in Algorand algorithm is honest majority of money. Moreover, some of the technical advancements like trivial computation, true decentralization, finality of payment, scalability and security are added in Algorand algorithm. Similarly, mechanism is resilient to Sybil and denial of service attacks. Hence, the proposed system used Algorand algorithm to detect deepfakes in financial transactions.

3.1 Pseudocode of Algorand Algorithm

Algorand: A secure and efficient distributed ledge

Begin

1. Select a random user
   1.1 prepare a block
   1.2 propagate block through gossiping
2. Select random committee with small number of users (~10 k)
   2.1 run Byzantine agreement on the block
   2.2 digitally sign the result
   2.3 propagate digital signatures

End

4 Results and Discussion

A working version of Algorand is developed in C++. For networking, Boost ASIO library is utilized and SHA-256 hash algorithm is used. Each user connects to four random peers using the indicated way, accepts connections from more peers, and engages in gossip with each of them. As a result, an average of 8 peers is obtained. Each user receives an "address book" file including their public key's IP address and port information. Users might either publish this knowledge through public bulletin board or engage in gossip about it in a real-world application. Parameters used in source code are mentioned in the Table 1.

The communication costs are determined by the expected committee size and the number of block proposers. As more users join the gossip network, message spreads more slowly. In Algorand's gossip network, each user is connected to a random peer, resulting in a random network graph. Users who are already coped up with the current ledger speed are not obliged to use the block history and matching certificates.

In order to support N shards, users store blocks or certificates whose round number is equal to their public key modulo N. In contrast to PoW and PoS, which require 120 and 10 MB of storage capacity respectively, PPoS requires only 130 KB of storage for every user to verify 1 MB of data.

Table 1 Parameter and its meaning

Parameter	Meaning
h	Honest users
$\lambda_{proposer}$	Block
λ_{step}	Committee members
$\lambda_{stepvar}$	BA* completion time variance estimation
$\lambda_{priority}$	Time to gossip sortition proofs
λ_{step}	Timeout for BA* step
λ_{block}	Timeout for receiving a block

Figure 1 shows the latency for one round with varying percentage of malicious insurers among 10,000 insurers. When the number of malicious insurers is 3000, latency increases due to increased verification steps (hard vote) and becomes stable.

Algorand's agreement time (BA), according to the results, is unaffected by block size and remain in the same (12 s) even for enormous blocks. To further increase throughput, the final step, which lasts about 6 s, can be pipelined with the next Algorand round. Due to the fixed execution time for BA and the linear growth in block propagation time (with block size), increasing block size enables one to commit more data in exchange for the fixed execution time for BA, thus maximizing throughput and network capability. Figure 2 shows this effect.

Block and transaction verification requires less computational work when using Proof of Stake. Because the machines of coin owners are used to verify the blocks, less computation is required (Table 2).

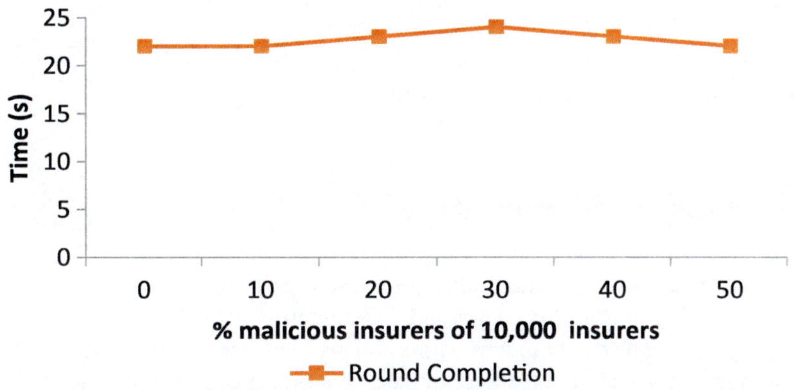

Fig. 1 Algorithm latency for a single round with 10,000 players

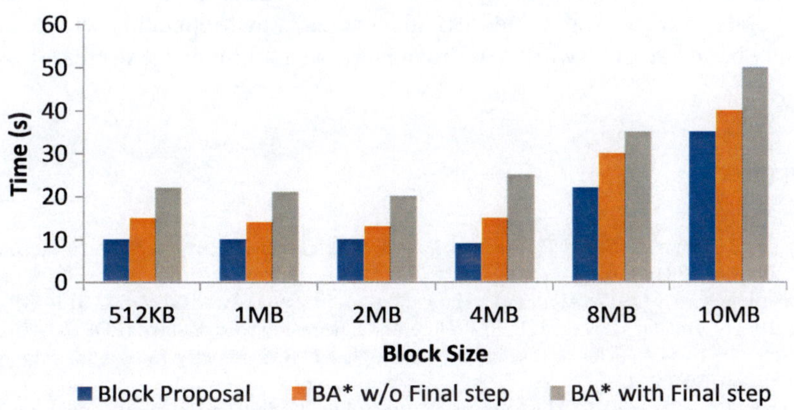

Fig. 2 Changes in latency for one round with different block size

Table 2 Consensus mechanisms and its parameter values

Consensus mechanism/ parameters	Transaction confirmation (finality)	Average time to generate block	Tolerated power of adversary	Block verification speed	Transactions per second (throughput)	Average energy consumption
Proof of work	8 min	10 min	< 25% of computing power	> 100 s	7–30	130 TWh (Bitcoin) 26 TWh (Ethereum)
Proof of stake	6 min	64 s	< 51% of stake	< 100 s	40	0.01 TWh
Delegated proof of stake	2 min	3 s	< 51% of validators	< 100 s	257	0.001 TWh
Leased proof of stake	1 min	2 s	< 51% of leasers	< 50 s	535	0.0015 TWh
Pure proof of stake	20 s	0.5 s	< 51% of committee size	< 25 s	875	0.0006 TWh

5 Conclusion and Future Work

New cryptocurrency, Algorand confirms transactions in less than a minute and has a very low forking probability. Algorand's architecture is based on the BA protocol and cryptographic sortition method. Algorand protects itself against targeted attacks by switching out participants at every phase. Algorand is a brilliant Proof of Stake blockchain platform, and see a lot of potential in using it as a framework for research purpose into a new consensus mechanism [20].

In the future, instead of an Algo-based financial stake, the stake would be reputation points. Another way to address adversaries is that reputation will be earned gradually but decreased swiftly (Additive increase, multiplicative decrease).

References

1. Marco S, Timur S, Bernd R, Damian B, Adversarial learning of deepfakes in accounting. arXiv:1910.03810 [cs.LG]
2. Chen J, Micali S (2017) Algorand technical report. https://arxiv.org/abs/1607.01341v9
3. Bachani V, Bhattacharjya A (2023) Preferential delegated proof of stake (PDPoS)—modified DPoS with two layers towards scalability and higher TPS. Symmetry 15(4). https://doi.org/10.3390/sym15010004
4. Kumar HU, RPSG (2019) Algorand: a better distributed ledger. In: 1st international conference on advances in information technology (ICAIT), pp 496–499. https://doi.org/10.1109/ICAIT47043.2019.8987305

5. Cristian L, Michela C, Andrea V, Udai Pratap R, Arvindbhai SK, Luca Z (2020) A survey on blockchain consensus with a performance comparison of PoW, PoS and pure PoS. Mathematics 8(10). https://doi.org/10.3390/math8101782
6. Dorri A, Jurdak R (2021) Tree-chain: a lightweight consensus algorithm for IoT-based blockchains. In: IEEE international conference on blockchain and cryptocurrency (ICBC), pp 1–9. https://doi.org/10.1109/ICBC51069.2021.9461098
7. Nakamoto, Bitcoin S (2008) A peer-to-peer electronic cash system
8. King S, Nadal S (2012) Ppcoin: peer-to-peer crypto-currency with proof-of-stake. Self-Publ Pap 19(1)
9. Daniel L (2014) Delegated proof-of-stake (DPoS). MBitshare whitepaper
10. Kwon J, Buchman E (2019) Cosmos whitepaper. In: A network of distributed ledgers
11. Leung D, Suhl A, Gilad Y, Vault ZN (2019) Fast bootstrapping for the Algorand cryptocurrency. In: Network and distributed systems security (NDSS)
12. Chen J, Gorbunov S, Micali S, Vlachos G (2018) Algorand agreement: super-fast and partition resilient byzantine agreement. IACR Crypto ePrint Arch 2018:377
13. Cohen S, Keidar I, Naor O (2021) Byzantine agreement with less communication: recent advances. SIGACT News 52(1):71–80
14. Gilad Y, Hemo R, Micali S, Vlachos G, Zeldovich N (2017) Algorand: scaling Byzantine agreements for cryptocurrencies. Cryptology ePrint Archive, Report 2017/454, Version 20170924:210956
15. Chen J, Micali S (2019) Algorand: a secure and efficient distributed ledger. Theoret Comput Sci 777:155–183. https://doi.org/10.1016/j.tcs.2019.02.001
16. Micali S (2016) Algorand: the efficient public ledger. https://arxiv.org/abs/1607.01341
17. https://attestiv.com/algorand-insurers-investigate-blockchain-corroborates/
18. https://www.algorand.com/resources/blog/algorand-announces-insurtech-use-case-attestiv
19. Zixiang N, Zhang M, Lu Y (2022) HPoC: a lightweight blockchain consensus design for the IoT. Appl Sci 12(24).https://doi.org/10.3390/app122412866
20. Kim T, Noh J, Cho S (2019) SCC: storage compression consensus for blockchain in lightweight IoT network. In: IEEE international conference on consumer electronics (ICCE), pp 1–4. https://doi.org/10.1109/ICCE.2019.8662032

Effective Ransomware Detection Method Using PE Header and YARA Rules

S. Hashwanth and S. Kirthica

1 Introduction

Nowadays, information security has become crucial [1]. The number of cyber-crimes has significantly increased in recent years [2]. Ransomware is one of the most serious risks (or) dangers to computer systems and data, and its prevalence has grown with the rise of digital currencies [3]. An illustration of a malicious program which encodes (or) encrypts data, prevents users from accessing the computer or their information (or) data, and then requests payment for the ransom is called ransomware. To prohibit a person or company from accessing files on a computer, ransomware is created [4]. Ransomware attacks are mainly focused on big tech giants where the data/information is more crucial aspect. These companies are more prone to pay a ransom because they are more likely to need immediate access to files. Without any predetermined targets, random attacks are launched over the Internet to infect as many systems as possible.

Recent ransomware attacks have had a significant negative impact on several businesses, crippled local services, and affected hospital's ability to provide vital services. In May 2021, Colonial Pipeline, a major US fuel pipeline operator, was hit by a ransomware attack that forced the company to shut down its operations. The attacker stole nearly 100 gigabytes of data from the company's server and demanded ransom in return. In this paper, ransomware is being detected with the help of the respective portable executable (PE) headers. In order to classify the application, initially, feature selection is done over the dataset. Feature selection is used to select

S. Hashwanth
Vellore Institute of Technology, Chennai, India

S. Kirthica (✉)
Department of Computer Science and Engineering, College of Engineering Guindy, Anna University, Chennai, India
e-mail: s.kirthica@gmail.com

J. K. Mandal et al. (eds.), *Proceedings of International Conference on Network Security and Blockchain Technology*, Lecture Notes in Networks and Systems 738,
https://doi.org/10.1007/978-981-99-4433-0_16

essential data and reduce the data ultimately leading to lesser computation [5]. Feature selection is nothing but selecting the attribute that is more than enough to compute the model and give more accurate results. Once the features are selected, the dataset is split into train and test data. Once the data is split, various machine learning models are trained over training data. Once the models are trained, it can be used to classify the legitimate and malware files. During live testing of the models, there are two ways for this feature extraction from the application [6]. One of the methods is through dynamic analysis. To extract the features, the dynamic approach needs to run the application. This method extracts features in accordance with the operations of the program as it runs. Many ransomware applications do not exhibit their true nature for a while after being launched, so they avoid detection and easily enters the system. During execution, how long the program should be watched over with this approach is unknown [6, 7]. Also, dynamic analysis cannot be done when the file is not executed. Thus, static analysis can be used to perform in this situation. Execution of the application is not required for static analysis. Recent studies have shown that static feature-based techniques can accurately identify ransomware [11].

In this paper, static analysis is used to get more accurate and precise results. The core objective of this proposed work is to detect the ransomware more efficiently and easily without any complex distributed modules. The proposed work has produced great accuracy and speed. Bitcoin is preferred by the attackers as a payment method since it is an anonymous payment system that hides their identity and location. So, along with the PE header classification, YARA rules are implemented to find whether a Bitcoin address is available in the file. This helps to significantly boost the model accuracy. Here, the final prediction result is provided to the user with the malware status of that application.

2 Related Works

Manavi et al. [8] discussed a method where an image based upon PE header is constructed, and CNN is used to extract and classify the features from those images. They performed static analysis to extract feature. Static analysis helps to extract the features without the program being executed. Their proposed method is complex, and the file is not scanned for crypto signatures.

Vinayakumar et al. [9] proposed a method using shallow and deep networks. They use the prevalence of API requests to distinguish ransomware from other malware families and benign malware. Simple MLP networks are utilized to save high computational costs; however, they perform less well than more complicated design choices when it comes to producing outcomes.

Homayoun et al. [10] discussed a dynamic strategy for ransomware imprisonment. They kept a collection of logs and used the collections to extract features. Finally, they classified the gathered features using CNN and LSTM. They demonstrated the existence of unique recurring patterns within various ransomware families. Since this approach is dynamic, during execution, how long the program should be watched over with this approach is unclear.

Vidyarthi et al. [11] suggested a technique for ransomware detection based on executable file's PE header. They retrieved features for each application by using a few headers field values from the executable files. They performed static analysis and unpacked packed applications. This approach was not efficient since they used lesser number of attributes for determining the final prediction result. Also, crypto signatures could have been detected for more precise results.

Hanqi et al. [12] proposed a method which includes transforming opcode sequences from ransomware samples to N-gram sequences. Then calculation of TF-IDF is performed for each N-gram to select feature N-grams. To extract features without program being executed, static analysis is performed. It is difficult to extract opcodes from files if the files are packed.

Bahrani et al. [13] proposed an approach that employs process mining to discover ransomware by first extracting the process model from the events logs, following which features are retrieved and combined with classification algorithms. The features used to classify are not sufficient, thus making it less efficient.

El-Kosairy et al. [14] proposed a method where using a deception system based on honey files and honeytokens, any intrusion or ransomware attempting to compromise private files is detected. This approach is easy to deploy but this only works with NTFS file systems and cannot work with the FAT32 file systems.

Rezaei et al. [15] proposed a method where different machine learning models are trained using information derived from the PE file structure and header to identify malware applications. They used static analysis, which is an advantage to extract features even when program is not executed. The accuracy of their proposed work is lesser and could have used more attributes to classify.

Manavi et al. [16] used LSTM network to process the executable file header and build a detection model. The approach performs well by consuming less time for delivering the results. Only con is that it is complex and gives less accuracy.

Belaoued et al. [17] suggested a technique for detecting malware in real-time that uses information stored in PE-optional header fields in 2016. To choose the most useful features, they combined the Chi-square and Phi coefficient feature selection algorithms. Finally, they trained several machine learning classifiers using the features they had obtained. This model only utilized the optional header section of the PE file for their research; however, this is insufficient to offer a thorough and useful model.

Vyas et al. [18] suggested a real-time system for detecting the network malware by investigated static features. The header and sections of PE files are mined for features, which are then reorganized into 4 categories. The 4 categories are file packing, DLL imports, imported functions, and file metadata. This proposed work uses 28 features which can be minimized in to give more accurate results or lesser computations.

By observing the previous works, it can be noticed that most of the previous proposed works undergo dynamic approach toward feature selection which is not efficient since the file needs to be executed for that [11]. In few cases, opcodes were taken for the detection of ransomware. In these works, it is difficult to get the opcodes if the file is packed [13]. Most research papers which have provided efficient results lack with respect to complexity of its architecture [9–11].

3 Proposed Method

The proposed method is divided into 3 modules—data pre-processing, model training, and static analysis. Further, YARA rules will be implemented to check for bitcoin address in the file. The dataset consists of the metadata from the PE file.

3.1 PE Structure

The portable executable file format is used by Windows executables, object code, and DLLs [15]. A header is followed by a number of sections in the PE file structure. The header contains details about the file itself. The next element is the PE file signature, which always contains the value "PE 0 0". The file header and optional header are then included. These headers provide crucial information. In Fig. 1, insight over PE file structure is shown.

The locations of additional significant executable file information are indicated in each row of this array. In fact, each file may contain a few tables, such as the resources table and import table. The size and address of a particular table are provided in each row of the Data Directory. The section table, which includes details on the program sections, comes as the final header section after the optional header. Each row describes a specific area of the PE file.

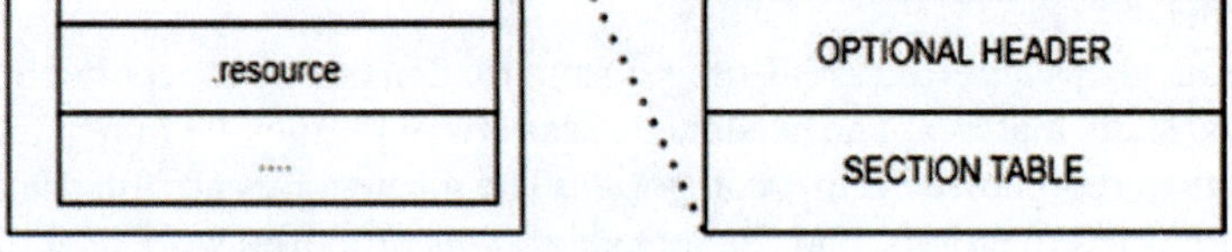

Fig. 1 PE file structure

3.2 Data Pre-processing

The dataset is imported with the help of panda's library. The dataset is visualized to get detailed insights over the data. Once the data is loaded, the first step is feature selection which is carried out to select essential features required for the computation [19]. Less resources are needed to carry out computations or actions when characteristics are minimized. Thus, computer can complete more work with less storage space and computation time.

In this paper, impurity-based feature importance is calculated using tree-based estimators, and this is then used to eliminate unnecessary features. After eliminating the unnecessary feature, the remaining features are used to get the most efficient and accurate results after classification. After feature selection, the attributes count is reduced from the 57 features to 14 features. These 14 features can be used for further model training to get more precise results.

3.3 Model Training

The final data from the previous step is used for training the model. The data is fed into different classification models to get the classification results. For classification, the data is split into train and test set. Once the data is divided into train and test set, the train set is then fed into each model to get the classification score of the record being a legitimate file or malware. Models included in the research are decision trees, gradient boosting, random forest, adaptive boosting, and Naïve Bayes, logistic regression, and KNN. Few of them are commonly used ones [20].

3.4 Static Analysis

Once the model is training, it can be tested against live applications. The applications to be classified are taken, and these applications undergo static analysis in order to get their respective PE headers. Static analysis, also known as static code analysis, examines a computer program to identify issues without running the application [21]. In order to understand the structure of the code and subsequently detect errors in it, static analysis is often conducted on the source code of the program using tools that turn the program into an abstract syntax tree (AST).

In this proposed work, PE headers are extracted from the file based on which the model predicts it as legitimate or malware. Windows 25 executables, and DLLs and object code all employ the portable executable (PE) file format. The Windows OS loader needs information from the PE file format, which is a data structure, to manage the wrapped executable code. PE headers are extracted using a Python package "pefile". A multi-platform Python module called "pefile" is used to parse and operate with portable executable (PE) files. Most of the data in the PE file headers, as well as all the metadata and data in the sections, are all accessible. Once the required features are taken out from the source application, the data is passed into the trained model so that the final classification output can be obtained whether the file is legitimate file or a malware.

3.5 YARA Rules Implementation

After successfully classifying the applications, further YARA rules are used to identify whether a file contains any Bitcoin address. Rules are written in order to check for any Bitcoin address presence in that file. A Bitcoin address is composed of 26–35 alphanumeric characters that start with 1 or 3.

4 Results and Discussion

4.1 Dataset

The dataset used in this project is taken from Kaggle [23]. The data in the dataset is obtained by conducting statistical analysis on 138,047 application files. The objective is to classify files as malware or legitimate. Legitimate files are software that does not behave like malware and are useful and harmless to the users. The dataset constitutes 41,323 binaries (exe, dll) which are legitimate and 96,724 malware files from virusshare.com. The statistical analysis extracted PE information and calculated the entropy of different sections of these files. Finally, the dataset consisted of 138,047 entries with 57 columns.

4.2 Evaluation Metrics

The proposed work is evaluated using the metrics: accuracy, precision, *F*-measure, and recall [22].

In order to calculate these metrics, the confusion matrix must be visualized as shown in Fig. 2, and the true (or) false positive and true (or) false negative values should be taken from the matrix. These values are used for the calculation of the metrics. In these metrics, true positive (TP) refers to the number of samples that are classified rightly as ransomware, true negative (TN) refers to the number of samples that are classified rightly as the legitimate files, false positive (FP) refers to the number of samples that are falsely classified as ransomware, and false negative (FN) refers to the number of samples that are falsely classified as legitimate files.

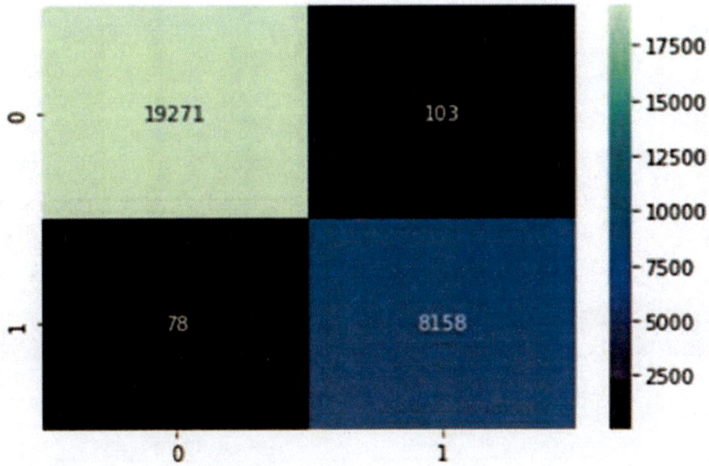

Fig. 2 Confusion matrix obtained using random forest model

4.3 Final Evaluation

After performing evaluation with the models, random forest turned out to give the highest prediction accuracy and score out of all the other models. Random forest gave a whopping accuracy of 99.3% in classification. The final score of the model is 99.34%. Compared to the previous works, the classification accuracy is the highest among them all as shown in Table 1. All the other evaluation metrics (recall, precision, and $F1$-score) are cross verified with the related works and visualized as shown in Fig. 3 and Table 2.

From this result, random forest is said to perform the best and give more accurate and precise results. Further, YARA rules are implemented to find whether the file contains any Bitcoin address or not. This helps to get an insight over the file to check for any Bitcoin address which will help to confirm with the ransomware.

Table 1 Final model classification scores

Model	Score
Logistic regression	70.17
KNN	98.97
Decision tree	99.05
Random forest	99.34
Naïve Bayes	70.17
Gradient boosting	98.88
AdaBoost	98.56

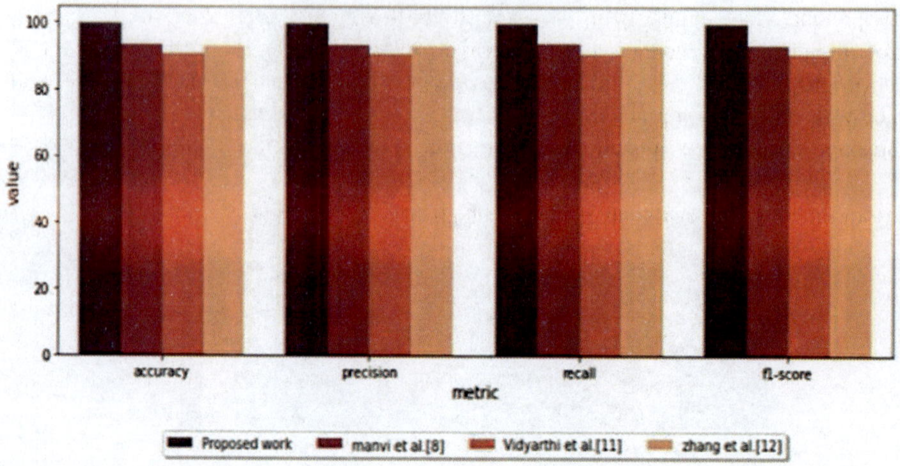

Fig. 3 Comparison of evaluation metrics of proposed work with the previous works

Table 2 Comparison of evaluation metrics of proposed work with related works

Metric Proposed work	Accuracy	Precision	Recall	*F*-score
Proposed method	99.34	99.59	99.46	99.53
Manvi et al. [8]	93.33	93.33	93.40	93.34
Vidyarthi et al. [11]	90.43	90.43	90.67	90.40
Zhang et al. [12]	92.75	92.75	92.78	92.74

4.4 Testing

The test bed is taken from dike dataset available in GitHub. This dataset is a labeled one which contains the benign PE and OLE files. The test bed consists of 750 benign files. And, also the model is tested against different setup files of many programs. Once the files are taken, it underwent static analysis in order to get those 14 features selected using feature selection during data pre-processing. After getting the required metadata, the data is tested with the loaded model (random forest). Then, the desired result is obtained. Further, YARA is compiled by importing YARA-Python, then the rule to find the bitcoin address is defined manually. The classified file is further process with this rule in order to check for presence of any Bitcoin address in that file. If presence of bitcoin is found, 1 is returned or else 0 is returned.

5 Conclusion

In this study, a ransomware classification system was created by analyzing structure and header of a PE file using static analysis. Static analysis helped us to extract the required data without the execution of the program. Random forests can handle large datasets with high dimensionality and can avoid overfitting by building multiple decision trees and aggregating their results. Random forest gave the highest accuracy. This model is stored and loaded during static analysis for classification of the application. Further, set of rules (YARA rules) were defined in order to check whether the file consists of any bitcoin address or not. The advantage of the proposed work over the previous studies is that the result gave a whopping accuracy of 99.34% which shows the efficiency of the proposed work. With this proposed work, the future impacts of ransomwares can be mitigated in large scale. Additional implementation of YARA rules in checking for Bitcoin address to get furthermore insights over the file. Future works are discussed in the below section.

The future work includes improvising the data to give a slightly higher performance. Further, model can be trained with more benign files in order to increase the accuracy of the project. Also, Bitcoin detection is not just enough to predict ransomware applications. In that case, the future work also comprises the implementation to find the crypto signatures in the executable file to provide a more reliable result.

References

1. Alkhudhayr F, Alfarraj S, Aljameeli B, Elkhdiri S (2019) Information security: a review of information security issues and techniques. In: 2019 2nd international conference on computer applications & information security (ICCAIS), pp 1–6. https://doi.org/10.1109/CAIS.2019.8769504
2. Humayun M, Niazi M, Jhanjhi NZ, Alshayeb M, Mahmood S (2020) Cyber security threats and vulnerabilities: a systematic mapping study. Arab J Sci Eng 1–19
3. Noorbehbahani F, Rasouli F, Saberi M (2019) Analysis of machine learning techniques for ransomware detection. In: 2019 16th international ISC (Iranian Society of Cryptology) conference on information security and cryptology (ISCISC), pp 128–133
4. Al-rimy BAS, Maarof MA, Shaid SZM (2018) Ransomware threat success factors, taxonomy, and countermeasures: a survey and research directions. Comput Secur 74:144–166
5. Sethi K, Chaudhary SK, Tripathy BK, Bera P (2018) A novel malware analysis framework for malware detection and classification using machine learning approach. In: Proceedings of the 19th international conference on distributed computing and networking—ICDCN '18, pp 1–4
6. Shijo PV, Salim A (2015) Integrated static and dynamic analysis for malware detection. Procedia Comput Sci 46:804–811. ISSN: 1877-0509
7. Sgandurra D, Munoz-Gonzalez L, Mohsen R, Lupu EC (2016) Automated dynamic analysis of ransomware: benefits, limitations and use for detection. arXiv Prepr. arXiv:1609.03020
8. Manavi F, Hamzeh A (2020) A new method for ransomware detection based on PE header using convolutional neural networks. In: 2020 17th international ISC conference on information security and cryptology (ISCISC), pp 82–87

9. Vinayakumar R, Soman KP, Velan K, Ganorkar S (2017) Evaluating shallow and deep networks for ransomware detection and classification. In: 2017 international conference on advances in computing, communications, and informatics (ICACCI), pp 259–265
10. Homayoun S, Dehghantanha A, Ahmadzadeh M, Hashemi S, Khayami R (2020) Know abnormal, find evil: frequent pattern mining for ransomware threat hunting and intelligence. IEEE Trans Emerg Topics Comput 8(2):341–351
11. Vidyarthi D, Kumar CRS, Rakshit S, Chansarkar S (2019) Static malware analysis to identify ransomware properties. Int J Comput Sci Issues 16(3):10–17
12. Zhang H, Xiao X, Mercaldo F, Ni S, Martinelli F, Sangaiah AK (2019) Classification of ransomware families with machine learning based on N-gram of opcodes. Future Gener Comput Syst 90:211–221
13. Bahrani A, Bidgly AJ (2019) Ransomware detection using process mining and classification algorithms. In :2019 16th international ISC (Iranian Society of Cryptology) conference on information security and cryptology (ISCISC), pp 73–77
14. El-Kosairy A, Azer MA (2018) Intrusion and ransomware detection system. In: 2018 1st international conference on computer applications & information security (ICCAIS), pp 1–7
15. Rezaei T, Hamze A (2020) An efficient approach for malware detection using PE header specifications. In: 2020 6th international conference on web research (ICWR), pp 234–239
16. Manavi F, Hamzeh A (2021) Static detection of ransomware using LSTM network and PE header. In: 2021 26th international computer conference, Computer Society of Iran (CSICC), pp 1–5
17. Belaoued M, Mazouzi S (2016) A chi-square-based decision for real-time malware detection using PE-file features. J Inf Process Syst 12(4):644–660
18. Vyas R, Luo X, McFarland N, Justice C (2017) Investigation of malicious portable executable file detection on the network using supervised learning techniques. In: 2017 IFIP/IEEE symposium on integrated network and service management (IM)
19. Benkessirat A, Benblidia N (2019) Fundamentals of feature selection: an overview and comparison. In: 2019 IEEE/ACS 16th international conference on computer systems and applications (AICCSA), pp 1–6
20. Baldwin J, Dehghantanha A (2018) Leveraging support vector machine for opcode density-based detection of crypto-ransomware. In: Cyber threat intelligence. Springer, pp 107–136
21. Hassen M, Carvalho MM, Chan PK (2017) Malware classification using static analysis-based features. In: 2017 IEEE symposium series on computational intelligence (SSCI), pp 1–7
22. Powers DM (2011) Evaluation: from precision, recall and *F*-measure to ROC, informedness, markedness and correlation
23. Jayanth D. PE-header data. https://www.kaggle.com/datasets/dasarijayanth/pe-header-data

Applied S P Integration Procedure for Enhanced Haphazardly Misplaced Values in Data Mining for Database Protection

Darshanaben Dipakkumar Pandya and Abhijeetsinh Jadeja

1 Introduction

From the original source of information, in the way of integration by the accessible methods, and the importance of the consequent variable reliant on every significance of the autonomous variable should be considered another time commencing the formula was developed with the importance of the autonomous variable in mind. If essential to be relevant method for each part significance for, and instance individual, the statistical significance of the changeable data, it is factual to declare principles for equivalent variable reliant happening an ambiguous value of the independent values by using a feasibly available assimilation type and from individual data such as be supposed to be done in one of the ways.

2 Related Work

Gaur and Dulawat [1] identified closer to the absence of attribute principals of mining approach and its related work in brief. Buck [2] stated the approach of estimating missing values for multivariate information suitable for use with an electronic computer was described in brief. Gaur and Dulawat [1] formed closer to the Lack of Principles of Attributes of the Mining Approach and its contents in brief. Sharma and Gaur [3] stated agile contiguous approach to handle strange block format that is

D. D. Pandya · A. Jadeja (✉)
Department of Computer Science, Shri C.J Patel College of Computer Studies (BCA), Sankalchand Patel University, Visnagar, India
e-mail: abjadeja.fca@spu.ac.in

D. D. Pandya
e-mail: ddpandya.fcs@spu.ac.in

J. K. Mandal et al. (eds.), *Proceedings of International Conference on Network Security and Blockchain Technology*, Lecture Notes in Networks and Systems 738,
https://doi.org/10.1007/978-981-99-4433-0_17

missing on data mining in brief. Lin and Brown [4] identified method of association of data based on anomalous values in detail.

3 Analysis of Method

Moreover new procedure related through changing absent regarding standards by artificially created data's. The new approach is based on replacing incorrect attribute values with values that were artificially made. This method is incredibly very useful for statistical characteristics. This procedure looks for a value that is very close to the element's accurate average as well as the value of the absent values' immediately preceding and following values. It is feasible to arbitrary obtain ideals as a table value for the directly integration table thanks to a process for creating the near values for absence value places. At these instant researches for belongings misplaced in the attribute start.

This current technique determines the closest fit method for absent data improvement. At this time, firstly evaluate the whole values from a database for lost value. Two variables, *A* and *B*, are currently listed as the dataset's year and value. For other qualities *B*, variable *A* for the year indication is fixed that include absent data values. *B* attributes are replaceable while A is constant for current investigation. By the beginning, look at the whole table database, with *A* and *B* variable. Now it is obvious that *B* is the variables that contains of an absent dataset value. An investigate pointer hardened to indicate out an absent data value place in the *B* variable, *B* [0] is the first value of the pointer and *B* [$n - 1$] is the preceding value or end one.

When a search pointer indicates a lost aspect for attribute, array the pointer will keep a copy of the variable. The explore pointer represents the EMPTY and space value in the feature. Those types of values recognized in the database, the technique for recovering lost values, and the pointer continuing to stay at the array subscript section. The current aspect or array subscript is indicated by (A_i). At this time it is obvious that (B_i) is near to (A_i). Above the data as (B_0) as of the lost value data and initial the succeeding data as (B_1), we regard as (A_0) to consider the data value of previous the aspect of A_i and (A_1) to consider the value of A_i's further actions. The iteration for "$i =$ ZERO up to $i = n - 1$" is currently being used.

$$B_0 = \text{value}(B_{i-2}) \quad (1.1)$$

B_0 second preceding data value from B_i

$$B_1 = \text{value}(B_{i-1}) \quad (1.2)$$

B_1 first preceding value from B_i

$$B_2 = \text{value}(B_{i+1}) \quad (1.3)$$

B_2 first succeeding value from B_i

$$B_3 = \text{value}(B_{i+2}) \tag{1.4}$$

B_3 second succeeding value from B_i.

$$A_0 = \text{value}(A_{i-2}) \tag{1.5}$$

A_0 second preceding element from A_i

$$A_1 = \text{value}(A_{i-1}) \tag{1.6}$$

A_1 first preceding element from A_i

$$A_2 = \text{value}(A_{i+1}) \tag{1.7}$$

A_2 first succeeding element from A_i

$$A_2 = \text{value}(A_{i+2}) \tag{1.8}$$

A_3 second succeeding element from A_i

$$A = \text{value}(A_i) \tag{1.9}$$

An equivalent response for the incomplete data subtype *Bi*.

Where $A_3, A_2, A_1, A_0, B_0, B_3, B_2, B_1$, and $A \neq$ "NULL".

Since subsequently phase set up the aspects inside this procedure.

$$\text{sum} = 0 \tag{1.10}$$

Now, the sum variable is initialized to zero, and then saving the outcome afterwards handing out, because that variable sum = zero is initially taken. After that we formulate calculations of sum for predictable value.

$$\text{Sum} = [(B_0 + B_3 + (B_0 * 2) + 3 * (B_1 + B_2)]/10 \tag{1.11}$$

Then after allocate sum to B estimated value.

$$B_{\text{est}} = \text{Sum} \tag{1.12}$$

Then after allocate expected in the absence of a result, therefore allocating is performed.

$$\text{value}(B_i) = B_{\text{est}} \tag{1.13}$$

$$i = i + 1$$

loop continues till i less than n.

4 Algorithm

The anticipated technique is related on changing an absent attribute values by useful numerical recovery manner. This technique is extremely more supportive for numerical attributes. Usually, this technique is look for of absent data and then after finding its data is changed through improved data the feature's random value absent dataset.

Starting: Here an array A and B are which is taken N size, N with 50 values, that process changes absent data using the improved records through information records. Similarly, predecessor is indicated by Prev variable of absent records. $A[I]$ and $B[I]$ are the first two arrays chosen in this instance. One aspect array that serves as a repository values for the table specific collection of data includes the variable I utilizes to index entries from 1 to N. Here are the algorithm's detailed steps in more clarity:

Attribute $A = \{A_1, \ldots, A_n\}$, $B = \{B_1, \ldots, B_n\}$
Where $A = A_{obs} + A_{mis}$
$A_{obs} = \{A_1, \ldots, A_k\}$ // element values observed
$A_{mis} = \{A_{k+1}, \ldots, A_n\}$ // element values missing
$B = B_{obs} + B_{mis}$
$B_{obs} = B_1, \ldots, B_k\}$ // element values observed
$B_{mis} = B_{k+1}, \ldots, B_n\}$ // Attribute Missing values
collection of data (A) = = collection of data (B)
Get $A = \{ A_1, \ldots, A_n\}$, $B = \{B_1, \ldots, B_n\}$ // Elements read
For i value 0 to n-1 do // Creation of iteration beginning element from beginning besides end
If (value (Bi) = = NULL) then
B_0 = value(B_{i-2}) // second preceding data of Bi
B_1 = value(B_{i-1}) // first preceding data of Bi.
B_2 = value(B_{i+1}) // first succeeding value of Bi.
B_3 = value(B_{i+2}) // second succeeding value of Bi..
A_0 = value(A_{i-2}) // second preceding element of Ai.
A_1 = value(A_{i-1}) // preceding first element of Ai.
A_2 = value(A_{i+1}) // value of first following Ai.
A_3 = value(A_{i+2}) //value of second following Ai.
A = value(Ai) // A is the subsequent value of absent value subscript Bi.
where A3, A2, A1, A0, B0, B3, B2, B1, A $\neq$ „NULL"
Sum = 0 // the variables Initialization

Sum = [(B_0 + B_3 + (B_0 * 2) + 3 * (B_1 + B_2))]/ 10 // Estimated value
B_{est} = Sum // final predicted data value
Value (Bi) = B_{est} // move calculated value
i = i + 1 // i value increases by 1
procedure continue till (i < n)
finish loop
stop.

5 Analysis for Outcome

Measurement of prevailing trend (mean): Here below list-1 table indicates emissions of the carbon dioxide as global using fossil fuels flaming by form of fuel oil, natural gas, and coal between 1960 and 2009. Coal, oil, and natural gas each provide an average of 2109, 2262, and 879 metric tons of carbon dioxide to the atmosphere, respectively, following missing values at the arbitrarily, the average considered through imperfect database lists are coal as 2125, oil for 2257 and natural gas for 900.

The planned technique is implemented relating to Table 1's datasets through replace absent data's. That is determined with correspond to coal and oil, natural gas values that are 2120, 2257, and 878 correspondingly. That is significant the average values received once changing absent data with planned strategy extremely near to the real average as provided.

(S.D) The Standard Deviation: Using the evaluation of outcome. It is of standard deviation. Seen once afterwards inference with absent values, the standard deviation data received are much related to the basis dataset's standard deviation. According to analysis of outcome calculations, we may propose that intended formula is proper for omitted data values data recovery with evaluation.

(C.V) Covariance Coefficient: Using the evaluation the outcome (CV) coefficient of difference that is initiated that follows evaluation with absent data, with data C.V. levels are not very high alter or vaguely deny which specifies that the sequence is consistent currently.

(ANOVA) Evaluation of Variability: In addition to the alternative, we must confirm the following hypothesis H0: $\mu 1 = \mu 2 = \mu 3$.

H1: as a minimum the two μ's are dissimilar (example as a minimum among the similarities cannot received).

We put up the below analysis of variance for all the data points to test this hypothesis:

(COAL)—Specific ANOVA One-Way

Table 1 Table for applied S P integration method haphazardly omitted values in dataset

S. No.	Year	Original data			Absent data values			Obtained values		
		Coal	Oil	Natural gas	Coal	Oil	Natural gas	Coal	Oil	Natural gas
		Million tons of carbon			Million tons of carbon			Million tons of carbon		
1	1960	1410	849	235	1410	849	235	1410	849	235
2	1961	1349	904	254	1349	904	254	1349	904	254
3	1962	1351	980	277	1351	980	–	1351	980	**270**
4	1963	1396	1052	300	1396	1052	300	1396	1052	300
5	1964	1435	1137	328	1435	–	328	1435	**1108**	328
6	1965	1460	1219	351	1460	1219	351	1460	1219	351
7	1966	1478	1323	380	1478	1323	380	1478	1323	380
8	1967	1448	1423	410	–	1423	410	**1464**	1423	410
9	1968	1448	1551	446	1448	1551	–	1448	1551	**435**
10	1969	1486	1673	487	1486	1673	487	1486	1673	487
11	1970	1556	1839	516	1556	1839	516	1556	1839	516
12	1971	1559	1946	554	1559	1946	554	1559	1946	554
13	1972	1576	2055	583	1576	2055	583	1576	2055	583
14	1973	1581	2240	608	–	2240	608	**1581**	2240	608
15	1974	1579	2244	618	1579	2244	618	1579	2244	618
16	1975	1673	2131	623	1673	2131	623	1673	2131	623
17	1976	1710	2313	650	1710	2313	650	1710	2313	650
18	1977	1766	2395	649	1766	–	–	1766	**2305**	**657**
19	1978	1793	2392	677	1793	2392	677	1793	2392	677
20	1979	1887	2544	719	1887	2544	719	1887	2544	719
21	1980	1947	2422	740	1947	2422	740	1947	2422	740
22	1981	1921	2289	756	–	2289	756	**1947**	2289	756
23	1982	1992	2196	746	1992	2196	746	1992	2196	746
24	1983	1995	2177	745	1995	2177	745	1995	2177	745
25	1984	2094	2202	808	2094	2202	808	2094	2202	808
26	1985	2237	2182	836	2237	2182	836	2237	2182	836
27	1986	2300	2290	830	2300	–	830	2300	**2246**	830
28	1987	2364	2302	893	2364	2302	893	2364	2302	893
29	1988	2414	2408	936	2414	2408	–	2414	2408	**911**
30	1989	2457	2455	972	2457	2455	972	2457	2455	972
31	1990	2409	2517	1026	2409	2517	1026	2409	2517	1026
32	1991	2341	2627	1069	2341	2627	1069	2341	2627	1069
33	1992	2318	2506	1101	2318	2506	1101	2318	2506	1101
34	1993	2265	2537	1119	2265	2537	1119	2265	2537	1119

(continued)

Table 1 (continued)

S. No.	Year	Original data			Absent data values			Obtained values		
		Coal	Oil	Natural gas	Coal	Oil	Natural gas	Coal	Oil	Natural gas
35	1994	2331	2562	1132	2331	2562	1132	2331	2562	1132
36	1995	2414	2586	1153	–	2586	–	**2862**	2586	**1158**
37	1996	2451	2624	1208	2451	2624	1208	2451	2624	1208
38	1997	2480	2707	1211	2480	2707	1211	2480	2707	1211
39	1998	2376	2763	1245	2376	–	1245	2376	**2697**	1245
40	1999	2329	2716	1272	2329	2716	1272	2329	2716	1272
41	2000	2342	2831	1291	2342	2831	1291	2342	2831	1291
42	2001	2460	2842	1314	2460	2842	1314	2460	2842	1314
43	2002	2487	2819	1349	–	2819	1349	**2517**	2819	1349
44	2003	2638	2928	1399	2638	–	1399	2638	**2915**	1399
45	2004	2850	3032	1436	2850	3032	1436	2850	3032	1436
46	2005	3032	3079	1479	3032	3079	1479	3032	3079	1479
47	2006	3193	3092	1527	3193	3092	1527	3193	3092	1527
48	2007	3295	3087	1551	3295	3087	1551	3295	3087	1551
49	2008	3401	3079	1589	3401	3079	1589	3401	3079	1589
50	2009	3393	3019	1552	3393	3019	1552	3393	3019	1552
Average		2109	2262	879	2125	2257	900	2120	2257	878
S.D		567.89	621.13	400.27	580.06	620.20	403.11	576.13	620.62	400.65
C.V		0.27	0.27	0.46	0.27	0.27	0.45	0.27	0.27	0.46

Global dataset on carbon dioxide emissions from the burning of fossil fuels 1960–2009, by fuel type (in million tons of carbon missing)
www.earth-policy.org

ANOVA						
Source of variation	SS	df	MS	*F*	*P*-value	*F* crit
Between groups	7225.026	2	3612.51312	0.011076	0.988986	3.061234
Within groups	45,335,990	139	326,158.202			
Total	45,343,215	141				

Table data: $F(2, 139)$ value 3.0718 at 5% level of significance and 4.7865 at 1% level of significance

(OIL)—Specific ANOVA One-Way

ANOVA						
Source of variation	SS	df	MS	*F*	*P*-value	*F* crit
Between groups	634.3473	2	317.1737	0.000907	0.999094	3.061234

(continued)

(continued)

ANOVA						
Source of variation	SS	df	MS	*F*	*P*-value	*F* crit
Within groups	48,614,912	139	349,747.6			
Total	48,615,547	141				

Table data: $F(2, 139) = 3.0718$ at 5% level of significance and 4.7865 at 1% level of significance

(Natural Gas)—Specific ANOVA One-Way

ANOVA						
Source of variation	SS	df	MS	*F*	*P*-value	*F* crit
Between groups	16,110.5	2	8055.251	0.051913	0.94943	3.061234
Within groups	21,568,237	139	155,167.2			
Total	21,584,348	141				

$F(2, 139) = 3.0718$ at the 5% level of significance and 4.7865 at the 1% level of significance, respectively

Result and Judgment: Given that (Calculated) *F* less than 3.0781, we accept H0 at a 5% degree of importance and draw the conclusion there being are no significant differences in mean values between the groups of (coal), (oil), and (gas).

6 Summary and Conclusion

In real world, there is no complete and supreme method for handling the values of absent attributes. A numerically suitable analysis which has proper mechanisms and assumptions for the absent data should be performed. The nearby fitting technique suggested is helpful with the statistical feature, through a variation inferior with the mean. That is the greatest idea to improve randomly absent data's with the database. Therefore it also is distinguished with the methods to arranging the values of absent data's must be selected separately, along with environment similarly nature of data.

References

1. Gaur S, Dulawat MS (2011) Closer to the absence of attribute principals of mining approach. Int J Adv Sci Technol 2(4)
2. The approach of estimating missing values for multivariate information suitable for use with an electronic computer is described by Buck SF (1960) Series B 2:302–306

3. Sharma S, Gaur S (2013) Agile contiguous approach to handle strange block format that is missing on data mining. Int J Adv Res Comput Sci 4(11):214–217
4. Lin S, Brown D (2003) Method of association of data based on anomalous values. In: The SIAM procedures international conference on data mining, San Francisco, CA, May 2003

DDoS Attack, a Threat to IoT Devices in the High-Speed Networks—An Overview

Pravir Chitre and Srinivasan Sriramulu

1 Introduction

Communication technology and the speed of communication over the network have been growing ever since communication technology and Internet have been invented. With the increasing speed and developments in the communication technology, new and new means of automation are being invented/developed. Among the latest development is the Internet of things, which is becoming more popular with the advancements in the network communication technology. The quest for more and more speed has given rise to the Wi-Fi-6, 5G mobile network, and high speed communication technology. World is now going for smart cities with the advent of this new technology i.e. Internet of things. The smart cities are bringing in the mission-critical applications as a part of automation that is using high-speed communication technologies in implementation of Automation Techniques as a part of smart city projects.

The next generation technologies are simplifying up-gradation and management of the high-speed network devices by implementing the new concept like software defined network (SDN), which in turn is providing additional techniques like network function virtualization (NFV) and network slicing (NS). Though these are some of the powerful features, the misuse of these features has also been designed by the miscreants.

The security threat to the SDN controller has been long identified, and the problem of securing the SDN controller is already being addressed. But with the use of Smart devices which are being developed with the inbuilt support of IoT technology, the

P. Chitre (✉) · S. Sriramulu
Galgotias University, Greater Noida, UP, India
e-mail: pravir.chitre_phd19@galgotiasuniversity.edu.in

S. Sriramulu
e-mail: s.srinivasan@galgotiasuniversity.edu.in

P. Chitre
Bhai Parmanand DSEU Shakarpur Campus II, Galgotias University, Delhi, India

J. K. Mandal et al. (eds.), *Proceedings of International Conference on Network Security and Blockchain Technology*, Lecture Notes in Networks and Systems 738,
https://doi.org/10.1007/978-981-99-4433-0_18

networks and SDN controllers are now facing new security threats. One of the major security threats has been identified in which the IoT devices are used to initiate the IoT-DDoS attack by the attackers without the knowledge of IoT device users by turning the IoT device as zombie or reflector. This is specially evident since the network servers faced DDoS attack where the innocent IoT end devices were used to initiate the DDoS attack using the Mirai Botnet [20]. In this paper, security threat due to poorly configured end point IoT devices is discussed. Also the measures that may help mitigate such security threats have been discussed.

2 Background

The high speed in the network communications is being achieved using the 5G mobile technology and the sixth generation Wi-Fi technology, as these are the vehicles in terms of achieving high speed in network communication. The 5G, or fifth generation, mobile phone system, and the sixth generation Wi-Fi, or Wi-Fi-6, which is also known as the IEEE 802.11ax standard, will coexist as these are complementary technologies. The 5G mobile technology is also enabling higher density and capacity for network devices. With the implementation of Wi-Fi-6, the data communication rates are increasing as Wi-Fi-6 offers Wi-Fi connectivity with the devices at much higher data transfer rate. This is also adding up to the increased data consumption [19]. While 5G is the preferred option for the outdoor network, Wi-Fi-6 is still the preferred indoor access network. With 5G and Wi-Fi-6 in place, data communication is advancing in terms of speed, latency, and device density [2, 7]. The next generation hardware is using the new technologies like:

- ***Software Defined Network (SDN)***: Software defined networking (SDN) is a networking architecture that makes networks more flexible and manageable. SDN centralizes management by decoupling the control plane from the data forwarding task in discrete networking devices. SDN uses software-based controllers or application programming interfaces (APIs) to direct traffic on a network in order to communicate with the underlying hardware infrastructure. The network control plane is physically separated from the forwarding plane, and a control plane is capable of controlling several devices [4].
- ***Network Function Virtualization (NFV)***: Network function virtualization (NFV) is a technology where the physical hardware devices like switches, routers, firewalls, load balancers, and others are replaced with software-based virtual devices which are highly scalable and are implemented using the technologies like virtual machines, thus forming a virtual network [5]. By implementing software solutions, NFV increases scalability and agility by allowing service providers to instantly deploy new network services and applications without the need for additional hardware resources [3].
- ***Network Slicing (NS)***: Network slicing is the process of dividing a single network connection into many virtual connections that deliver various amounts of resources

to different types of traffic using network virtualization. The network slice is a logically segregated, self-contained, independent, and secure portion of the network that targets distinct services with varying speed, latency, and reliability needs [6].

High-speed networks are being deployed using technologies like SDN, NFV for a variety of applications that will support the next generation smart city. As smart cities are growing, new technology like IoT is being used to automate the mission-critical and high risk applications. Together with activities that require quick responses, such as self-driving cars, automation is being used in applications like numerous crucial services including distribution of water and electrical supplies and traffic control, and others [1, 24]. The IoT devices are entering the households with the use of smart and IoT-enabled devices.

This implementation of applications that are mission-critical and response-sensitive is producing new security issues, and these challenges are driving the need for a higher level of security protection. For instance: If a smart city that makes use of 5G services for essential services like supplying the city with electricity or water and has a security system flaw due to which a hacker succeeds in getting the access of control of these essential services, then they might be able to tamper the services or even shut down these services, which could have devastating effects on the smart city [1, 24].

Major Security Threats

In the era of 5G and high-speed networks when the new concepts like, i.e., Internet of things (IoT) are gaining popularity and the new projects like smart city and implementation of mission-critical applications are being executed, the security threats on the next generation network are getting higher and higher. This is because with IoT and smart city like applications, lots of small devices are getting installed for controlling various gadgets and appliances. These devices are designed for low battery consumption and are being implemented with SDN-based application. Due to their small size and low-power consumption, these devices' designs provide new vulnerabilities and can be readily compromised by hackers who then use them as weapons to launch various forms of attacks on mission-critical applications and other targets. The most popular of the attacks/vulnerability is denial of service and/or distributed denial of service attack. As per "DDoS Attack Statistics, Facts and Figures For 2022" from Web portal PixelPrivacy [8–10], DDoS attack is one of the most popular online weapon that is used by the hackers. As mentioned in Web portal pixelprivacy, "Cisco estimates that the total number of distributed denial of service attacks will double from the 7.9 million attacks experienced in 2018 to 15.4 million attacks in 2022" [8]. This security threat has already been experienced by the world with the DDoS attack on the Security Blog of Brian Krebs on September 20, 2016 [20, 26] which was initiated by weaponizing the CCTV cameras and other IoT devices. More such attacks are being faced by the world.

Denial of Service (DoS) and Distributed Denial of Service (DDoS)

DoS attacks have the potential to deplete an opponent's network resources. DoS is a type of security attack that reduces a network's availability. A DoS attack can be

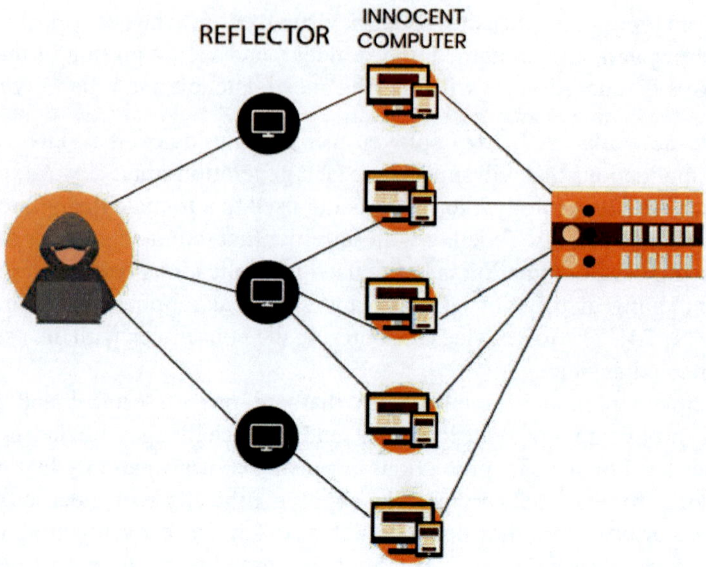

Fig. 1 Attacker is using innocent machines to perform a DDoS assault on the victim system [11]

launched using jamming or flooding so that the target is overloaded with requests and the service's or resource's usage is severely compromised. It is possible to identify and mitigate the attack when there is a single attacker which is DoS attack. But when the similar type of attack is carried out using multiple resources, then the attack is called as DDoS attack. In such an attack it is difficult to identify the attacker and to mitigate the attack. In case of DDoS attack, attacker rather than directly attacking, identifies and uses the weak devices to launch the attack. Attacker uses the remote handlers like Botnets, reflectors and controls the devices remotely to initiate and manage the attack on the target system/device/network/resource. A DDoS model is shown in Fig. 1.

The prevention of DoS or DDoS attacks is not possible. The only solution is to identify and detect these attacks. If these attacks can be detected, then these attacks may be stopped or prevented. The best defense toward the DDoS attack is to detect the attack, as soon as possible and to mitigate it. Sooner the attack can be detected, earlier it can be mitigated and lesser the damage due to the attack.

DoS and DDoS attacks have recently become a serious concern to many Websites and could pose a big threat to operators as the deployment of large number of devices in 5G wireless networks has been increasing. These connected devices are normally designed to be smaller in size, and to manage with the smaller size, these devices may be designed to be either less secured or with the advent of SDN, devices and are secured and controlled by the SDN controller. Attacker can gain control of the SDN controller, thereby gaining the control and access to the controlled devices. The attacker can use these devices to launch the attack and hides itself, behind these smaller devices. A DoS/DDoS assault can be categorized as either a device/user

DoS/DDoS attack or a network infrastructure DoS attack, depending on the attacker's target. DoS attacks can target the battery, memory, disc, CPU, radio, actuators, and sensors in devices [1, 24].

DDoS attacks are, according to Norton, "*one of the most powerful weapons on the internet*". Denial of service attacks can strike at any time, affecting any aspect of a Website's operations or resources, and resulting in major service disruptions and financial losses. DDoS assaults used to be a source of amusement, but research suggests that they're increasingly being used by hackers to make money or to cause disruption for political reasons [9].

Why there is increase in DDoS attacks
As it has been observed that there in explosive growth of DDoS attacks in the world of Internet and Web. DDoS attacks can be considered to be asymmetrical warfare. Hackers can more easily build the firepower needed for a DDoS assault, thanks to millions of vulnerable IoT devices. Manufacturers of IoT devices who are looking to cut costs, frequently overlook security features. This omission causes broad harm and stymies IoT growth in the long run. Once an IoT device is installed, it is challenging to upgrade its security [12, 13].

Types of DDoS attacks
DDoS assaults can be divided into three categories: volume-based, protocol-based, and application-layer attacks. SYN attacks are the most prevalent attacks among the many varieties (about 94% of all types of DDoS Attacks) [14]. Other common DDoS attacks includes ICMP (Ping) Flood, UDP Flood, etc. Earlier, it was the trend of DDoS attacks that would last for long time. As per the analysis done by Kaspersky, the longest known DDoS attack was of about 509 hours [9].

3 Discussion

Security Challenges in SDN
The networking technology known as software defined network (SDN) offers innovation in 5G networks and the next generation of network hardware. The primary cause of security challenges is SDN. In older technology, the hardware was in charge of controlling the network; hardware upgrades could not be made simply by updating the firmware. In case of any technological changes required in the network or the network devices, the only solution was to replace the old devices with devices supporting the new technology.

With SDN, majority of technological up-gradations are possible simply by software configuration or software up-gradation, and the need for replacement of existing hardware which may be a costly affair, that may not be required. The network devices can be remotely controlled and managed by software, thanks to the innovative SDN technology. Moreover, the network controller contains centralized network intelligence.

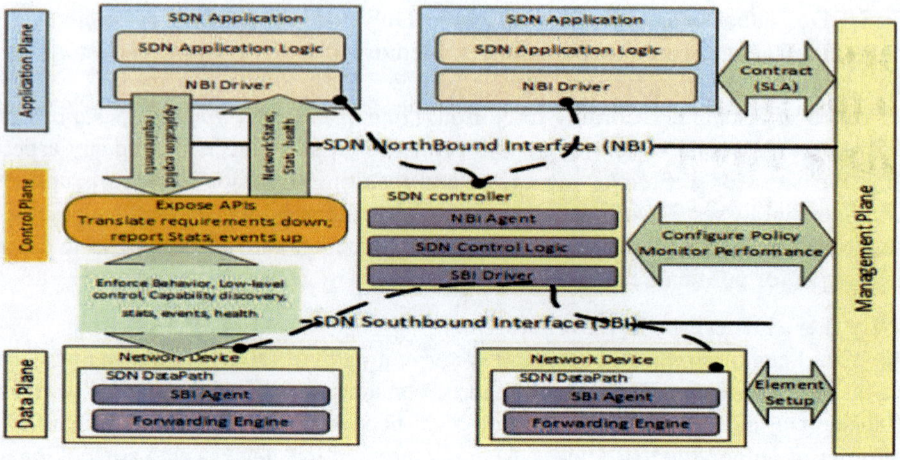

Fig. 2 SDN reference architecture [15]

Applications based on SDN are used to implement the network functionalities of next generation network hardware (including 5G and high speed network). As a result, designing and establishing network architecture as required is simple. It is still simpler for an attacker or hacker to begin an attack because they must concentrate on the primary network controller [24]. A hacker or attacker can send malicious code to the network controller, where it will start to wreak havoc on the entire network. So, keeping such malicious code out of the system becomes essential for safeguarding the Network.

The availability of open APIs in network equipment is one of the key security vulnerabilities that applications might provide to the network that can alter or change network behavior. The majority of 5G features will be developed as applications depending on how quickly they can be supplied, updated, and modified. NFV will make application-based services possible and introduce them into the networking domains [24]. Hence, safeguarding the network from anomalies brought on by applications will be crucial. Figure 2 illustrates how the control plane in SDN functions as a centralized decision-making unit. The controller may be heavily targeted for network intrusion or malicious operations due to its crucial role. In this case, it is important to keep in mind that since the hacker or attacker will attempt to assault the SDN Controller's core control plane, it will be easier to mitigate the attack there.

Distributed Denial of Service Attack in SDN

In recent years, a DDoS assault has been one of the most serious security concerns to the SDN network. In addition to preventing authorized users from using and accessing network resources, it has the power to fully destroy the network. As a result, defending the SDN network against DDoS attacks is critical [16, 17].

In order to establish zombie host groups that meet their attack needs, attackers hack many hosts and unite them. These zombie hosts bombard the target with an enormous number of useless data packets, consuming a lot of its bandwidth and

CPU resources, in the process. The target host will become unresponsive and unable to process real data packets when it receives much more data packets than it can handle. DDoS attacks are favored by attackers because they are easy to carry out. The switch receives attack packets and matches each one with a flow entry when the controller is assaulted. The flow table entry in the flow table cannot be matched because the attack packet is invalid. The packet is wrapped by the switch and sent as a packet-in message to the controller. The controller then decides on the direction of the data packet. Attackers flood the controller with attack packets, causing packet-in messages to continuously flow in and use up a lot of the controller's resources. The controller is therefore unable to process legitimate traffic data and becomes inefficient, or gets flawed. This demonstrates that the security issues associated with SDN should not be overlooked [16, 17].

Use of IoT devices as remote devices to Launch attack

The attacker and hacker have been using the attack on SDN controller as it is deduced that the attack will be very much destructive for the services which the said SDN has been in use. Most of the SDN implementations have now understood the need for securing controller. Since the SDN controllers are secured, there is no way to target the SDN controller. The hackers have been trying to use some different techniques for the attack. They seem to have found the possible loop hole in the end IoT devices. Hackers/attackers are trying to use the vulnerabilities [12] of the IoT devices, which are the part of network that is controlled by SDN controllers-based network. This is evident from the Mirai Botnet attack on the Security Blog of Brian Krebs that happened on September 20, 2016 and in Dyn DDoS attack that happened on October 21, 2016 [22, 23, 26]. Most of these IoT devices have following vulnerabilities:

- Most of these devices are small in size and are designed to consume less power. So, these devices are designed with scaled downed version of embedded operating system as a part of firmware so that it can run with the limited resources.
- These devices have hard coded password that is not been set. Thus, the devices have weak authentication mechanism.
- Devices comes with insecure Telnet support, so it is easy to install a Malware.
- The traffic between the controller and the device is normally not encrypted.
- As reported by the authority on video surveillance IPVM [25], CCTV can be a soft target.
- Open JTAG interface in the chip that gives physical access to the chip. (JTAG is a special interface added to the chip. With the help of these pins the test probe can be connected to the chip) [18]. It is used to inspect chip interconnects and printed circuit boards (PCBs). To determine how a chip responds to different commands, hackers use JTAG and debugging tools.

The Mirai botnet is a worm that would replicate by itself. It would infect and attack the vulnerable IoT devices. It is called as botnet as the infected devices would be then controlled by the remote command and control server. These server would tell the infected device to target which site. This would be possible as these bots would have replication module and the attacking modules. Later, lots of similar bots surfaced

Table 1 Some of the incidences of Mirai like attacks [22]

Date	Target	Type of attack
20 September 2016	Website of Computer security journalist Brian Krebs using Mirai and BASHLITE, a malware	DDoS attack
March 2018	A new variant of Mirai named "OMG", target vulnerable IoT devices	Truned IoT devices into Proxy Servers
May to June 2018	A Mirai variant "Wicked" targetting Netgear Routers and CCTV DVRs	Locate vulnerable IoT devices
July 2018	13 variants of Mirai detected to target Android devices	Target vulnerable IoT devices
21 October 2016	Multiple major DDoS attack on DNS service provider DYN using IoT devices	DDoS attack

and have been used for attacking the target network by the attacker [21]. Some of the popular attacks using the IoT devices is listed in Table 1.

Security Solution in SDN

A global perspective of the network is provided by the logically centralized control plane of SDN, which also makes it possible to configure network components in real-time. The SDN architecture makes network forensics, security policy changes, and security service insertion possible by supporting extremely proactive and reactive security monitoring, traffic analysis, and reaction systems.

By continuously collecting and accumulating data from network resources, states, and flows, SDN enables quick threat identification. With adjustments to the flow table, the SDN design offers traffic redirection for data analysis, policy change, and reconfiguring the network. SDNs' programmability makes it possible to change security policies on the fly without needing to individually configure each piece of hardware. Mis-configurations and policy conflicts across numerous networks are less likely as a result of this automation. Due to the network's high visibility, standardized network security standards may be established. As a result, security services like firewalls and intrusion detection systems (IDS) may be applied to specific traffic in line with generally recognized security standards.

The above measures may be enough so as to secure the SDN controller and the network, but in view of the new challenges those are added due to the vulnerabilities of IoT devices, which are used by hackers and attackers to launch their attacks on to the target victim network resources, the above measures may not be enough for securing the network from vulnerabilities due to vulnerable IoT devices.

Possible Solution to IoT-DDoS Security Threats

There have been research in the area of IoT-DDoS, and some of the solution have been proposed include the use of edge computing [26]. In this paper, the authors have proposed to implement a Show Net at the edge devices and also to implement

Shadow Server to detect and mitigate the IoT-DDoS attack. While in another paper [27], the authors have proposed to implement flow guard at the edge devices, an edge defense mechanism to detect the IoT-DDoS attacks. But in both the solutions, such mechanism has to be implemented in each edge device. This can be herculean task as the IoT devices may be located geographically apart and implementing such edge mechanism may not be possible in big network.

Following care may be taken to secure the IoT devices:

1. For your IoT devices, choose a strong operating system along with powerful development tools like *Windriver* Helix or *VMware Liota*. They also make it easier to apply security upgrades.
2. Implement more stringent authentication measures. A strong and unique password must be created by users, or default passwords should be updated. To protect IoT devices, public key authentication could be used.
3. The telnet support may be disabled on the device and a strong password with strong authentication mechanism may be followed, may be by using public key encryption, by enabling SSH instead of Telnet.
4. Verify that the correct control server is being used by the devices. This can be done by guarding the IP addresses of the control servers and limiting access to them. A reliable control server should be used to validate IoT firmware changes. Also by implementing the defense mechanism at the Edge devices.
5. To prevent hackers from accessing your IoT devices, make sure the JTAG interface is encrypted.
6. Ensure the security of both your own servers and the IoT control servers against such assaults. A number of services are provided, including an advanced threat solution that provides traffic visibility, security information, and situational awareness across the whole network. Better threat identification and incident response are made possible by real-time insights, visualization, and forensics.

4 Conclusion

The 5G network, 6G network, and high-speed wireless network are the future. But need for speed brings various risks. Biggest risk that is troubling the connected world is the threat of various cybersecurity attacks. As has been discussed, the next generation network will employ a number of technologies, including software defined networking, network function virtualization, and network slicing. Additionally, the use of connected devices for the Internet of things will expose the network to security risks like DDoS attacks. These new generation networks must be protected from these risks by law. Since now the techniques are being developed to secure such SDN controllers. It is the need of hour to secure the networks and network resources by securing the end point devices from being misused for launching attacks like DDoS attack and other type of attacks. One of the security mechanisms that may be considered is implementation of edge computing in the edge devices, but this technique is in nascent stage and security measure which are required to be developed, so that the world can enjoy the speed of life that will take us to the future.

Acknowledgements The authors would like to express sincere gratitude to the guide and mentor Dr. Srinivasan Sriramulu for his sincere efforts and all his support. The author would also like to thank Dr. Sampath Kumar, Dr. Naresh Kumar and all the Faculty members of Galgotias University for their support and guidance without which this paper would never have happened.

References

1. Fang D, Qian Y, Hu R (2018) Security for 5G mobile wireless networks. IEEE Access 6:4850–4874
2. Mantas G, Komninos N, Rodriguez J, Logota E, Marques H (2015) Security for 5G communications. In: Fundamentals of 5G mobile networks, pp 207–220
3. A smart city solution with 5G mMTC technology. https://www.gigabyte.com/Solutions/mmtc#:~:text=Example%20applications%20can%20include%20waste,charging%20stations%20for%20electric%20vehicles
4. Software defined network. In: Software-defined networking. Cisco. https://www.cisco.com/c/en/us/solutions/software-defined-networking/overview.html
5. What is NFV? In: Red Hat—we make open source technologies for the enterprise. https://www.redhat.com/en/topics/virtualization/what-is-nfv
6. 5G network slicing. https://www.ericsson.com/en/network-slicing
7. 5G phase I release 15 documentation released by 3GPP and TSG. https://www.3gpp.org/release-15
8. DDoS attack statistics, facts and figures for 2022. https://pixelprivacy.com/resources/ddos-attack-statistics-report/
9. DDoS attack statistics and facts for 2018–2022. https://www.comparitech.com/blog/information-security/ddos-statistics-facts/
10. 32 remarkable DDoS statistics for 2022. https://www.softactivity.com/ideas/ddos-statistics/
11. What is a DDoS attack? How to prevent DDoS attacks? https://www.testbytes.net/blog/ddos-attack/
12. Puri D. DDoS attacks using IoT devices follow The Manchurian Candidate model. https://www.networkworld.com/article/3128372/ddos-attacks-using-iot-devices-follow-the-manchurian-candidate-model.html
13. Galov N. 39 Jaw-dropping DDoS statistics to keep in mind for 2022. https://webtribunal.net/blog/ddos-statistics/
14. DDoS attacks. https://www.imperva.com/learn/ddos/ddos-attacks/
15. Hakiri A, Berthou P (2015) Leveraging SDN for the 5G networks: trends, prospects and challenges. CoRR. abs/1506.02876. http://arxiv.org/abs/1506.02876
16. Fan C, Kaliyamurthy N, Chen S, Jiang H, Zhou Y, Campbell C (2022) Detection of DDoS attacks in software defined networking using entropy. Appl Sci 12. https://www.mdpi.com/2076-3417/12/1/370
17. Mahrach S, Haqiq A (2020) DDoS flooding attack mitigation in software defined networks. Int J Adv Comput Sci Appl 11
18. JTAG. https://www.jtag.com/jtag-interface/
19. Why India's 5G users are experiencing high data consumption, choppy connectivity. https://www.techcircle.in/2022/11/11/why-india-s-5g-users-are-experiencing-high-data-usage-choppy-connectivity
20. Heightened DDoS threat posed by Mirai and other Botnets. https://www.cisa.gov/news-events/alerts/2016/10/14/heightened-ddos-threat-posed-mirai-and-other-botnets

21. Inside the infamous Mirai IoT Botnet: a retrospective analysis. https://blog.cloudflare.com/inside-mirai-the-infamous-iot-botnet-a-retrospective-analysis/
22. Mirai (malware). https://en.wikipedia.org/wiki/Mirai_(malware)%23:%7E:text%3DThe%20Mirai%20botnet%20was%20first,Krebs%27%20website%2C%20an%20attack%20on
23. How IoT is making DDoS attacks more dangerous? https://insights2techinfo.com/how-iot-is-making-ddos-attacks-more-dangerous/
24. Chitre P, Sriramulu S (2023) Analysis and evaluation of security and privacy threats in high speed communication network. In: Gupta D, Khanna A, Bhattacharyya S, Hassanien AE, Anand S, Jaiswal A(eds) International conference on innovative computing and communications. Lecture notes in networks and systems, vol 471. Springer, Singapore. https://doi.org/10.1007/978-981-19-2535-1_39
25. The authority on physical security technology. https://ipvm.com/about?from=quick-links
26. Bhardwaj K, Miranda JC, Gavrilovska A (2018) Towards IoT-DDoS prevention using edge computing. In: USENIX workshop on hot topics in edge computing
27. Jia Y, Zhong F, Alrawais A, Gong B, Cheng X (2020) FlowGuard: an intelligent edge defense mechanism against IoT DDoS attacks. IEEE Internet Things J 7(10):9552–9562. https://doi.org/10.1109/JIOT.2020.2993782

Dual Image-Based Watermarking Scheme Using Interpolation

Swarup Kumar Bhunia, Pabitra Pal, and Debasis Giri

1 Introduction

Information security has been a valuable issue with the development of the Internet in modern society. Today's information (i.e., messages, audio, and video) hiding techniques become important fields with digital watermarking. This information can be altered by unauthorized users when data is transmitted or stored for illegal use. This illegal use affects human real-life applications. So anybody wants to prevent this type of illegal use with the help of watermarking scheme, which is used to protect against unauthorized use. A dual watermarked images (WIs) combines two host images and a watermark image. It embeds secret watermark bits stream in each host image (HI) pixel and generates an embedded dual image, i.e., watermarked embedded image. The modified dual image is transmitted to the receiver. The original image can be extracted using the reversible watermarking technique and stored successfully. At this work, first a $(N + 1) \times (N + 1)$ host image is generated from a $(2N) \times (2N)$ color image using interpolation algorithm. This interpolation algorithm increases the size of the host image (HI), i.e., $(2N + 1) \times (2N + 1)$ host image and watermark (W) embedded in each host image. This technique improves picture quality and increases

Pabitra Pal and Debasis Giri contributed equally to this work.

S. K. Bhunia
Department of Commerce, Rishi Bankim Chandra Evening College affiliated to West Bengal State University, North 24 Parganas, Naihati 743165, West Bengal, India

P. Pal (✉)
Department of Computer Applications, Maulana Abul Kalam Azad University of Technology, Haringhata, Nadia 741249, West Bengal, India
e-mail: pabipaltra@gmail.com

D. Giri
Department of Information Technology, Maulana Abul Kalam Azad University of Technology, Haringhata, Nadia 741249, West Bengal, India

J. K. Mandal et al. (eds.), *Proceedings of International Conference on Network Security and Blockchain Technology*, Lecture Notes in Networks and Systems 738,
https://doi.org/10.1007/978-981-99-4433-0_19

data memory capacity. In this work, an interpolation-based watermarking scheme was used to obtain a highly perceptible watermarked image with a high embedding capacity. Thus data hiding using the interpolation method increases the ability to hide the data and enhances the picture quality. This method enlarges the overall pixel count, conveying a hidden watermark bitstream. It restricts highest pixels deflection.

The remaining work is divided into the following sections. The state-of-the-art works have been reviewed in Sect. 2. The used techniques are introduced in Sect. 3 follows. Next, the experimental results and discussions are elaborated in Sect. 4. Lastly, the conclusion is given in Sect. 5.

2 Related Work

Several study papers have recently been published on several digital picture platforms. Jana et al. [1] published two types of symmetry keys for data hiding and restoration introducing Hamming code. Jung et al. [2] suggested a data hiding method to provide authentication of embedding for forgery attacks. Yao et al. [3] presented an overall structure for a position-based reversible data masking scheme. This pattern has been suggested to build integration capacity. Jung and Yoo et al. [4] introduced RDH techniques using Average neighbor Interpolation (NMI), providing high storage data capacity. Pal et al. [5] represented a watermarking method of double image by applying the quorum function and interpolation technique for better data security. Hassan et al. [6] provide an efficient RDH technique based on interpolation optimization that achieves a good PSNR value. Parah et al. [7] developed a new RDH process for eHealth applications including high capacity of data hiding. Yao et al. [3] proposed a general framework for a two-image-based RDH scheme with a shiftable position. Darwish et al. [8] proposal for a new model for protecting image copyrights based on the fusion of consecutive and segmented watermarks in double images. Chowdhuri et al. [9] proposal for a new steganographic method for authentication and detection of manipulation in double images. Mohammad et al. [10] developed a data stash technique using image interpolation to enhance the ability to integrity. Yalman et al. [11] proposed an RDH scheme with R-weighted coding and image interpolation.

This publication presented a watermarking scheme of double images using interpolation to produce better embedding capacity and improve better quality of the watermark images. It also protects digital documents from unauthorized alteration. It is impossible to extract images without a secret key.

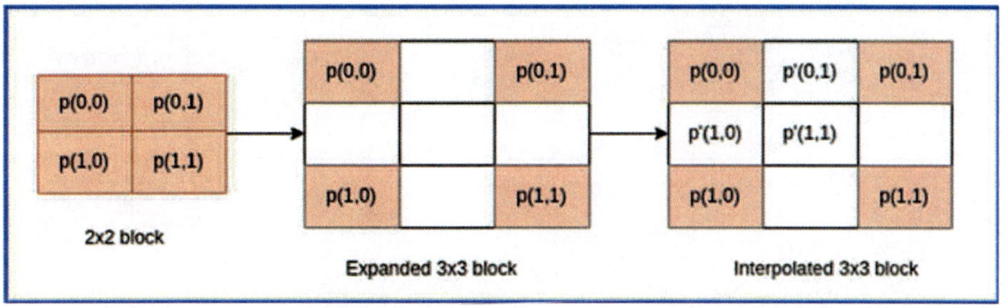

Fig. 1 Block diagram of interpolation

3 Contribution of Watermarking Method

In the current work, the embedding and extraction methods have been discussed and elaborated in the Pre-processing Phase, Embedding Phase, and Extraction Phase, respectively.

3.1 Pre-processing Phase

Here, at first, the Host image is considered, and then the division of the image is (2×2) blocks. These (2×2) blocks are interpolated into (3×3) blocks using the interpolation Eq. 1.

$$\hat{\alpha}\,(0,1) \equiv \left[\frac{\alpha\,(0,0) + \alpha\,(0,1)}{4} + \frac{\alpha\,(1,0) + \alpha\,(1,1)}{6}\right] \tag{1}$$

$$\hat{\alpha}\,(1,0) \equiv \left[\frac{(\alpha(0,0) + \alpha(1,0)}{4} + \frac{\alpha(0,1) + \alpha(1,1)}{6}\right]$$

$$\hat{\alpha}\,(1,1) \equiv \left[\frac{\alpha(0,0) + \alpha(1,0) + \alpha(0,1) + \alpha(1,1)}{4}\right]$$

Every (2×2) block is expanded in a block (3×3) by entering a blank line and column. Then the empty space will be filled by introducing an interpolation algorithm. The dilated block will be computed by Eq. 1. For each (2×2) block shown in Fig. 1, $\alpha(0,0), \alpha(0,1), \alpha(1,0), \alpha(1,1)$ are the earliest pixel, and an interpolated pixel is denoted by $(\hat{\alpha})$. The important part is that two empty pixels (except for the corner pixels, which remain unchanged) within a dilated images are computed according to X-axis and Y-axis adjacent blocks.

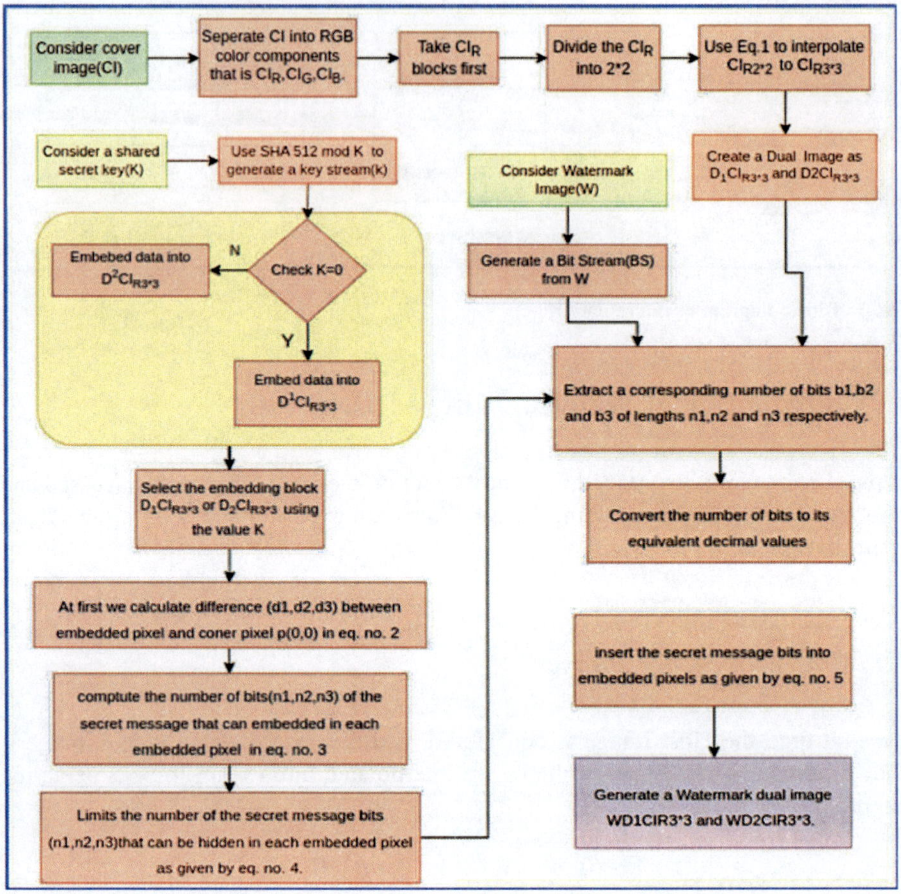

Fig. 2 Block diagram of embedding process

3.2 Embedding Phase

The detailed embedding algorithm is discussed here. The basic block diagram of the embedding scheme is shown in Fig. 2.

At first, a color cover image is chosen for the watermarking process. Then repeat horizontally to create $\mathrm{CI}_{(2\times2)}$ overlapping blocks of each of the four adjacent cells, which is then considered a cover image (CI). After that, the cover image (CI) was divided into RGB color components called CI_R, CI_G, and CI_B. The Red block (CI_R) is considered first for the embedding process. After that each $\mathrm{CI}_{R_{2\times2}}$ has been interpolate to (3×3) blocks named as $\mathrm{ICI}_{R_{3\times3}}$ by using interpolation rule given in Eq. 1. A dual illusion of the cover image is generated and named as $D^1\mathrm{ICI}_{R_{3\times3}}$ and $D^2\mathrm{ICI}_{R_{3\times3}}$. Now a shared secret key K is considered and generates a 512-bit key stream (κ) using the SHA-512 hash algorithm. Now based on the κ value, a randomization of the block chosen has been provided. Either block is chosen based on the K's value. The data embedding starts by embedding data into the interpolated portion only and selecting interpolated pixels is performed in zigzag form according to X-axis and Y-axis approaches.

The details algorithmic description has been illustrated in the following:

Step 1: The data has been embedded based on the payload computed between the color shading embedded pixels and the pixel $\alpha(0, 0)$. Equation 2 protects the process of overflowing by the upper bound of 255 integrated pixels which corresponds to the maximum pixel values in the images.

$$\begin{aligned}\beta(0,1) &\equiv \text{Minimum}\left[\alpha(0,1) - \alpha(0,0), (255 - \alpha(0,1))\right]\\ \beta(1,0) &\equiv \text{Minimum}\left[\alpha(1,0) - \alpha(0,0), (255 - \alpha(1,0))\right]\\ \beta(1,1) &\equiv \text{Minimum}\left[\alpha(1,1) - \alpha(0,0), (255 - \alpha(1,1))\right]\end{aligned} \tag{2}$$

Step 2: now Eq. 3 has been used to count the bits number of the watermark bitstream that can be embedded in each pixel.

$$\begin{aligned}\gamma(0,1) &\equiv \left\lfloor \log_2(\beta(0,1)) \right\rfloor\\ \gamma(1,0) &\equiv \left\lfloor \log_2(\beta(1,0)) \right\rfloor\\ \gamma(1,1) &\equiv \left\lfloor \log_2(\beta(1,1)) \right\rfloor\end{aligned} \tag{3}$$

Step 3: In his step, the number of hidden watermark bits in each embedded picture element is restricted by Eq. 4. This step is very important to restrict the highest deformation in a unique picture element to an admissible size. There is a restriction of the values of $\gamma(i, j)$ to the highest value of n. The calculation of values is given below.

$$\begin{aligned}\gamma(0,1) &\equiv \text{Minimum}\left(\left\lfloor \log_2(\beta(0,1)) \right\rfloor, n\right)\\ \gamma(1,0) &\equiv \text{Minimum}\left(\left\lfloor \log_2(\beta(1,0)) \right\rfloor, n\right)\\ \gamma(1,1) &\equiv \text{Minimum}\left(\left\lfloor \log_2(\beta(1,1)) \right\rfloor, n\right)\end{aligned} \tag{4}$$

Here $n = 1, 2, 3, 4, 5, 6, 7, 8$. If one select n is equal to 1, then Embedding results show high image quality and low embedding, but if $n = 9$, then results show a higher embedding capacity corresponding to the image quality.

Step 4: Take an appropriate bit lengths B_1, B_2 and B_3 of the lengths $\gamma(0, 1)$, $\gamma(1, 0)$, and $\gamma(1, 1)$, from the watermark bits stream and is converted into the corresponding decimal values $D1$, $D2$, and $D3$.

Step 5: Insert the watermark bits into embedded pixels by Eq. 5.

$$\begin{aligned}\hat{\alpha}(0,1) &\equiv \alpha(0,1) + D1\\ \hat{\alpha}(1,0) &\equiv \alpha(1,0) + D2\\ \hat{\alpha}(1,1) &\equiv \alpha(1,1) + D3\end{aligned} \tag{5}$$

The process of embedding hidden watermarks continues in each pixel block until the end of the secret bitstream is reached or all blocks are covered. This is a very important point because it basically depends on the hidden watermark bitstream size and the numeric value n, have to be provided with the watermark image. To activate the target page extract the secret watermark bitstream. This process has been identified by adding 3 bytes and concatenated bits stream to get the whole secret bits stream. These 3 bytes are integrated using the above algorithm and applying $n \equiv 8$.

3.3 Extraction Phase

The extraction of watermarks and the restoration of images is the reverse algorithm of the data integration process. Bit extraction and image retrieval are blind because they generated the secret bit flow of the integration watermark image to get the original cover image. At first, dividing the watermarked embedded image into non-overlapping blocks (3×3) in the same procedure and flow of the embedding process. The dual image has been chosen using the SHA-256 and generated the original cover image by using the reverse procedure of embedding.

It is noted that the corner pixel $\alpha(i, j)$ will remain the same during the entire process of image block expansion, insertion, and image recovery. The important factor is that the secret bits stream is embedded only with the pixel $\hat{\alpha}\ (i, j)$ and will remain two empty pixels. The image recovery process is described below.

Here is the value of $k \equiv 1$, So we choose WD2CIR3*3.

Step 1: At first, the values of the original cover image will be generated from the embedding image by Eq. 6.

$$
\begin{aligned}
\alpha\,(0,1) &\equiv \left\lfloor \frac{\alpha\,(0,0)+\alpha\,(0,2)}{4} + \frac{\alpha\,(2,0)+\alpha\,(2,2)}{6} \right\rfloor \\
\alpha\,(1,0) &\equiv \left\lfloor \frac{\alpha\,(0,0)+\alpha\,(2,0)}{4} + \frac{\alpha\,(0,2)+\alpha\,(2,2)}{6} \right\rfloor \\
\alpha\,(1,0) &\equiv \left\lfloor \frac{\alpha\,(0,0)+\alpha\,(2,0)+\alpha\,(0,2)+\alpha\,(2,2)}{4} \right\rfloor
\end{aligned}
\tag{6}
$$

Step 2: The next step is to generate a decimal number of hidden watermark bits in each and every pixel $\hat{\alpha}\ (i, j)$ by Eq. 7.

$$
\begin{aligned}
D_1 &\equiv \hat{\alpha}\,(0,1) - \alpha\,(0,1) \\
D_2 &\equiv \hat{\alpha}\,(1,0) - \alpha\,(1,0) \\
D_3 &\equiv \hat{\alpha}\,(1,1) - \alpha\,(1,1)
\end{aligned}
\tag{7}
$$

Step 3: Then, to determine the number of hidden watermark bits in each pixel $\hat{\alpha}\ (i, j)$ by Eq. 8.

$$\begin{aligned}\beta(0,1) &\equiv \text{Minimum}\left[(\alpha(0,1)-\alpha(0,0)),(255-\alpha(0,1))\right]\\ \beta(1,0) &\equiv \text{Minimum}\left[(\alpha(1,0)-\alpha(0,0)),(255-\alpha(1,0))\right]\\ \beta(1,1) &\equiv \text{Minimum}\left[(\alpha(1,1)-\alpha(0,0)),(255-\alpha(1,1))\right]\end{aligned} \tag{8}$$

Step 4: After determining the number of bits, Eq. 9 will help to determine the maximum length of bits.

$$\begin{aligned}\gamma(0,1) &\equiv \text{Minimum}\left(\left\lfloor \log_2(\alpha(0,1)-\alpha(0,0))\right\rfloor, n\right)\\ \gamma(1,0) &\equiv \text{Minimum}\left(\left\lfloor \log_2(\alpha(1,0)-\alpha(0,0))\right\rfloor, n\right)\\ \gamma(1,1) &\equiv \text{Minimum}\left(\left\lfloor \log_2(\alpha(1,1)-\alpha(0,0))\right\rfloor, n\right)\end{aligned} \tag{9}$$

Step 5: In this step decimal values obtained in **Step 2** are converted into binary equivalent B_1, B_2, B_3 with the length obtained in **Step 4**.
Step 6: after the binary equivalent values have been found, The values are concatenated to obtain watermark secret bits $W \equiv B_1B_2B_3$.
Step 7: For creating the original cover image, There must be changed the pixel values $\hat{\alpha}(0,1)$, $\hat{\alpha}(1,0)$, $\hat{\alpha}(1,1)$ with the pixel values $\alpha(0,1)$, $\alpha(1,0)$, $\alpha(1,1)$ in **Step 1** accordingly.
Step 8: This process continues until the secret watermark bits are generated.

4 Experimental Results and Discussion

This section describes the experimental results. The different types of the metrics evaluation process are PSNR, BER, Q-index, Structural Similarity Index Measurement, Correlation-coefficient, Normalize correlation coefficient (NCC), Universal image quality index (UIQI) and Standard Deviation (SD). These types of metrics have been introduced to analyze the measurement of this algorithm technique. To deploy the watermark algorithm, a standard database of images such as USC-SIPI image database was used for providing an exclusive test bed of colorful images of sized (512×512). The experiments have been executed in MATLAB R2016b using an image processing toolbox and Windows 10 operating systems.

4.1 Analysis

For two ($M \times N$) images, the peak-to-noise signal (PSNR) measurement measures the standard of the reconstructed image. The PSNR metrics are mostly used to determine the standard of the image. It is the ratio between the maximum signal strength to the maximum error signal strength. The PSNR can be described by Eq. 10.

Table 1 Experimental results with respect to various parameter

Image database	Image	Payload	PSNR	Q-index	Capacity (bits)
USC-SIPI	Lena	2	44.52	0.912	1,966,080
	Baboon	2	44.36	0.910	1,966,080
	Tiffany	2	44.31	0.902	1,966,080
	Pepper	2	44.37	0.910	1,966,080
	Airplane	2	44.41	0.907	1,966,080

Table 2 Results on number of image versus PSNR value

Database	Cover	No. of images	PSNR
USC-SIPI	(513 × 513)	10	44.51
		25	44.41
		50	44.38

Table 3 Different types of metrics

Image dataset	Image	MSE	BER	NCC	Q-index	PSNR	SSIM
USC-SIPI	Lena	2.32	0.01198	0.912	0.912	44.52	0.898
	Baboon	2.31	0.01207	0.907	0.910	44.36	0.901
	Tiffany	2.11	0.01212	0.902	0.902	44.31	0.879
	Pepper	2.15	0.01224	0.903	0.911	44.37	0.897
	Average	2.13	0.01192	0.912	0.907	44.41	0.899

$$\text{PSNR} = 10\log_{10}\frac{225^2}{\text{MSE}},\ \text{MSE} = \frac{\sum_{i=1}^{N}\sum_{j=1}^{M}[\text{CI}(i,j) - \text{DCI}(i,j)]^2}{N \times M}, \quad (10)$$

CI(i, j) is represented as the cover image pixel and DCI(i, j) is represented as the watermark image. PSNR $\geq$ 30 dB, represent good quality visual application to human and always a quality algorithm exist with higher PSNR values.

After evaluation, the available technique shows that the PSNR value is 44 dB after 1,966,080 bits are embedded in the cover image. This means that it can easily measure a standard image quality after integrating 1,966,080 bits of stash information. Then the result of Q-index, the value is 0.908 and this value is close to 1.

In Table 1 shows the testing values of PSNR, Payload, Q-Index, capacity bits. After adding 1,966,080 watermark bits to the Lena image, the results show a 2 bpp payload with a Q-index of 0.908 and a PSNR of 44.39 dB.

Table 2 shows different databases of images of various PSNR values. The result of the experiment generates that the average value of PSNR is 44.39 dB according to the image count 10, 25, 50, respectively.

Table 4 Experimental results on algorithmic complexity

Algorithms	[3]	[6]	[1]	[2]	Proposed algorithm
Times	0.72	0.521	1.32	6.16	0.56

In this proposed technique, There are compared with four simple pictures that is Lena, Baboon, Tiffany, Pepper were taken from the database of USC-SIPI and the testing result are shown in Table 3. The result shows that after embedding the maximum 1,966,080 watermark bits stream the value of PSNR is 44.39 dB (approx).

The complexity of this algorithm is very needful in the application of real-time. The run time of this method has been verified and compared to a few current techniques. The examination query is in Table 4.

5 Conclusion

The proposed technique elaborated the watermarking scheme of dual cover image by introducing interpolation and also applied a Secure Hash Algorithm. Firstly there must construct a dual interpolation image. Then the selection of a double interpolated image follows the generation of the key. When the value of key is equal to 1, the watermark bits hide in the 2nd interpolated matrix, otherwise in the remaining interpolated matrix. This proposed technique increases the data storage capacity compared with previous work. This scheme used the SHA-512 encryption algorithm for better security of data. This proposed technique reduces the distortion in the interpolation expansion step. This technique is vulnerable, and therefore no robustness experiments against attacks have been performed. This technique is appropriate for the authentic transmission of different experimental images of the medical departments, different satellite images of the army, and similar applications from different important departments that require precise coverage image overlay.

References

1. Jana B, Giri D, Kumar Mondal S (2018) Dual image based reversible data hiding scheme using (7, 4) hamming code. Multimed Tools Appl 77(1):763–785
2. Jung K-H, Yoo K-Y (2009) Data hiding method using image interpolation. Comput Stand Interfaces 31(2):465–470
3. Yao H, Qin C, Tang Z, Zhang X (2018) A general framework for shiftable position-based dual-image reversible data hiding. EURASIP J Image Video Process 2018(1):1–10
4. Jung K-H (2018) A survey of interpolation-based reversible data hiding methods. Multimed Tools Appl 77(7):7795–7810
5. Pal P, Jana B, Bhaumik J (2019) Watermarking scheme using local binary pattern for image authentication and tamper detection through dual image. Secur Privacy 2(2):59

6. Hassan FS, Gutub A (2021) Efficient image reversible data hiding technique based on interpolation optimization. Arab J Sci Eng 46(9):8441–8456
7. Parah SA, Ahad F, Sheikh JA, Loan NA, Bhat GM (2017) A new reversible and high capacity data hiding technique for e-healthcare applications. Multimed Tools Appl 76(3):3943–3975
8. Darwish SM, Al-Khafaji LDS (2020) Dual watermarking for color images: a new image copyright protection model based on the fusion of successive and segmented watermarking. Multimed Tools Appl 79(9):6503–6530
9. Chowdhuri P, Pal P, Jana B (2018) A new dual image-based steganographic scheme for authentication and tampered detection. Inf Technol Appl Math ICITAM 699:163
10. Mohammad AA, Al-Haj A, Farfoura M (2019) An improved capacity data hiding technique based on image interpolation. Multimed Tools Appl 78(6):7181–7205
11. Yalman Y, Akar F, Erturk I (2010) An image interpolation based reversible data hiding method using r-weighted coding. In: 2010 13th IEEE international conference on computational science and engineering. IEEE, pp 346–350

Network Security and Its Applications

Congestion Control Enhancement in TCP

Vishwanath Chikkareddi, Vinaykumar Chikaraddi, Santosh Chinchali, and Chidanand Kusur

1 Introduction

Computer networks play major role when communication between computers is considered. To make the communication possible in all physical conditions, OSI layer is developed. OSI layers separates the functionalities that needed in communication steps. One of the seven layers of OSI layers is Transport layer. Transport layers deals with end-to-end communication. End-to-end communication is between sender and receiver, Transport layers ignores all the other activities that needed to make this communication possible and relays the duty of routers and message passing through different networks to layers below it. Transport layer majorly utilizes two types of mode, TCP and UDP. TCP is connection oriented while UDP is connection less. TCP guarantees delivery of packets in orderly fashion. As the application in modern world require more network utilization, chances of network getting congested is increase.

A. *Background*

While there are different ways to detect congestion, different method suitable for each method is developed. TCP at its common algorithm to tackle congestion problem uses TCP Reno [1]. An improvement on this TCP veno [2] is suggested which is also combination of a different algorithm TCP vegas and native TCP Reno. Reno, veno uses packet loss as signal for congestion. Similarly algorithm exists for other methods. Approaches so far used are packet loss-based, delay-based, and hybrid. Another

V. Chikkareddi (✉) · S. Chinchali · C. Kusur
B.L.D.E.A's V.P. Dr. P.G. Halakatti College of Engineering and Technology, Vijayapura, India
e-mail: ise.vishwanath@bldeacet.ac.in

S. Chinchali
e-mail: cse.santoshchinchali@bldeacet.ac.in

V. Chikaraddi
Department of BCA, BLDEA's A.S. Patil College of Commerce (Autonomous), Vijayapura, India

J. K. Mandal et al. (eds.), *Proceedings of International Conference on Network Security and Blockchain Technology*, Lecture Notes in Networks and Systems 738,
https://doi.org/10.1007/978-981-99-4433-0_20

approach that is used is to make use of explicit feedback from the network to achieve high performance. It essentially asks for congestion signals from router, i.e., communication with next layer for taking decision. Both layers, Transport and Network layer working together gives better results than previous methods mentioned.

B. *Motivation*

Congestion control in TCP is a crucial task. There are many congestion control techniques exists. The aim of these methods is to prevent congestion and also to maximize the use of bandwidth available. Though packet loss is one of the indication that congestion has occurred it cannot be useful if network is lossy such as wireless network which are more prone to packet loss due to variable bandwidth or interference [1]. Delay-based approaches use queue delay as indication of congestion. Mainly RTT is used for estimating the delay. But variation in delay cannot be always considered due to presence of congestion [2]. Hybrid approach make use of both above approaches as input to determine congestion. Though this method works well for most of the applications, it is challenging to deliver quality performance for the upcoming new generation of high-capacity wireless networks like LTE and WiMax [3]. To deliver quality performance, network's explicit feedback is employed by most of the protocols. The majority of these either have performance issues as a result of inadequate congestion feedback or require more feedback bits than are provided in the IP header. Lately, protocols that combine the explicit congestion notification (ECN) marks of many packets and use these estimations to direct the multiplicative increase, additive increase, multiplicative decline (MI-AI-MD) window adaptation have gained attention [4–8, 10, 11].

2 TCP-FI

The congestion control window is directly adjusted via the additive increase/multiplicative decrease (AIMD) technique by TCP-FIT [9]. When a packet loss event occurs, the AIMD infuse exponential reduction with linear expansion of the congestion control window C. General terms for the AIMD's congestion control window C adjustment equation include,

$$\text{For every RTT: } C \leftarrow C + x,$$
$$\text{For every Loss: } C \leftarrow C - y \cdot C.$$

Here x and y represent increasing and decreasing factors of AIMD, respectively. In TCP Reno, $x = 1$ and $y = 1/2$. To achieve N times throughput of the TCP Reno, the MulTCP sets $x = N$ and $y = 1/2N$. In the TCP-FIT, C is decreased by a factor of $y = 2/(3N + 1)$ instead of $y = 1/2N$, when a packet loss occurs. This helps to keep the throughput of TCP-FIT flows, exactly N times of the TCP Reno with similar network parameters. The congestion control window C adjustment equation for TCP-FIT can be given as,

$$\text{Each RTT: } C \leftarrow C + N,$$
$$\text{Each Loss: C} \leftarrow C - (2 \cdot C/3N + 1).$$

In modification of TCP-FIT, this RTT-based calculation is removed and data from network layer is taken into consideration. Thus, provides more accurate calculation. The calculation done for N in TCP-FIT is for estimation of Queue delay in overall network. It is therefore better to get value of such parameter directly from layer itself rather than calculating depending on amount of time taken by packet to reach through network on multiple successions.

3 Modification of TCP-FIT

This protocol model works in two separate parts, first being the part of TCP-FIT and later is for better performance. The TCP-FIT, is inspired from mulTCP. It is a concept of running parallel virtual TCP connection under one TCP. Since these are Virtual Connections window size is changed in main TCP as if actual TCP connections are running. It is observed that running any number of connections doesn't linearly increase throughput. After a certain threshold throughput degrades. For the native implementation authors have used queuing delay as the measure for number of TCP connections. As the number of connection increases queue space becomes the bottleneck and hence degradation starts.

The other part tries to remove the error that may be caused in TCP-FIT, as TCP-FIT uses RTT for estimation. Using time quantity for estimation in a network with high BDP, gives outdated information to source. Hence it is better that router itself provides the data needed for estimation. This gives better approximation of load at router. Exata contains base queue model, which provide numerous utility functions and variables, among them queue length, queue delay, and average queue length are used in this model. The path from source to destination is calculated with the help of routing table at each node. Entries in routing table corresponding to destination are accumulated and average queue lengths for respective interfaces is used. The min average queue is among all the interfaces is provided to Transport layer. Later while deciding window size TCP-FIT can use this size for estimation.

As one can see TCP-FIT though achieving good throughput can do better if number of connections are adjusted accurately. BMCC uses similar technique to find if the node congested by the help of ECN markings and estimating load factor. Proposed system takes the merits of cross layer communication as done in BMCC to estimate load factor and improve utilization. The same is depicted in Fig. 1.

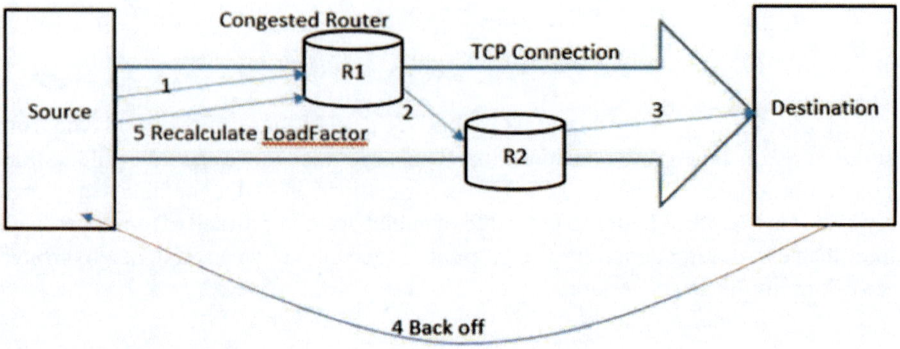

Fig. 1 Modified TCP-FIT

4 Results and Analysis

In this section, results of TCP-FIT and Reno is compared upon a common scenario. The metric used for same are mentioned below. The objective of this comparison is to show the difference in results due to enhancement done in TCP-FIT over Reno. For the testing of the protocol, scenario as depicted in Fig. 2 is used. This scenario is consisted of 12 nodes connected to each other through different network. Three nodes namely 1, 3, and 10 are transmitting FTP generic traffic, causing TCP packets in network. Link connecting nodes are having different bandwidth allocated. Further, to make sure that congestion occurs, the queue size and also the number of queues allocated to each interface of each node is kept to 2048 and 1, respectively.

Since TCP-FIT increases window depending on number of packets which are sent but not delivered. The very increase in window size doesn't increase congestion but only make use of available bandwidth. Therefore, the average jitter as in Fig. 4 is lower than that of Reno.

Till the congestion occurs in network, TCP-FIT and Reno works exactly in same way. But after congestion occurred TCP-FIT tries to send more number of packets to increase throughput.

Thus, Figs. 3 and 4 show more number of packets being sent by TCP-FIT.

In Fig. 5 average hop count of each protocol is compared. Though there is not much variation (0.02 metric), it's due to same topology being tested, average hop count, however, decreases slightly in modified TCP-FIT, as no of packets sent from source to destination are increased and thus more no of packets dropped in same model.

In Fig. 6 for Reno is minimum, and as that of Base TCP-FIT is maximum. The reason for being so is that, length of queue is directly proportional to the amount of data sent. Thus, in Reno as after occurrence of congestion less number of window of TCP hence the less no of packets in network are sent. Whereas in TCP-FIT, the window size updates exponentially till queue reaches its capacity as depicted by RTT. Also, in modified TCP-FIT node itself provides data of Q-delay and thus

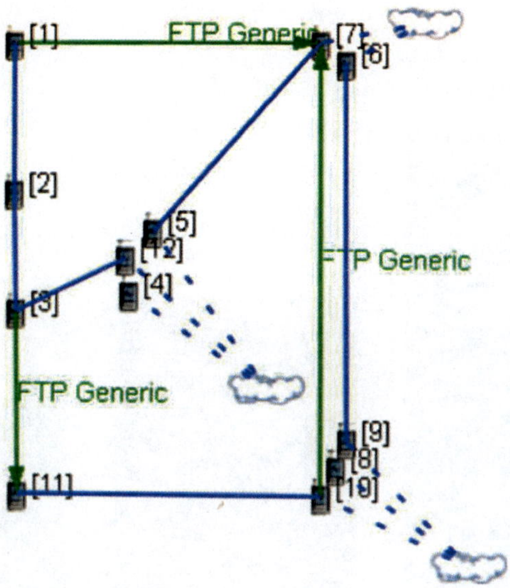

Fig. 2 Scenario used for experiment

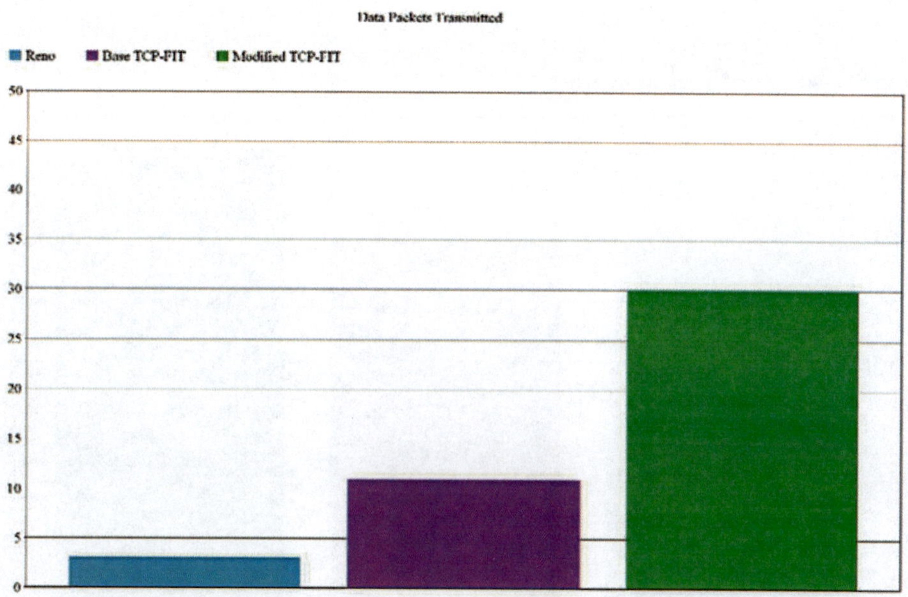

Fig. 3 Data packets transmitted

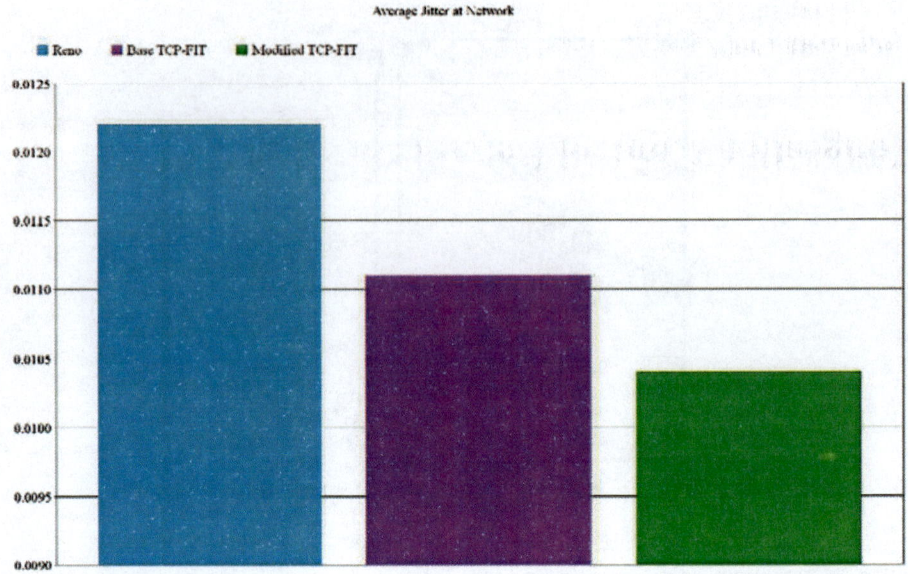

Fig. 4 Average jitter at network

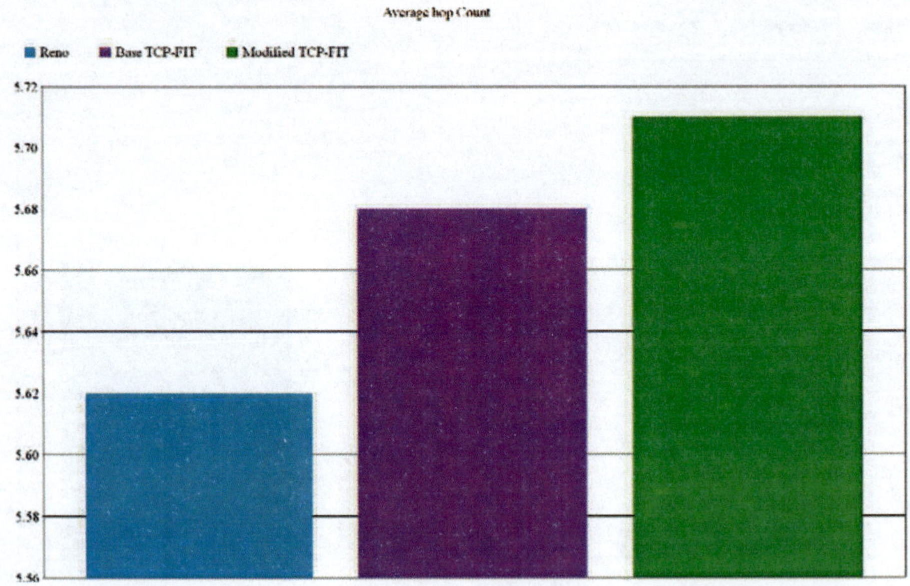

Fig. 5 Average hop count

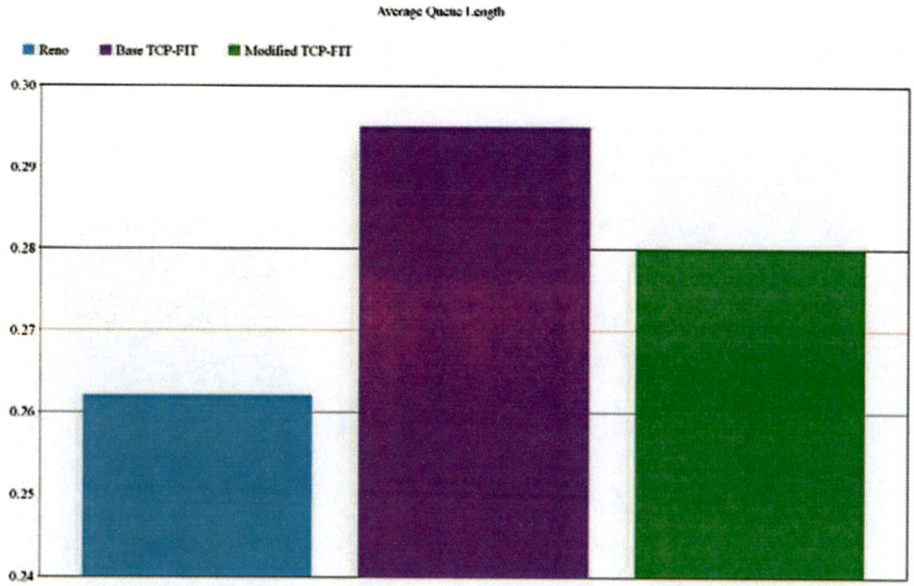

Fig. 6 Average queue length at network

current status of queue, number of packets sent by modification is comparatively more appropriate than TCP-FIT.

The reason for which, average delay shown in Fig. 7, says Reno has experience more delay is that, as congestion occurs, for doing fast recovery from congestion, Reno decreases window size and then increase by one till next congestion is detected. Therefore, number of segments going out of transport is more than that of TCP-FIT-based protocol which increases window size exponentially.

As window size increase the number of control and data packets sent for the TCP-FIT is also increases. Thus modified TCP-FIT sending more number of control packets shows more overhead than the rest two protocols in Fig. 8.

Figures 9 and 10 show the comparisons of behavior of window sizes in protocols mentioned. Since window size in modified TCP-FIT is more, and also due to changes in back off parameters of TCP-FIT, it sends more no of packets than Reno and TCP-FIT. Windows sizes are in native TCP are controlled in AIMD manner, where after congestion discovery windows will be updated to minimum value and then try to recover until it reaches threshold. Whereas in TCP-FIT the factor is always manipulated N times, if increasing window is not working toward recovery from congestion then window is reduced by factor of N, the same factor used in increase. In modified version of TCP-FIT values from network layer are taken into consideration.

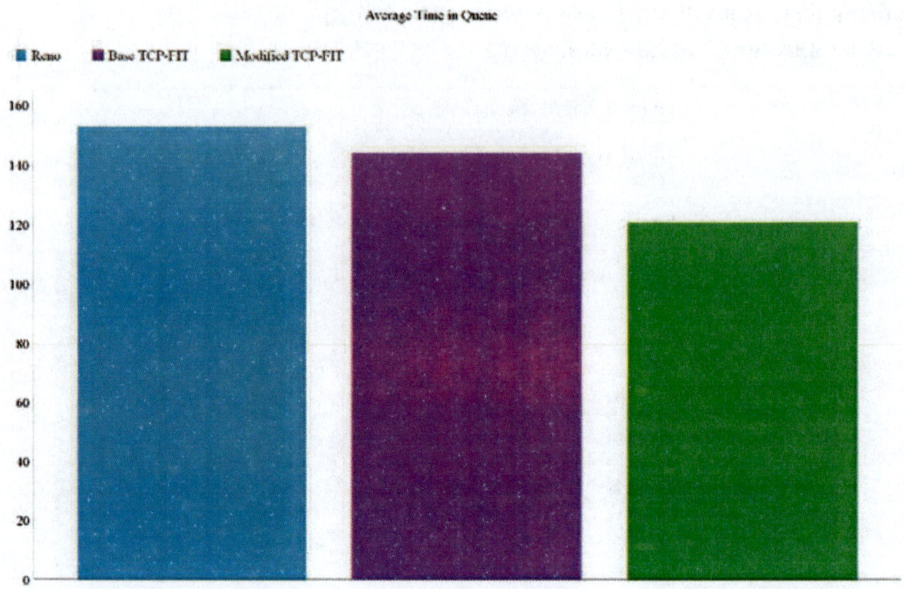

Fig. 7 Average time in queue

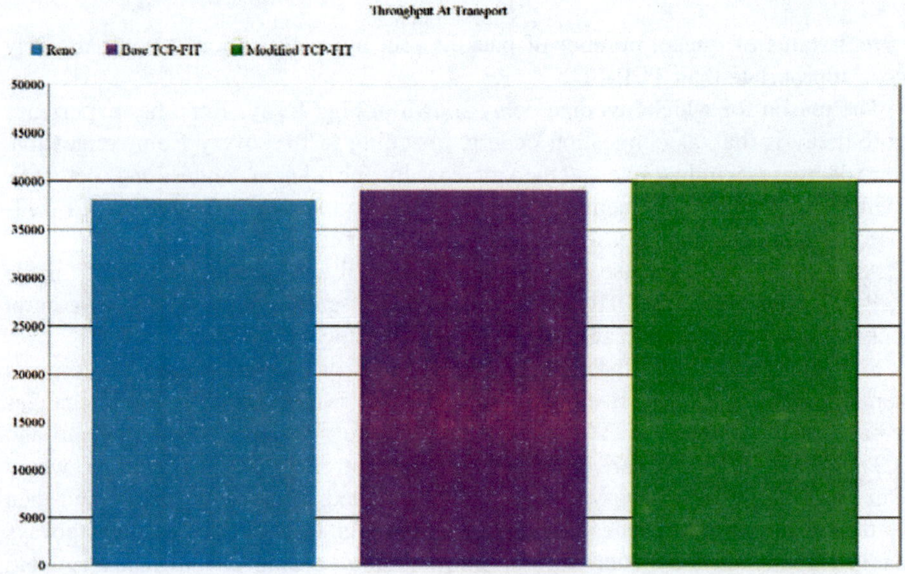

Fig. 8 Throughput at transport

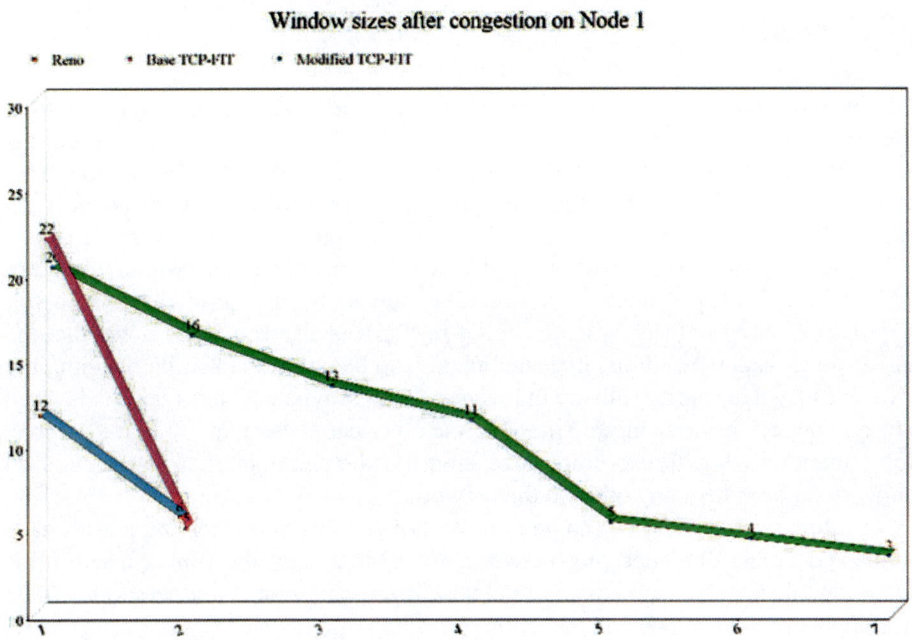

Fig. 9 Window comparison on Node 1

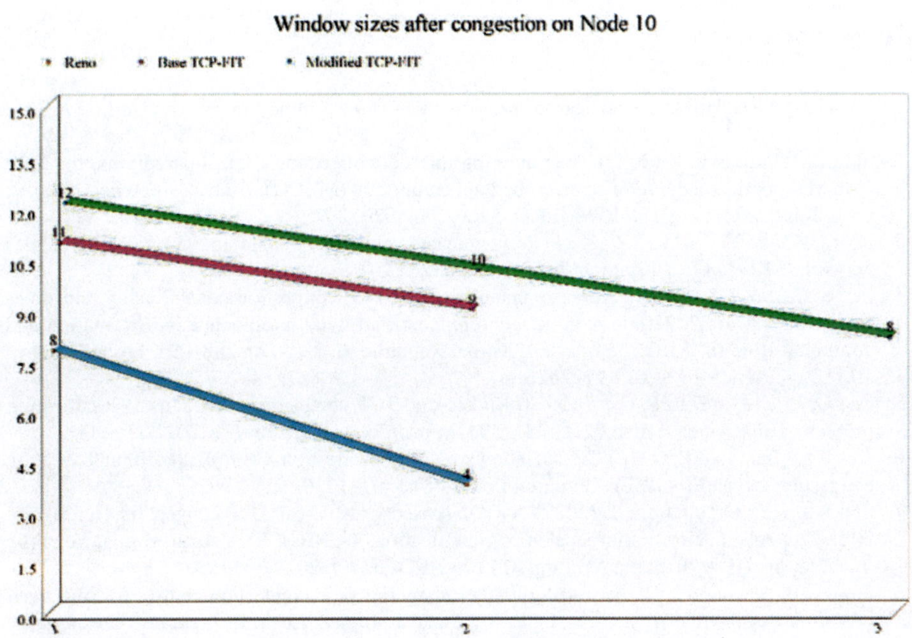

Fig. 10 Window comparison on Node 10

5 Conclusion

It is been observed that by the use of intermediate network devices, such as router, the congestion level can be estimated. The Dissertation concerns are based on this very idea and proposes a method to use this advantage for better estimation. This dissertation provides idea of using queue delay as measure of the network congestion, which significantly improves results obtained from previous methods as TCP-FIT. After all, whenever the congestion of TCP is to be controlled, only window size can be helpful, as it the parameter which deciding how loaded the network is. The better approximation will always give the better result. It is also should be noted that the queue delay experienced on congested node is to be considered while deciding the window size. Because any other window more than in sync with this queue delay will always congest the node more. Similarly, the other parameters, in future if available, for estimation, should also emphasize more on the status of such parameters on congested nodes than any other in the network.

In future work, TCP-FIT can be improved, if sending and receiving parties have more knowledge of underlying network. As BMCC uses the Link capacity, this Dissertation suggest involvement of network layer, such more parameters should be used for improved congestion detection.

References

1. Postel J (1981) Transmission control protocol, RFC 793 [online]. http://www.ietf.org/rfc/rfc793.txt
2. Tafa Z, Milutinovic V (2022) The emerging internet congestion control paradigms. In: 2022 11th Mediterranean conference on embedded computing (MECO), Budva, Montenegro, 2022, pp 1–5. https://doi.org/10.1109/MECO55406.2022.9797207
3. Kim GH, Seo WK, Cho YZ (2021) Performance evaluations of TCP in 5G mmWave cellular network. J Korean Inst Commun Inf Sci 46:2237–2250
4. Seo S-J, Cho Y-Z (2022) Fairness enhancement of TCP congestion control using reinforcement learning. In: 2022 international conference on artificial intelligence in information and communication (ICAIIC), Jeju Island, Korea, Republic of, 2022, pp 288–291. https://doi.org/10.1109/ICAIIC54071.2022.9722626
5. Sun G, Li C, Ma Y, Li S, Qiu J (2023) End-to-end TCP congestion control as a classification problem. IEEE Trans Reliab 72(1):384–394. https://doi.org/10.1109/TR.2022.3172335
6. Jiang X, Jin G (2015) CLTCP: an adaptive TCP congestion control algorithm based on congestion Level. Proc IEEE Commun Lett 19(8):1307–1310
7. Seo S-J, Kim G-H, Cho Y-Z (2022) A DQN-based CUBIC for TCP congestion control. In: 2022 27th Asia Pacific conference on communications (APCC), Jeju Island, Korea, Republic of, 2022, pp 419–420. https://doi.org/10.1109/APCC55198.2022.9943650
8. Francis B, Narasimhan V, Nayak A (2012) Enhancing TCP congestion control for improved performance in wireless networks. In: Li XY, Papavassiliou S, Ruehrup S (eds) Ad-hoc, mobile, and wireless networks. ADHOC-NOW 2012. Lecture notes in computer science, vol 7363. Springer, Berlin, Heidelberg
9. Wang J, Wen J, Zhang J, Han Y (2011) TCP-FIT: an improved TCP congestion control algorithm and its performance. In: Proceedings of IEEE on INFOCOM, pp 2894–2902

10. Srivastava A, Fund F, Panwar SS (2022) Coexistence of delay-based TCP congestion control: challenges and opportunities. In: 2022 IEEE international workshop technical committee on communications quality and reliability (CQR), Arlington, VA, USA, 2022, pp 43–48. https://doi.org/10.1109/CQR54764.2022.9918593
11. Athuraliya S, Li VH, Low SH, Yin Q (2001) REM: active queue management. Teletraffic Sci Eng 4. https://doi.org/10.1016/S1388-3437(01)80172-4

Classification of Bruteforce Attacks Using Convolution Neural Network

Srikakulapu Bhavitha, S. Kranthi, and Adapaka Sai Kishore

1 Introduction

One of the issues with the current common occurrence of computer networks is the abundance of cyber-attacks that occur on a daily basis. These attacks have resulted in cyberwarfare and cyberterrorism, both of which can be extremely aggravating to deal with. If this is successful, data loss, hardware degradation, and server degradation may occur.

To address these issues, one approach used to prevent such cyber-attacks is the Intrusion Detection System (IDS). To extract features from network monitoring data and categorize the type of attack, we proposed a deep learning model. We used a subset of the CIC-IDS-2018 dataset to evaluate our model.

S. Bhavitha (✉) · S. Kranthi · A. Sai Kishore
Department of Information Technology, Velagapudi Ramakrishna Siddhartha Engineering College, Vijayawada, Andhra Pradesh, India
e-mail: bhavithasrikakulapu@gmail.com

S. Kranthi
e-mail: kranthiit@vrsiddhartha.ac.in

A. Sai Kishore
e-mail: asaikishore2002@gmail.com

J. K. Mandal et al. (eds.), *Proceedings of International Conference on Network Security and Blockchain Technology*, Lecture Notes in Networks and Systems 738,
https://doi.org/10.1007/978-981-99-4433-0_21

2 Preliminaries

This section defines the key terms and concepts used in this paper.

2.1 *Threat*

A potentially harmful action or event enabled by a vulnerability that has an unfavorable impact on a computer system or application.

2.2 *Attack*

An attack is the act of gaining unauthorized access that may result in a huge damage and lead to data loss.

2.3 *Bruteforce Attack*

Bruteforce attacks on networks are very common because they are used to break into accounts that have weak username and password combinations.

2.4 *Deep Learning*

Machine learning is superset of deep learning. Deep learning is a multi-layered neural network which generally consists of two or more layers. Large amounts of data is fed to the deep learning models to allow it to learn from the data similar to that of the human brain. There are additional hidden layers that helps in optimizing and improving the accuracy of the model.

2.5 *CNN Algorithm*

A CNNs, or convolutional neural networks, are data-processing deep learning neural networks. It is a multi-layered feed-forward neural network built by stacking many unseen layers on top of one another in a specific order.

2.6 Data Pre-processing

The process of formatting the raw data to make it more effective is known as Data Pre-Processing. This formatted data can be used for analysis by various classification model.

3 Related Work

This section gives the brief information about the resources that provided a base for the proposed research of intrusion detection and attack classification methods. Various modern algorithms are used in the research publications for the extraction of features and classification of the network attacks. The study of the research and the review papers helped in proposing a deep learning model for the classification of attacks and to use an effective network traffic dataset. Thus, we require to present a model with better accuracy that runs at a high speed.

Bagaa [1] proposed a security framework that helps in anomaly detection using the artificial intelligence. Various machine learning models are used in the NFV and SDN for providing an aggregated security policy.

Sahu [2] suggested various pre-processing techniques to normalize and reduce data volume using systematic and planned techniques like PCA. Various co-relation factors are also identified for developing a balanced dataset.

Akhil Reddy [3] presented a classification model to detect and classify network attacks using artificial neural networks. They evaluated this model using KDD cup 99 network traffic dataset.

Ubale [4] the SDN framework was used to discuss the DoS network attack. They conducted a survey to raise awareness about the vulnerability associated with the SDN.

Karatas [5] six different machine learning models were used to implement the CSE-CICIDS2018 dataset. The models used here are Decision tree, Random forest, KNN, Adaboost, Linear discriminate analysis, and Gradient boosting.

Amir Ali [6] the intruders are identified using a three-tire system IDPS by validating the user, packet, and flow information. The packets are classified as malicious and normal using this system. OMNeT++ environment is used for evaluation of the model. Features like failure rate, intrusion detection rate, throughput and traffic load are provided to view the performance of the developed model.

Thakkar and Lohiya [7] this study discovered that the underlying dataset needs to be updated in order to identify the various attacks in the field of intrusion detection with improved performance. Furthermore, the pattern of carrying out various attacks simulates the requirement for datasets with realistic network scenarios. CIC-IDS-2017 and CSE-CICIDS-2018 datasets have been introduced to meet the IDS dataset resembling the realistic network traffic.

4 System Architecture

The network traffic dataset is preprocessed to filter the noisy data, encode the label to numeric format and remove the null values. The data is augmented by resampling to ensure proper distribution of data. The preprocessed dataset is further classified as train and test data. The train data is sent to the proposed convolution neural network. The test data is passed onto the trained deep learning model and the model is evaluated using performance metrics like accuracy (Fig. 1).

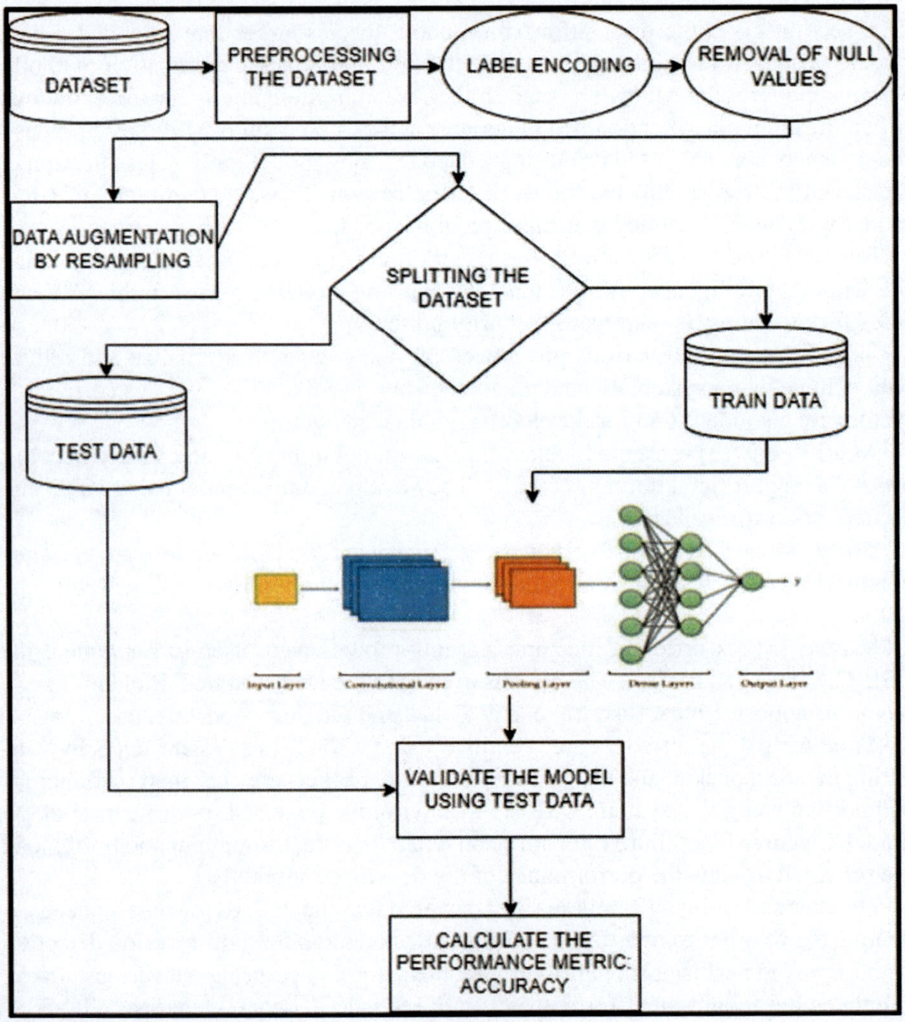

Fig. 1 System architecture

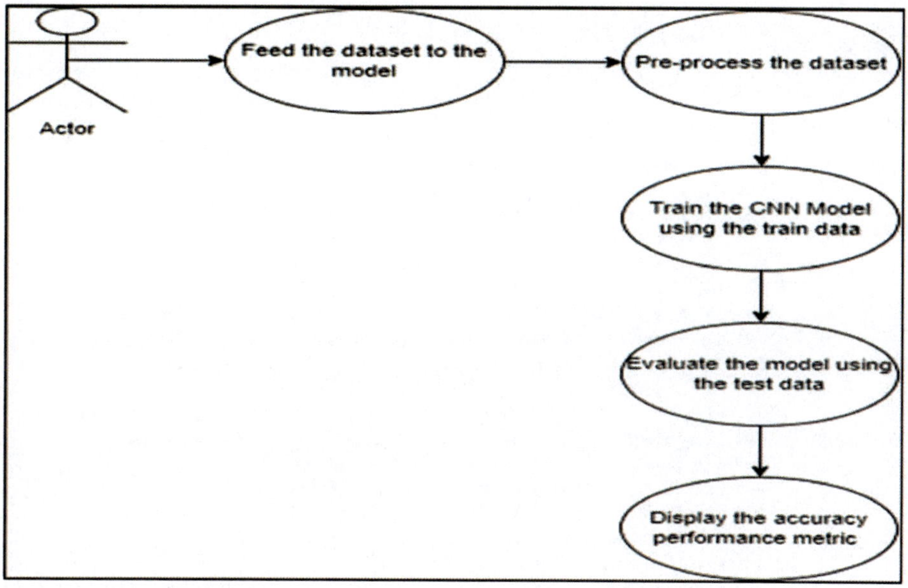

Fig. 2 Design methodology

5 Design Methodology

In this methodology (Fig. 2)

- The model is fed by the user from a pre-defined dataset.
- The file is preprocessed by removing the null values.
- The CNN model is then used to classify the data.
- The model is then evaluated using the test data.
- The performance metric, accuracy is displayed.

6 Dataset Description

- We have a huge amount of data entries (1 million+).
- There are a total of 80 columns belonging to each sample (Fig. 3).
- Labels present in the selected portion of the dataset are shown in Fig. 4.

```
# check the columns in data
network_data.columns

Index(['Dst Port', 'Protocol', 'Timestamp', 'Flow Duration', 'Tot Fwd Pkts',
       'Tot Bwd Pkts', 'TotLen Fwd Pkts', 'TotLen Bwd Pkts', 'Fwd Pkt Len Max',
       'Fwd Pkt Len Min', 'Fwd Pkt Len Mean', 'Fwd Pkt Len Std',
       'Bwd Pkt Len Max', 'Bwd Pkt Len Min', 'Bwd Pkt Len Mean',
       'Bwd Pkt Len Std', 'Flow Byts/s', 'Flow Pkts/s', 'Flow IAT Mean',
       'Flow IAT Std', 'Flow IAT Max', 'Flow IAT Min', 'Fwd IAT Tot',
       'Fwd IAT Mean', 'Fwd IAT Std', 'Fwd IAT Max', 'Fwd IAT Min',
       'Bwd IAT Tot', 'Bwd IAT Mean', 'Bwd IAT Std', 'Bwd IAT Max',
       'Bwd IAT Min', 'Fwd PSH Flags', 'Bwd PSH Flags', 'Fwd URG Flags',
       'Bwd URG Flags', 'Fwd Header Len', 'Bwd Header Len', 'Fwd Pkts/s',
       'Bwd Pkts/s', 'Pkt Len Min', 'Pkt Len Max', 'Pkt Len Mean',
       'Pkt Len Std', 'Pkt Len Var', 'FIN Flag Cnt', 'SYN Flag Cnt',
       'RST Flag Cnt', 'PSH Flag Cnt', 'ACK Flag Cnt', 'URG Flag Cnt',
       'CWE Flag Count', 'ECE Flag Cnt', 'Down/Up Ratio', 'Pkt Size Avg',
       'Fwd Seg Size Avg', 'Bwd Seg Size Avg', 'Fwd Byts/b Avg',
       'Fwd Pkts/b Avg', 'Fwd Blk Rate Avg', 'Bwd Byts/b Avg',
       'Bwd Pkts/b Avg', 'Bwd Blk Rate Avg', 'Subflow Fwd Pkts',
       'Subflow Fwd Byts', 'Subflow Bwd Pkts', 'Subflow Bwd Byts',
       'Init Fwd Win Byts', 'Init Bwd Win Byts', 'Fwd Act Data Pkts',
       'Fwd Seg Size Min', 'Active Mean', 'Active Std', 'Active Max',
       'Active Min', 'Idle Mean', 'Idle Std', 'Idle Max', 'Idle Min', 'Label'],
      dtype='object')
```

Fig. 3 Columns in the dataset

Fig. 4 Label

```
Benign            667626
FTP-BruteForce    193360
SSH-Bruteforce    187589
Name: Label, dtype: int64
```

7 Implementation and Results

We'll show how we implemented our proposed system in the following phases.

- Importing libraries.
- The libraries we are using are:
  - NumPy
  - Pandas
  - Matplotlib
  - Scikit-learn
  - Keras
  - TensorFlow
  - Loading the Data
  - First step is to load the available data into our memory.
- For making a proper understanding of dataset we are using, we will perform a brief Exploratory Data Analysis (EDA). The EDA is sub-divided into:
  - Data Visuals
  - Data Understanding
  - Data Analysis
- We conclude that:
  - We have a huge amount of data entries.
  - There are a total of 80 columns belonging to each sample.
  - There are missing values in our data, which need to be filled or dropped for proper modeling.
- Data pre-processing plays an important part because data may not be completely clean and may contain missing or null values, this is an important step in the data science process. In this step, we will check to see if our data contains any null or missing values.
- The data's label feature contains three labels: Benign, BruteForceFTP, and BruteForceSSH. These are all in string format. We need to convert them into numbers so that our neural network can understand their representations.
- Creating data for CNN.
- We followed the steps below to apply a convolutional neural network to our data:
  - Separate the data for each label.
  - Create a matrix representation for the labels in the numeric format.
  - Make equal distribution of data among all the labels using data resampling.
  - Create the target and the predictor variables.
  - The dataset is splitted for the training and the testing purposes.
  - Create multidimensional data for providing input to the CNN.
  - Use the CNN model.
  - Visualization of Results (CNN).

- Let's make a graphical visualization of results obtained by applying CNN to our data (Fig. 5).

- The model is evaluated using validation data and test data.
- The results obtained are shown in Figs. 6 and 7.
- The validation accuracy and loss obtained are shown in Fig. 8.

```
Model: "sequential"
_________________________________________________________________
 Layer (type)                Output Shape              Param #
=================================================================
 conv1d (Conv1D)             (None, 72, 64)            448

 batch_normalization (BatchN  (None, 72, 64)           256
 ormalization)

 max_pooling1d (MaxPooling1D  (None, 36, 64)           0
 )

 conv1d_1 (Conv1D)           (None, 36, 64)            24640

 batch_normalization_1 (Batc  (None, 36, 64)           256
 hNormalization)

 max_pooling1d_1 (MaxPooling  (None, 18, 64)           0
 1D)

 conv1d_2 (Conv1D)           (None, 18, 64)            24640

 batch_normalization_2 (Batc  (None, 18, 64)           256
 hNormalization)

 max_pooling1d_2 (MaxPooling  (None, 9, 64)            0
 1D)

 flatten (Flatten)           (None, 576)               0

 dense (Dense)               (None, 64)                36928

 dense_1 (Dense)             (None, 64)                4160

 dense_2 (Dense)             (None, 3)                 195

=================================================================
Total params: 91,779
Trainable params: 91,395
Non-trainable params: 384
```

Fig. 5 Model summary

```
# check the model performance on test data
scores = model.evaluate(X_test, y_test)
print("%s: %.2f%%" % (model.metrics_names[1], scores[1] * 100))

188/188 [==============================] - 1s 6ms/step - loss: 0.0515 - accuracy: 0.9962
accuracy: 99.62%
```

Fig. 6 Accuracy metric

```
scores = model.evaluate(X_test, y_test)
print("%s: %.2f%%" % (model.metrics_names[0], scores[0]))

188/188 [==============================] - 1s 7ms/step - loss: 0.0515 - accuracy: 0.9962
loss: 0.05%
```

Fig. 7 Loss metric

8 Comparative Analysis

The performance of the previous models are compared with the proposed deep learning model. It is observed that the proposed model provided us with maximum accuracy ensuring that the CNN model is more efficient when compared to traditional classification models (Fig. 9).

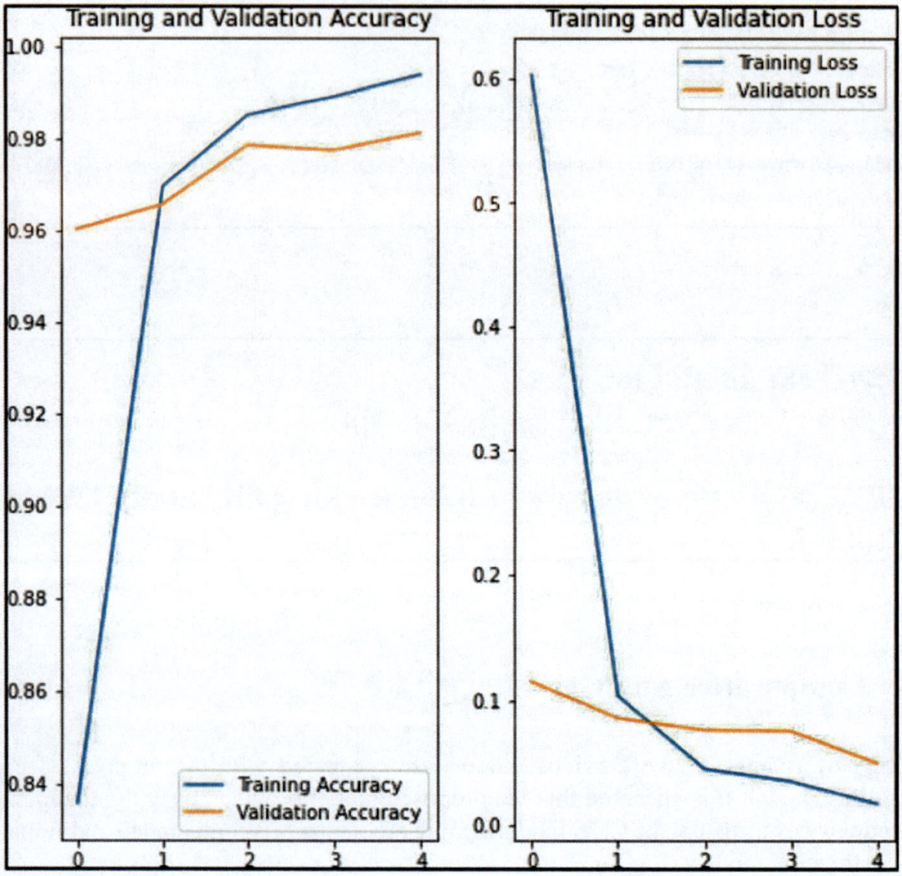

Fig. 8 Plotted results of CNN

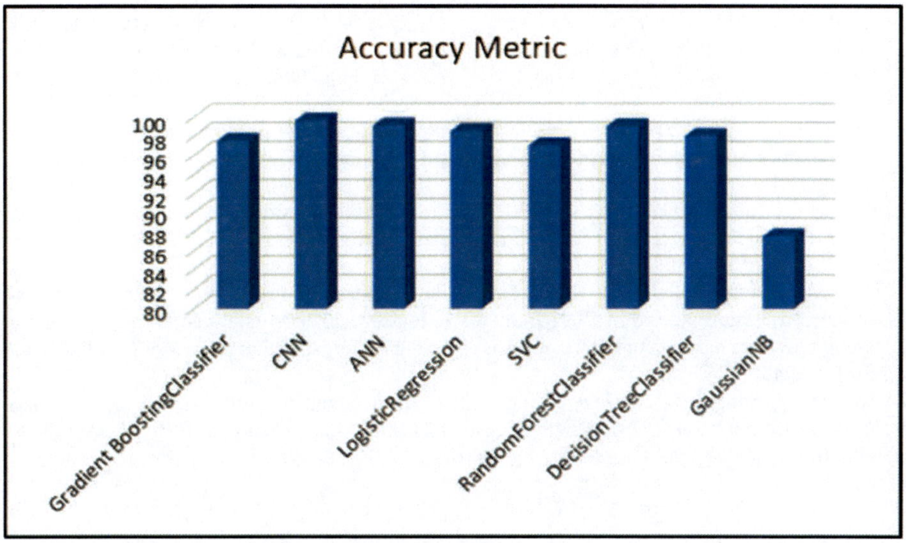

Fig. 9 Comparison of various ML and DL algorithms

9 Conclusion

The suggested model has been developed successfully to detect and classify the Bruteforce attacks in the given dataset. Its performance is measured using the accuracy metric and the result obtained is 99.62% which is more when compared to other traditional machine learning models. The loss obtained is about 0.0515.

10 Future Work

The model can be extended in future using the LSTM techniques to reduce the run time. The entire dataset can be used for classification of multiple network attacks along with the Bruteforce attacks.

References

1. Bagaa M, Taleb T, Bernabe JB, Skarmeta A (2020) A machine learning security framework for IoT systems. IEEE Access 8:114066–114077. https://doi.org/10.1109/ACCESS.2020.2996214
2. Sahu A, Mao Z, Davis K, Goulart AE (2020) Data processing and model selection for machine learning-based network intrusion detection. In: 2020 IEEE international workshop technical committee on communications quality and reliability (CQR). IEEE, pp 1–6. https://doi.org/10.1109/CQR47547.2020.9101394

3. Akhil Reddy D, Puneet V, Siva Rama Krishna S, Kranthi S (2022) Network attack detection and classification using ANN algorithm. In: 2022 6th international conference on computing methodologies and communication (ICCMC), pp 66–71. https://doi.org/10.1109/ICCMC53470.2022.9753934
4. Ubale T, Jain AK (2020) Survey on DDoS attack techniques and solutions in software-defined network. In: Handbook of computer networks and cyber security. Springer, pp 389–419. https://doi.org/10.1007/978-3-030-22277-2_15
5. Karatas G, Demir O, Sahingoz OK (2020) Increasing the performance of machine learning-based IDSs on an imbalanced and up-to-date dataset. IEEE Access 8:32150–32162. https://doi.org/10.1109/ACCESS.2020.2973219
6. Dahiya S, Siwach V, Sehrawat H (2021) Review of AI techniques in development of network intrusion detection system in SDN framework. In: 2021 international conference on computational performance evaluation (ComPE), pp 168–174. https://doi.org/10.1109/ComPE53109.2021.9752430
7. Thakkar A, Lohiya R (2020) A review of the advancement in intrusion detection datasets. Procedia Comput Sci 167:636–645. ISSN: 1877-0509. https://doi.org/10.1016/j.procs.2020.03.330

Selective Text Encryption Using RSA for E-governance Applications for Pdf Document

Subhajit Adhikari and Sunil Karforma

1 Introduction

The exchange of data or information is now quite frequent in e-governance applications. Textual data, like legal data and the personal data of citizens, flows from different departments in e-governance. If there is any form of leakage during transit, security properties like confidentiality will not be preserved. The confidentiality of sensitive data is to be checked during transmission from the sender to the receiver. To remove threats to confidentiality and other security parameters, the technique of encryption is widely used. Traditional encryption systems can be divided into two subcategories: symmetric and asymmetric methods [1]. But in recent studies, there have been various proofs available to disqualify the applicability of the symmetric key concept in terms of textual information encoding. So, as a consequence, the asymmetric key concept is a good choice for encryption of textual data. With a different view point, it can also be stated that the encoding methods can be of two types: encoding with a selective portion and encoding with the whole portion of the original text. Both the two methods have its benefits and drawbacks. Full encryption methods are not suitable for resource constrained environment [2]. Considering the method of whole text encoding, it is obvious that it must consume the more

S. Adhikari (✉)
Assistant Professor, BSH Department, Institute of Engineering and Management, University of Engineering and Management, Kolkata, India
e-mail: Subhajit.adhikari@iem.edu.in

Research Scholar, Department of Computer Science, University of Burdwan, Burdwan, India

S. Karforma
Dean(Science) Faculty, Department of Computer Science, The University of Burdwan, Burdwan, India
e-mail: skarforma@cs.buruniv.ac.in

J. K. Mandal et al. (eds.), *Proceedings of International Conference on Network Security and Blockchain Technology*, Lecture Notes in Networks and Systems 738,
https://doi.org/10.1007/978-981-99-4433-0_22

computation time than selective encoding, but the speedup factor is also a major factor [3]. In selective encoding, the speed of encryption is much higher for huge amounts of data produced from different sources maintaining same level of security of whole text encryption method. In our proposed method, we consider the benefits of both the asymmetric key method and the selective encoding approach to design a robust and secure encryption system. So, regular expressions are used to select the segment of textual data, given a text as user input, and then RSA cryptography is implemented to encrypt the selected segment of text. In our research study, 1024 bit RSA is used for strongest encryption process. The cryptosystem RSA is very famous for its class of algorithms in asymmetric key cryptography [4]. The steps of RSA algorithm has already defined in [5]. In our research study, the predefined function $rsa.encrypt(Org\_msg, Pub\_key)$ of 1024 bits in Python-RSA module [6] as pure Python-RSA implementation for encryption is taken for the experiment. In decoding step, $rsa.decrypt(Enc\_msg, Priv\_key)$ is used to decode the orginal text, where Org_msg depicts original message, Pub_Key depicts public key of the receiver and $Priv\_Key$ depicts the private key of the receiver. The message is encoded and decoded with the 'utf8' format before encrypting process and after decrypting process respectively.

2 Our Contribution

Selective encryption in the context of text encryption is very rare. Our main contribution is that some portion of the data must be untraceable, even if the attacker manages to extract the rest of the unencrypted data. Assume the PAN or the Aadhaar number is important information of citizen that must be kept private. Whenever an Income Tax Return form is generated by the authority, the PAN number is added to it. If the attacker can obtain the PAN number, he or she can obtain all the legal information pertaining to a particular citizen. Our aim is to encode only the PAN, while the rest of the document will not be encoded. So, RSA with a 1024-bit encoding technique is implemented. We combine the benefits of selective encryption and an asymmetric key algorithm to design our new encoding technique. We chose the selective encoding method by search to reduce the time required by traditional whole text encryption. The asymmetric key encoding scheme is then used to achieve the highest level of security while maintaining the data's confidentiality. Our method can be extended and applied to secure medical records and sensitive data generated by wireless and IOT devices.

3 Literature Review

The purpose of the research study [1] is to introduce a novel selective significant data encryption algorithm, where a significant amount of uncertainty is added to data as it is encrypted. This algorithm takes help of the concept of natural language processing and extracts the data from the whole text. There are four steps to the selective encryption technique studied in this study. First step is to removing special characters, secondly tokenization fetches all words available in the message l, after that the words signifies termination have been removed. Lastly, encryption process is applied to the keywords to leaving the common words as it is. Both encrypted keywords and plain common words are sent to the network. In recent times, a research [2] is carried out considering selective encryption for image and audio data in resource constrained environment in terms of low memory, low computation capacity and low power requirements. Also, selective encoding technique is evaluated in association with metrics like tenability, degradation of visual effect, cryptographic security, encryption ratio, compression friendliness, format compliance and error tolerance. The categorization of selective encoding is also done based on pre-compression, in-compression and post-compression approaches. The selective significant data encryption [3] approach for text data encryption was introduced in the previous study. This method chooses just relevant data from the entire message in terms of the whole message's keywords, which gives the data encryption procedure enough uncertainty. This improves speed and cuts down on the overhead associated with encryption. The symmetric key encoding technique is used to carry out the encryption process. The Blowfish algorithm is employed for this. A comparative study of the proposed technique, the full encoding scheme, and the toss of a coin method is also included in terms of proportion of encoded text and computation time. In this study of a selective encoding scheme[7], they provide an innovative AES-Rijndael-based encryption technique for medical data. Firstly, a selector component is depicted that allows the method to be implemented on a variety of platforms, with the required size of input, count of rounds. In the second phase, the compression process of original picture is done with the Huffman algorithm to decrease the size of the picture and encryption time of AES method by more than half. And thirdly, the simulation time of AES algorithm is kept minimum with the concept of loop unrolling and methods of merging in proposed algorithm. Experimental study proves that this novel selective encoding scheme cut down the average execution time by 35% comparing to traditional AES scheme. Previously, a modified RSA [8] method has been presented with improved security for message encryption. By identifying three factors of n instead of two, makes the proposed encrypting model more difficult for an attacker to guess by the process of factorization. Thus the security is raised by two levels. Finding a public key and a private key as a result of the second modulus x being used in place of the modulus n being passed is challenging since only using these keys makes it feasible to encrypt or decode messages while maintaining message secrecy. The time to produce the keys of the encoding system is less than the traditional RSA cryptographic method. In this article, a new selective encryption technique[9]

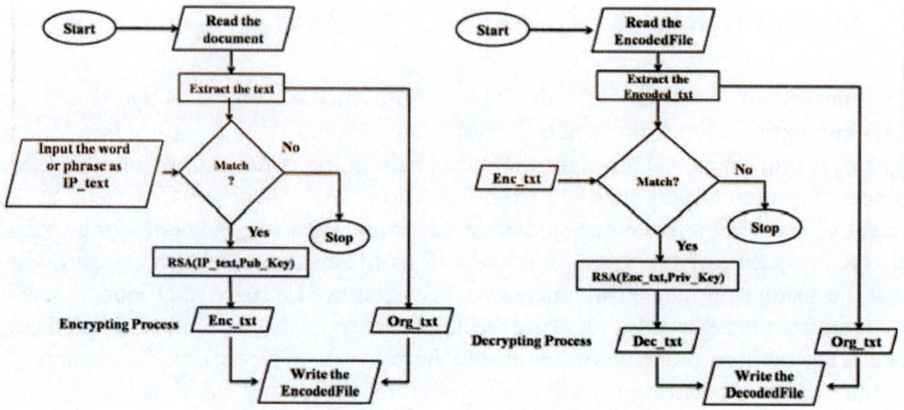

Fig. 1 Block diagram of encryption and decryption process

is demonstrated that employs a safe, index-based chaotic sequence to encrypt only the chosen compressed video frames from each set of images. Simulation results and statistical analysis have done based on quality analysis, keyspace metric, psnr analysis, mean-square-error analysis and computation time analysis and it is found effective and efficient rather than traditional AES and RC5 encoding algorithms. The concept of the CMYK color model [10] has already been used to create a unique encoding and decoding approach with four keys for conversion from text to image. This approach encrypts data faster in terms of text characters. In order to prevent the mathematical factorization of n from leading to the factors p and q, the modified RSA algorithm [11] incorporates the removal of the large prime number n from the key. A one-digit number serves as the initial message in this experiment. According to the analytical report, the suggested approach encrypts and decrypts faster than a conventional RSA strategy. To address the issue of slow key decryption or slow key transmission, an improved method of homomorphic encryption based on Chinese remainder theorem with a Rivest-Shamir-Adleman [12] method was developed, utilizing multiple keys. It performs the cipher text decoding better than standard RSA for documents.

4 Proposed Algorithm

The proposed algorithm is depicted in a block diagram in the Fig. 1.

4.1 Encrypting and Decrypting Procedure

The process of encrypting and decrypting schemes are given below.

Algorithm 1: Encryption Procedure

Input: OriginalPDF, Input text as IP_txt
Output: PDF as encodedFile
1.Read the text lines from the document in *Org_txt*.
2.Take input the word or phrase to be searched and saved into *IP_txt*.
3.Loop
4. If *IP_txt*== *Org_txt* then
5. Compute rsa.encrypt(*IP_txt*, *R_pubKey*) and save it to *Enc_txt*.
6. Add a special symbol "??" To the end of *Enc_txt* and save it to *FinEnc_txt*.
7. Write *FinEnc_txt* as string to a encoded file as "encodedFile".
8. Else
9. *Org_txt* as string to a encoded file "encodedFile".
10. EndIf
11.Untill End of File.
12.Stop.

Algorithm 2: Decryption Procedure

Input: PDF as encodedFile
Output: OrginalPDF as DecodedFile
1. Read the text from the encodedFile.
2. Separate the encoded string in "EncSting" using special symbol "??" from the original text *Org_txt*.
3. Compute rsa.decrypt(*EncSting*, *R_privKey*) and save it to *Dec_txt*.
4. Write *Dec_txt* to the file DecodedFile .
5. Write *Org_txt* to the file DecodedFile.
6. Stop.

---

5 Implementation Example

The experiment has been conducted in Intel 3rd gen processor computer having 1.70 GHz cpu speed, 500GB HDD and RAM of 4GB capacity. The software Pycharm of version 2020.2 is utilized for the experiment along with Matlab R2016b for statistical analysis. Different standard pdf documents are collected from the web sources [13–15]. In the following example, the content of the pdf document is considerd for analysis irrespective of the position and layout and font of the pdf document. The content "July 4, 1776" is selected from second line of text the for encrypting and decrypting process. The process of selective encoding mechanism is applied to the selected part "July 4, 1776" and the encrypted form of the text is written to the encoded pdf file. The content of encoded pdf file is shown in Fig. 2 in the middle .The decrypting process converts the encoded content back to the original text "July 4, 1776" and written to a new decoded pdf file. The content of decoded pdf file is shown in Fig. 2 in bottom part.

Declaration of Independence

IN CONGRESS, July 4, 1776.

The unanimous Declaration of the thirteen united States of America,

When in the Course of human events, it becomes necessary for one people to dissolve the political bands which have connected them with another, and to assume among the powers of the earth, the separate and equal station to which the Laws of Nature and of Nature's God entitle them, a decent respect to the opinions of mankind requires that they should declare the causes which impel them to the separation.

Declaration of Independence

IN CONGRESS, b"~\x87\x97>@%\xf6KP\x08\xd9\xa9\xefQ\x1e\xd9\xf6\xf3E\x

The unanimous Declaration of the thirteen united States of America,

When in the Course of human events, it becomes necessary for one people to dissolve the political bands which have connected them with another, and to assume among the powers of the earth, the separate and equal station to which the Laws of Nature and of Nature's God entitle them, a decent respect to the opinions of mankind requires that they should declare the causes which impel them to the separation.

Declaration of Independence

IN CONGRESS, July 4, 1776

The unanimous Declaration of the thirteen united States of America,

When in the Course of human events, it becomes necessary for one people to dissolve the political bands which have connected them with another, and to assume among the powers of the earth, the separate and equal station to which the Laws of Nature and of Nature's God entitle them, a decent respect to the opinions of mankind requires that they should declare the causes which impel them to the separation.

Fig. 2 Original text, encrypted text and decrypted text

6 Analysis of Security Parameters

The dataset is composed of three standard pdf documents. The extracted portion of the text is named "Data1","Data2" and "Data3", respectively. As for example the "Data1" consists of the text "July 4, 1776". As for example the "Data2" consists of the text "SEMPRONIO". As for example the "Data3" consists of the text "Contents".

6.1 Study of Key Space

Study of keyspace considers the number of changing variables used for computation. The high value of this metric discards any type of attacks that are bruteforce in nature. The standardization made with IEEE floating-point value consideration, is that the accuracy of double variables is approximately 10^{-15} with the bit capacity 64. We have four double variables as p,q,e and d. So, the keyspace value is about $10^{60} \approx 2^{199.31569}$. So, our scheme of encrypting and decrypting text is constituted to give protection about all attakcs made in bruteforce approach considering this large keyspace.

Table 1 Study of entropy

Content	Original	Encrypted
Data1	3.251629	4.3741
Data2	2.947702	3.16992
Data3	2.5	4.54205

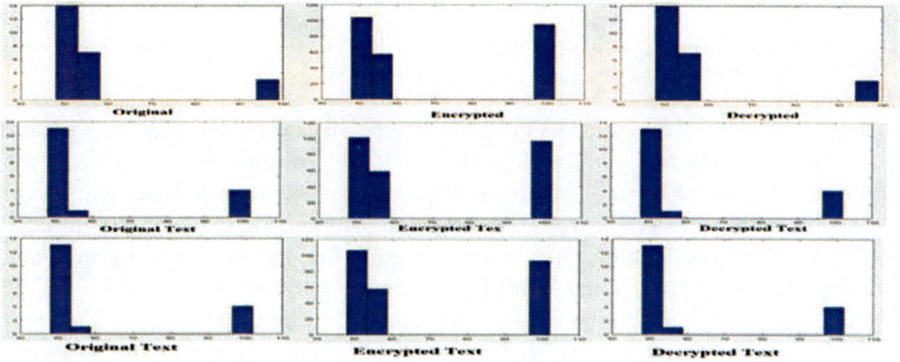

Fig. 3 Study of histogram of Data1, Data2 and Data3

6.2 Entropy

The term is first uttered by the famous mathematician Shannon as a metric to measure uncertainty. It has been applied in the domain of information processing [16]. The value of a text with a lower probability of the occurrence of an event retains more information, and thus it has a higher information entropy [17]. As a consequent, suppose "Data security" has less probability of appearance than the sentence "Data security is applicable to different fields". The metric entropy of a sentence represents indicates how much information it contains [18]. The study of entropy can be depicted as the Eq. 1 given below [19]

$$H(P) = \sum_{i=0}^{255}[\text{Prob}(X_i) \times \log(\frac{1}{\text{Prob}(X_i)})] \tag{1}$$

In the above equation $Prob(X_i)$ represents the probability of existence of symbol X_i

From the above Table 1, the encrypted text has more entropy value than original text. The higher value of entropy makes the encrypting and decrypting scheme very hard to crack.

Table 2 Study of avalanche effect

Content	Avalanche_Effect
Data1	0.51956947
Data2	0.5112414467
Data3	0.5341796875

6.3 *Histogram Analysis*

Each letter or symbol that appears in the message "Msg" is shown by a histogram. If the spread of the letters or symbols is uniform, the encrypting technique is also insurmountable in the face of statistical assaults [20]. The histogram plot of the ciphered text should differ drastically from the histogram of the plain text and should be as evenly distributed as is humanly feasible, meaning that the likelihood of any value existing is the same [10]. In the above Fig. 3, the histogram of original, encoded and decoded text is depicted taking conversion to ASCII format. For the encrypted text, the histogram representation is uniform in terms of vertical bars than the histogram of original text.

6.4 *Avalanche Effect*

A feature of an encryption method known as the "avalanche effect" causes changes in multiple bits of the encoded text when one bit of the original text is changed [21]. Avalanche impact should be 0.5 under ideal circumstances [22]. The Eq. 2 of avalanche effect is depicted below. In the equation "CTEXT" represents cipher text.

$$\text{Avalanche Effect} = \frac{\text{Number of Bits Flipped in Ctext}}{\text{Number of Bits in Ctext}} \tag{2}$$

From the above Table 2, the conclusion can be made easily that our proposed technique crossed the ideal range of the avalanche effect value, depcting a good encrypting system property.

6.5 *Plaintext Sensitivity*

The study of plaintext sensitivity depicts that a small moderation in the original content in terms of a bit can create a rapid change in the encoded content. The original text is "July 4, 1776" is changed to "July 4, 1777" to compute the plaintext sensitivity and the result is given in the above Fig. 4. As a consequent, the above

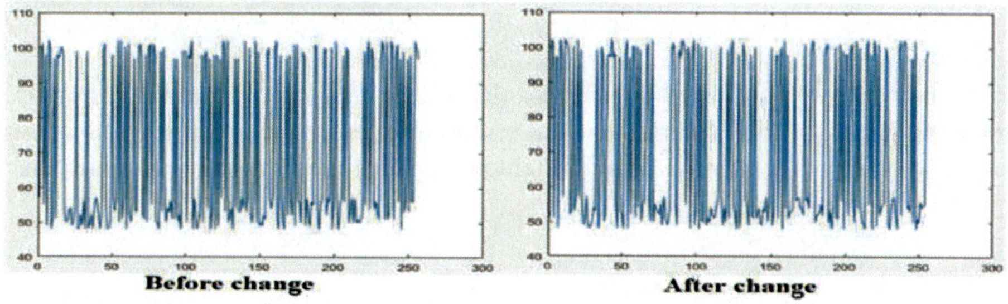

Fig. 4 Study of plaintext sensitivity

Table 3 Study of encryption and decryption time

Content	Enc_Time(Sec)	Dec_Time(Sec)
Data1	0.0004920	0.0133413
Data2	0.000353	0.027816
Data3	0.00031	0.009625

Table 4 Comparison result of proposed text encoding with others

	Entropy	Avalanche	Enc_Time(Sec)
Proposed method	4.3741	0.51956947	0.0133413
[3]	NA	NA	1500
[8]	NA	NA	0.0050
[10]	6.92	NA	0.000001
[11]	NA	NA	21980
[12]	NA	NA	0.412

two encoded images are totally different before and after the encoding process. So, only one-digit change in the original string make a huge change in cipher text.The correlation between two cipher files is -0.0276. This low value of correlation means there is no relation between two encoded files.

6.6 Computation Time Analysis

In the Table 3, the computation time for encoding and decoding text file is given in seconds. The time analysis satisfies that our method consumes less cpu time and can be incorporated not only in e-governance application but also in resource limited environment.

From the above Table 4, it is very clear that existing methods of text encryption lack in detailed statistical anlysis in terms of metrics like entropy and avalanche effects and only present required encryption time. Our method has high value of entropy, ideal value of avalanche effect with low encryption time. Also, our propsed method of encoding text constists of detailed study of statistical metrics which proves the robustness against different attaks. The important metrics like plaintext sensitivity and histogram study have also been included in our research study to qualify as a good cryptosystem.

7 Conclusion and Future Scope

Our research study provides the text data security in e-governance applications. The asymmetric approach of encoding text is discussed in this paper using 1024 bit RSA cryptographic algorithm. The confidentiality property of data is guaranteed by our proposed method along with high security features. Government documents and Legal documents can be secured using our proposed encoding scheme. Important selected data like account number, PAN and Aadhaar of any citizen can be encrypted using proposed method and added in the government documents. Attacker may find the document but unable to decrypt the selected part of the content which leads to an unsuccessful attempt of data theft. The security analysis report proves the robustness of our method against different attacks causing security threats. Also, the proposed model of encrypting and decrypting specific part of the content fetched from pdf document takes less time than whole text encoding. As a consequence, the applicability of our encrypting method rises for resource limited environment. As of now, the method is implemented for text in pdf document but can also be applied for multimedia content like image and video. In future, chaotic functions may be incorporated to introduce more randomness in the encoding and decoding technique. The encoding scheme can also be extended with the elliptic curve cryptography. The proposed method of encryption can be done with any length and in any position, but in the context of "Selective Encryption", a small portion of the whole text is taken for experiment.

References

1. Kushwaha A, Sharma HR, Ambhaikar A (2018) Selective encryption using natural language processing for text data in mobile ad hoc network. In: Modeling, simulation, and optimization. Springer, Cham, pp 15–26
2. Massoudi A, Lefebvre F, De Vleeschouwer C, Macq B, Quisquater JJ (2008) Overview on selective encryption of image and video: challenges and perspectives. Eurasip J Inf Secur 2008(1):179290
3. Kushwaha A, Sharma HR, Ambhaikar A (2016) A novel selective encryption method for securing text over mobile ad hoc network. Procedia Comput Sci 79:16–23
4. Kota CM, Aissi C (2022) Implementation of the RSA algorithm and its cryptanalysis. In: 2002 GSW

5. Shawkat SA (2007) Enhancing steganography techniques in digital images. Faculty of Computers and Information, Mansoura University Egypt-2016
6. https://stuvel.eu/python-rsa-doc/usage.html , Accessed 06 Dec 2022
7. Oh JY, Yang DI, Chon KH (2010) A selective encryption algorithm based on AES for medical information. Healthc Inf Res 16(1):22–29
8. Jaju SA, Chowhan SS (2015) A modified RSA algorithm to enhance security for digital signature. In: 2015 international conference and workshop on computing and communication (IEMCON). IEEE, pp 1–5
9. Batham S, Yadav VK, Mallik AK (2014) ICSECV: an efficient approach of video encryption. In: 2014 seventh international conference on contemporary computing (IC3). IEEE, pp 425–430
10. Noor NS, Hammood DA, Al-Naji A, Chahl J (2022) A fast text-to-image encryption-decryption algorithm for secure network communication. Computers 11(3):39
11. Minni R, Sultania K, Mishra S, Vincent DR (2013) An algorithm to enhance security in RSA. In: 2013 fourth international conference on computing, communications and networking technologies (ICCCNT). IEEE, pp 1–4
12. Abid R, Iwendi C, Javed AR, Rizwan M, Jalil Z, Anajemba JH, Biamba C (2021) An optimised homomorphic CRT-RSA algorithm for secure and efficient communication. Pers Ubiquitous Comput 1–14
13. https://www.kaggle.com/code/gauravduttakiit/working-with-pdf-files/data , Accessed 06 Dec 2022
14. https://www.kaggle.com/datasets/paretogp/examples-exams-pdf , Accessed 06 Dec 2022
15. https://www.bl.uk/collection-metadata/downloads , Accessed 06 Dec 2022
16. Xu W, Pan Y, Chen X, Ding W, Qian Y (2022) A novel dynamic fusion approach using information entropy for interval-valued ordered datasets. IEEE Trans Big Data
17. Xu H, Lv Y (2022) Mining and application of tourism online review text based on natural language processing and text classification technology. Wireless Commun Mob Comput
18. Khurana A, Bhatnagar V (2022) Investigating entropy for extractive document summarization. Expert Syst Appl 187:115820
19. Lin H, Wang C, Cui L, Sun Y, Zhang X, Yao W (2022) Hyperchaotic memristive ring neural network and application in medical image encryption. Nonlinear Dyn 110(1):841–855
20. Hagras T, Salama D, Youness H (2022) Anti-attacks encryption algorithm based on DNA computing and data encryption standard. Alexandria Eng J 61(12):11651–11662
21. Gamido HV, Sison AM, Medina RP (2018) Modified AES for text and image encryption. Indonesian J Electr Eng Comput Sci 11(3):942–948
22. Ghadirli HM, Nodehi A, Enayatifar R (2019) An overview of encryption algorithms in color images. Sig Process 164:163–185

Malware Analysis Based on Malicious Web URLs

Ritam Ghosh and Soumen Kanrar

1 Introduction

The growing significance of the World Wide Web (WWW) provides a medium for malware to compromise individuals and organizations. It gains remote access by embedding malicious Web URLs with scripts, exploits, and executable files. An uniform resource locator (URL) refers to a unique resource identifier on the Internet. The attackers try to manipulate protocols and modify the URL structure to deceive the targeted users. Then, users are redirected to the forged Web page, targeted by malicious unintentional downloads, phishing, social engineering, adware, and spam. That leads to financial loss, sensitive information disclosure, and data extortion. Malicious Web URLs are classified into four types: malware, spoofing, phishing, and defacement. Malware is malicious software and is a code planted on victims' devices to obtain unauthorized access. It is associated with system files that can contaminate government or corporate Websites and cloud systems. In spoofing, victims' personal information, such as usernames, passwords, and credit/debit card information, is compromised. As the victims believe they are communicating with a legitimate Website, they are actually communicating with replicated Websites with similar structural designs and functionalities. Those are hosted on attacker servers. Phishing is achieved by sending deceptive emails to gain sensitive information. Unlike spoofing, phishing emails contain fraudulent Web links that compromise end-user information. Defacement involves altering a trusted Website's legitimate content and structure by injecting malicious scripts. Attackers gain unauthorized access to replace

R. Ghosh
ACM Student Member, Kolkata, India
e-mail: ritamg@acm.org

S. Kanrar (✉)
Amity University, Jharkhand, India

Vlenzor Technologies Pvt. Ltd., Kolkata 60, India

J. K. Mandal et al. (eds.), *Proceedings of International Conference on Network Security and Blockchain Technology*, Lecture Notes in Networks and Systems 738,
https://doi.org/10.1007/978-981-99-4433-0_23

the legitimately hosted Website with their malicious one, which leads to phishing, code injection, and cross-site scripting. To detect and identify these malware-based malicious Web URLs, security researchers have designed and developed defence techniques such as static analysis, dynamic analysis, hybrid analysis, blacklisting-based analysis, and heuristic-based analysis against malware-based malicious Web links. Static analysis crawls and inspects Web links using mathematical-statistical features [1, 2]. The dynamic analysis uses the cuckoo sandbox and Yara signatures to analyze networks and behaviour to detect suspicious malware scripts [3]. The hybrid analysis is a combination of both static and dynamic, where evasion chances reduce due to continuous crawling and monitoring of the behaviour of the suspicious scripts present in Web URLs. In blacklisting-based methods, URLs are scanned against a list of predefined malicious Web URLs. Whereas heuristic-based methods use known patterns and signatures to scan suspicious Web URLs. Unfortunately, blacklisting can be evaded by modifying the URL string structure, and heuristic-based blacklisting fails for Web URLs that are unknown to the scanning module. So according to the current trend, detecting malware-based malicious Web URLs is based on a combination of signatures, pattern matching, and behavioural analysis [4]. In this case, advanced machine learning or deep learning techniques are used to detect and classify malware [5]. Our research work uses advanced classifiers to classify malicious Web URLs based on features and behaviours. From static and dynamic behaviours, features are segregated and extracted. Then, it passes through multiple classifiers such as random forest, decision tree, extra tree, Gaussian Naive Bayes, neighbours, SGD, and AdaBoost to detect and analyze malware through malicious Web URLs in terms of accuracy, macro average, weighted average, precision, recall, F1-score, and support. The paper is organized as follows: Sect. 2 describes related works and provides a literature overview on malicious URLs. Section 3 denotes the proposed methodology and technicalities. Section 4 provides results and performance analysis. Section 5 outlines our conclusion and future scope.

2 Related Works

In recent years, malware-detecting methods based on malicious URLs have continuously evolved from traditional methods such as patterns or signatures, which apply only to known malicious URLs. Modern methods are using advanced algorithms and classifiers for behavioural analysis. Yong Shi presented a machine learning-based technique called extreme learning machine (ELM) for malicious domain detection [6]. This system uses single-hidden-layer feedforward networks (SLFNs) and module detection problems as SLFNs with a 94% detection rate and 96% accuracy. Hung Le proposed an end-to-end deep learning framework called URLNet [7]. URLNet uses both character-level and word-level embedding using convolutional neural network (CNN). It learns and detects sequential words with 99% accuracy and a higher execution time. Justin Ma explored statistical methods to classify based on lexical and host-based features to detect malicious Web URLs [1]. Host-based feature extraction is time-consuming but achieves 95–99% accuracy. Charlie Curtsinger proposed a low-overhead mechanism for detecting and preventing JavaScript-based malware known as ZAZZLE [8] which uses a Bayesian classifier and achieves a low false-positive rate of 0.0003%. Yongjie Huang presents a deep learning approach named Phishing Net [4] which uses CNN for character-level and recurrent neural network (RNN) for word-level feature extraction. Ability to detect newly emerging malicious URLs with 97% accuracy. Martino Trevisan proposed generative adversarial networks for malicious URL classification [9]. It is generated results based on benign datasets (Checkpoint) with a precision of 0.9900 and malware datasets (Tidserv) with a precision of 0.5600. Sungjin Kim presents a YARA-based malicious Webpage machine learning detection model called WebMon [3]. Rokkathapa and Kanrar proposed an approach for predicting the malware attacks [10]. It detects covert payloads by tracing linked URLs to confirm the legitimacy of the Websites. WebMon is 7.6 times more efficient than conventional Webpage detection tools, with a 98% detection rate. Ignacio Arnaldo proposed automated learning system using deep learning mechanism which periodically gets updated with a blacklisting rate of 9.36% and anticipate with previously marked malicious domains [11]. Cloud computing has a number of advantages, but there are existing issues with confidentiality and integrity [12]. Whereas the blockchain uses an attribute-based cryptosystem to transfer the digital data [13], more robust algorithms are required for the secure transportation of voice data samples over the Internet during real-time testing [14].

3 Methodology

Our proposed malicious URL detection system aims to analyze classify and detect Web-based URLs as malware or benign. The proposed system architecture is shown in Fig. 1, which is primarily categorized into three stages: data processing, training, and testing.

Data Processing

To detect malicious URLs and label them as malicious or benign, initially, we need to crawl the Web to extract URLs and normalize them. After this, indexing is required to structure and categories them into benign, malware, defacement, and phishing. Then, the features are extracted through lexical scanning of the Web strings. Finally, preprocessed data is split into two subsets. 80% for training to fit the model, and 20% for testing to evaluate the fitted model.

Feature Extraction and Selection

Features are extracted from crawled Web-based URLs to properly classify and detect Web-based URLs as malware or benign. The primary feature groups used for malicious Web URL detection are listed as follows:

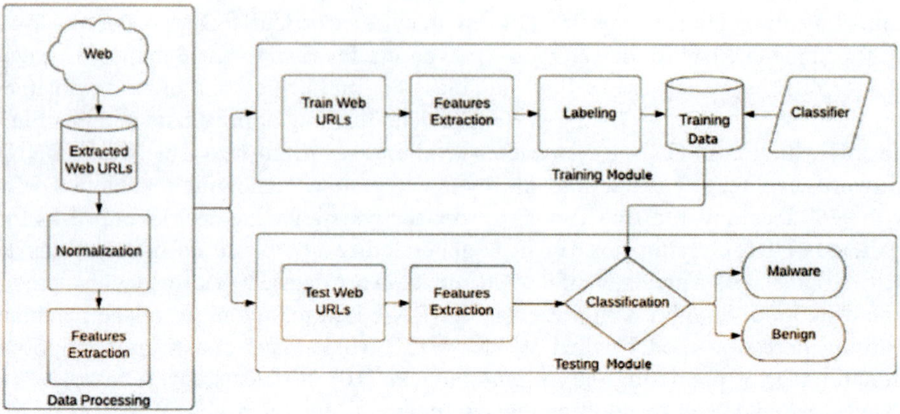

Fig. 1 System architecture

a. Lexical features contain special characters such as ('@', '?', '-', '=', '.', '#', '%', '+', '$', '!', '*', ',', '//') found in a URL and presented in Fig. 2. Also includes Web URL length, primary domain length, average path length, and average token length.
b. Reputation-based features examine the reputation of a Website and provide indexing; such Web indexes are Alexa, Google, Yahoo, and Baidu.
c. Entropy-based features measure the presence of randomness or uncertainty in Web URLs. Higher entropy means a higher randomness factor. It is used to categorize normal or abnormal Websites.
d. Content-based features are acquired by crawling or downloading the whole Web page. Sensitive content is mostly present on malicious Websites. To count sensitive content, the tokenized approach is used; such content areas are 'login', 'sign in', 'signup', 'confirm', 'account', and 'secure'.
e. Host-based features are extracted from the host attributes, which indicate the location, identity, technology, and degree of impact of malicious servers. As per the mentioned host attributes, censorship, protocols, and regulations are standardized.

	url	type	Category	url_len	domain	@	?	-	=	.	#	%	+	$	!	*	,	//
0	br-icloud.com.br	phishing	2	16	br-icloud.com.br	0	0	1	0	2	0	0	0	0	0	0	0	0
1	mp3raid.com/music/krizz_kaliko.html	benign	0	35	mp3raid.com	0	0	0	0	2	0	0	0	0	0	0	0	0
2	bopsecrets.org/rexroth/cr/1.htm	benign	0	31	bopsecrets.org	0	0	0	0	2	0	0	0	0	0	0	0	0
3	http://garage-pirenne.be/index.php?option=com_...	defacement	1	84	garage-pirenne.be	0	1	1	4	2	0	0	0	0	0	0	0	1
4	http://adventure-nicaragua.net/index.php?optio...	defacement	1	235	adventure-nicaragua.net	0	1	1	3	2	0	0	0	0	0	0	0	1
5	http://buzzfil.net/m/show-art/ils-etaient-loin...	benign	0	118	buzzfil.net	0	0	16	0	2	0	0	0	0	0	0	0	1
6	espn.go.com/nba/player/_/id/3457/brandon-rush	benign	0	45	espn.go.com	0	0	1	0	2	0	0	0	0	0	0	0	0
7	yourbittorrent.com/?q=anthony-hamilton-soulife	benign	0	46	yourbittorrent.com	0	1	2	1	1	0	0	0	0	0	0	0	0
8	http://pashminaonline.com/pure-pashminas	defacement	1	40	pashminaonline.com	0	0	1	0	1	0	0	0	0	0	0	0	1
9	allmusic.com/album/crazy-from-the-heat-r16990	benign	0	45	allmusic.com	0	0	4	0	1	0	0	0	0	0	0	0	0

Fig. 2 Lexcial feature extraction

f. Domain-based features include 'Who Is' (WHOIS) domain information, which ranks among domain name, domain expiry, domain registry, server section, and 'domain length'.

Training Module

The training Web URL data must be subjected to feature extraction during the training phase to properly determine malware or benign URLs. Then, the extracted data is labelled accordingly and passed through classifiers where multiple classifying algorithms are being used, namely decision trees, extra trees, K-nearest neighbours, Gaussian Naive Bayes, random forests, AdaBoost, and stochastic gradient descent classifiers. The model will be used in the testing phase based are on its performance.

In training module, feature extraction undergoes through multiple levels of extraction. Initially, all the data are sorted according to lexical features. Then, the data is matched with respect to the standard URL structure and further sorted into normal and abnormal URL category referred to Fig. 3.

After abnormal URL sorting, URL is sorted to identify the availability of SSL/TLS encrypted communication channel in form of hyper text transfer protocol (HTTP) or HTTP secure (HTTPS) referred to Fig. 4.

After security channel (http or https) wise sorting, we need to proceed with matching legitimate URL shortening services refered in Fig. 5. After security channel (http or https) wise sorting, we need to proceed with standard IP address structure currently available and in use referred to Fig. 6. Finally, we are able to generate a correlated heat map with proper labels, after passing our data through multiple layer of extraction processes referred to Fig. 7, which will be trained using mutiple classifer to achieve our results.

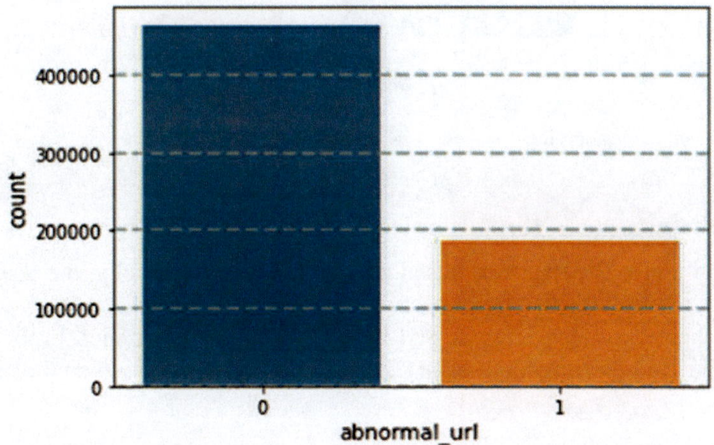

Fig. 3 Abnormal URL sorting

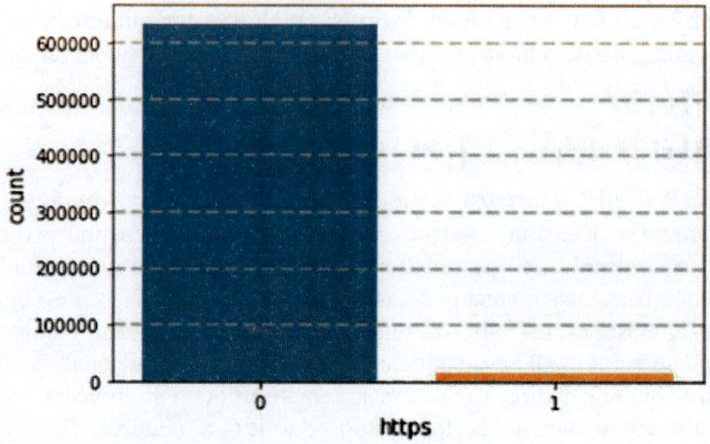

Fig. 4 HTTP or HTTPS filtering

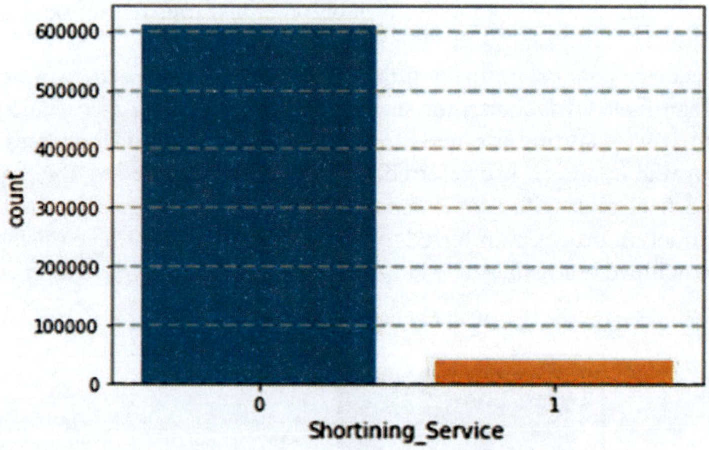

Fig. 5 URL shortining service

Testing Module

The testing phase is performed based on trained results. Initially, the testing Web URLs pass through the feature extraction process.

Then, these features are processed by the classifiers mentioned in the training module to determine whether the URL is malware or benign based on the prediction Table 1.

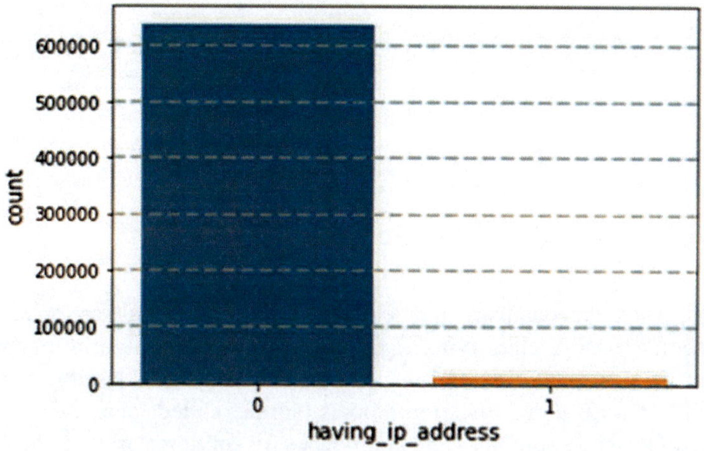

Fig. 6 IP address-based filtering

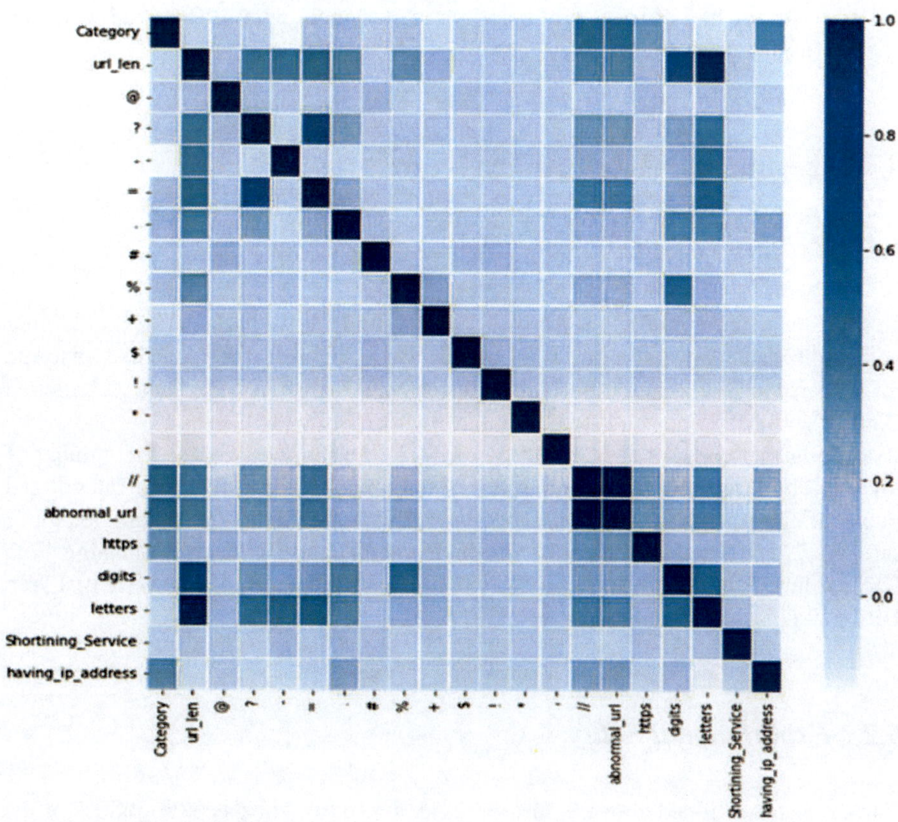

Fig. 7 Generated heat map

Table 1 Prediction table

Confusion matrix		Predicted values	
		Malicious	Benign
Real values	Malicious	TP	FN
	Benign	FP	TN

4 Classification Algorithm Selection

In this work, a novel mechanism for detecting and classifying malware or benign well-known machine learning classifying algorithms is used. Those algorithms are suitable as they utilize our new features selected for detecting and classifying malicious Web-based URLs. The classification models (simply called 'classifiers') provide a wide range of possible outcomes and approaches for solving classification problems. Decision trees, extra trees, K-nearest neighbours, Gaussian Naive Bayes, random forests, AdaBoost, and stochastic gradient descent classifiers are included to explore the performance and efficiency of using these classifiers with different adjustments of features and parameters.

5 Experimental Result Analysis

5.1 Experimental Dataset

To train and test our model's performance and efficiency, we use a dataset containing both legitimate and malicious web URLs. We have used the Canadian Institute for Cybersecurity dataset from University of New Brunswick [15] and A Labelled Dataset with Malicious and Benign IoT Network Traffic [16].

This dataset consists of 651,191 Web URLs (the sample size, i.e. the number of rows or observations), then the number of columns, which are further categorized into four classes: benign, malware, phishing, and defacement.428103 are benign or safe, 32,520 are malware, 96,457 are defacements, and 94,111 are phishing Web URLs referred to in Fig. 8. It has mainly two columns, compromising the actual Web URLs and classes of malicious Web URLs.

5.2 Experimental Setup

The dataset mentioned above is further divided into two subsets. About 80% of the dataset is used for training, and 20% of the dataset is used for testing. Multiple classifying algorithms repeatedly runs with different parameter settings. Software Environment: We have used Anaconda and Collab platforms with Python version

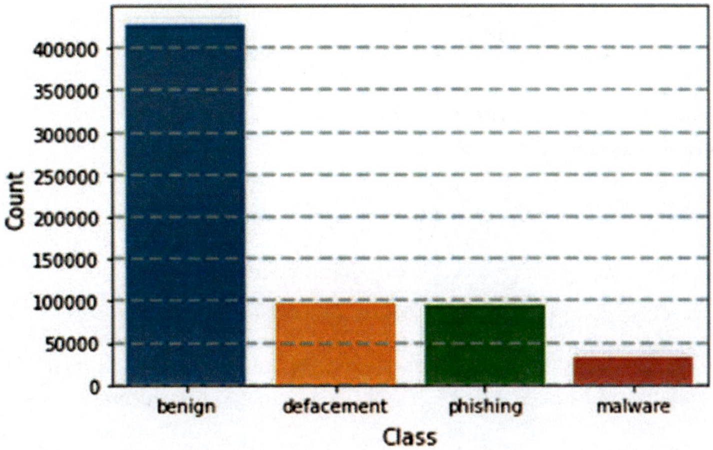

Fig. 8 Data categorization based on class

3.6. Hardware configuration is built with 16 GB of RAM, an Intel(R) Core(TM) i7-9750HF, and a 2.60 GHz processor.

5.3 Experimental Evaluation

In our experiment, accuracy, precision, confusion matrix, recall, and F1-score are used as performance metrics to evaluate and analyze performance as presented in Eq. (1). Where true positives (TPs) represent correct malicious URL prediction, true negatives (TNs) represent correct benign URL prediction, false positives (FPs) represent incorrect malicious URL prediction, and false negatives (FNs) represent incorrect benign URL prediction. Accuracy is defined as the ratio of correct predictions (TP + TN) among total samples (TP + TN + FP + FN).

$$\text{Accuracy} = \frac{\text{TP} + \text{TN}}{\text{TP} + \text{TN} + \text{FP} + \text{FN}}$$

Precision: Defined as the ratio of correctly labelled samples (TP) over all correctly labelled samples (TP + FP).

$$\text{Precision} = \frac{\text{TP}}{\text{TP} + \text{FP}}$$

Confusion Matrix: Evaluate the accuracy of a classifier with predicted and actual classifications.

$$\text{Confusion matrix} = \begin{bmatrix} \text{TN FP} \\ \text{FN TP} \end{bmatrix}$$

The recall is defined as the ratio of correctly labelled (TP) over all correctly labelled predicted data (TP + FP),

Recall = $\frac{\text{TP}}{\text{TP+FN}}$

F1-score: Defined as harmonic mean of precision and recall.

$$\text{F1} = \frac{2 * \text{Precision} * \text{Recall}}{\text{Precision} + \text{Recall}} \quad (1)$$

Various confusion matrices are presented based on the models described in Table 2 and are presented in Figs. 9, 10, 11, 12, 13, 14, and 15, respectively.

Table 2 Experimental results

Model	Accuracy %	Precision %	Recall %	F1-score %
Decision tree	90	92	97	94
Random forest	91	92	98	95
AdaBoost	82	84	98	90
K-neighbours	89	91	96	93
SGD	81	82	99	90
Extra trees	91	92	98	95
Gaussian NB	79	85	92	88

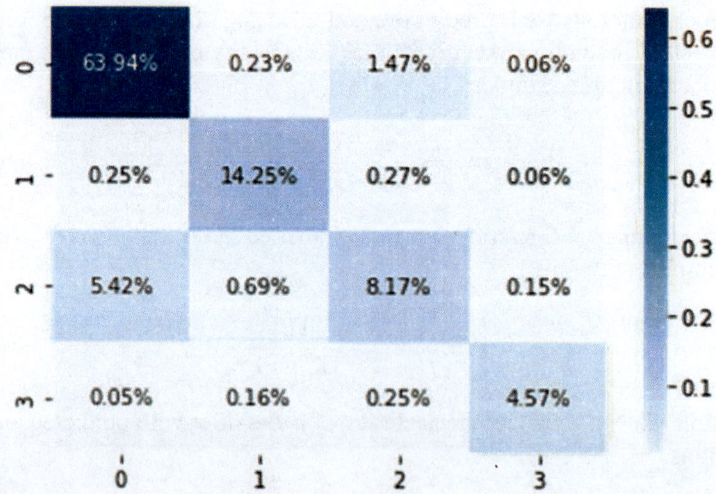

Fig. 9 Decission tree confusion matrix

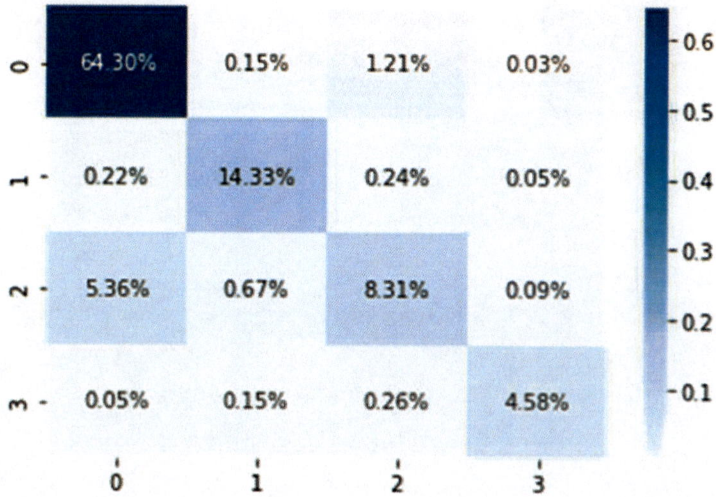

Fig. 10 Random forst confusion matrix

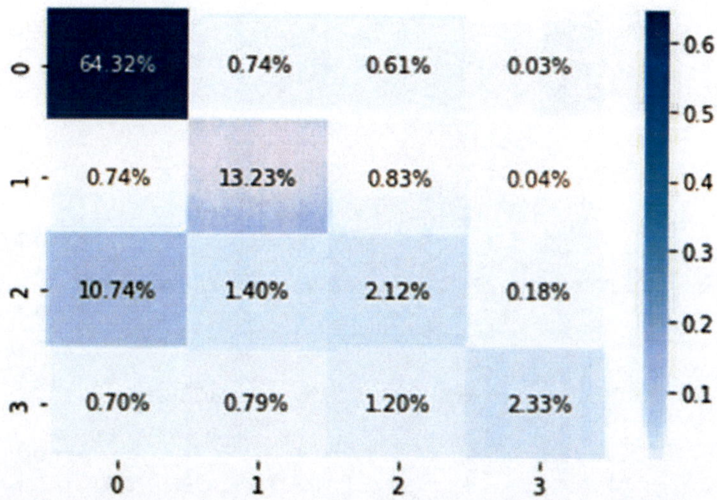

Fig. 11 AdaBoost confusion matrix

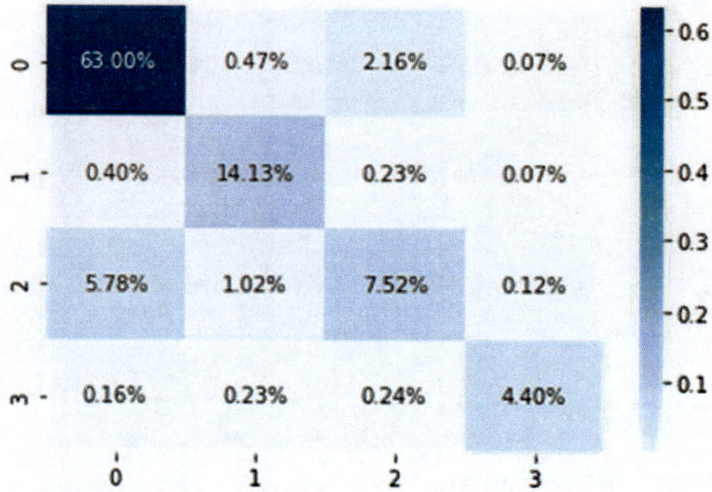

Fig. 12 K-neighbours confusion matrix

Fig. 13 SGD confusion matrix

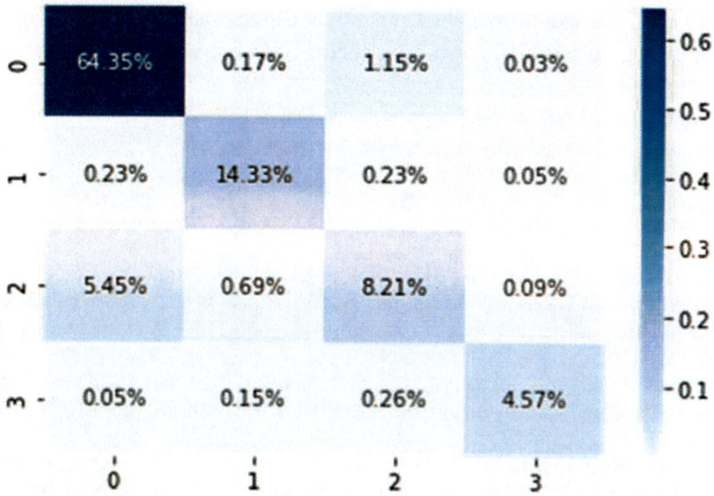

Fig. 14 Extra tree confusion matrix

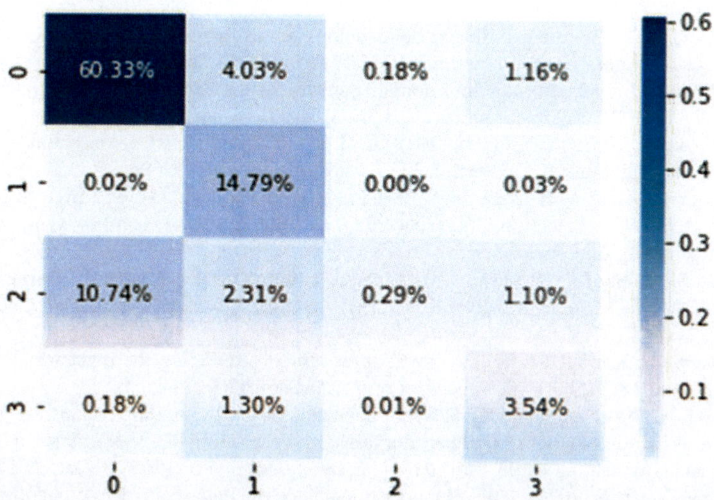

Fig. 15 Gaussian Naive Bayes confusion matrix

6 Conclusion

In this paper, a mechanism for detecting and classifying malware or benign Web-based URLs is presented. The results in Table 2 have shown the effectiveness of the proposed model. In this study, we have developed an approach based on multiple classifiers to detect URLs compromised by benign content, malware, phishing, and defacement. Web URLs are crawled and lexical and content-based features extracted to effectively identify benign or malware. In further steps, we are planning to seek more datasets, combine them with our system, and design a Web application interface. So it can be implemented by information security systems or used by researchers for further development.

References

1. Saul JKL, Savage S, Voelker MG (2009) Beyond blacklists: learning to detect malicious web sites from suspicious URLs. In: Proceedings of the 15th ACM SIGKDD international conference on Knowledge discovery and data mining. Paris, pp 1245–1254. https://doi.org/10.1145/1557019.1557153
2. Wang S, Chen Z, Yan Q, Ji K, Peng L, Yang B, Conti M (2020) Deep and broad URL feature mining for android malware detection. Inf Sci 513:600–613. https://doi.org/10.1016/j.ins.2019.11.008
3. Kim S, Kim J, Nam S, Kim D (2018) WebMon: ML- and YARA-based malicious webpage detection. Comput Netw 137(4):119–131. https://doi.org/10.1016/j.comnet.2018.03.006
4. Huang Y, Yang Q, Qin J, Wen W (2019) Phishing URL detection via CNN and attention-based hierarchical RNN. In: Proceeding of 13th IEEE international conference on big data science and engineering. New York, pp 112–119. https://doi.org/10.1109/TrustCom/BigDataSE.2019.00024
5. Kanrar S (2023) Machine learning model development using computational neurology. Smart Innov Syst Technol 313. https://doi.org/10.1007/978-981-19-8669-7_14
6. Shi Y, Chen G, Li J (2018) Malicious domain name detection based on extreme. Neural Process Lett 48(3):1347–1357. https://doi.org/10.1007/s11063-017-9666-7
7. Le H, Pham Q, Sahoo D, Hoi CHS (2018) URLNet: learning a URL representation with deep learning for malicious URL detection 1–13. https://doi.org/10.48550/arXiv.1802.03162
8. Curtsinger C, Livshits B, Zorn B, Seifert C (2011) ZOZZLE: fast and precise in-browser javascript malware detection. In: Proceeding of 20th USENIX security symposium. San Francisco, pp 1–16
9. Trevisan M, Drago I (2018) Robust URL classification with generative adversarial networks. ACM SIGMETRICS Perform Eval Rev 46(3):143–146. https://doi.org/10.1145/3308897.3308959
10. Rokkathapa E, Kanrar S (2019) A novel approach for predicting the malware attacks. Int J Comput Appl 181(45):30–32. https://doi.org/10.5120/ijca2019918585
11. Arnaldo I, Arun A, Kyathanahalli S, Veeramachaneni K (2018) Acquire, adapt, and anticipate: continuous learning to block malicious domains. In: Proceeding IEEE ınternational conference on big data (Big Data). Seattle, pp 1891–1898. https://doi.org/10.1109/BigData.2018.8622197
12. Verma G, Kanrar S (2022) A novel model to enhance the data security in cloud environment. Multiagent Grid Syst 18(1):45–63. https://doi.org/10.3233/MGS-220361
13. Verma G, Kanrar S (2022) Secure digital documents sharing using blockchain and attribute based cryptosystem. Multiagent Grid Syst 18(3–4):365–379. https://doi.org/10.3233/MGS-221361
14. Kanrar S (2022) Robust threshold selection for environment specific voice in speaker recognition. Wireless Pers Commun 126(4):3071–3092. https://doi.org/10.1007/s11277-022-09852-2
15. UNB dataset. https://www.unb.ca/cic/datasets/index.html
16. Aposemat IoT-23. https://www.stratosphereips.org/datasets-iot23

Asymptotic Diffusion Analysis of a Queueing System $M^X/G/1$ with Collisions and Unreliable Servers in the Process of Communication

R. Vanalakshmi, S. Maragathasundari, B. Balamurugan, M. Kameswari, and C. Swedheetha

1 Introduction

1.1 Communication Network System

A communication network is one that explains how ideas are transmitted between two locations. Information is transmitted and received during communication. The data producer, the channel and the data consumer are important components of data. A data transmission is a theory of systems that depicts the information transfer between two stations, the transmitter, and the receiver. The term "stream" refers to the way that signals or information from data transfer travel from one target node to another. Signals must first pass-through multiple stages of filtering before being communicated via a communication network. These processes include signal portrayal, signal creation, signal encrypting, and signal activation (Figs. 1 and 2).

R. Vanalakshmi · S. Maragathasundari (✉) · M. Kameswari
Kalasalingam Academy of Research and Education, Chennai, Tamilnadu 626126, India
e-mail: maragatham01@gmail.com

B. Balamurugan
Velammal Institute of Technology, Chennai, Tamilnadu 601204, India

C. Swedheetha
Vaigai College of Engineering, Madurai, Tamilnadu 625122, India

J. K. Mandal et al. (eds.), *Proceedings of International Conference on Network Security and Blockchain Technology*, Lecture Notes in Networks and Systems 738,
https://doi.org/10.1007/978-981-99-4433-0_24

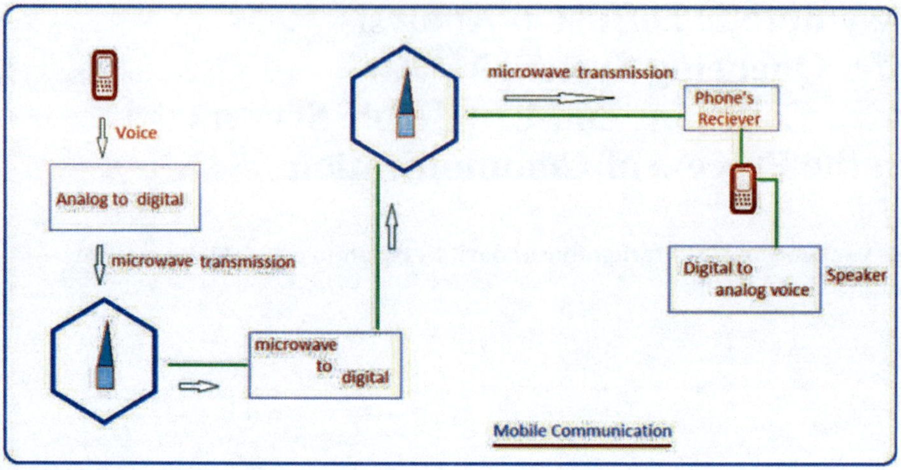

Fig. 1 Mobile communication

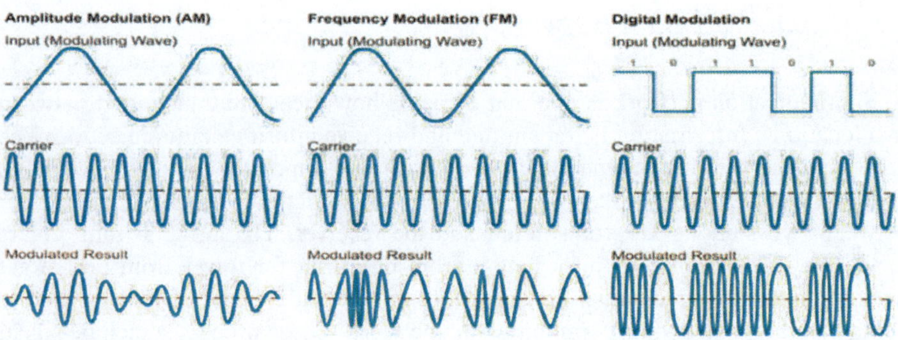

Fig. 2 The modulating signal mix up with carrier to form modulated signal

1.2 Following is an Overview of the Wireless Transmission Platform's Queuing Framework

The following queuing parameters are used to clarify the process used in communication research. (i) Standby server (ii) Necessary Stages of service (iii) Compulsory vacation (iv) Setup time stage (v) Service interruption and Repair process.

Stage 1 Service:

Modulation: Synchronization, which involves incorporating data into an electromagnetic carrier signal, transforms data into radio waves. A stable waveform with a constant amplitude, and frequency is referred to as a carrier signal. For optical fiber and even semiconducting phenomena like spin, characteristics can be added by changing the carrier's amplitude, frequency, phase, and polarization. Direct current can be manipulated, mainly by switching it on and off as in a virtual current loop interface, because it can be thought as a corrupted carrier wave with a fixed amplitude and frequency of 0 Hz. Electronic signals including radio waves, lasers and optical waves, and computer networks are routinely modulated. Base band modulation is what happens when there is no carrier or when a storage device is disconnected from a remote system. In Amplitude modulation, the signal carrier's intensity is differed to display the data of been appended to the signal. Next, in case of Frequency modulation the carrier wave form's frequency is diverse to portray the spread of the data. Phase modulation changes the process of the reference wave to represent changes in data frequency. Also, Quadrature amplitude modulation encodes two or more snippets in a single transmission using two AM carriers (Figs. 3 and 4).

Signal Transmission: A transmission medium in the context of telecommutes and Internet technologies can either refer to a wire network across a multi modal medium, such as a radio channel, or to a physical data transfer, like a cable. One or more senders convey an information signal—such as a digital bit stream—to one or more receivers via a channel. To be transported through space, an information signal needs to pass through some sort of channel or medium. These "communication channels" use cable and broadcast media. A channel is a hypothetical channel model with specific error properties that is used in information theory. According to this broader perspective, a data store is also a communication medium that may be used to send and receive information signals as well as transmit them over time. A signal carries the information through the channel.

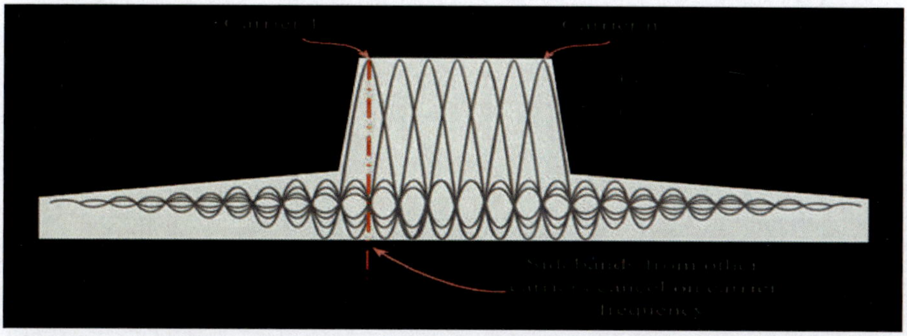

Fig. 3 Carrier synchronization is meant by cancel side band carrier frequencies with transmitted carrier signal

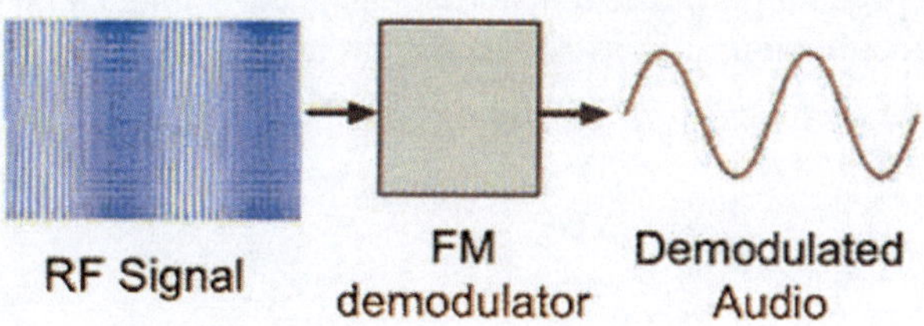

Fig. 4 The RF signal frequency gets mix up with carrier signal and gets demodulated

Optional Vacation (An optional maintenance work is carried out after completion of Stage 1 service):
Demodulates must complete several maintenance activities after the second phase of service is accomplished, comprising error detection and correction, pulse compression, rake receiver, carrier restoration, clock recovery, bit slip, frame synchronization, and acquired spectral response indication.

Carrier Recovery: For the objective of harmonic demodulation, a transmission equipment is a device used to calculate and account for time—frequency discrepancies between the carrier wave of a received signal and the recipient's local oscillator. The carrier frequency oscillators of the sender and receiver in an ideal communications system would be exactly matched in frequency and phase, allowing for flawless coherent decoding of the modulated base band signal. On the other hand, sharing a carrier oscillator between transmitters and receivers is uncommon. Typically independent of the transmitting systems, communications receiver systems each had their own synthesizers with frequency and phase aberrations and instabilities.

Setup Time Stage (In this step, well before work required for the Phase two service is completed.)

Instantaneous amplitude, phase, and frequency are important to information exchange and are used in many applications of signals processing. The transmission needs to be modified in order to carry information. Amplitude, phase, and frequency modulations are only a few of the various types of modifications that can be used. Data is contained as fluctuations in the intensity of a message signal in frequency modulation. Extraction of the modulated signal's envelope is required for demodulation of an amplitude-modulated signal. The instantaneous phase can be extracted after creating an analytical signal by applying the Hilbert transform to a phase modulated signal of type $x(t)$. The instantaneous phase of the signal must be subtracted from this linear offset in order to produce the signal that is modulated with information.

Stage 2 service: Demodulation: It is the procedure of isolating the initial information-carrying signal from a carrier wave. The information content of a modulated carrier wave is retrieved using an electrical device known as a demodulation. Demodulates come in a variety of forms, just as there are various types of modulation. A demodulation's signal output may contain binary data or pictures. Demodulates exist in a variety of shapes and are frequently used in conjunction with radio receivers, but they are also a part of many other systems. There are numerous demodulation techniques, which differ depending on how base band signal properties like intensity, speed, or phase are transmitted in the carrier signal. With a signal that exhibits linear modulation, such as AM, a synchronous detector is utilized. A PM demodulation, on the other hand, is required to demodulate a signal produced by angular modulation.

Breakdown—Interference: The best way to define interference is as the impact of undesired signals or background noise on the receipt of a desired signal. Interference can reduce the quality of the music or image the device produces, completely block reception, or simply temporarily disrupt a transmission. Electrical equipment and radio transmitters are the two main sources of interference. All signal-transmitting communication systems have the capacity to produce interference. Amateur radios, radio and television stations, and mobile phones are some of these systems. Power lines or electrical appliances in your home could be the source of electrical interference. In signal analysis, distortion is the alteration of a signal's original configuration. It alludes to the alteration of a signal's structure in a transmission medium or electronic device.

Steps to overcome these interference: Repair Activity:

Transmit in different places: Spaced apart transmitters can both use the same frequency concurrently. The physical distance between the places ensures that the signals will attenuate and not interfere at authorized power levels, therefore this does not provide a problem. A station would interfere with neighboring stations if it went over its permitted power limit.

Transmit at different frequencies: If two emitters use separate frequencies, they can reach the same location and broadcast simultaneously. Local television and radio stations, mobile phones, and most of the other wireless communication use this approach. Although transmit power is constrained in license-free bands, radios and their processors must nevertheless anticipate and manage interference.

Transmit at different times: If two transmitters transmit at different times, they can encompass the same region and use the same frequency. For example, a point-multi point network could be setup in which distant radios could only transfer when given "authorization" by the base station radio. Intervention would be eliminated because it would only grant access to one at a time. Another option is to allow remote radios to transmit when they do not detect a signal from another radio in the area.

The following section provides a mathematical description of the communication queuing system as a queuing description (problem). The queuing extra variable technique is used to fix the issue, and Sect. 3's performance measurements are acquired in relation to it.

In Sect. 4, a comprehensive study of the performance measurements of the queuing system created for the communication process is done using numerical analysis.

The study is programmed using Python, MATLAB, and R Tool. The simulated study of the communication process is thoroughly covered in this section.

2 Mathematical Framework of the Model (Communication System—A Queuing Approach)

2.1 Description of the Model in Diagram

Description of the above Proposed model

The current study provides a mounting method to address the considerable queuing problem that many systems face, including maintenance work, service stages, preparation work, interruptions, and repair operations. Changes in probability at different system levels can affect the protective device. The method a system takes to strategy has completely transformed into a line issue. The approach includes offering a variety of administrations. After the first administration stage is complete, customers line up for the second level of service. The server takes a brief break to finish the required scheduled maintenance if no clients are awaiting the first level of service. In this sense, the term "vacations" refers to the server maintenance chores that must be completed during that period to assure server uptime or prevent server downtime. This maintenance effort helps the system as much as it can to deliver a smooth service. During the time of vacation, to avoid the lengthen of the queue, a standby server is introduced in the place of the existing server to provide service to all the arriving customers. The server enters a setup phase after the vacation is ended, which is when all pre-processing tasks essential for the second phase of service will be finished. Also, server failure occurs during setup and cannot be prevented. As a result, the server starts the repair procedure straight away. Once the issue has been resolved, service is again made available. All of the clients waiting in line are then given the second step of service. After finishing all service procedures, customers finally leave the system (Fig. 5).

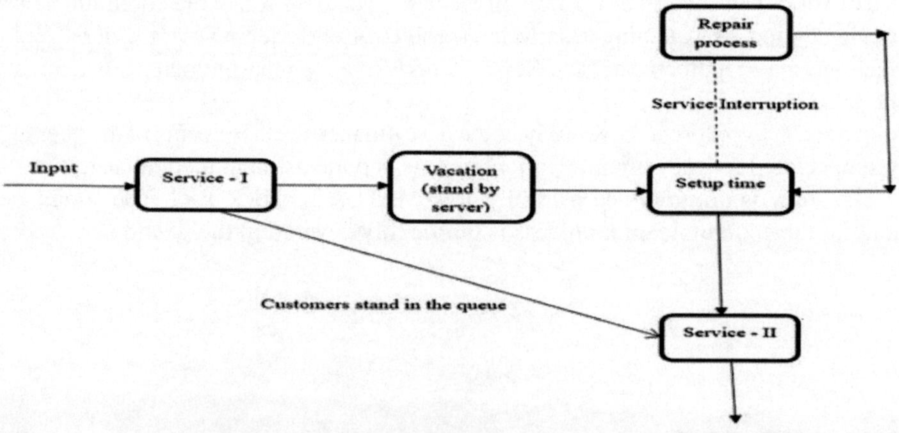

Fig. 5 Queuing model in the process of communication system

The vacation procedure (maintenance work) has only been mentioned in the queueing literature so far after the service has been completed, assuming there are no consumers in the queue. Nonetheless, the vacation procedure is established in between the services according to the proposed model. There is a potential that an interruption will happen while the system is running, and in some situations, it will be visible beforehand. In that situation, the vacation procedure (maintenance work) is done in between services to prevent a severe downtime of the system.

A useful tool for analyzing the common occurrence of standing in line is the queuing theory. The mathematical analysis of how waiting lines or queues arise, function, and get congested is known as queuing theory. The non-Markovian queuing model has been described by numerous writers using a variety of parameters. Two are more than stage of duty throughout this time period, according to some authors. The server then goes through a repair process. For a non-Markovian queuing model, some writers have increased the optional vacation and setup time stages. Asitha K. S. and Khoo. H. L. have thoroughly examined the Incentivized Travel and Mobile Application with Multiple Policy Intervention [1]. In a batch arrival queuing model, Ayyappan and Karpagam [2] thoroughly evaluated the service stage with the breakdown and repair process. Bwambale et al. [3] developed a framework using mobile phone data for trip planning modeling. Ghahramani M et al. investigated the methods, uses, and difficulties of mobile phone data [4]. Ghose et al. [5] assessed the study on environmental marketing based on mobile mobility apps. Through a queuing technique, Gillis et al. [6] investigated how to measure delays at signalized junctions. Fabisiak and Szyjewski [7] worked at a company that utilized mobile phone technology. Kim and Sohn [8] investigated the analysis of travel behaviors using data from mobile phones. Mobile usage in travel has been explained by Linton and Kwortnik [9] using a queuing technique. Maragathasundari [10] does a good job of investigating the two periods of service. Maragathasundari S and Dhanalakshmi investigate the queueing approach problem in mobile adhoc networks [11]. The optional sorts of service—longer vacations, reneging processes, and service interruptions followed by phases of repair—are the focus of a study by Maragathasundari and Manikandan [12]. Maragathasundari and Joy [13] examined the extra services that required service disruption and reconfiguration. Maragathasundari et al. [14] took into account the non-Markovian queue with breakdown, repair procedure, and standby server. Maragathasundari and Srinivasan [15] presented a non-Markovian batch arrival queueing model with several service stages, restricted admissibility, feedback service, and three possible vacations. Maragathasundari et al. [16] considered the concept of limited client admittance in non-Markovian queues. Alkhazaali et al. [17] have studied mobile communication using 5G technology. Liu et al. [18] looked into the Performance Study of Queueing Systems with a Particular Service Disruption. Rama Devi et al. [19] looked at a study of an M/M/1 queuing system with two-phase service. In order to account for server outages, maintenance, breaks, and backup servers, Srinivas Chakravarthy et al. [20] created a non-Markovian queueing model. Vanalakshmi and Maragathasundari [21] created a study on a non-Markovian queueing model with two stages of service, a breakdown, repair process, and an optional vacation. An analysis of a queuing model with two phases of service, setup time stage, and optional vacation was done by Maragathasundari et al. [22]. Using a computational framework, Xu et al. [23] has finished a computational framework using fine-grained mobile phone data. Mobile phone GPS data in urban ride-sharing was alluded to by Zhang et al. [24]

2.2 *Description of the Model's Notations, Assumptions and Parameters*

Clients enter the framework in varying-sized groups and follow a Poisson distribution with an arrival rate $\lambda_{\varsigma} > 0$. The service is made available to everyone follows a general distribution.

Notations	Description	Distribution function and density function
$E^{(i)}(x)$	**Stages of Service**: The number of parallel service stations that are functioning at once is referred to as "service channels." A customer must stop by a specific number of service stations in a series in order to receive full servicing. The stages of service are used to describe this. The conditional probability is denoted by $\mu_{e_i}(x)$, can be determined using the formula $\mu_{e_i}(x) = \frac{c_i^*(x)}{1-C_i^*(x)}$ and $c_i^*(x) = \mu_{e_i}(x)e^{-\int_0^x \mu_{e_i}(t)dt}$, $i = 1, 5$	$C_i^*(x)$ $c_i^*(x)$
$F_n(x)$	**Compulsory Vacation**: Employees or machines can take a predetermined period off from work, usually to relax and recharge. Most devices or enterprises have a vacation time limit $\mu_f(x) = \frac{c_2^*(x)}{1-C_2^*(x)}, c_2^*(x) = \mu_f(x)e^{-\int_0^x \mu_f(t)dt}$	$C_2^*(x)$ $c_2^*(x)$
$G_n(x)$	**Setup time stage of service:** The amount of time required to setup a machine so that it is ready to perform a task is known as setup time. Reducing setup time for small production runs is crucial to facilitating just-in-time production $\mu_g(x) = \frac{c_3^*(x)}{1-C_3^*(x)}, c_3^*(x) = \mu_g(x)e^{-\int_0^x \mu_g(t)dt}$	$C_3^*(x)$ $c_3^*(x)$
ξ	**Standby server**: The standby server is available while the vacation procedure is in progress. Another server is quickly deployed in the system to supply the service to all incoming clients once the normal server has entered the maintenance phase. The standby server rate is displayed as ξ	— —

(continued)

(continued)

Notations	Description	Distribution function and density function
$G_n(x)$	**Repair Process**: As during setup time stage, the service is interrupted. Break down arrival rate is θ. Immediately the server is sent into repair process. The conditional density function is given by $\mu_h(x) = \frac{c_4^*(x)}{1-C_4^*(x)}, c_4^*(x) = \mu_h(x)e^{-\int_0^x \mu_h(t)dt}$	$C_4^*(x)$ $c_4^*(x)$

2.3 *Description of Governing Equation Using the Birth and Death Process*

Only birth and death transitions can be states in a birth–death process, a particular sort of continuous time Markov process. When a birth occurs, the procedure switches from state i to state i + 1. Like this, when death occurs, the process shifts from stage i to state i-1. Input is considered to be the birth and the output (Exit) is related to death.

The following governing equations for the specified model are framed based on the analysis of the birth and death processes mentioned above.

Governing Equations

$$\frac{\mathrm{d}}{\mathrm{d}x}E_n^{(1)}(x) + \left(\lambda + \mu_{e_1}(x)\right)E_n^{(1)}(x) = \lambda \sum_{r=1}^{n} w_r E_{n-r}^{(1)}(x), n > 1 \tag{1}$$

$$\frac{\mathrm{d}}{\mathrm{d}x}E_0^{(1)}(x) + \left(\lambda + \mu_{e_1}(x)\right)E_0^{(1)}(x) = 0 \tag{2}$$

$$\frac{\mathrm{d}}{\mathrm{d}x}F_n(x) + \left(\lambda + \mu_f(x) + \xi\right)F_n(x)$$

$$= \lambda \sum_{r=1}^{n} w_r F_{n-r}(x) + \xi F_{n-r}(x), n > 1 \tag{3}$$

$$\frac{\mathrm{d}}{\mathrm{d}x}F_0(x) + \left(\lambda + \mu_f(x) + \xi\right)F_0(x) = \xi F_1(x) \tag{4}$$

$$\frac{\mathrm{d}}{\mathrm{d}x}G_n(x) + \left(\lambda + \mu_g(x) + \theta\right)G_n(x) = \lambda \sum_{r=1}^{n} w_r G_{n-r}(x) \tag{5}$$

$$\frac{\mathrm{d}}{\mathrm{d}x}G_0(x) + \left(\lambda + \mu_g(x) + \theta\right)G_0(x) = 0 \tag{6}$$

$$\frac{\mathrm{d}}{\mathrm{d}x}H_n(x) + (\lambda + \mu_h(x))H_n(x) = \lambda \sum_{r=1}^{n} w_r H_{n-r}(x) \tag{7}$$

$$\frac{\mathrm{d}}{\mathrm{d}x}H_n(x) + (\lambda + \mu_h(x))H_n(x) = 0 \tag{8}$$

$$\frac{\mathrm{d}}{\mathrm{d}x}E_n^{(2)}(x) + \left(\lambda + \mu_{e_2}(x)\right)E_n^{(2)}(x) = \lambda \sum_{r=1}^{n} w_r E_{n-r}^{(2)}(x), n > 1 \tag{9}$$

$$\frac{\mathrm{d}}{\mathrm{d}x}E_0^{(2)}(x) + \left(\lambda + \mu_{e_2}(x)\right)E_0^{(2)}(x) = 0 \tag{10}$$

Initial and Boundary Conditions

$$E_n^{(1)}(0) = \int_0^\infty H_n(x)\mu_h(x)\mathrm{d}x + \int_0^\infty E_n^{(2)}(x)\mu_{e_2}(x)\mathrm{d}x + \lambda w_{r+1}A \tag{11}$$

$$F_n(0) = \int_0^\infty E_n^{(1)}(x)\mu_{e_1}(x)\mathrm{d}x \tag{12}$$

$$G_n(0) = \int_0^\infty F_n(x)\mu_f(x)\mathrm{d}x \tag{13}$$

$$H_n(0) = \theta \int_0^\infty G_n(x)\mathrm{d}x = \theta G_{n-1}(x) \tag{14}$$

$$E_n^{(2)}(0) = \int_0^\infty G_n(x)\mu_g(x)\mathrm{d}x \tag{15}$$

$$\lambda A = \int_0^\infty H_0(x)\mu_h(x)\mathrm{d}x + \int_0^\infty E_0^{(2)}(x)\mu_{e_2}(x)\mathrm{d}x \tag{16}$$

3 Methodological View of Resolving the Queuing Problem

The supplemental variable method, a technique from queuing theory, can be used to determine the stable probability of non-Markov queues. David Cox and David George Kendall made the introduction. Assume that the accumulation's outcome is distributed using a common probability density function to demonstrate how to use additional variables. Second, the waiting size process $N(t)$ cannot be altered since the Markov property is absent. As a result, we introduce X as a variable that is created at random (t). It displays how long the consumer received service at time t. We demonstrate the transformation of the two-dimensional random variable $(N(t);$ $X\ (t))$ into a Markov process by including extra variables in the state description.

The probability generating function of queue size, denoted as $Y\ q\ (z)$, is produced by using the additional variable technique on the equations previously described. This function is used as the basis for determining the various queue performance measurements for the particular queuing problem.

$$L_q^*(z) = \left[\left[\frac{1 - C_1^*(a)}{a}\right] + \left[\frac{1 - C_2^*\left(a + \xi - \frac{\xi}{z}\right)}{a + \xi - \frac{\xi}{z}}\right] C_1^*(a) + \left\{C_1^*(a) C_3^*(a + \theta)\left[\frac{1 - C_2^*\left(a + \xi - \frac{\xi}{z}\right)}{a + \xi - \frac{\xi}{z}}\right]\right\} \left[1 + \xi z\left(\left[\frac{1 - C_4^*(a)}{a}\right] + C_4^*(a)\left[\frac{1 - C_5^*(a)}{a}\right]\right)\right]\right]$$

We derive the performance metrics from the queue size's probability generating function as follows.

3.1 Metrics of the Queuing Framework Designed in Communication Process

(a) *Time Spent Idle and the Operational Mode*

To find C, the constraint that normalizing, $L_q(z) + A = 1$ is employed.
L. Hospital's rule is used as a result of $L_q(z)$'s indeterminacy.
Hence, we get

$$A = \frac{\lambda E(c_5)\left(1 - \xi\left(\frac{1 - C_3^*(\theta)}{\theta}\right)\right) + 2\left[1 + \{-\xi\lambda E(c_1) + \xi(-\lambda + \xi)E(c_2) - \lambda\xi E(c_4) - \lambda\xi E(c_5)\}\left(\frac{1 - C_3^*(\theta)}{\theta}\right) - \xi C_3^{*\prime}(\theta)\right]}{\lambda\left[\frac{1 - C_3^*(\theta)}{\theta}\right] + 2\left[1 + \{-\xi\lambda E(c_1) + \xi(-\lambda + \xi)E(c_2) - \lambda\xi E(c_4) - \lambda\xi E(c_5)\}\left(\frac{1 - C_3^*(\theta)}{\theta}\right) - \xi C_3^{*\prime}(\theta)\right]}$$

(b) *Utilization Factor*

$$\rho = \frac{\lambda\left[\frac{1-C_3^*(\theta)}{\theta}\right] + \lambda E(c_5)\left(1-\xi\left(\frac{1-C_3^*(\theta)}{\theta}\right)\right)}{\lambda\left[\frac{1-C_3^*(\theta)}{\theta}\right] + 2\left[1+\{-\xi\lambda E(c_1)+\xi(-\lambda+\xi)E(c_2)-\lambda\xi E(c_4)-\lambda\xi E(c_5)\}\left(\frac{1-C_3^*(\theta)}{\theta}\right)-\xi C_3^{*\prime}(\theta)\right]}$$

(c) *Length of the Queue* ($\boldsymbol{L_q}$)*:* L' Hospital's rule is used to determine the typical queue length in a stable equilibrium.

$$L_q = \lim_{z\to 1}\frac{D'(z)N''(z) - N'(z)D''(z)}{2(D'(z))^2}$$

$$N'(1) = \lambda\left[\frac{1-C_3^*(\theta)}{\theta}\right]$$

$$N''(1) = 2\lambda\left[E(c_1)E(c_2) + \left(\frac{1-C_3^*(\theta)}{\theta}\right)(\xi(E(c_4)+E(c_5)))\right.$$
$$\left. + (\lambda E(c_1)+\xi E(c_2))\left(\frac{1-C_2^*(\theta)}{\theta}\right) - C_3^{*'}(\theta)\right]$$

$$D'(1) = \lambda E(c_5)\left(1-\xi\left(\frac{1-C_3^*(\theta)}{\theta}\right)\right)$$
$$+ 2[1+\{-\xi\lambda E(c_1)+\xi(-\lambda+\xi)E(c_2)-\lambda\xi E(c_4)-\lambda\xi E(c_5)\}$$
$$\left.\left(\frac{1-C_3^*(\theta)}{\theta}\right) - \xi C_3^{*'}(\theta)\right]$$

$$D''(1) = 2\lambda E(c_5)\left[1-\{\xi\lambda(E(c_1)+E(c_4)+E(c_5))+\xi E(c_2)(-\lambda+\xi)\}\left(\frac{1-C_3^*(\theta)}{\theta}\right) - C_3^{*'}(\theta)\right]$$
$$+2-\xi\lambda^2E(c_1^2)+\lambda E(c_1)E(c_2)(-\lambda+\xi)+\lambda(E(c_4)+E(c_5))+\lambda[E(c_1)+E(c_4)+E(c_5)]$$
$$-E(c_2)(-\lambda+\xi)+(-\lambda+\xi)+(-\lambda+\xi)[\lambda(E(c_1)+E(c_4)+E(c_5)-E(c_2)(-\lambda+\xi))]$$
$$-2\xi\lambda E(c_4)-\lambda\xi E(c_4)\{\lambda(E(c_1)+E(c_5))-(-\lambda+\xi)E(c_2)\}+\lambda^2E(c_4^2)$$
$$+\lambda E(c_5)(E(c_1)+E(c_4)-(-\lambda+\xi)E(c_2))+\lambda^2E(c_5^2)\left(\frac{1-C_3^*(\theta)}{\theta}\right)$$
$$+\xi C_3^{*'}(\theta)[\lambda(+E(c_4)+E(c_5))+(-\lambda+\xi)E(c_2)]-C_3^{*''}(\theta)(-\lambda+\xi)$$
$$-\xi C_3^{*'}(\theta)[\lambda(E(c_1)+E(c_4)+E(c_5))+(-\lambda+\xi)E(c_2)]$$

$$D''(1) = 2\lambda E(c_5)\left[1-\{\xi\lambda(E(c_1)+E(c_4)+E(c_5))+\xi E(c_2)(-\lambda+\xi)\}\left(\frac{1-C_3^*(\theta)}{\theta}\right) - C_3^{*'}(\theta)\right]$$
$$+2-\xi\lambda^2E(c_1^2)+\lambda E(c_1)E(c_2)(-\lambda+\xi)+\lambda(E(c_4)+E(c_5))+\lambda[E(c_1)+E(c_4)+E(c_5)]$$
$$-E(c_2)(-\lambda+\xi)+(-\lambda+\xi)+(-\lambda+\xi)[\lambda(E(c_1)+E(c_4)+E(c_5)-E(c_2)(-\lambda+\xi))]$$
$$-2\xi\lambda E(c_4)-\lambda\xi E(c_4)\{\lambda(E(c_1)+E(c_5))-(-\lambda+\xi)E(c_2)\}+\lambda^2E(c_4^2)$$
$$+\lambda E(c_5)(E(c_1)+E(c_4)-(-\lambda+\xi)E(c_2))+\lambda^2E(c_5^2)\left(\frac{1-C_3^*(\theta)}{\theta}\right)$$
$$+\xi C_3^{*'}(\theta)[\lambda(+E(c_4)+E(c_5))+(-\lambda+\xi)E(c_2)]-C_3^{*''}(\theta)(-\lambda+\xi)$$

$$-\xi C_3^{*'}(\theta)\left[\lambda\left(E(c_1)+E(c_4)+E\left(c_5\right)\right)+(-\lambda+\xi)E(c_2)\right]$$

(d) *Waiting Time Performance Measures*

Calculations of the system length, a customer's anticipated arrival time within the framework, and their average wait time in line are all possible. Using Little's method.

$W_q = \frac{L_q}{\lambda_c}$, $W = \frac{L}{\lambda_c}$, $L = L_q + \rho$. The model is thoroughly analyzed and justified in the following part using a numerical illustration technique with the programmed R Tool, MATLAB, and Python.

4 Numerical Structural View of the Queuing Problem in the Communication System

The service time is distributed exponentially in this case.

$$\begin{aligned}
\lambda &= 2, \xi = 0.5, \theta = 2, \mu_{e_1} = 3, \mu_f = 4, \mu_{e_2} = 5, \mu_g \\
&= 6, \mu_h = 7, E(C_1) = \frac{1}{\mu_{e_1}}, E(C_2) = \frac{1}{\mu_f}, E(C_4) \\
&= \frac{1}{\mu_h}, E(C_5) = \frac{1}{\mu_{e_2}}, E\left(C_1^2\right) = \frac{2}{\mu_{e_1}^2}, E\left(C_2^2\right) \\
&= \frac{2}{\mu_h^2}, E\left(C_5^2\right) = \frac{2}{\mu_{e_2}^2}, C_3^*(\theta) = \frac{\mu_g}{\mu_g + \theta}, \\
C_3^{*'}(\theta) &= \frac{-\mu_g}{\left(\mu_g + \theta\right)^2}, C_3^{*'}(\theta) = \frac{2\mu_g}{\left(\mu_g + \theta\right)^3}
\end{aligned}$$

Case 1: Arrival rate $\lambda = 2, 5, 10, 15, 20$(Figs. 6 and 7).

Case 2: Breakdown rate $\theta = 2, 4, 6, 8, 10$ (Figs. 8 and 9).

Case 3: Standby server $\xi = 1, 1.5, 2, 2.5, 3$ (Figs. 10 and 11).

Table 1 shows how the defined queuing method is impacted by the client delivery ratio. The rate of entry (arrival) rises as the line length does. More people are using the platform now, and there are longer lineups. Pre-processing is completed during setup, which lowers the load factor and decreases the amount of downtime experienced as queues develop. Table 2 now includes an explanation of service outages. Under practical circumstances, this is inexplicable and happens randomly. The queue size and client wait times do not grow when a service is suspended. This is due to the

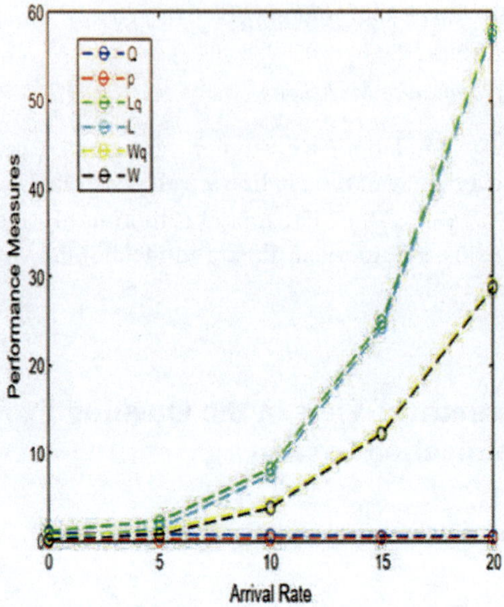

Fig. 6 The graphical portrayal of arrival rate

```
import pandas as pd
import matplotlib.pyplot as plt

data = pd.read_csv("Arrival Rate.csv")
x = data.M
y1 = data.Q
y2 = data.P
y3 = data.Lq
y4 = data.L
y5 = data.Wq
y6 = data.W
plt.plot(x,y1,color="blue",label="Q")
plt.plot(x,y2,color="green", label="P")
plt.plot(x,y3,color="black", label="Lq")
plt.plot(x,y4,color="red", label="L")
plt.plot(x,y5,color="cyan", label="Wq")
plt.plot(x,y6,color="magenta", label="W")
plt.xlabel("Arrival Rate")
plt.ylabel("Performance Measures")
plt.legend()
plt.title("Arrival Rate Effectives")
plt.show()
```

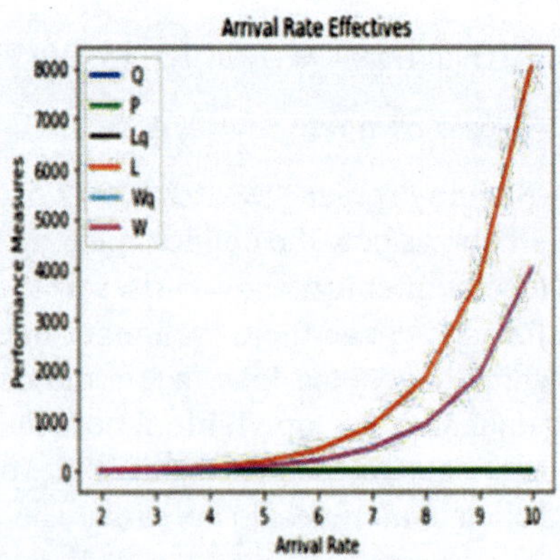

Fig. 7 The arrival rate is represented graphically and through Python programming

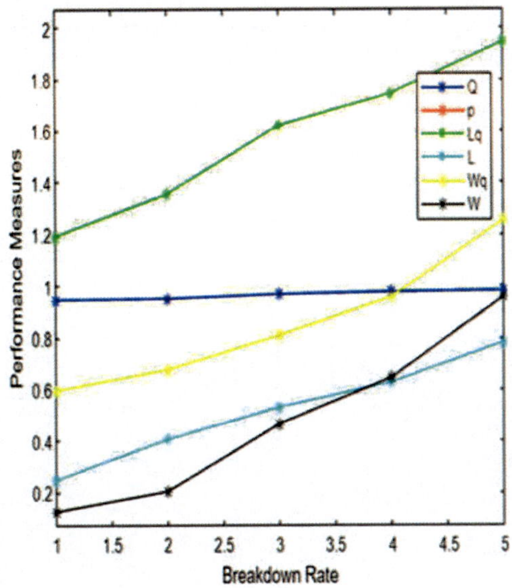

Fig. 8 The graphical portrayal of breakdown rate

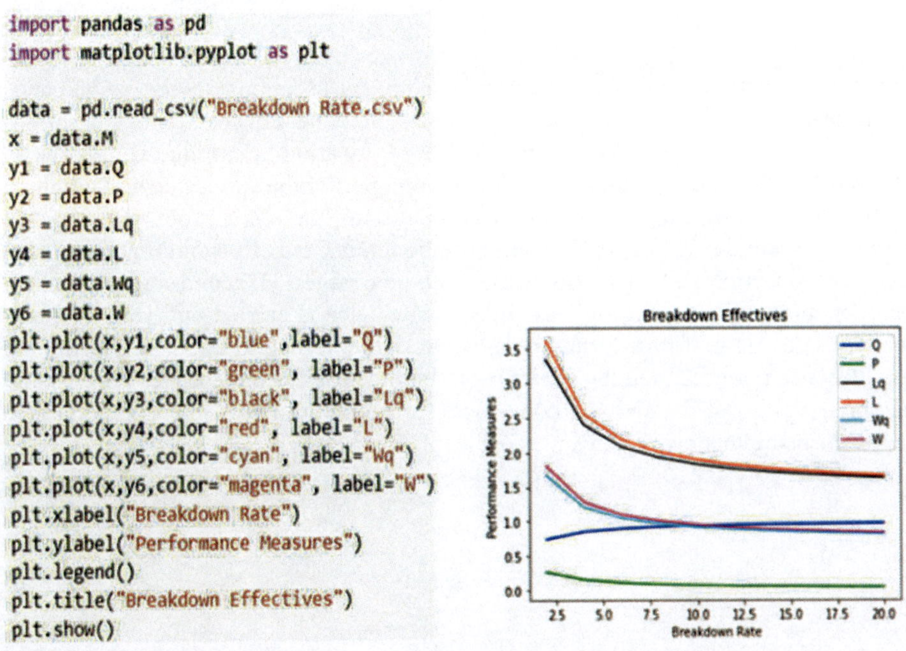
```
import pandas as pd
import matplotlib.pyplot as plt

data = pd.read_csv("Breakdown Rate.csv")
x = data.M
y1 = data.Q
y2 = data.P
y3 = data.Lq
y4 = data.L
y5 = data.Wq
y6 = data.W
plt.plot(x,y1,color="blue",label="Q")
plt.plot(x,y2,color="green", label="P")
plt.plot(x,y3,color="black", label="Lq")
plt.plot(x,y4,color="red", label="L")
plt.plot(x,y5,color="cyan", label="Wq")
plt.plot(x,y6,color="magenta", label="W")
plt.xlabel("Breakdown Rate")
plt.ylabel("Performance Measures")
plt.legend()
plt.title("Breakdown Effectives")
plt.show()
```


Fig. 9 The breakdown rate is represented graphically and through Python program

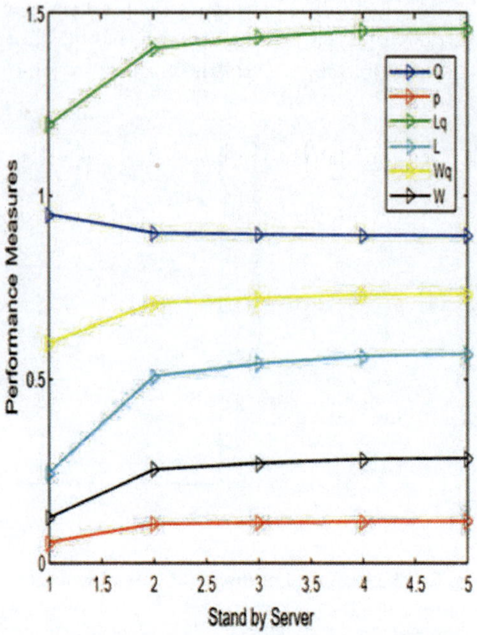

Fig. 10 The graphical portrayal of standby server

fact that presets are typically used during setup. By executing server maintenance after the service has concluded, failure can absolutely be reduced. Table 3 describes the concept of a backup server in the system. A vacation is introduced in between the initial stage of service and setup. The second part of the service should continue without interruption. In this communication system, there is a probability that the setup and second stage of service will both be interrupted. To avoid that, a specific maintenance work is finished prior to the setup time stage and second stage of service. As a result, service is interrupted while maintenance is carried out. But, as maintenance is being performed, the line's length will continue to increase. During the maintenance period, a standby server is placed to partially compensate. This backup server serves all of the arriving consumers. Although it takes some time, doing so helps to ensure that every consumer is completely satisfied.

```
import pandas as pd
import matplotlib.pyplot as plt

data = pd.read_csv("Stand by server.csv")
x = data.M
y1 = data.Q
y2 = data.P
y3 = data.Lq
y4 = data.L
y5 = data.Wq
y6 = data.W
plt.plot(x,y1,color="blue",label="Q")
plt.plot(x,y2,color="green", label="P")
plt.plot(x,y3,color="black", label="Lq")
plt.plot(x,y4,color="red", label="L")
plt.plot(x,y5,color="cyan", label="Wq")
plt.plot(x,y6,color="magenta", label="W")
plt.xlabel("Stand by Server")
plt.ylabel("Performance Measures")
plt.legend()
plt.title("Stand by Server Effectives")
plt.show()
```

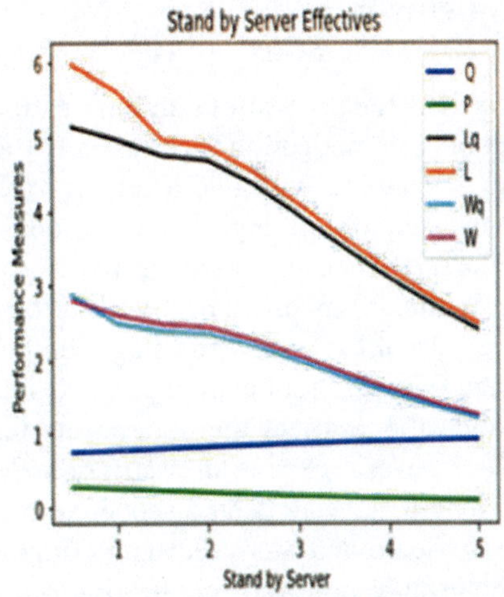

Fig. 11 The standby server is represented graphically and through Python programming

Table 1 Arrival rate λ = 2, 5, 10, 15, 20

λ_S	R	ρ	L_q	L	W_q	W
2	0.8387	0.1613	0.8796	1.1876	0.4568	0.6965
5	0.7979	0.2021	1.4963	2.2942	0.7482	1.1471
10	0.6934	0.3066	7.5108	8.2042	3.7554	4.1021
15	0.6295	0.3705	24.2955	24.9249	12.1477	12.4625
20	0.5864	0.4136	57.3890	57.9753	28.6945	28.9877

Table 2 Breakdown rate $\theta = 2, 4, 6, 8, 10$

θ	R	ρ	L_q	L	W_q	W
2	0.9458	0.0542	0.2458	1.1916	0.1229	0.5958
4	0.9500	0.0500	0.4041	1.3541	0.2020	0.6770
6	0.9666	0.0334	0.5289	1.6191	0.4645	0.8095
8	0.9750	0.0250	0.6257	1.7432	0.6445	0.9574
10	0.9800	0.0200	0.7805	1.9436	0.9256	1.2561

Table 3 Standby server $\xi = 1, 1.5, 2, 2.5, 3$

ξ	R	ρ	L_q	L	W_q	W
1	0.8953	0.1047	0.5073	1.4026	0.2536	0.7013
1.5	0.8914	0.1086	0.5405	1.4319	0.2703	0.7160
2	0.8888	0.1112	0.5610	1.4498	0.2805	0.7249
2.5	0.8877	0.1123	0.5671	1.4548	0.2836	0.7274
3	0.8734	0.1266	0.5768	1.4674	0.2882	0.7291

5 Conclusion

In this article, the process of communication is explained in detail. The system in the concept of communication is completely transformed to a batch arrival mathematical concept of queuing with two levels of service is covered in this article. During the period of vacation, all significant system maintenance tasks can be finished. As a result, the server can minimize service disruptions. All pre-processing tasks necessary for all significant maintenance work to be finished during the break are completed with the help of the setup time stage. In the numerical analysis part, it is thoroughly examined how different characteristics, such as standby servers, break down arrival rates, and others, affect the queuing system. The probability generating function, as well as several execution metrics, including the mean queue length and mean system length, have all been discovered. Numerous every day and industrial systems can be modeled and analyzed using queuing models. Additionally, they portray a more believable option of congestion problems, which is why system developers, manufacturing engineers, and administrators prefer them.

References

1. Asitha KS, Khoo HL (2020) Incentivised travel and mobile application as multiple policy intervention for mode shift. KSCE J Civ Eng 24(10):3074–3091
2. Ayyappan G, Karpagam S (2018) An M[X]/G (a, b)/1 queueing system with breakdown and repair, stand-by server, multiple vacation and control policy on request for re-service. Mathematics 6(101):1–18
3. Bwambale A, Choudhury CF, Hess S, Iqbal MS (2020) Getting the best of both worlds: a framework for combining disaggregate travel survey data and aggregate mobile phone data for trip generation modelling. Transportation 48(5):2287–2314
4. Ghahramani M, Zhou M, Wang G (2020) Urban sensing based on mobile phone data: approaches, applications, and challenges. IEEE/CAA J Automatica Sinica 7(3):627–637
5. Ghose A, Kwon HE, Lee D, Oh W (2019) Seizing the commuting moment: contextual targeting based on mobile transportation apps. Inf Syst Res 30(1):154–174
6. Gillis D, Gautama S, Van Gheluwe C, Semanjski I, Lopez AJ, Lauwers D (2020) Measuring delays for bicycles at signalized intersections using smartphone GPS tracking data. ISPRS Int J Geo Inf 9(3):174
7. Szyjewski G, Fabisiak L (2018) A study on existing and actually used capabilities of mobile phones technologies. Procedia Comput Sci 126:1627–1636

8. Kim H, Sohn D (2020) e urban built environment and the mobility of people with visual impairments: analysing the travel behaviours based on mobile phone data. J Asian Architect Build Eng 19(6):731–741
9. Linton H, Kwortnik RJ (2019) Mobile usage in travel: bridging the supplier-user gap. Int J Contemp Hosp Manag 31(2):771–789
10. Maragathasundari S (2018) An examination on queuing system of general service distribution with an establishment time and second discretionary administration. Int J Appl Comput Math 4(3):1–12
11. Maragathasundari S (2018) Dhanalakshmi: Mobile adhoc networks problem a queueing approach. Int J Commun Netw Distrib Syst 21(4):475–495
12. Maragathasundari S, Manikandan P (2020) A study on the performance measures of the non-Markovian model of optional types of service with extended vacation, reneging process and service interruption followed by phases of repair. Int J Process Manage Benchmarking 10(4):520–249
13. Maragathasundari S, Joy MC (2017) Queueing model of optional type of services with service stoppage and revamp process in web hosting. Int J Knowl Manage Tourism Hospitality 1(2)
14. Maragathasundari S, Radha S (2019) A study on the investigation of mathematical modelling in non-Markovian queue. AIP Conf Proc 2177:20042–20049. https://doi.org/10.1063/1.5135217
15. Maragathasundari S, Srinivasan S (2017) Analysis of non-Markovian batch arrival queueing model with multi stages of service of restricted admissibility, feedback service and three optional vacations in production and manufacturing. Int J Math Oper Res 11(3):285–309
16. Maragathasundari S, Vanalakshmi R, Somasundaram RS (2020) A study on the concept of restricted admissibility of customers in non-Markovian queues. J Crit Rev 7(19):5006–5011
17. Alkhazaali NH, Aljiznawi RA, Jabbar SQ, Kadhim DJ (2017) Mobile communication through 5G technology (challenges and requirements). Int J Commun Netw Syst Sci 10(5B):1–5
18. Liu P, Jiang T, Cha X (2020) Performance analysis of queueing systems with a particular service interruption discipline. Discrete Dyn Nat Soc 1:13. Article ID 1847512
19. Rama Devi VN, Rao A, Chandan K (2019) Analysis of a M/M/1 queueing system with two-phase, N-policy, server failure and second optional batch service with customers impatient behaviour. IOP Conf Ser J Phys 1344(012015):1–10
20. Srinivas Chakravarthy R (2020) Shruti, Rakhee Kulshrestha: a queueing model with server breakdowns, repairs, vacations and backup server. Oper Res Perspect 7(100131):1–13
21. Vanalakshmi R, Maragathasundari S, Dhanalakshmi KS (2021) Queuing system in VLSI physical design. Math Comput Simul 201:755–768
22. Vanalakshmi R, Maragathasundari S, Kishore Eswar S (2021) Queuing system behaviour in thermo pack process. J Phys Conf Ser 1850(012047):1–13
23. Xu Y, Li J, Xue J, Park S, Li Q (2021) Tourism geography through the lens of time use: a computational framework using fine-grained mobile phone data. Ann Assoc Am Geogr 111(5):1–25
24. Zhang H, Chen J, Li W, Song X, Shibasaki R (2020) Mobile phone GPS data in urban ride-sharing: an assessment method for emission reduction potential. Appl Energ 269:115038–115048

The Development of a Tool for the Detection of Cotton Wool Spots, Haemorrhage, and Exudates Using Multi-resolution Analysis

Yogesh Rajput, Sonali Gaikwad, Rajesh Dhumal, and Jyotsna Gaikwad

1 Introduction

The World Health Organization (WHO) estimates that there are 347 million people living with diabetes around the globe, and that more than 80% of diabetes-related fatalities take place in various nations. According to projections made by the WHO, diabetes would rank as the seventh greatest cause of death in the year 2030. Diabetes can induce a condition called diabetic retinopathy, which damages the retina by causing blood or fluid to flow from blood vessels in the retina. The diabetic retinopathy may be broken down into two stages: early and advanced. The first kind of diabetic retinopathy is called non-proliferative diabetic retinopathy (NPDR), while the second type is called proliferative diabetic retinopathy (PDR). Cotton wool spots are the name given to the yellow and white dots. They are brought on by microinfarcts that occur in the retinal nerve fibre layer. The axoplasm of exploded retinal ganglion cell axons is extruded like toothpaste from the cell. You should be on the lookout for Patches that look like cotton wool scattered over the optic disc and along the temporal vascular arcades. Exudates is the term that's used to describe those golden specks. These lipid remnants are the result of serous fluid leaking out of capillaries that have been damaged. And retinal haemorrhage is a condition of the eye in which bleeding occurs into the tissue that is located on the back wall of the eye and is responsible for retentive vision. A retinal haemorrhage can be caused by hypertension, retinal

Y. Rajput (✉) · R. Dhumal
Symbiosis Institute of Geoinformatics (SIG), Symbiosis International (Deemed University) (SIU), Model Colony, Pune, Maharashtra, India
e-mail: yogeshrajput128@gmail.com

S. Gaikwad
Shree Shivaji Science and Arts College, Chikhli, Maharashtra, India

J. Gaikwad
Deogiri College, Aurangabad, Maharashtra, India

J. K. Mandal et al. (eds.), *Proceedings of International Conference on Network Security and Blockchain Technology*, Lecture Notes in Networks and Systems 738,
https://doi.org/10.1007/978-981-99-4433-0_25

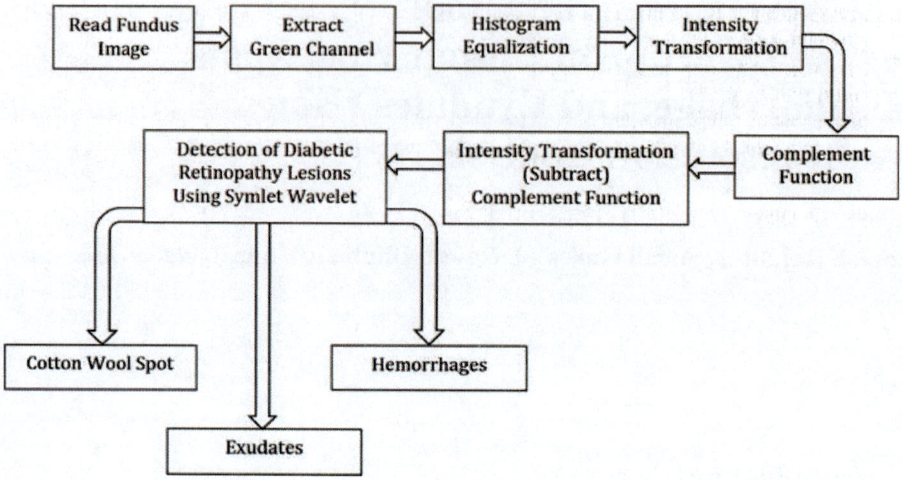

Fig. 1 Workflow for detection of cotton wool spot, exudates and haemorrhages

vein occlusion, which refers to the blocking of a retinal vein, or diabetes mellitus, which causes the formation of tiny blood vessels in the retina that are fragile and easily destroyed. Shaking the head, especially in very young newborns, or receiving a significant blow to the head can also cause retinal haemorrhages [1]. Shaking can also cause retinal haemorrhages. Blood vessels are enhanced and segmented by utilising Gabor wavelet and multilayered thresholding, respectively. This computer-aided technique for the early diagnosis of DR was proposed by Usman M. Akram and others. After that, they localised the optic disc by using a thresholding and an average filter, and then determined the border of the optic disc by using a Hough transform and edge detection. After the blood vessels and optic disc (OD) have been separated from one another, a hybrid fuzzy classifier is used to identify dark and bright lesions [2] (Fig. 1).

Rupa V. L. and P. S. Kulkarni details the process of extracting a variety of elements from fundus pictures, including exudates, microaneurysms, optic Disc, macula, blood vessels, and textural attributes such as entropy, amongst other things. In addition, the genetic algorithm and the multilayer feed forward neural network are utilised for the categorization of diabetic retinopathy lesions. The work that is being suggested has a primary emphasis on detecting and classifying [3]. This technique achieves a sensitivity of 80% while also maintaining a specificity of 83%. There are a number of different lesions that manifest themselves, including microaneurysms, haemorrhages, cotton wool patches, and exudates. Exudates have a tendency to gather in a ring around the location of the diseased vessel and seem as yellowish-white deposits with well-defined borders, while cotton wool patches are also present. Since it has a clearly defined border, exudates are simpler to distinguish from the backdrop than cotton wool spots are. This is because of the difference in texture. In order to identify these lesions, the cotton wool patches and exudates need to be separated from the background using the appropriate method. Therefore, the purpose of this research is to suggest refining the edge in order to ease the segmentation procedure for cotton wool spots and exudates by reducing ramp width [4].

2 Methodology

In the proposed approach, the fundus pictures first undergo preprocessing, which aims to eliminate noise from those images. Remove the optical disc from consideration. In order to get rid of the optical disc, we extract the green channel from the RGB image. This is done since the green channel has a higher intensity than the red and blue channels. After that, perform histogram equalisation so that the image is improved. After that, perform the intensity transformation so that the optic disc is brought to the forefront. After that, perform the function of the complement. The OD may then be removed by deleting the complement image and replacing it with an intensity modified image. In addition to this, remove the mask from the fundus picture. The process of extracting the fundus mask begins with the removal of the red channel from the RGB picture, followed by the use of binarization with a threshold. Following the completion of the preprocessing step, we use the symlet wavelet algorithm to extract diabetic retinopathy lesions. Wavelet analysis is notable for having the critical attribute of flawless reconstruction. This refers to the process of reassembling a deconstructed signal or picture into its original form without any information being lost in the process. In the process of wavelet transformation, there are a few different fundamental functions that might be utilised as the mother wavelet. Because the mother wavelet is responsible for producing all wavelet functions that are utilised in the transformation through translation and scaling, it is the mother wavelet that dictates the properties of the wavelet transform that is produced. In order to make good use of the wavelet transform, it is necessary to take into consideration the specifics of the application (Fig. 2).

Following figure shows the fundus image mask and removal of optic disc (OD) (Figs. 3 and 4).

Under consideration and select an acceptable mother wavelet. The shapes of the wavelets and their capacity to perform signal analysis in a certain context are taken into consideration when selecting the wavelets to use. Orthogonal and biorthogonal wavelet families are the two primary groups that may be identified from one another. The Daubechies, Coiflet, and Symlet wavelet families are all considered to be orthogonal [5]. The extraction of diabetic retinopathy lesions is accomplished with the help of symlet wavelet. Including but not limited to cotton wool spot, exudates, and haemorrhages.

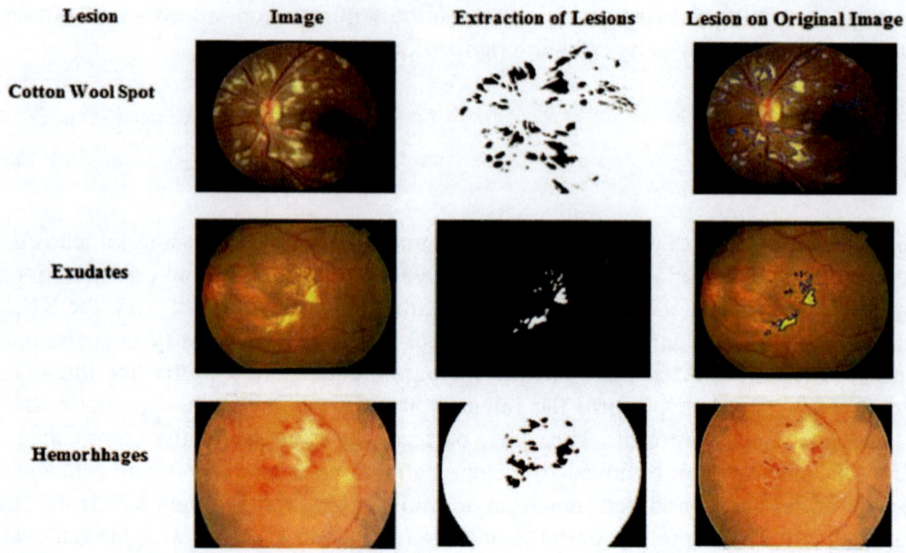

Fig. 2 Extracted diabetic retinopathy lesions

3 Result

Long-term diabetes and fluctuating blood glucose levels can develop diabetic retinopathy, which is now the most prevalent cause of vision loss globally. It has developed into a serious issue among people of working age that requires speedy response to prevent future eyesight loss. Develop a graphical user interface tool for the diagnosis of diabetic retinopathy complications such cotton wool spots, haemorrhages, and exudates (EX). Digital image processing techniques and wavelet decomposition with the help of symlet wavelet are utilised by our team in the process of detecting diabetic retinopathy lesions. The graphical user interface (GUI) tool was designed with MATLAB 2013a. Use certain online databases in addition to the local fundus image database that was created by Dr. Manoj Saswades for the purpose of evaluating this method. The specifics of the databases are shown in the Table 1.

Following the extraction of diabetic retinopathy lesions, statistical analysis is performed by calculating the mean, the variance, the standard deviation, and the correlation. The statistical approach for cotton wool spot is presented in the following Table 2.

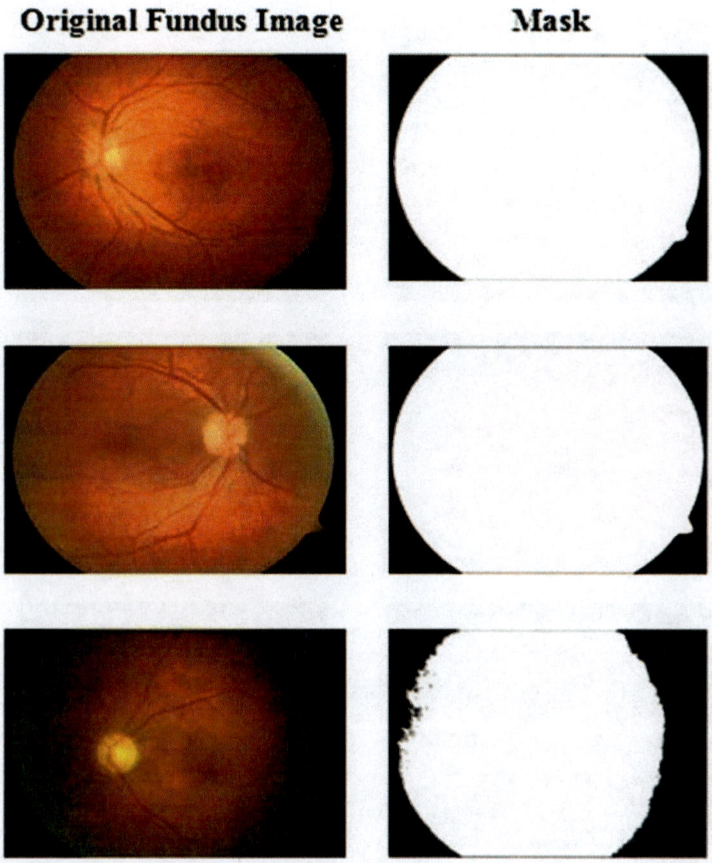

Fig. 3 Fundus mask

3.1 Statistical Operation on Cotton Wool Spots

$$M.(x). = \frac{491.7}{30} = 16.39$$

$$M.(y) = \frac{502.7}{30} = 16.76$$

$$\text{Var.}(x) = \frac{\sum(x - \overline{X})}{N} = \frac{475.31}{30} = 15.85$$

$$\text{Var.}(y) = \frac{\sum(y - \overline{Y})}{N} = \frac{485.94}{30} = 16.19$$

$$\text{Std.}(x) : \sqrt{\text{Variance}(x)} = \sqrt{15.85} = 3.99$$

$$\text{Std.}(y) : \sqrt{\text{Variance}(y)} = \sqrt{16.19} = 4.03$$

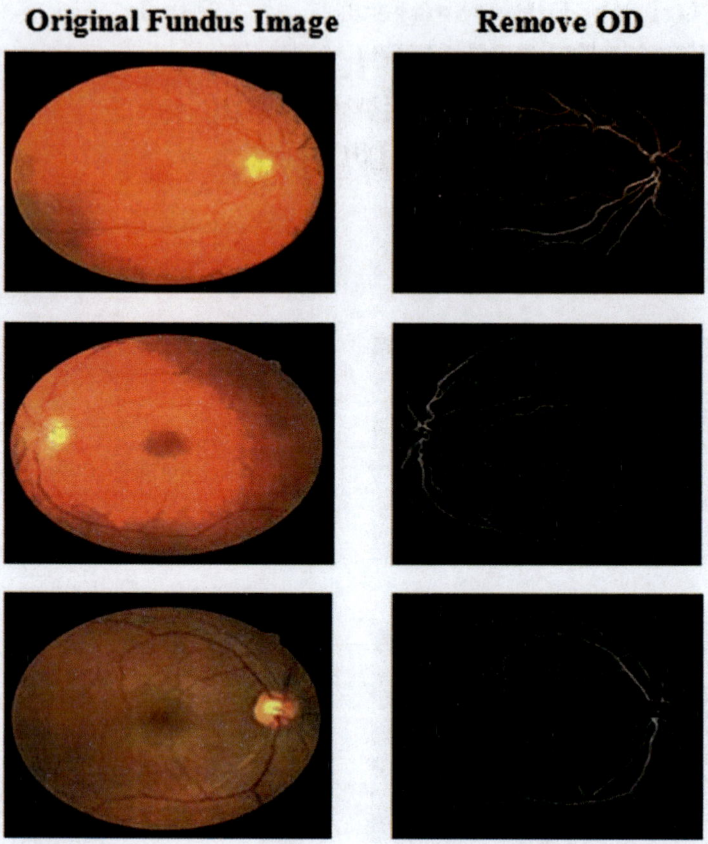

Fig. 4 Optic disc removal from fundus image

Table 1 Fundus image database

Sr. No	Name of fundus database	Total images
1	HRF (Diabetic retinopathy) [21]	15
2	HRF (Glaucoma) [21]	15
3	Diarect DB 1 [22]	89
4	DRIVE [23]	40
5	STARE [23]	402
6	Saswade (Local)	500

Correlation:
Where

$$\sum(x - \overline{X}) = 475.31,$$
$$\sum(y - \overline{Y}) = 485.94$$

Table 2 Statistical techniques on diabetic retinopathy lesions

Sr. No	Manual counting of CWS (x)	CWS by algorithm (y)	Manual counting of haemorrhages (x)	Haemorrhages by algorithm (y)	Manual counting of exudates (x)	Exudates by algorithm (y)
1	612	912	120,070	120,080	35	42
2	984	985	110,070	110,071	47	49
3	932	932	131,600	131,602	102	106
4	889	905	145,410	145,414	31	31
5	204	204	138,320	138,320	68	68
6	795	795	98,868	98,869	18	18
7	891	891	80,622	80,627	41	41
8	138	138	90,982	90,989	96	96
9	137	147	113,110	113,110	89	89
10	688	688	104,120	104,120	48	48
11	474	474	98,650	98,650	21	21
12	100	100	82,427	82,428	48	48
13	136	136	121,290	121,290	302	309
14	143	143	116,040	116,040	37	37
15	149	149	80,753	80,753	66	66
16	31	31	75,480	75,480	65	65
17	88	89	87,198	87,198	39	39
18	320	322	193,540	193,540	154	155
19	734	734	109,350	109,350	8	8
20	817	817	128,770	128,770	37	37

$$\sum (x - \overline{X})^2 = 225919.60,$$
$$\sum (y - \overline{Y})^2 = 236137.69$$

$$r = \frac{475.31 * 485.94}{\sqrt{225919.60 * 236137.69}}$$

$$r = \frac{230972.15}{230971.51} = 1$$

The value of the coefficient, denoted by r, might fall anywhere between $+1$ and -1. If one of the variables has a value of 0, it means that there is no connection between the other two variables. If the value is larger than zero, this shows that there is a positive link between the two variables; this indicates that the value of the other variable will also increase whenever the value of the first variable increases. A number that is less than zero shows an inverse relationship; this means that as the value of one variable increases, the value of the other variable decreases. This is shown by the fact that an inverse relationship is denoted by a number that is less than zero.

3.2 Statistical Operation on Haemorrhages

$$M.(x) = \frac{110105.37}{30} = 3670.17$$

$$M.(y) = \frac{110106.77}{30} = 3670.2$$

$$\text{Var.}(x) = \frac{\sum(x - \overline{X})}{N} = \frac{106435.2}{30} = 3547.84$$

$$\text{Var.}(y) = \frac{\sum(y - \overline{Y})}{N} = \frac{106436.57}{30} = 3547.89$$

$$\text{Std.}(x) : \sqrt{\text{Variance}(x)} = \sqrt{3547.84} = 59.57$$

$$\text{Std.}(y) : \sqrt{\text{Variance}(y)} = \sqrt{3547.89} = 59.57$$

Correlation
Where

$$\sum(x - \overline{X}) = 106435.2,$$

$$\sum(y - \overline{Y}) = 106436.57,$$

$$\sum(x - \overline{X})^2 = 11328451799.1,$$

$$\sum(y - \overline{Y})^2 = 11328749819.6.$$

$$r = \frac{106435.2 * 106436.57}{\sqrt{11328451799.1 * 11328749819.6}}$$

$$r = \frac{11328597615.27}{11328597615.22} = 1$$

3.3 Statistical Operation on Exudates

$$M.(x) = \frac{52.86666667}{30} = 1.77$$
$$M.(y) = \frac{54.2}{30} = 1.81$$

$$\text{Var.}(x) = \frac{\sum(x - \overline{X})}{N} = \frac{51.11}{30} = 1.71$$
$$\text{Var.}(y) = \frac{\sum(y - \overline{Y})}{N} = \frac{52.40}{30} = 1.75$$
$$\text{Std.}(x) : \sqrt{\text{Variance}(x)} = \sqrt{1.71} = 1.31$$
$$\text{Std.}(y) : \sqrt{\text{Variance}(y)} = \sqrt{1.75} = 1.33$$

Correlation:

$$r = \frac{\sum(x - \overline{X})\sum(y - \overline{Y})}{\sqrt{\sum(x - \overline{X})^2 \sum(y - \overline{Y})^2}} \quad (1)$$

where

$$\sum(x - \overline{X}) = 51.11,$$
$$\sum(y - \overline{Y}) = 52.40,$$
$$\sum(x - \overline{X})^2 = 2612.24,$$
$$\sum(y - \overline{Y})^2 = 2745.76$$
$$r = \frac{51.11 * 52.40}{\sqrt{2612.24 * 2745.76}}$$

$r = \frac{2678.17}{2678.17} = 1$

3.4 K-Means Clustering

K-means is a clustering technique. Clustering algorithms are unsupervised approaches for dividing a larger dataset into more manageable groups. The moniker that can be used to a priori label unsupervised data indicates that they do not originate from clearly defined categories. In machine learning, the challenge of unsupervised learning is to look for latent structure in unlabelled data. Since there are no error or reward signals to aid in the identification of workable solutions, learners are taught

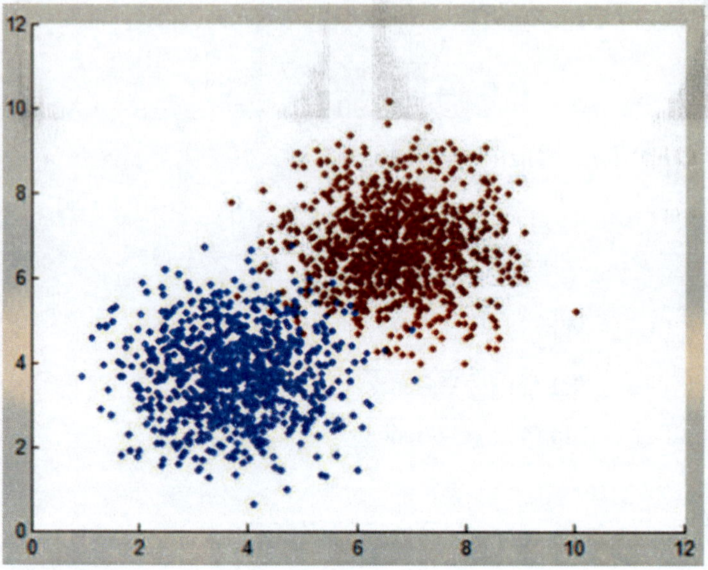

Fig. 5 K-means clustering

using examples without labels. Now, supervised learning and reinforcement learning may be distinguished from unsupervised learning (Fig. 5).

3.5 *ROC Curve*

The area under the receiver operating characteristic curve (AUC-ROC) is a performance assessment that can be used for classification problems using a variety of threshold settings. The ROC is a probability curve, and the AUC is the degree of separability, sometimes known as a measure of separability. It indicates the degree to which the model is able to differentiate between different classes. The higher the area under the curve (AUC), the more accurately the model can predict that 0 classes will be 0 and 1 classes will be 1. By analogy, a higher AUC indicates that the model is more able to differentiate between patients who have the disease and those who do not have the disease (Fig. 6).

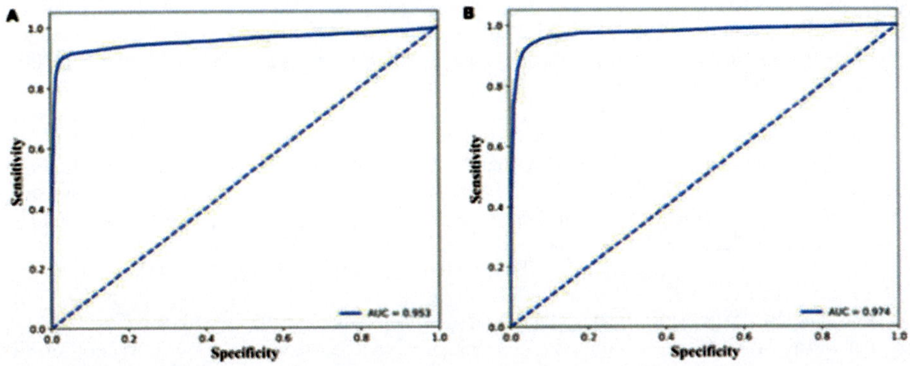

Fig. 6 ROC curve for performance analysis

4 Conclusion

In order to assist in the diagnosis of diabetic retinopathy lesions, we make use of digital image processing techniques as well as symlet wavelet. After the lesions have been removed, we then make use of statistical approaches such as computing the mean, standard deviation, variance, and correlation. Additionally, we have designed a diagnostic tool for diabetic retinopathy that features a graphical user interface (GUI). This tool is used to locate lesions caused by diabetic retinopathy. Which is of great use to the ophthalmologist in reaching a diagnosis of the illnesses when it comes to the matter at hand. When evaluated via the lens of statistical methodology, the proposed algorithm demonstrated a success rate of 94%. Working with a dataset that is both balanced and multimodal could be the focus of work to be done in the future. The second thing is to combine deep neural networks with techniques such as supervised learning and unsupervised learning [6–21].

References

1. American Academy of Ophthalmology: Protecting Sight. Empowering Lives—American Academy of Ophthalmology. (n.d.). https://www.aao.org/
2. Akram UM, Khan SA (2011) Automated detection of dark and bright lesions in retinal images for early detection of diabetic retinopathy. Springer Science+Business Media, LLC
3. Rupa VL, Kulkarni PS (2013) Automatic diagnosis of diabetic retinopathy by hybrid multilayer feed forward neural network. Int J Sci Eng Technol Res (IJSETR) 2(9), Sept 2013
4. Yazid H, Arof H, Mokhtar N (2010) Edge sharpening for diabetic retinopathy detection. In: IEEE conference on cybernetics and intelligent systems
5. Kumari S, Vijay R (2012) Effect of symlet filter order on denoising of stillimages3. Adv Comput Int J (ACIJ) 3(1), Jan 2012. https://doi.org/10.5121/acij.2012.3112
6. Singh K(2022) Diabetes in India. Diabetes. https://www.diabetes.co.uk/global-diabetes/diabetes-in-india.html
7. ROC 1 (n.d.-b). http://www.vassarstats.net/roc1.html

8. Patwari MB, Manza RR, Saswade M, Deshpande N (2012) A critical review of expert systems for detection and diagnosis of diabetic retinopathy. Ciit Int J Fuzzy Syst Feb 2012. FS022012001 ISSN 0974-9721, 0974-9608
9. Rajput YM, Manza RR, Patwari MB, Deshpande N (2013) Retinal blood vessels extraction using 2D median filter. In: Third national conference on advances in computing (NCAC-2013), 5th to 6th March 2013. School of Computer Sciences, North Maharashtra University, India
10. Rajput YM, Manza RR, Patwari MB, Deshpande N (2013) Retinal optic disc detection using speed up robust features. In: National conference on computer and management science [CMS-13], April 25–26, 2013, Radhai Mahavidyalaya, Auarngabad-431003 (MS India)
11. Patwari MB, Manza RR, Rajput YM, Saswade M, Deshpande NK (2013) Review on detection and classification of diabetic retinopathy lesions using image processing techniques. Int J Eng Res Technol (IJERT). 2(10), ISSN: 2278-0181, Impact Factor 1.76, Oct 2013
12. Patwari MB, Manza RR, Rajput YM, Deshpande NK, Saswade M (2013) Extraction of the retinal blood vessels and detection of the bifurcation points. Int J Comput Appl (IJCA), 18 Sept, 2013. ISBN: 973-93-80877-61-7
13. Patwari MB, Manza RR, Rajput YM, Saswade M, Deshpande NK (2013) Calculation of retinal blood vessels tortuosity by using image processing techniques and statistical techniques. In: 2nd international conference on system modeling and advancement in research trends (SMART) Department of Computer Applications, TMIMT, Teerthanker Mahaveer University, Academic Journal online (AJO), Int J Trends Comput Sci 2(11), ISSN: 7462-8452
14. Patwari MB, Manza RR, Rajput YM, Saswade M, Deshpande NK (2013) Detection and counting the microaneurysms using image processing techniques. Int J Appl Inf Syst (IJAIS) 6(5):11–17, Nov 2013. Published by Foundation of Computer Science, New York, USA. ISSN: 2249-0868, Vol 6 Number 5, Oct 2013
15. Patwari MB, Manza RR, Rajput YM, Saswade M, Deshpande NK (2014) Automatic detection of retinal venous beading and tortuosity by using image processing techniques. Int J Comput Appl (IJCA), Feb 2014. ISBN: 973-93-80880-06-7
16. Patwari MB, Manza RR, Rajput YM, Saswade M, Deshpande NK (2014) Personal Identification algorithm based on retinal blood vessels bifurcation. In: 2014 international conference on intelligent computing applications, 2014 IEEE, 978-1-4799-3966-4/14 ©. https://doi.org/10.1109/ICICA.2014.51
17. Telander DG, Small KW, Browning DJ (2010) Genetics and diabetic retinopathy. In: Browning D (eds) Diabetic retinopathy. Springer, New York. https://doi.org/10.1007/978-0-387-85900-2_2,2010
18. Lu L, Jiang Y, Jaganathan R, Hao Y (2018) Current advances in pharmacotherapy and technology for diabetic retinopathy: A systematic review. J Ophthalmol1–13. https://doi.org/10.1155/2018/1694187
19. Martins B, Amorim M, Reis F, Ambrósio AF, Fernandes R (2020) Extracellular vesicles and microrna: putative role in diagnosis andtreatment of diabetic retinopathy. Antioxidants 9:1–26. https://doi.org/10.3390/antiox9080705
20. Calderon GD, Juarez OH, Hernandez GE, Punzo SM, De La Cruz ZD (2017) Oxidative stress and diabetic retinopathy: development and treatment Eye (Basingstoke) 31:1122–30
21. Lehrstuhl für Mustererkennung F-A-UE-N (2022) Contact, high-resolution fundus (HRF) image database. Available at: https://www5.cs.fau.de/research/data/fundus-images/ (Accessed: 7 Mar 2022)
22. Lehrstuhl für Mustererkennung F-A-UE-N (2022) Contact, high-resolution fundus (HRF) image database. Available at: https://www5.cs.fau.de/research/data/fundus-images/ (Accessed: 10 Apr 2022)
23. Structured analysis of the Retin The STARE Project. Available at: https://cecas.clemson.edu/~ahoover/stare/ (Accessed: 7 Sep 2022)
24. Drive—Grand challenge grand. Available at: https://drive.grand-challenge.org/ (Accessed: 7 Sept 2023)

Enhancement of Data Security for Cloud Computing with Cryptography Techniques

Govinda Giri, Kunal Chakate, Dirun Reddy, Prachi Mohite, Mebanphira Cajee, Snehal Bhosale, and Sonali Kothari

1 Introduction

Cloud computing has become increasingly popular as a way for individuals and organizations to store and process data. However, the use of cloud computing also raises concerns about data security. Here are some measures that can help ensure data security for cloud computing:

i. **Strong Passwords**: It is important to use strong passwords that are difficult to guess, contain a mix of upper and lowercase letters, numbers and special characters.
ii. **Multi-factor Authentication**: Multi-factor authentication (MFA) adds an extra layer of security by requiring users to provide more than one form of authentication to access their data. This can include a combination of passwords, security tokens, biometric data, or other forms of authentication.
iii. **Encryption**: Encryption is the process of converting plain text data into a coded language that can only be read by authorized users with the proper decryption key. It is important to ensure that data is encrypted both during transmission and when it is stored in the cloud.
iv. **Access Control**: Access control is the process of determining who is authorized to access data and what level of access they have. This can be achieved through the use of access control policies and permissions.

G. Giri · K. Chakate · D. Reddy · P. Mohite · M. Cajee · S. Kothari
Department of Computer Science and Engineering, Symbiosis Institute of Technology (SIT), Symbiosis International (Deemed University) (SIU), Lavale, Pune, Maharashtra, India

S. Bhosale (✉)
Department of E&TC, Symbiosis Institute of Technology (SIT), Symbiosis International (Deemed University) (SIU), Lavale, Pune, Maharashtra, India
e-mail: snehal.bhosale@sitpune.edu.in

J. K. Mandal et al. (eds.), *Proceedings of International Conference on Network Security and Blockchain Technology*, Lecture Notes in Networks and Systems 738,
https://doi.org/10.1007/978-981-99-4433-0_26

v. **Data Backup and Recovery**: Data backup and recovery are critical to ensuring that data is not lost in the event of a security breach or other incident. Cloud providers should offer regular backups of data and provide tools for recovering data in case of a disaster.
vi. **Regular Audits and Monitoring**: Regular audits and monitoring can help identify potential security vulnerabilities and ensure that the cloud provider is adhering to security policies and procedures.
vii. **Vendor Selection**: It is important to carefully select a cloud provider that has a strong track record in security and offers the necessary security features.

By implementing these measures, individuals and organizations can help ensure that their data is secure in the cloud computing environment.

An evolving virtual distributed system called cloud computing incorporates the concepts of connection, virtualization, power processing, and storage. A huge pool of resources, storage media, and sharing media is made available through communication over a wide network, or Internet cloud. It makes it easier to provide on-demand services. This will make it easier for end users to adhere to the concepts of isolation, flexibility, security, and dispersion [1].

The biggest challenge in the cloud is security, making it the biggest barrier to the growth of IT-based businesses that provide consumers with on-demand services. The application phase, network phase, authentication or authorization phase, information storage phase, and virtualization phase are all phases where these security vulnerabilities may be seen. These difficulties or dangers continue to stand in the way of cloud computing's full success. Many businesses and individuals store their data in cloud databases, making user privacy and security a top priority. Important data shouldn't be lost or altered as it moves from one location to another over the network. Therefore, it is crucial to guarantee the Confidentiality (C), Integrity (I), and Availability (A) of user information. The unauthenticated user attempting to access the data belonging to the authorized user is another factor [2, 3].

1.1 Related Work

To combat these dangers, we can use encryption methods on cloud servers. A single encryption technique, however, is insufficient to provide data security and manage access control procedures in a cloud computing environment when a user's access is terminated. These methods are applied to encryption for data security. Complete data encryption can become highly costly in terms of memory and processing time. Therefore, it would be preferable if we first separated our sensitive data and then used encryption technologies to remedy this issue [4].

Services are distributed throughout all servers, users, and people in the cloud environment. Since security is the most important concern in data processing and transmission since the original data form can be accessed, mishandled, and destroyed, cloud providers struggle to ensure file protection. Security in the cloud is a serious problem in the context of cloud computing.

To safeguard the cloud environment, several research initiatives are being offered. To get over the security issue and get Confidentiality, Integrity and Availability (CIA) property, cryptography is utilized. The best way to ensure the security of data transmission and storage is through cryptography. Classical symmetric and asymmetric have various limitations [5].

1.2 Importance, Relevance, and Timeliness of Enhancement of Data Security for Cloud Computing

Firstly, cloud computing has become a fundamental part of modern-day business operations. Companies, organizations, and individuals now rely on cloud services to store, process, and access their data. The sheer amount of data being generated and transmitted over cloud networks means that securing this data is critical. Cryptography techniques can ensure that sensitive data remains confidential and protected against unauthorized access, data breaches, and cyberattacks.

Secondly, the relevance of cryptography techniques in enhancing data security for cloud computing cannot be overstated. As more businesses move their data to the cloud, the need for robust security measures becomes more pressing. Cryptography techniques such as encryption, decryption, and hashing provide a level of data protection that is necessary in today's digital age.

Finally, the timeliness of enhancing data security for cloud computing with cryptography techniques is crucial. Cybercriminals are constantly developing new methods to breach security measures and steal data. In recent years, there have been numerous high-profile data breaches that have cost companies millions of dollars in damages and lost business. The use of cryptography techniques in cloud computing can provide an extra layer of security to protect against these threats.

In summary, the importance, relevance, and timeliness of enhancing data security for cloud computing with cryptography techniques cannot be understated. The increasing reliance on cloud services for storing and accessing data means that security measures must be put in place to protect against cyberthreats. The use of cryptography techniques can provide the necessary level of security needed to protect sensitive data in the cloud.

2 Literature Review

There was a discussion of consumer-facing cloud computing challenges. The two key security concerns are information security and secrecy preservation. To safeguard the data, other methods, such as Airavat by Roy et al. (2020), had been offered.

The strategy employed proved impractical and depended on several model execution techniques. The AES approach, which ensures the security of the users' information present on the cloud servers, was utilized to protect information systems in the cloud in [6].

An overview of numerous currently utilized approaches is presented in the literature study. In [7], Thabit et al. (2020) discuss the examination of security difficulties in cloud computing, including problems, threats, and strategies for mitigating assaults. In [8], a study by Verma et al. (2020) examined how cryptography can ensure cloud computing data security [8].

Data backup and recovery are provided by Deerthana et al. [9], which also use cryptography and steganography to improve security. Cloud computing security was developed by Mohammed et al. (2021) in the direction of homomorphic encryption [10]. Numerous keys are generated and the size of the key is selected in the paper by Vishwamitra et al. (2020) [11] using cryptographic techniques (Symmetric Cryptography and Asymmetric Cryptography). Then, it will be compared using various criteria, including a database, the size of the key, functionality, execution time, and the memory required for data cryptographic operations.

Agrahari et al. [12] have used the algorithms like AES, RSA, and D-H, among others. These algorithms may be used to circumvent security issues in order to assure data security via the cloud. These algorithms are compatible with all Internet protocols that use IPv4 and IPv6, and they can defend against any kind of assault. The study by Budhani et al. [4] has explored the significant issue of data security in cloud computing, which might hinder deployment.

The security concerns that might occur with each cloud computing service have been addressed (SaaS, PaaS, and IaaS). Discussion of security measures in cloud-enabled systems, such as firewalls, virtual private networks, and encryption [13].

3 Methodology

In our methodology, we have used algorithms like AES, 3DES, and Steganography. Advanced Encryption Standard (AES) uses a substitution permutation technique, AES is made up of a 128-bit block and keys of various sizes, including 128, 192, and 256 bits. It is made up of a cipher that defines the number of transformation rounds required to convert plaintext to ciphertext.

It is necessary that the sender and receiver possess the same secret key. Three categories of information are distinguished: confidential, secret, and top secret. The system's Confidential and Secret levels can be protected by keys of any length. A key with 192 bits or 256 bits must be used for top-secret material. 256-bit keys have fourteen rounds, while 128-bit keys have ten rounds and 192-bit keys have twelve rounds. A round is made up of various processing steps that transform plaintext into ciphertext. These include substitutions, transpositions, and mixes. Four operations make up AES encryption, and InvSubBytes, which is made up of SubBytes, Shift

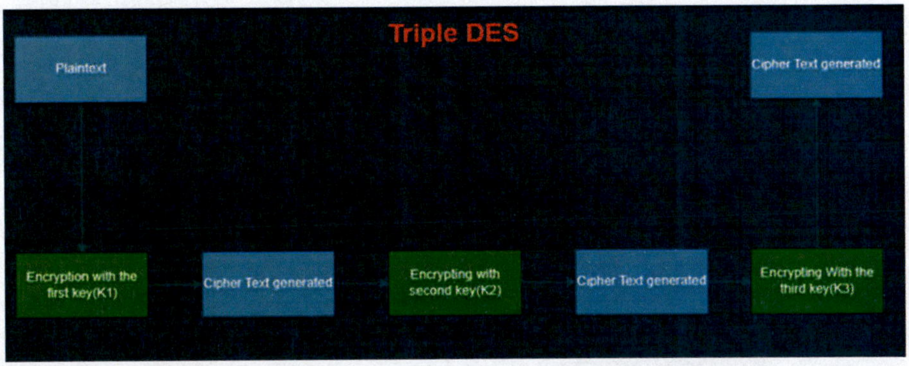

Fig. 1 Triple DES

Rows, Mix Columns, and Xor Round keys. The decryption of AES consists of two opposite functions, InvShiftRows, and InvMixColumns.

3.1 Encryption Process

A more sophisticated variant of DES and 3DES encrypts each data block three times with DES as shown in Fig. 1. It uses a 168-bit key that is permuted into 16 subkeys and 64-bit data/plain text with 48 rounds. The 16 subkeys are each 48 bits long. There are eight distinct numbers that contain S-boxes.

The outcome of the previous step is then encrypted once again with K3. Ciphertext is the result of the last K3 step. In a reverse way, the encrypted text is decrypted. In the final step, K1 is utilized for decryption while K3 and K2 are used for encryption. Systems based on 3DES are more secure than those based on DES, but because the process must be done three times, they are often slower. In a reverse way, the encrypted text is decrypted. K2 and K3 are used for encryption and decryption, respectively.

The last stage of encryption uses K1 for decryption. Systems based on 3DES are more secure than systems based on DES, but because the process must be repeated three times, they are often slower. The process is shown in Fig. 2.

3.2 Decryption Process

During the decryption stage, the whole encryption process is reversed. The DES algorithm encrypts plaintext with key K1 at the beginning of the process. Key K2 is then used by the DES algorithm to decode the output generated at the previous step. The process is shown in Fig. 3.

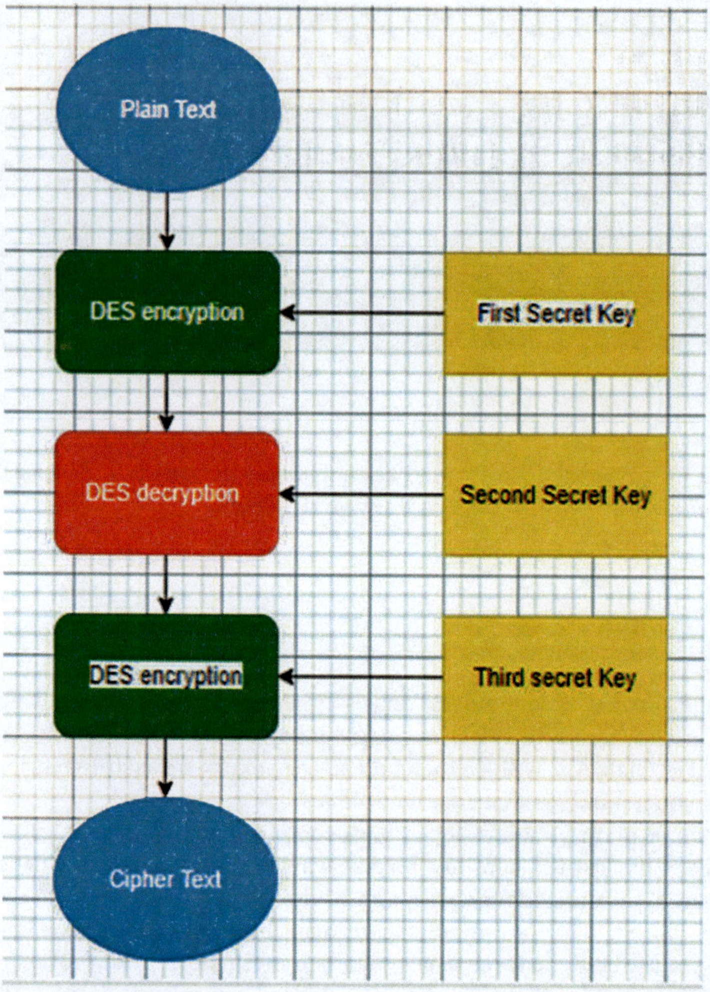

Fig. 2 Encryption process

4 Steganography

Steganography is an approach to concealing secret data within a non-secret file or message to avoid detection. The sensitive data is subsequently removed from the usual file or communication at its destination, avoiding discovery. It can be combined with encryption to obscure or safeguard data. The steganography method may be applied to text, photos, video files, or audio files.

i. **Text Steganography**: The concealed data is encoded into each letter of each word.

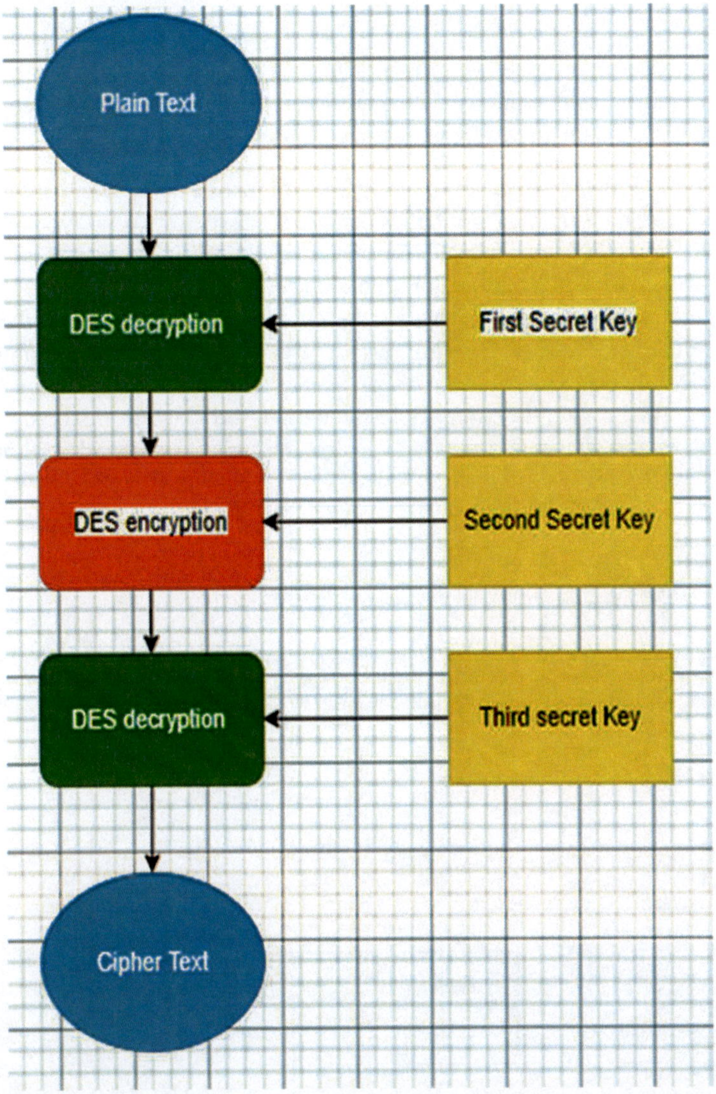

Fig. 3 Decryption process

ii. **Image Steganography**: It comprises hiding data behind a picture of another item as a cover. Image steganography relies on pixel intensities to disguise data.
iii. **Audio and Video Steganography**: As in here the audio is endowed in an audio signal and changes the sequence of the audio and the data is invisible in the case of video.

5 Results

To get the required results, the workflow is shown in Fig. 4. Result analysis contains security analysis. It analyzes several security characteristics, including:

1. Data Confidentiality: In order to analyze it, it must be compared with other data encrypted by DES or AES, both of which use the same key to encrypt and decrypt data. We propose a system that employs three levels of security to prevent any attempts to access personal data in the cloud. In this case, the owner of the data ensures the confidentiality of the data by knowing the key for three algorithms.
2. Authentication: Through registration, the user sets a password to complete the authentication process.
3. Integrity: As a result, the integrity of the data in the cloud is ensured.
4. Encryption and decryption were made stronger using the hybrid method.

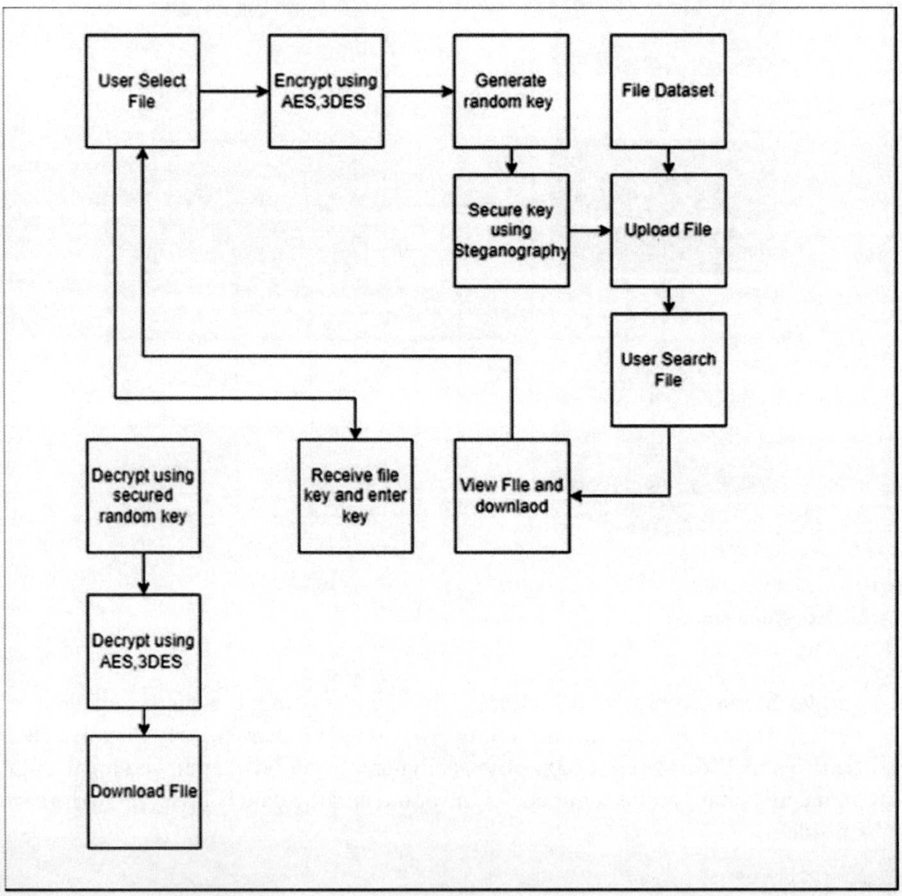

Fig. 4 Project block diagram

5. In comparison to other techniques, AES and 3DES are used as integral part of the system, resulting in higher performance.
6. Cloud computing is secure when AES, 3DES, and Steganography are used together.

6 Conclusion and Future Scope

Cloud computing is a rapidly increasing technology that is being used by the current world, but it still has certain open areas that are affecting its robust features. According to the results of our survey, people have severe worries about the open nature of privacy and security. We have analyzed the cloud's characteristics and identified specific categories of dangers that must be addressed. Security is determined by how a cloud service provider allows his clients to access and register with his cloud network.

We wish to incorporate stateful apps in the hybrid cloud computing model's application scope going forward. Security and privacy issues for cloud storage will endure if adaptable techniques are not used in daily life to enhance client and user experiences. Another constraint is to categorize privacy and security issues according to the frequency of attacks. In cloud privacy models, which lack flexibility and control over security and privacy precautions that protect users' sensitive data, individuals are not prioritized. Introducing an updated key distribution technique that would provide clients access to the encryption key without involving the cloud provider.

References

1. Shirole BS, Vishwamitra LK (2020) Review paper on data security in cloud computing environment. In: 2020 9th International conference system modeling and advancement in research trends (SMART), pp 79–84. https://doi.org/10.1109/SMART50582.2020.9337115
2. Mudawi NA, Beloff N, White M (2020) Issues and challenges: cloud computing e-government in developing countries. Int J Adv Comput Sci Appl 11(4). https://doi.org/10.14569/IJACSA.2020.0110402
3. Tan CB, Hijazi MHA, Lim Y, Gani A (2018) A survey on proof of retrievability for cloud data integrity and availability: cloud storage state of-the-art, issues, solutions and future trends. J Netw Comput Appl. https://doi.org/10.1016/j.jnca.2018.03.017
4. Joshi M, Budhani S, Tewari N, Prakash S (2021) Analytical review of data security in cloud computing. In: 2nd International conference on intelligent engineering and management (ICIEM), pp 362–366. https://doi.org/10.1109/ICIEM51511.2021.9445355
5. Van Der Ham J (2021) Toward a better understanding of cybersecurity. Digit Threat Res Pract 2(3), Article 18. https://doi.org/10.1145/3442445
6. Roy I, Setty STV, Vitaly AK, Witchel SE (2010) Airavat: security and privacy for MapReduce. In: Conference: proceedings of the 7th USENIX symposium on networked systems design and implementation, NSDI (2010), 28–30 Apr 2010, San Jose, CA, USA
7. Thabit F, Al-ahdal A, Alhomdy S, Jagtap S (2020) Exploration of security challenges in cloud computing: issues, threats, and attacks with their alleviating techniques. J Inf Comput Sci 10:35–59

8. Verma V, Tiwari AK (2020) Analytic study of data security in cloud computing using cryptography. Int J Technol 10(1):88–92. https://doi.org/10.5958/2231-3915.2020.00017.6
9. Deerthana K, Devi Saranya B, Jayamala R (2016) Enhancing security using cryptography and steganography and providing data backup and recovery in cloud. Int J Eng Res Technol (IJERT) NCICCT 4(19)
10. Mohammed SJ, Basheer D (2021) From cloud computing security towards homomorphic encryption: a comprehensive review. TELKOMNIKA Telecommun Comput Electron Control 19(4):1152–1161. https://doi.org/10.12928/TELKOMNIKA.v19i4.16875
11. Albugmi A, Alassafi MO, Walters RJ, Wills G (2016) Data security in cloud computing. In: Conference: 2016 fifth international conference on future generation communication technologies (FGCT), vol 1. IEEE, At, Luton, UK. https://doi.org/10.1109/FGCT.2016.7605062
12. Agrahari V (2020) Data security in cloud computing using cryptography algorithms. Int J Sci Dev Res (IJSDR) 5(9). www.ijsdr.org
13. Meinardi M, Clayton T (2020) How to manage and optimize costs of public cloud IaaS and PaaS, p 70

A Novel Approach of Network Security Using Genetic Algorithm

Arkojeet Bera, Debarpito Sinha, Soumyadip Maity, and Soumya Paul

1 Introduction

In the era of evolving Internet and cascading computer networks, network security has undoubtedly become one of the most burning issues to address. Cryptography plays a crucial role in any interconnected network environment to provide "security and privacy." It is crucial for preserving data confidentiality, authenticity, and integrity. The fundamental concepts of cryptography are encryption and decryption [1]. There are effectively two techniques in cryptography: symmetric key cryptography or public key cryptography, as it uses the same key for encryption and decryption, whereas asymmetric key cryptography or private key cryptography uses different keys for the same purpose. The cryptographic algorithm takes the plain text as input and converts it to cipher text, i.e., scrambled and unreadable. Genetic algorithms are a class of optimization algorithms inspired by the process of natural selection and genetics. As it does not utilize natural numbers directly, it is safe to say that the genetic algorithm is secure [2]. In the general sense, the genetic algorithm involves two essential functions: crossover and mutation. Crossover is a process that involves taking two parent solutions from the current population and combining them to produce one or more offspring solutions. Mutation, on the other hand, involves making small random changes to an individual solution to introduce new variation into the population and prevent the algorithm from getting stuck in local optima.

The approach discussed in this paper extensively uses genetic algorithms and mathematical and bitwise operations as core components for the cryptosystem.

A. Bera (✉)
National Institute of Technology Karnataka, Surathkal, Mangalore 575025, Karnataka, India
e-mail: arkojeetphy007@gmail.com

D. Sinha · S. Maity
Ramakrishna Mission Vidyamandira, Belur Math, Howrah 711202, West Bengal, India

S. Paul
St. Mary's Technical Campus Kolkata, Barasat 700126, West Bengal, India

J. K. Mandal et al. (eds.), *Proceedings of International Conference on Network Security and Blockchain Technology*, Lecture Notes in Networks and Systems 738,
https://doi.org/10.1007/978-981-99-4433-0_27

2 Related Work

Mittal [3] proposed a cryptosystem using matrix manipulations, substitution algorithms, and genetic operators.

Sindhuja [4] discussed and approached using a two-point crossover mechanism using two random points selected from the two parents, and the bits between them are swapped to produce the child chromosome.

Dutta [5] proposes a cryptosystem using the concept of genetic algorithms with pseudorandom function.

3 Proposed Heuristic

3.1 Heuristic for Encryption and Decryption

Step 1: Take the text string, key string, and resizer parameter (k) as input.
Step 2: Call encryption algorithm.\\ 3.1.1
Step 3: Cipher text generated.
Step 4: Call decryption algorithm.\\ 3.1.2
Step 5: Plain text retrieved.
Step 6: Stop.

3.1.1 Encryption Heuristic

Step 1: Convert each element (i.e., character) of the text list (say L) into their corresponding ASCII value and calculate the length of the text list (n).
Step 2: Use the formula $n + x = 0 \pmod{4k}$, $k \in N$ and find the value of x.
Step 3: Increase the size of the list by x and write the value $(127 - x)$ at the last index of the resized list and fill the remaining empty index positions with random numbers from 10 to 127.
Step 4: Convert each element (i.e., character) of the key list into their corresponding ASCII value and apply XOR operation index-wise with the text list (L) and store the values back to L.
Step 5: Divide the list into 2 equal halves by the strategy of odd index and even index, i.e., even indexed list (say E) and odd indexed list (say O).
Step 6: Calculate the new length of the acquired lists (E and O), i.e., $n = n/2$, and apply 2-point crossover at $(n/4)$th (inclusive) and $(3n/4)$th (exclusive) indexes of E and O.
Step 7: Divide the newly obtained lists (E and O) after 2-point crossover into yet another 2 equal parts, $evenH_1$, $evenH_2$ from E and $oddH_1$, $oddH_2$ from O.
Step 8: Cross-merge $evenH_1$ with $oddH_2$ (say Arr_1) and $evenH_2$ with $oddH_1$ (say Arr_2).

Step 9: Convert Arr_1 and Arr_2 into their nearest square matrix (say Mat_1 and Mat_2) and fill the first blank space with the ASCII value of $\wedge$ (i.e., 94).
Step 10: Fill the remaining blank spaces of both the matrices (Mat_1 and Mat_2) with random values between 0 and 94.
Step 11: Take transpose of both the matrices, $Mat_1 = (Mat_1)^{\mathrm{T}}$ and $Mat_2 = (Mat_2)^{\mathrm{T}}$.
Step 12: Convert both the matrices into their respective lists (say L_1 and L_2).
Step 13: Apply fixed point mutation at $(n/2)$th and $(n-2)$th indexes of list L_1 and L_2 by doing XOR operation with the respective indexes of the key list.
Step 14: Merge L_1 and L_2 and convert the obtained list into string format to get the final cipher text.
Step 15: Stop.

3.1.2 Decryption Heuristic

Step 1: Take the cipher text and convert it into list format.
Step 2: Convert each element (i.e., character) of the list into their corresponding ASCII value.
Step 3: Split the list into 2 equal halves, let the first half be L_1 and the second half be L_2.
Step 4: Apply fixed point mutation at $(n/2)$th and $(n-2)$th indexes of list L_1 and L_2 by doing XOR operation with the respective indexes of the key list.
Step 5: Convert L_1 and L_2 into their nearest square matrices (say Mat_1 and Mat_2).
Step 6: Take transpose of both the matrices, $Mat_1 = (Mat_1)^{\mathrm{T}}$ and $Mat_2 = (Mat_2)^{\mathrm{T}}$.
Step 7: Convert both the matrices into their respective lists (say L_1 and L_2).
Step 8: Start traversing from the end of both the lists (L_1 and L_2), if 94 is encountered then slice both the lists up to the index position of 94. Else carry on with L_1 and L_2.
Step 9: Split L_1 and L_2 into their respective odd and even index lists (i.e., $evenL_1$, $oddL_1$, $evenL_2$, and $oddL_2$).
Step 10: Merge the above-obtained lists by maintaining proper symmetry with the encryption technique, i.e., $evenL_1$ with $evenL_2$ (say Arr_1) and $oddL_1$ with $oddL_2$ (say Arr_2).
Step 11: Apply 2-point crossover at $(n/4)$th (inclusive) and $(3n/4)$th (exclusive) indexes of Arr_1 and Arr_2.
Step 12: Arr_1 is the even index list and Arr_2 is the odd index list, hence merge them accordingly, to get the list L.
Step 13: Convert each element (i.e., character) of the key list into their corresponding ASCII value and apply XOR operation index-wise with the text list (L) and store the values back to L.
Step 14: Slice list L by length $(127 - L[n-1])$ from the end.
Step 15: Convert the elements (integer) of list L into their corresponding characters as per the ASCII numbers. The plain text in list format will be obtained.
Step 16: Convert the list into string format. It will be the original plain text.
Step 17: Stop.

4 Example Illustration and Result Analysis

4.1 Encryption

Let's consider an example of plain text—"Hello There."

Now the first step requires to convert the whole text string into a list structure and calculate the length of the list (say n) as shown below,

[0]	[1]	[2]	[3]	[4]	[5]	[6]	[7]	[8]	[9]	[10]	
H	e	l	l	o		T	h	e	r	e	**Len(n) = 11**

Let this list be L. Now convert each element (character) of the list L into its corresponding ASCII value and store it back into their corresponding positions in L.

	[0]	[1]	[2]	[3]	[4]	[5]	[6]	[7]	[8]	[9]	[10]
L:	72	101	108	108	111	32	84	104	101	114	101

Now it is required to resize the list L into its nearest multiple of $4K$, $K \in N$, where $4K > 11$. Let the value of K for this example be 2. Hence, we require to resize the list L into a nearest multiple of 8 greater than 11. In order to achieve this a resizer variable (say x) is required, which will determine up to how much length the list L needs to be increased. This can be achieved by the following formula:

$$\mathbf{n + x = 0\ (mod\ 4K),\quad K \in N}$$

$$\mathbf{11 + x = 0\ (mod\ 8)}$$

Thus, the value of x can be calculated and, in this case, it will be 5. Now it is also required to fill up the blank spaces list caused due to resizing. At the last index of the list L the value $(127 - x)$, i.e., in this case will be $(127 - 5) = 125$ will be stored. And remaining blank indexes will be filled by random numbers from 10 to 127. Hence, the new list and its length (n) will be,

[0]	[1]	[2]	[3]	[4]	[5]	[6]	[7]	[8]	[9]	[10]	[11]	[12]	[13]	[14]	[15]	
72	101	108	108	111	32	84	104	101	114	101	102	103	104	105	122	**n=16**

Now consider a string input for the key. As it a symmetric key cryptography scheme the following key will used for both encryption and decryption.

The key also needs to be stored in list structure after input. Let the key string for this particular example be,

Key = list("qwertyuiopasdfghjklzxcvbnm")

Convert each element (character) of the key list into their respective ASCII value. Now, perform XOR operation between the corresponding elements of key list and list L index by index and store the value back in list L.

	[0]	[1]	[2]	[3]	[4]	[5]	[6]	[7]	[8]	[9]	[10]	[11]	[12]	[13]	[14]	[15]
L:	72	101	108	108	111	32	84	104	101	114	101	102	103	104	105	122
	⊕	⊕	⊕	⊕	⊕	⊕	⊕	⊕	⊕	⊕	⊕	⊕	⊕	⊕	⊕	⊕
Key:	113	119	101	114	116	121	117	105	111	112	97	115	100	102	103	104
	[0]	[1]	[2]	[3]	[4]	[5]	[6]	[7]	[8]	[9]	[10]	[11]	[12]	[13]	[14]	[15]

↓

	[0]	[1]	[2]	[3]	[4]	[5]	[6]	[7]	[8]	[9]	[10]	[11]	[12]	[13]	[14]	[15]
L:	57	18	9	30	27	89	33	1	10	2	4	21	3	14	14	18

Now divide the list L into Even indexed list (list containing the elements of even index of list L) and Odd indexed list (list containing the elements of odd index of list L) as follows:

	[0]	[1]	[2]	[3]	[4]	[5]	[6]	[7]	
Even:	57	9	27	33	10	4	3	14	**n=8**
Odd:	18	30	89	1	2	21	14	18	**n=8**
	[0]	[1]	[2]	[3]	[4]	[5]	[6]	[7]	

Apply two-point crossover at $(n/4)$th (inclusive) and $(3n/4)$th (exclusive) indexes, i.e., in this case 2nd (inclusive) and 6th (exclusive) indexes of list Even and list Odd as shown,

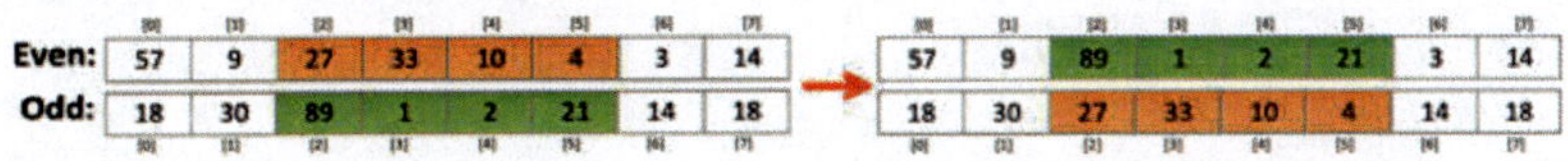

Now, divide the list Even and list Odd in yet another two equal parts respectively (by dividing them from the middle). Let the acquired lists be $evenH_1$, $evenH_2$ for list Even and and $oddH_1$, $oddH_2$ for list Odd as shown:

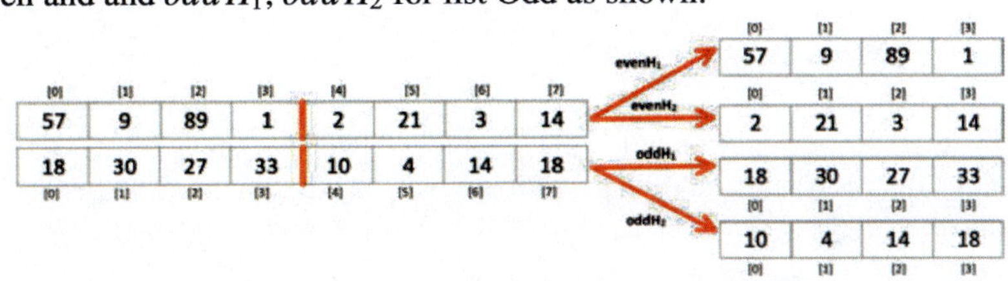

Now, cross-merge the elements of elements of list $evenH_1$ with $oddH_2$ (let the resultant list be Arr_1) and list $evenH_2$ with $oddH_1$ (let the resultant list be Arr_2) as shown below:

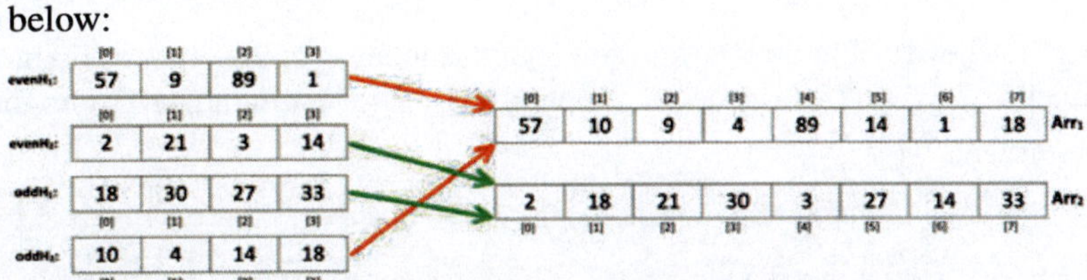

Now convert Arr_1 and Arr_2 into their nearest square matrix greater than or equal to their length, i.e., in this case the nearest square number greater than or equal to 8 is 9. Hence, the dimensions of the square matrix(s) will be (3×3). Start filling elements of the lists linearly column-wise starting from row 1 in their respective matrices (say Mat_1 and Mat_2). If the elements of the lists (Arr_1 and Arr_2) exhausts and the matrix is not yet filled, then fill the first blank space with the ASCII value of $\wedge$, i.e., 94 and the rest of the blank spaces with random numbers between 0 and 93. Else (i.e., if the elements of the lists fill their respective matrices) leave the lists as it is and proceed.

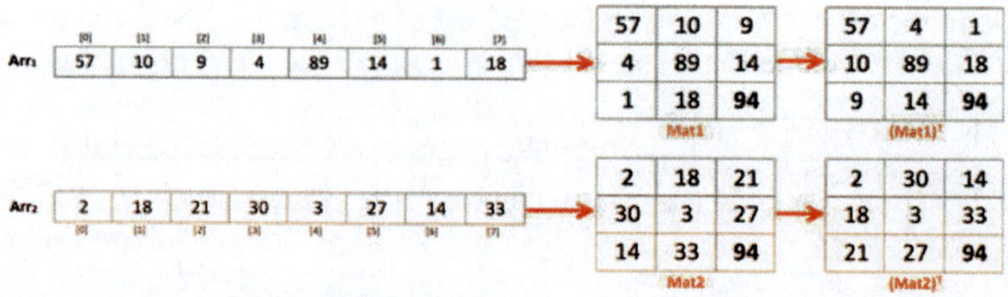

Take transpose of both the matrices, $Mat_1 = (Mat_1)^T$ and $Mat_2 = (Mat_2)^T$, i.e., $A_{ij} = A_{ji}$. Convert Mat_1 and Mat_2 into their corresponding list representation (say L_1 and L_2),

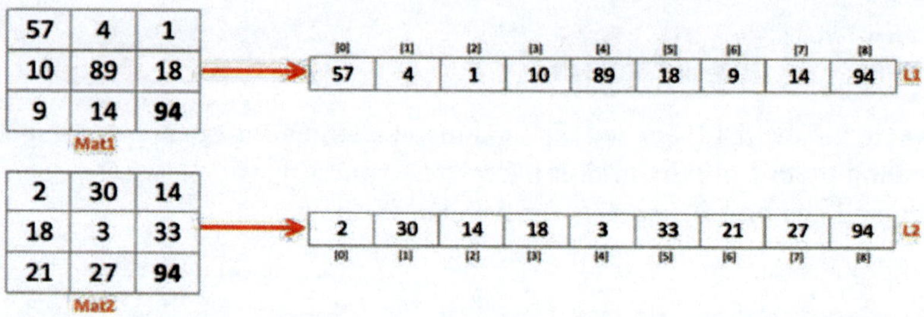

Apply fixed point mutation at $(n/2)$th and $(n-2)$th indexes of list L_1 and L_2 by doing XOR operation with the respective indexes of the key list. The resultant will be as follows:

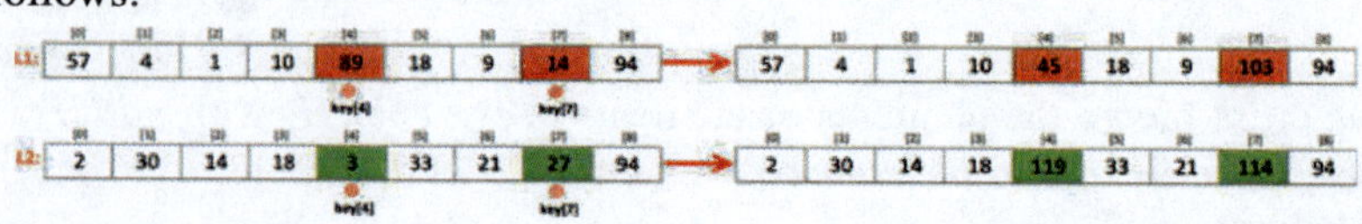

Merge the newly obtained lists L_1 and L_2 lists mutation merge the lists maintaining the sequence L_1 first L_2 last, i.e., L_1.append(L_2). The resultant list will be the final Cipher list (say L).

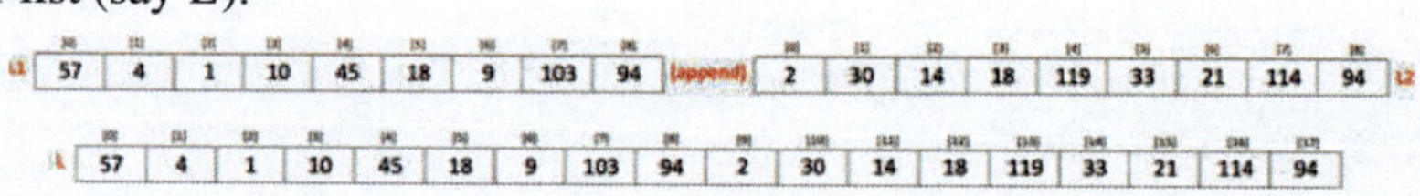

After changing each element (integer) of L in to their corresponding character value according to ASCII values, the following can be obtained:

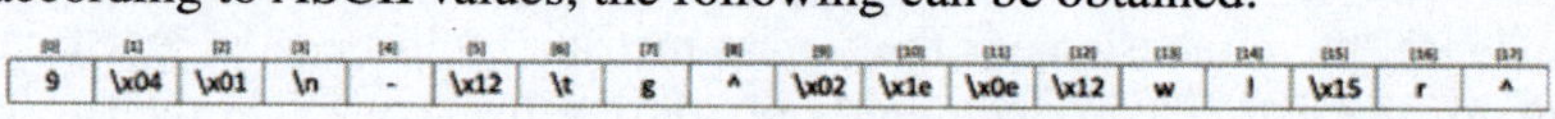

By converting this list into string format, the following cipher text is obtained:

9♦☺

-↕ g^●▲♫↕w!§r^

4.2 Decryption

The following cipher text was obtained during encryption:

9♦☺

-↕ g^● ▲♫↕w!§r^

After converting this cipher text into list structure, the following list (say L) can be obtained:

Now convert each element (character) of the list L into its corresponding ASCII value and store it back into their corresponding positions in L.

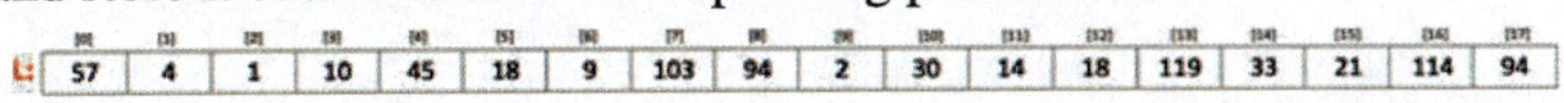

Split the list L into two equal halves from the middle. Let Arr_1 for 1st half and Arr_2 for 2nd half.

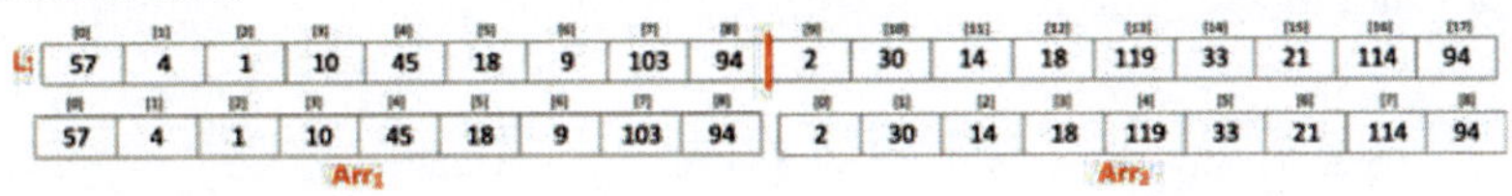

Apply fixed point mutation at $(n/2)$th and $(n-2)$th indexes of list Arr_1 and Arr_2 by doing XOR operation with the respective indexes of the key list. The resultant will be as follows:

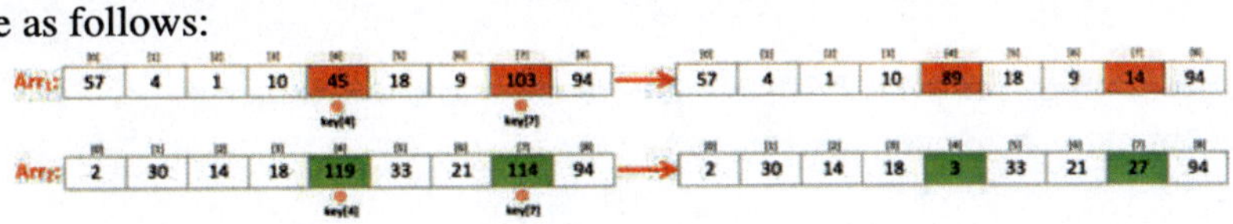

Now convert Arr_1 and Arr_2 into square matrices (will always be square matrices) by start filling elements of the lists linearly column-wise starting from row 1 in their respective matrices (say Mat_1 and Mat_2) as shown:

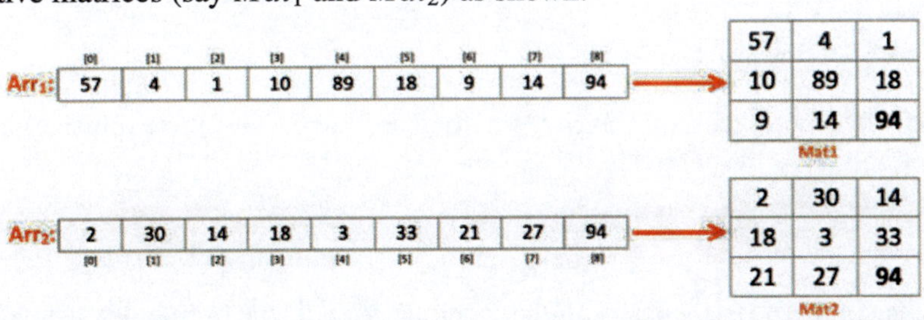

Take transpose of both the matrices, $Mat_1 = (Mat_1)^T$ and $Mat_2 = (Mat_2)^T$, i.e., $A_{ij} = A_{ji}$. Convert Mat_1 and Mat_2 into their corresponding list representation (say L_1 and L_2):

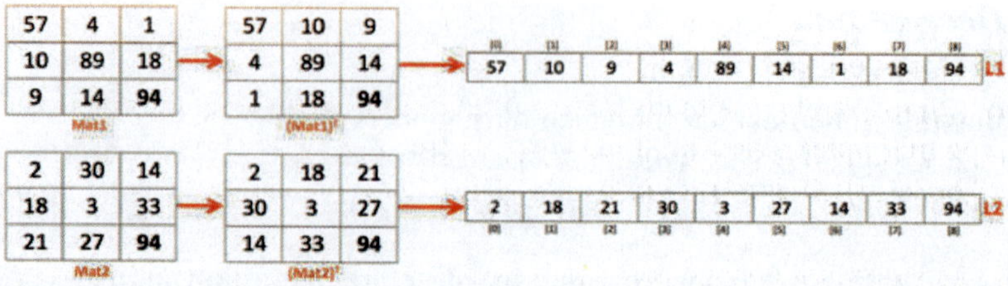

Now start traversing from the end of both the lists (L_1 and L_2), if 94 is encountered then slice both the lists up to the index position of 94. Else carry on with L_1 and L_2. In this case 94 is at the last index, hence the last index from both the lists will be sliced off.

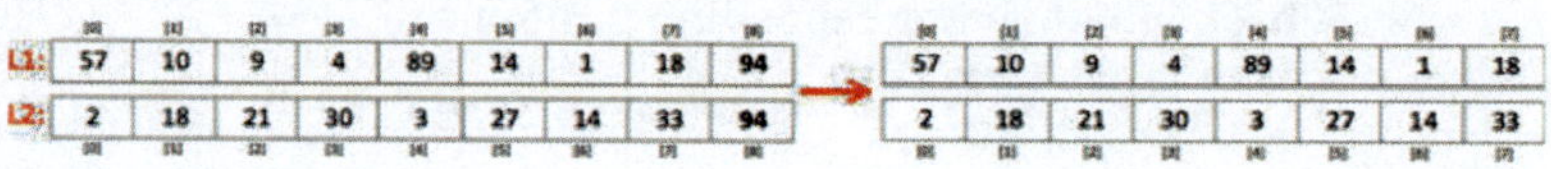

Split L_1 and L_2 into their respective odd and even index lists (i.e., $evenL_1$, $oddL_1$ for list L_1 and $evenL_2$, $oddL_2$ for list L_2) as follows:

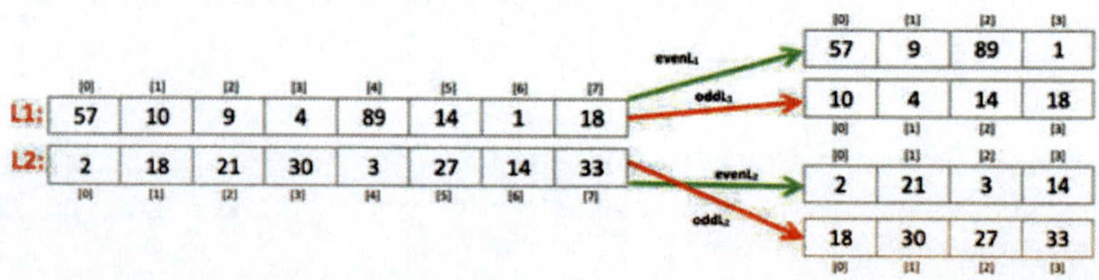

Merge the above-obtained lists by maintaining proper symmetry with the encryption technique, i.e., $evenL_1$ with $evenL_2$ (say Arr_1) and $oddL_1$ with $oddL_2$ (say Arr_2),

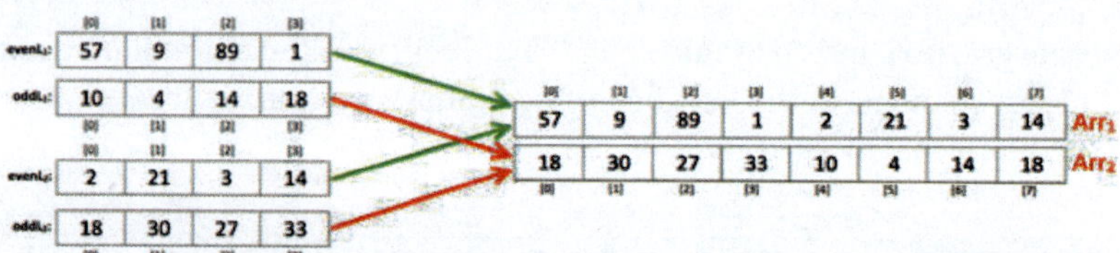

Apply two-point crossover at $(n/4)$th (inclusive) and $(3n/4)$th (exclusive) indexes of Arr_1 and Arr_2, respectively.

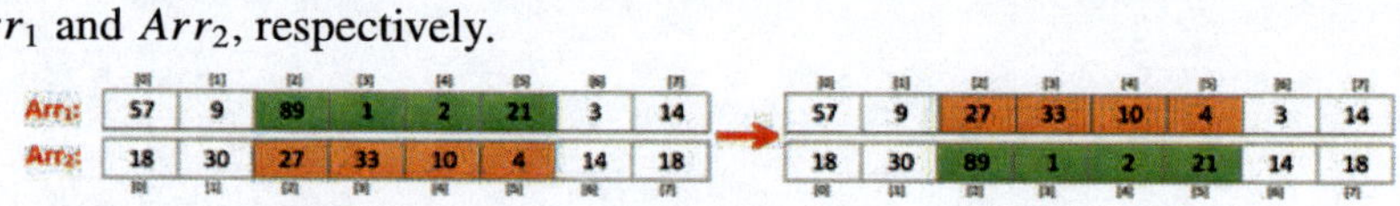

When compared to the encryption technique, Arr_1 is the even index list and Arr_2 is the odd index list hence merge them accordingly to get the list L.

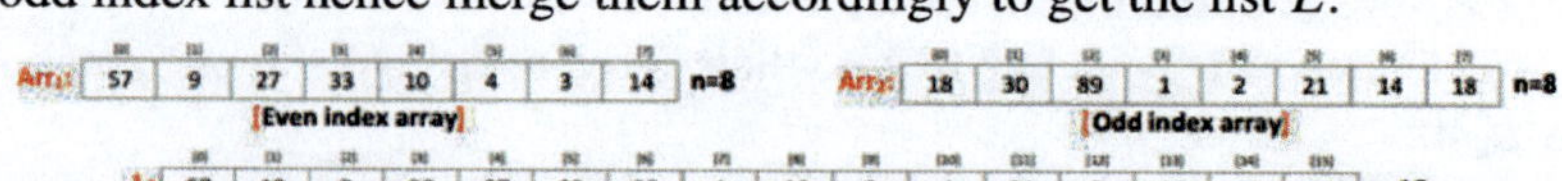

Convert each element (character) of the key list into their respective ASCII value. Now, perform XOR operation between the corresponding elements of key list and List L index by index and store the value back in list L.

	[0]	[1]	[2]	[3]	[4]	[5]	[6]	[7]	[8]	[9]	[10]	[11]	[12]	[13]	[14]	[15]	
L:	57	18	9	30	27	89	33	1	10	2	4	21	3	14	14	18	
	⊕	⊕	⊕	⊕	⊕	⊕	⊕	⊕	⊕	⊕	⊕	⊕	⊕	⊕	⊕	⊕	
Key:	113	119	101	114	116	121	117	105	111	112	97	115	100	102	103	104	
	↓																
L:	72	101	108	108	111	32	84	104	101	114	101	102	103	104	105	122	n=16

Slice list L by length $(127 - L[n-1])$ from the end, i.e., in this case $(127 - 122) = 5$. Hence slice L by 5 indexes from the end.

[0]	[1]	[2]	[3]	[4]	[5]	[6]	[7]	[8]	[9]	[10]	
72	101	108	108	111	32	84	104	101	114	101	n=11

Convert the elements (integer) of list L into their corresponding characters as per the ASCII number. The plain text in list format will be obtained.

[0]	[1]	[2]	[3]	[4]	[5]	[6]	[7]	[8]	[9]	[10]
H	e	l	l	o		T	h	e	r	e

Convert this list into String format. The original plain text has been retrieved: **"Hello There."**

5 Time Complexity of Proposed Cryptosystem

Let the function $f(u) = v$, such that v is a perfect square number greater than or equal to u. For example, $f(8) = 9$, $f(12) = 16$.

Now, let n = length of input plain text and x = resizer parameter.

$$f\left(\frac{n+x}{2}\right) = Y\,(\text{let})$$

Hence, the Time Complexity = $O(2Y) \approx O(Y)$.

6 Results

See Figs. 1, 2, 3, 4 and 5.

7 Conclusion

Network security is a crucial aspect of modern-day communication systems that heavily rely on long haul transmission using both guided and unguided media. In this work, a cryptosystem using genetic algorithm in has been proposed. A symmetric

```
Original Plain Text:
Hello There
Inside textToascii()
[72, 101, 108, 108, 111, 32, 84, 104, 101, 114, 101]

Inside resizer()
Text re-size parameter is 5
[72, 101, 108, 108, 111, 32, 84, 104, 101, 114, 101, 102, 103, 104, 105, 122]

Inside applyKey()
[57, 18, 9, 30, 27, 89, 33, 1, 10, 2, 4, 21, 3, 14, 14, 18]
```

Fig. 1 Intermediate cipher

```
Even: [57, 9, 27, 33, 10, 4, 3, 14]
Odd:  [18, 30, 89, 1, 2, 21, 14, 18]

Inside two_pointCrossover()
[57, 9, 89, 1, 2, 21, 3, 14]
[18, 30, 27, 33, 10, 4, 14, 18]

Inside crossMerge()
[57, 9, 89, 1] [2, 21, 3, 14]
[18, 30, 27, 33] [10, 4, 14, 18]
[57, 10, 9, 4, 89, 14, 1, 18]
[2, 18, 21, 30, 3, 27, 14, 33]
```

```
Transpose(Matrix_1):
[57, 4, 1]
[10, 89, 18]
[9, 14, 94]
-----------------------
Transpose(Matrix_2):
[2, 30, 14]
[18, 3, 33]
[21, 27, 94]

List(Matrix_1): [57, 4, 1, 10, 89, 18, 9, 14, 94]
List(Matrix_2): [2, 30, 14, 18, 3, 33, 21, 27, 94]
```

Fig. 2 Final cipher (i)

```
Inside MatrixOperations()
Matrix Dimensions are 3 x 3

Matrix_1:
[57, 10, 9]
[4, 89, 14]
[1, 18, 94]
-----------------------
Matrix_2:
[2, 18, 21]
[30, 3, 27]
[14, 33, 94]
```

```
Inside Mutation()
[57, 4, 1, 10, 45, 18, 9, 103, 94]
[2, 30, 14, 18, 119, 33, 21, 114, 94]


Inside AsciiToText()
Cipher Text in List Format:
['9', '\x04', '\x01', '\n', '-', '\x12', '\t', 'g', '^', '\x02',
 '\x1e', '\x0e', '\x12', 'w', '!', '\x15', 'r', '^']

The Cipher text is:
9 A
-    g^  w! r^
```

Fig. 3 Final cipher (ii)

```
Acquired Cipher Text is:
9 A
-   g^  w! r^

Inside textToascii()
[57, 4, 1, 10, 45, 18, 9, 103, 94, 2, 30, 14, 18, 119,
 33, 21, 114, 94]

Inside Mutation()
[57, 4, 1, 10, 89, 18, 9, 14, 94]
[2, 30, 14, 18, 3, 33, 21, 27, 94]

Inside MatrixOperations()
Matrix Dimensions are 3 x 3

Matrix_1:
[57, 4, 1]
[10, 89, 18]
[9, 14, 94]
------------------------
Matrix_2:
[2, 30, 14]
[18, 3, 33]
[21, 27, 94]
------------------------
Transpose(Matrix_1):
[57, 10, 9]
[4, 89, 14]
[1, 18, 94]
------------------------
Transpose(Matrix_2):
[2, 18, 21]
[30, 3, 27]
[14, 33, 94]
```

```
List(Matrix_1): [57, 10, 9, 4, 89, 14, 1, 18, 94]
List(Matrix_2): [2, 18, 21, 30, 3, 27, 14, 33, 94]

After Slicing:
List(Matrix_1): [57, 10, 9, 4, 89, 14, 1, 18]
List(Matrix_2): [2, 18, 21, 30, 3, 27, 14, 33]

Inside rev_crossMerge()
[57, 9, 89, 1] [10, 4, 14, 18]
[2, 21, 3, 14] [18, 30, 27, 33]

Merging the lists by maintaining proper symmetry with
 encryption Technique:
[57, 9, 89, 1, 2, 21, 3, 14]
[18, 30, 27, 33, 10, 4, 14, 18]

Inside two_pointCrossover()
[57, 9, 27, 33, 10, 4, 3, 14]
[18, 30, 89, 1, 2, 21, 14, 18]

Inside odd_evenMerge()
[57, 18, 9, 30, 27, 89, 33, 1, 10, 2, 4, 21, 3, 14,
 14, 18]
```

Fig. 4 Intermediate text

```
Inside applyKey()
[72, 101, 108, 108, 111, 32, 84, 104, 101, 114, 101, 102, 103, 104, 105, 122]

Slicing the Text to Original Length:
[72, 101, 108, 108, 111, 32, 84, 104, 101, 114, 101]


Inside AsciiToText()
Plain Text in List format:
['H', 'e', 'l', 'l', 'o', ' ', 'T', 'h', 'e', 'r', 'e']

The Original Plain text is:
Hello There
```

Fig. 5 Actual text

cryptosystem heuristic backed by extensive use of mathematical functions along with genetic algorithm has been implemented to address various threats such as cyber-attacks, hacking, and data breaches.

References

1. Stinson DR (2005) Cryptography: theory and practice. Chapman and Hall/CRC
2. Mitchell M (1998) An introduction to genetic algorithms. MIT Press
3. Mittal A, Gupta RK (2019) Encryption and decryption of a message involving genetic algorithm. Int J Eng Adv Technol (IJEAT) l 9(2):2249–8958
4. Sindhuja K, Devi SP (2014) A symmetric key encryption technique using genetic algorithm. Int J Comput Sci Inf Technol 5(1):414–416
5. Dutta S, Das T, Jash S, Patra D, Paul P (2014) A cryptography algorithm using the operations of genetic algorithm & pseudo random sequence generating functions. Int J 3(5):325–330

Mathematical Model for Improving Cloud Load Balancing Using Scheduling Algorithms

Prathamesh Vijay Lahande and Parag Ravikant Kaveri

1 Introduction

Cloud is a network that provides on-demand services to the end user over the Internet [1, 2]. Among several services, computing is an essential service provided by the cloud, so users can opt to process and execute their tasks on the cloud computing environment rather than on their local machines [1, 3]. By using limited resources, the cloud uses resource scheduling algorithms and executes the tasks on the cloud Virtual Machines (VMs) by trying to maintain a smooth flow of high task execution and maintain its on-demand availability characteristics [4, 5]. However, the cloud faces major issues and challenges with respect to improper load balancing when the requests are high, and the cloud resources are not available [2]. With accurate load balancing, the cloud can provide better cost and time for processing and executing tasks, otherwise will output limited results. Hence, it becomes essential to study the resource scheduling algorithms with respect to load balancing. The resource scheduling algorithms considered for this study are Max–Min (MX–MN), Minimum Completion Time (MCT), and Min–Min (MN–MN). The tasks are executed in a simulation environment and results are compared with respect to the load-balancing mechanism. The reinforcement learning (RL) [5] mechanism is proposed at the end to improve the load balancing and resource scheduling mechanisms. The research paper is organized as follows: Sect. 2 provides the literature review. Section 3 includes the experimental setup. Section 4 includes the mathematical model for load balancing. Section 5 includes the empirical analysis, followed by the conclusion in Sect. 6.

P. V. Lahande · P. R. Kaveri (✉)
Symbiosis Institute of Computer Studies and Research, Symbiosis International (Deemed University), Pune, India
e-mail: parag.kaveri@sicsr.ac.in

P. V. Lahande
e-mail: prathamesh.lahande@sicsr.ac.in

J. K. Mandal et al. (eds.), *Proceedings of International Conference on Network Security and Blockchain Technology*, Lecture Notes in Networks and Systems 738,
https://doi.org/10.1007/978-981-99-4433-0_28

2 Literature Review

The load-balancing problem is an NP-hard problem. Several researchers have designed and presented their work to address and solve this load-balancing issue. The researchers have proposed a hybrid task scheduling algorithm made from the MX–MN, MN–MN, and genetic algorithm to reduce the makespan and enhance the load balances between the resources [6]. The major use of this hybrid algorithm is the genetic algorithm which decides which tasks should use MX–MN and which ones should use MN–MN. This paper proposed a load-balancing virtual network functions deployment scheme to balance the physical machine loads to save the deployment and migration costs [3]. To improve the poor resource scheduling performance, the researchers have proposed a fuzzy iterative algorithm for the cloud computing environment [7]. The experimental results of this algorithm depict that it can improve the load-balancing mechanism and management efficiency of the cloud platform. This paper presents a load-balancing mechanism applied to a software-defining network in a cloud environment [8]. To balance the load in the host machines, a strategy is presented based on machine learning for Virtual Machine replacement [9].

The researchers have provided a load-balancing scheme to the proposed three-layer mobile hybrid hierarchical peer-to-peer model to balance the load with increasing mobile edge computing servers and query loads [10]. A load-balanced service scheduling approach is presented in this paper, which considers load balancing when scheduling requests to resources using classifications such as important, real time, or time tolerant [11]. This approach also considers the rate of failure of resources to provide better reliability to all the requests. A scheduling method named dynamic resource allocation for the purpose of load balancing is proposed to improve the load imbalance, which affects the scheduling efficiency and resource utilization [4]. The researchers have proposed a load-balancing algorithm to dynamically balance the cloud load in the present work [12]. A novel load-balancing task scheduling algorithm is proposed by combining the dragonfly and firefly algorithms [13]. The researchers have proposed scheduling algorithms for the heterogenous cloud environment to balance the load and time of resources [14].

The researchers have provided a mechanism to ensure that each node in the cloud is appropriately balanced [15]. A hybrid soft computing-inspired technique is introduced to achieve an optimal load of the VMs by tutoring the cloud environment [16]. This technique gives better load-balancing results when compared to other existing algorithms. To minimize the load balancing and overall migration overhead, the researchers have proposed a load-balancing method to provide a probable assurance against resource overloading with VM migration [17]. The researchers have presented two algorithms to distribute the cloud physical resources to obtain load-balancing consolidated systems with minimum power, memory, and time or processing [18]. The researchers have contributed three algorithms to improve the load-balancing mechanism [19]. The researchers have proposed a combination of the swarm intelligence algorithm of an artificial bee colony with the heuristic algorithm to improve the load balancing and makespan and minimize the makespan [20].

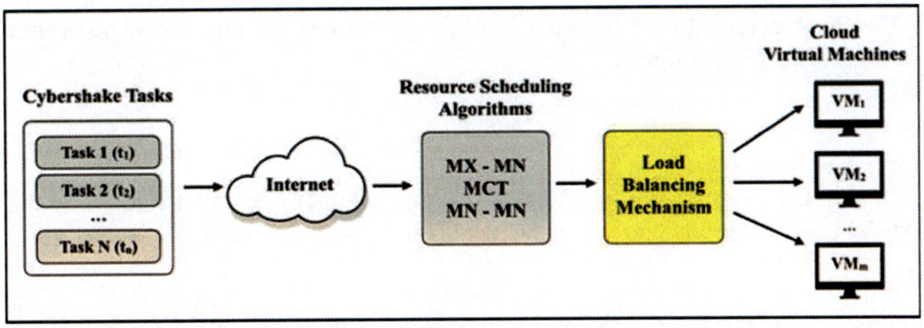

Fig. 1 Architecture of the conducted experiment

3 Experimental Setup

An experiment was conducted in the WorkflowSim environment where the Cybershake tasks are processed and executed on the cloud VMs in four scenarios. The Cybershake 30, 50, 100, and 1000 tasks are processed and executed in each scenario on the cloud VMs in the series: 5, 10, 15, ..., and 50. All these tasks are submitted to the cloud for processing and execution. The cloud accepts these tasks and places them in the ready queue, which contains a list of all the tasks waiting to be assigned a VM. The respective resource scheduling algorithm selects the tasks from this ready queue, performs the load-balancing mechanism, and appropriately schedules it to the target VM. Figure 1 depicts the architecture of the conducted experiment.

4 Mathematical Model for Load Balancing

This section includes the mathematical model for load balancing using the resource scheduling algorithms MX–MN, MCT, and MN–MN.

The task set of the Cybershake dataset used for the experiment is as follows:

$$\text{Task set} = T_1 + T_2 + T_3 + \cdots + T_n.$$

The Virtual Machine (VM) set can be depicted as follows:

$$\text{VM set} = \text{VM}_1 + \text{VM}_2 + \text{VM}_3 + \cdots + \text{VM}_m.$$

Every VM processes and executes tasks, and the count of tasks that the VM has processed and executed is as follows:

$$\text{VM}(x) = \sum_{i=0}^{m} [\text{VM}_i = x].$$

The ideal expected load for each VM 'm' processing 'n' number of tasks in a certain scenario is as follows:

$$\text{Average}(A) = \frac{n}{m}.$$

To measure load balance of a VM 'VM_i,' the amount of deviation caused by the VM 'VM_i' from the ideal expected average load 'A' can be calculated as

$$\text{Deviation}(\text{VM}_i) = \text{Absolute}\left(\sum_{i=0}^{m}[\text{VM}_i = x] - A\right).$$

Load balancing using MX–MN scheduling algorithm: The MX–MN algorithm first finds the minimum execution time (ET) from all the tasks. Later, it chooses the task 'T' with maximum ET from them, depicted as follows:

$$\text{MX--MN}(T) = \text{Max}\left[\text{Min}\left[\sum_{i=0}^{n}\text{E.T.}(T_i)\right]\right].$$

Load balancing using MCT scheduling algorithm: The MCT algorithm schedules the task 'T' based on the expected Minimum Completion Time (CT) among all the tasks, which can be depicted as follows:

$$\text{MCT}(T) = \text{Min}\left[\sum_{i=0}^{n}\text{C.T.}(T_i)\right].$$

Load balancing using MN–MN scheduling algorithm: The MN–MN algorithm first finds the minimum execution time (ET) from all the tasks. Later, it chooses the task 'T' with minimum ET from them. The below equation depicts the same:

$$\text{MN--MN}(T) = \text{Min}\left[\text{Min}\left[\sum_{i=0}^{n}\text{E.T.}(T_i)\right]\right].$$

5 Empirical Analysis of the Results and Their Implications

This section includes the detailed empirical analysis of the results of the experiment conducted with respect to the load-balancing mechanism along with their implications, which is further divided into four sub-sections: Sections 5.1, 5.2, 5.3, and 5.4

include the empirical analysis with respect to load balancing for Cybershake 30, 50, 100, and 1000 tasks, respectively.

5.1 *Scenario 1: Empirical Analysis with Respect to Cybershake 30 Tasks*

Table 1 depicts the deviation percentage of algorithms for Cybershake 30 tasks.

Figure 2 depicts the deviation graph of algorithms for Cybershake 30.

Table 1 Deviation percentage of load balancing for Cybershake 30

VMs	MX–MN	MCT	MN–MN
5	56.7742	42.5806	61.9355
10	37.276	56.6308	91.0394
15	37.276	56.6308	91.0394
20	37.276	56.6308	91.0394
25	37.276	56.6308	91.0394
30	37.276	56.6308	91.0394
35	37.276	56.6308	91.0394
40	37.276	56.6308	91.0394
45	37.276	56.6308	91.0394
50	37.276	56.6308	91.0394

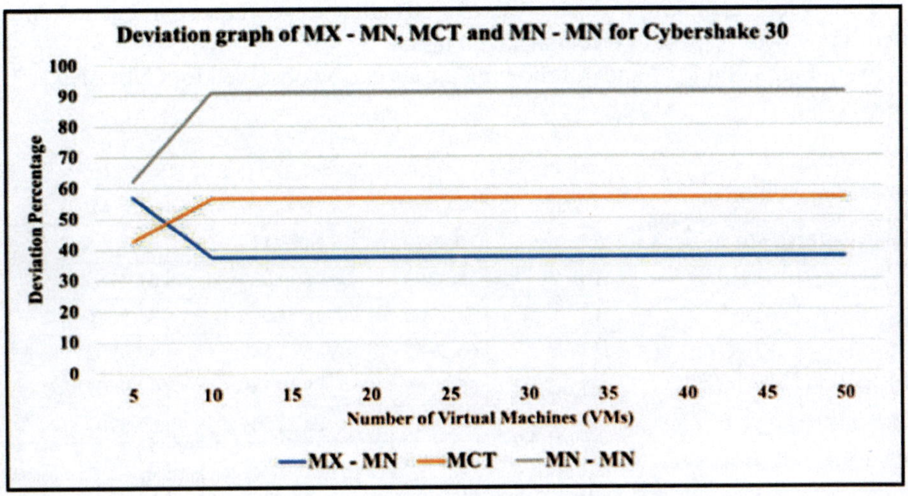

Fig. 2 Deviation graph for resource scheduling algorithms for Cybershake 30

Table 2 Empirical analysis of load balancing with respect to Cybershake 30 tasks

Parameter	MX–MN	MCT	MN–MN
Linear regression equation	$y = -1.60x + 45.0$	$y = 0.76x + 51.01$	$y = 1.58x + 79.39$
Regression line slope	− 1.6035	0.7664	1.5875
Y-intercept	45.07	51.011	79.398
Relationship	Negative	Positive	Positive
R^2 value	0.2727	0.2727	0.2727
VM–LB analysis	↑ VM = ↑ LB	↑ VM = ↓ LB	↑ VM = ↓ LB
Performance	MX–MN ≈ MCT ≈ MN–MN		

Table 2 depicts the empirical analysis of load balancing for the resource scheduling algorithms with respect to Cybershake 30 tasks.

From Fig. 2, Tables 1 and 2, following points can be observed for Cybershake 30 tasks:

- ↑ VM = ↑ LB: As VMs increase, load balancing improves for MX–MN.
- ↑ VM = ↓ LB: As VMs increase, load balancing degrades for MCT and MN–MN.
- Performance (MX–MN) ≈ Performance (MCT) ≈ Performance (MN–MN).

5.2 *Empirical Analysis with Respect to Cybershake 50 Tasks*

Table 3 depicts the deviation percentage of algorithms for Cybershake 50 tasks.

Figure 3 depicts the deviation graph of algorithms for Cybershake 50.

Table 4 depicts the empirical analysis of load balancing for the resource scheduling algorithms with respect to Cybershake 50 tasks.

From Fig. 3, Tables 3 and 4, following points can be observed for Cybershake 50 tasks:

Table 3 Deviation percentage of load balancing for Cybershake 50

VMs	MX–MN	MCT	MN–MN
5	29.8039	49.4118	33.7255
10	58.0392	69.8039	61.1765
15	61.1765	80.0000	81.5686
20	66.1765	87.2549	87.9902
25	66.1765	87.2549	87.9902
30	66.1765	87.2549	87.9902
35	66.1765	87.2549	87.9902
40	66.1765	87.2549	87.9902
45	66.1765	87.2549	87.9902
50	66.1765	87.2549	87.9902

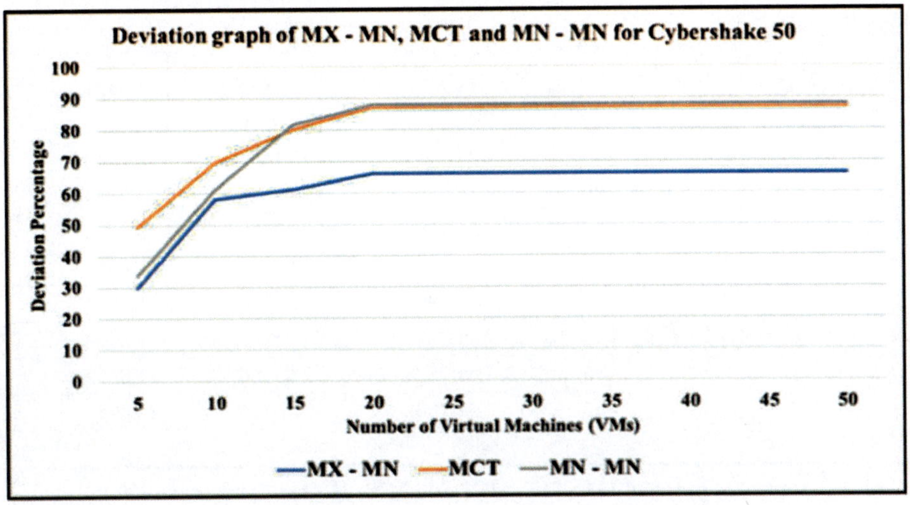

Fig. 3 Deviation graph for resource scheduling algorithms for Cybershake 50

Table 4 Empirical analysis of load balancing with respect to Cybershake 50 tasks

Parameter	MX–MN	MCT	MN–MN
Linear regression equation	$y = 2.48x + 47.58$	$y = 3.02x + 64.36$	$y = 4.29x + 55.63$
Regression line slope	2.4807	3.0244	4.292
Y-intercept	47.582	64.366	55.634
Relationship	Positive	Positive	Positive
R^2 value	0.4343	0.5398	0.5171
VM–LB analysis	$\uparrow$ VM = $\downarrow$ LB	$\uparrow$ VM = $\downarrow$ LB	$\uparrow$ VM = $\downarrow$ LB
Performance	MX–MN > MN–MN > MCT		

- $\uparrow$ VM = $\downarrow$ LB: As VMs increases, load balancing degrades for all algorithms.
- Performance (MX–MN) > Performance (MN–MN) > Performance (MCT).

5.3 *Empirical Analysis with Respect to Cybershake 100 Tasks*

Table 5 depicts the deviation percentage of algorithms for Cybershake 100 tasks.

Figure 4 depicts the deviation graph of algorithms for Cybershake 100.

Table 6 depicts the empirical analysis of load balancing for the resource scheduling algorithms with respect to Cybershake 100 tasks.

From Fig. 4, Tables 5 and 6, following points can be observed for Cybershake 100 tasks:

- $\uparrow$ VM = $\downarrow$ LB: As VMs increase, load balancing degrades for all algorithms.

Table 5 Deviation percentage of load balancing for Cybershake 100

VMs	MX–MN	MCT	MN–MN
5	18.2178	41.1881	11.0891
10	40.5941	38.8119	39.0099
15	20.5941	50.165	48.1848
20	44.9505	54.7525	58.9109
25	56.7129	72.7129	76.7525
30	64.7808	77.5106	81.8953
35	64.7808	77.5106	81.8953
40	64.7808	77.5106	81.8953
45	64.7808	77.5106	81.8953
50	64.7808	77.5106	81.8953

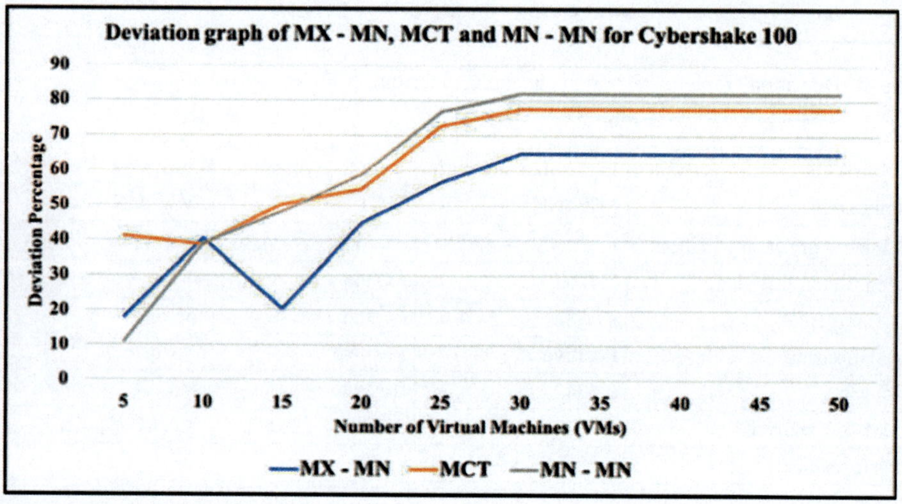

Fig. 4 Deviation graph for resource scheduling algorithms for Cybershake 100

Table 6 Empirical analysis of load balancing with respect to Cybershake 100 tasks

Parameter	MX–MN	MCT	MN–MN
Linear regression equation	$y = 5.31x + 21.26$	$y = 4.89x + 37.59$	$y = 7.15x + 25.00$
Regression line slope	5.3143	4.8945	7.1521
Y-intercept	21.268	37.599	25.006
Relationship	Positive	Positive	Positive
R^2 value	0.7459	0.8175	0.7725
VM–LB analysis	↑ VM = ↓ LB	↑ VM = ↓ LB	↑ VM = ↓ LB
Performance	MX–MN > MN–MN > MCT		

Table 7 Deviation percentage of load balancing for Cybershake 1000

VMs	MX–MN	MCT	MN–MN
5	2.5175	3.3567	1.9581
10	2.9571	11.6684	7.9321
15	7.1795	21.2055	8.9644
20	3.5665	20.2698	10.9091
25	4.9791	22.6654	13.8982
30	19.374	28.2918	13.4333
35	21.3387	38.2818	17.8622
40	34.6554	37.2828	15.3247
45	6.9842	41.7317	17.0541
50	4.5795	43.2808	20.4956

- Performance (MX–MN) > Performance (MN–MN) > Performance (MCT).

5.4 Empirical Analysis with Respect to Cybershake 1000 Tasks

Table 7 depicts the deviation percentage of algorithms for Cybershake 1000 tasks.

Figure 5 depicts the deviation graph of algorithms for Cybershake 1000.

Table 8 depicts the empirical analysis of load balancing for the resource scheduling algorithms with respect to Cybershake 100 tasks.

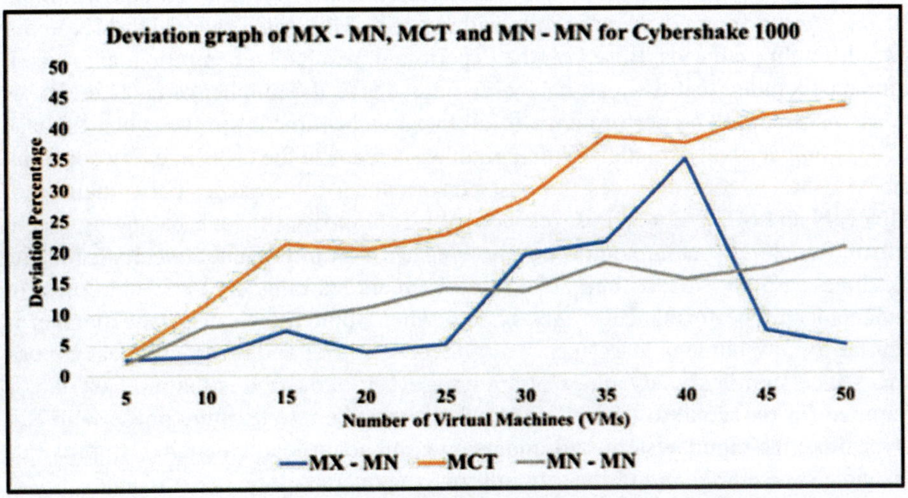

Fig. 5 Deviation graph for resource scheduling algorithms for Cybershake 1000

Table 8 Empirical analysis of load balancing with respect to Cybershake 1000 tasks

Parameter	MX–MN	MCT	MN–MN
Linear regression equation	$y = 1.52x + 2.41$	$y = 4.30x + 3.14$	$y = 1.71x + 3.35$
Regression line slope	1.5263	4.3019	1.7145
Y-intercept	2.4185	3.1431	3.3536
Relationship	Positive	Positive	Positive
R^2 value	0.1855	0.9494	0.8966
VM–LB analysis	$\uparrow$ VM = $\downarrow$ LB	$\uparrow$ VM = $\downarrow$ LB	$\uparrow$ VM = $\downarrow$ LB
Performance	MX–MN > MN–MN > MCT		

From Fig. 5, Tables 7 and 8, following points can be observed for Cybershake 1000 tasks:

- $\uparrow$ VM = $\downarrow$ LB: As VMs increase, load balancing degrades for all algorithms.
- Performance (MX–MN) > Performance (MN–MN) > Performance (MCT).

6 Conclusion

Load balancing plays a critical role in improving cloud performance. With an enhanced load-balancing technique, the performance of the cloud will be enhanced; otherwise, the cloud faces increased downtime with improper load balancing. Therefore, it becomes vital to study the resource scheduling algorithms with respect to the load-balancing mechanism. In this research paper, the Cybershake 30, 50, 100, and 1000 tasks were processed and executed in the WorkflowSim environment utilizing the resource scheduling algorithms MX–MN, MCT, and MN–MN under four different scenarios. Based on the experiment, results, and empirical analysis, it can be concluded that the load balancing degrades as the number of tasks increases in each scenario. The performance of all these algorithms is similar when the task size is comparatively smaller. With a gradual increase in the tasks in every scenario, the MX–MN algorithm gives the best performance, followed by performances of MN–MN and MCT algorithms, respectively. To improve the load-balancing mechanism, the cloud system should be provided with an intelligence mechanism. The machine learning (ML) technique of reinforcement learning (RL) is well known for enhancing the performance of any system when applied to it. RL's mechanism is similar to how humans learn, i.e., using feedback, trial and error, and past experiences. The significant advantage of applying RL to the cloud is that no past data is required for the cloud to learn. The cloud will first be in a learning phase with RL. Over time, the cloud system will understand and adapt load balancing, improve its resource scheduling, and ultimately improve the overall cloud performance.

References

1. Armbrust M, Fox A, Griffith R, Joseph AD, Katz R, Konwinski A, Lee G, Patterson D, Rabkin A, Stoica I, Zaharia M (2010) A view of cloud computing. Commun ACM 53(4):50–58
2. Dillon T, Wu C, Chang E (2010) Cloud computing: issues and challenges. In: 2010 24th IEEE international conference on advanced information networking and applications
3. Wang YC, Wu SH (2022) Efficient deployment of virtual network functions to achieve load balance in cloud networks. In: 2022 23rd Asia-Pacific network operations and management symposium (APNOMS)
4. Chhabra S, Singh AK (2021) Dynamic resource allocation method for load balance scheduling over cloud data center networks. J Web Eng
5. Vengerov D (2007) A reinforcement learning approach to dynamic resource allocation. Eng Appl Artif Intell 20(3):383–390
6. Aref IS, Kadum J, Kadum A (2022) Optimization of max-min and min-min task scheduling algorithms using G.A in cloud computing. In: 2022 5th International conference on engineering technology and its applications (IICETA)
7. Bin D, Yu T, Li X (2021) Research on load balancing dispatching method of power network based on cloud computing. J Phys Conf Ser 1852(2):022047
8. Omer YAH, Mohammedel-Amin MA, Mustafa ABA (2021) Load balance in cloud computing using software defined networking. In: 2020 International conference on computer, control, electrical, and electronics engineering (ICCCEEE)
9. Ghasemi A, Toroghi Haghighat A (2020) A multi-objective load balancing algorithm for virtual machine placement in cloud data centers based on machine learning. Computing 102(9):2049–2072
10. Duan Z, Tian C, Zhang N, Zhou M, Yu B, Wang X, Guo J, Wu Y (2022) A novel load balancing scheme for mobile edge computing. J Syst Softw 186:111195
11. Alqahtani F, Amoon M, Nasr AA (2021) Reliable scheduling and load balancing for requests in cloud-fog computing. Peer Peer Netw Appl 14(4):1905–1916
12. Negi S, Rauthan MMS, Vaisla KS, Panwar N (2021) CMODLB: an efficient load balancing approach in cloud computing environment. J Supercomput 77(8):8787–8839
13. Neelima P, Reddy ARM (2020) An efficient load balancing system using adaptive dragonfly algorithm in cloud computing. Clust Comput 23(4):2891–2899
14. Lin W, Peng G, Bian X, Xu S, Chang V, Li Y (2019) Scheduling algorithms for heterogeneous cloud environment: main resource load balancing algorithm and time balancing algorithm. J Grid Comput 17(4):699–726
15. Kaviarasan R, Harikrishna P, Arulmurugan A (2022) Load balancing in cloud environment using enhanced migration and adjustment operator-based monarch butterfly optimization. Adv Eng Softw 169:103128
16. Negi S, Panwar N, Rauthan MMS, Vaisla KS (2021) Novel hybrid ANN and clustering inspired load balancing algorithm in cloud environment. Appl Soft Comput 113:107963
17. Yu L, Chen L, Cai Z, Shen H, Liang Y, Pan Y (2020) Stochastic load balancing for virtual resource management in datacenters. IEEE Trans Cloud Comput 8(2):459–472
18. Nashaat H, Ashry N, Rizk R (2019) Smart elastic scheduling algorithm for virtual machine migration in cloud computing. J Supercomput 75(7):3842–3865
19. Saber W, Moussa W, Ghuniem AM, Rizk R (2021) Hybrid load balance based on genetic algorithm in cloud environment. Int J Electr Comput Eng (IJECE) 11(3):2477
20. Kruekaew B, Kimpan W (2020) Enhancing of artificial bee colony algorithm for virtual machine scheduling and load balancing problem in cloud computing. Int J Comput Intell Syst 13(1):496

Blockchain Technology and IoT

Securing Farm Insurance Using a Private-Permissioned Blockchain Driven by Hyperledger Fabric and IPFS

Nishat Tasnim Haque, Zerin Tasnim, Ananya Roy Chowdhury, and Saha Reno

1 Introduction

The majority of the population in Bangladesh depends on farming. Livestock farming is an integral element of our country's agriculture production system, which provides multiple sources of food, employment opportunities, agricultural development, and other services. However, it is concerning that a larger percentage of them are not protected by a safe insurance system that will offer financial security or compensation for damages. As a result, they are unable to recover from the immense loss they experience in the event of incidents like fires or natural disasters. An insurance management system that records crucial data about a farm, its owner, loss history, etc., is needed to ensure that livestock farmers can claim refunds in case of mishaps or unforeseen events. And on account of this, it will be effortless to provide the insurance they require based on their circumstances and to identify fraudulent insurance claimants.

Various authors have provided numerous ways to secure insurance management; however, the constraints in their systems prevent real-world use of these approaches. Raikwar and his co-authors suggested a permissioned blockchain-based approach [8] for insurance processes using Hyperledger Fabric v1.0.0-beta. Golanf v1.8 was used to create the chaincode. However, their database is not encrypted, and each transaction cannot have its own set of endorsing peers since smart contracts are not implemented at the transactional level. Additionally, the confirmation time will increase with node count, making the network slower. The authors of "NEO smart contract for drought-based insurance" [7] developed a system for South-East Asian (SEA) countries where small households experienced insufficient crop production due to heavy drought. The system was built on the NEO blockchain technology and runs on NeoVM. Given that this system's consensus protocol lacks a fallback mechanism, the

N. T. Haque · Z. Tasnim · A. R. Chowdhury · S. Reno (✉)
Bangladesh Army International University of Science and Technology, Cumilla, Bangladesh
e-mail: reno.saha39@gmail.com

J. K. Mandal et al. (eds.), *Proceedings of International Conference on Network Security and Blockchain Technology*, Lecture Notes in Networks and Systems 738,
https://doi.org/10.1007/978-981-99-4433-0_29

consensus must be forcibly reset when the code cannot reach consensus. NEO consensus relies on bookkeeper nodes that stop trying to come to an understanding when the time difference between the timestamp of the previous block and the present time exceeds 10 min. "Application of smart contracts based on the Ethereum blockchain for the purpose of insurance services" [1] proposed an Ethereum blockchain-based solution for insurance processes. They created a decentralized crypto-token based on the ECR20 smart-token standard. The problem with their approach is that they automated the handling of insurance claims using a public, permission-free blockchain, which could cause problems with real-time confirmation. On top of that, Ethereum employs a computationally intensive consensus process known as proof-of-work (POW), which has a significant scalability issue that restricts it to about 20 transactions per second.

In this study, we developed a system that handles applicants' information for insurance companies in the safest and most practical way possible using the hyperledger fabric blockchain. It's a special form of a private blockchain that enables the transaction of an asset or its state to be consented upon, maintained, or observed by only the members of a permissioned group [5]. Despite the fact that it is built on the idea of a distributed or shared ledger, each participant must be verified. This assists in concealing sensitive data and prevents data manipulation (i.e., creating new data, updating old data, or deleting existing data). Hyperledger fabric does not require consensus techniques like the computationally demanding proof of work (POW) or proof of stake (POS) used in blockchain networks because it is a private and permissioned network [2]. Scalability, transaction speed, and overall network performance are all boosted by doing this [9]. Last but not least, hyperledger fabric gives users the option to alter the entire blockchain's underlying infrastructure. Our proposed system offers

(i) Information access control within a permissioned group. Outsiders are not permitted to change the data; only the allowed participants can add new assets, along with updating, deleting, and searching the existing information.
(ii) An insured is only permitted to read his/her transaction information in order to prevent data manipulation.
(iii) Insurance companies can look up information about an insured by the asset ID, which is a unique identification number that prevents them from claiming insurance to which they are not entitled.

2 Literature Review

Over the past several years, blockchain research has gained popularity. Numerous people have introduced their outstanding efforts to address significant issues regarding insurance with a blockchain-based solution. Loukil and others proposed CioSy, a collaborative blockchain-based insurance system for monitoring and processing insurance transactions [6]. Through the use of three smart contracts, they created a community of insurers, acquired insurance from the insurer, submitted claims in accordance with the insurance contract, and automated the processing of claims and refund payments. This system, though, works best for goods with extremely low risk expectations. Premiums for claims in the event of sudden catastrophes, like natural disasters, may be woefully inadequate.

The MIStore is a medical insurance storage system [12] by Lijing and his co-authors that achieves several unique features, including decentralization, secure data storage, threshold, efficient verification, and efficient homomorphic computation. The tamper-resistance of the blockchain provides users with great trust; and because it is decentralized, users may connect with one another without the use of intermediaries. The system's drawbacks include that it utilized the Ethereum blockchain approach, which presented challenges with scalability due to its multipurpose ledger and complex programming language.

In [11], by Iman and others, a new framework is presented for securing a cyber-product using blockchain technology. To share the risk of insurance, they have applied for crowdfunding through a sealed-bid auction process. The binding property ensures that a commitment cannot be opened to another value, whereas hiding demands that a commitment not reveal any information about the committed value. They used the Ethereum-SHA3 hash function (Keccak256) as the commitment scheme for their implementation. The drawback of this work is that the price of Ether has fluctuated dramatically in the past, so beginners may find it risky to invest in this currency.

A framework for using smart contracts and storing them in blockchain for insurance contracts is presented in [3] by Abid and his co-authors. The proof of authority (PoA) consensus algorithm is used to verify the transactions, and it rejects any invalid transaction requests. Any incorrect endorsement will be obvious and detectable; the algorithm operates on both private and public networks. A drawback of smart contracts is that they significantly limit the volume of transactions that the network can handle in a given second.

A solution [10] by Rui and others proposed that prevents specific sorts of fraud in the field of vehicle insurance was created using blockchain and smart contracts. The double dipping fraud arrangement is the use case for the created Ethereum-based blockchain system to prevent insurance fraud. Drawback is that they are now in the prototyping stage of their suggested solution. It has been tested using data from hypothetical insurance companies, clients, and vehicles, and it has shown to be effective. However, expanding the solution's capabilities and developing a finished product are required.

Nishant and others have developed a blockchain-based crop insurance for Indian farmers [4]. Due to the fact that latest blockchain-based frameworks are built on the idea of smart contracts, they offer a tamper-proof environment where a transaction is only carried out after receiving validated data, and no bad user is able to alter the system. When the instances were upgraded, the throughput rose to 220. When the CPU load was 70%, RAM consumption reached 80%. Drawback is, since this system doesn't directly employ any cryptocurrencies like Bitcoin, system failure in unstable cryptocurrency markets may be more likely.

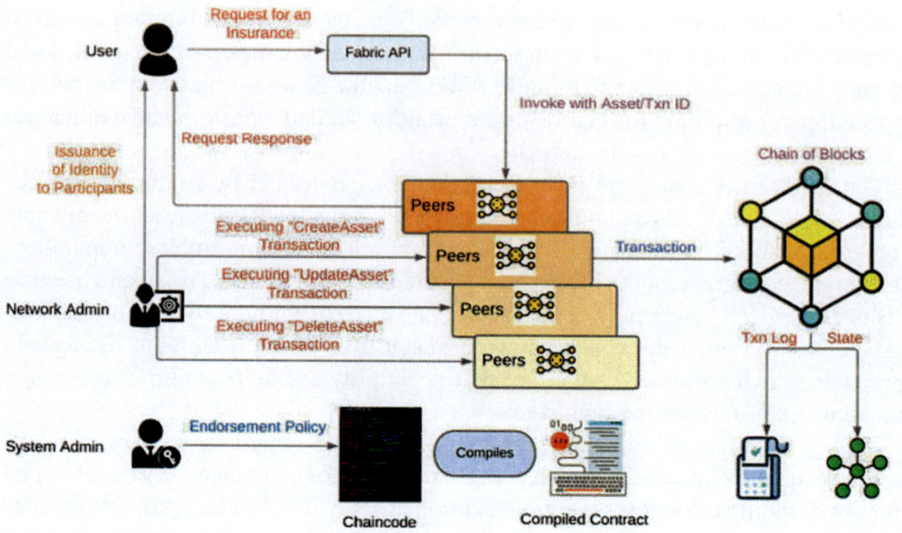

Fig. 1 Basic architecture of the proposed system

Our system overcomes the aforementioned shortcomings as: (i) it does not require any transaction fee, (ii) has better throughput rate (0.021 s per transaction), (iii) supports querying the ledger using SQL and (iv) has a robust access control module.

3 Methodology

Our chaincodes, written in JavaScript, have been installed in Hyperledger Fabric 2.3.1. Only those users who are identified in the system's asset identification field can access the document information. There must be an electronic copy of every document that is connected to the information provided by the insured. So, IPFS stores the images of the documents by cryptographically hashing them, creating a unique content identifier (CID) for each image. Additionally, they are linked to the appropriate assets. By doing this, we can instantly detect any altered or destroyed images, given that the accompanying asset's hash will likewise change. Any time an asset is created, added to, updated, or removed from a block, a transaction involving that asset is recorded. Nobody can effectively alter or modify the information provided in the system in this way without notifying others. The main concept of our system is illustrated in Fig. 1.

3.1 Resources of the Network

Any object that has information is a resource in the Hyperledger Fabric and is called an asset. Details included within an asset can be seen as key value pairs, which signify the value of anything, from the tangible to the intangible. With hyperledger fabric, we can manipulate these assets via chaincode-dependent transactions. In our work, each farm is considered an asset, and for each of these assets, there are 11 attributes which correspond to different relevant and unique information about the farms.

An asset of our private blockchain, which is built on hyperledger fabric, is characterized by these attributes. A new asset ID is constructed each time information regarding a new farm is provided, permitting the ledger to be queried for that person's farm information. Only by performing transactions is it possible to modify and update this information in future. Participants in our system fall into one of three categories: (i) system admin, (ii) network admin, and (iii) user. The users in this scenario are farm owners who may only request information updates and read information about their insurance transactions. Admin has broader access and control over system resources than users have, which is covered in more detail in the following subsections. The configuration setup is done by Peer Admin; similarly identities to new participants are appointed by Network Admin. The relation between assets and participants is depicted in Fig. 2.

3.2 Initiating and Processing Transactions

A private blockchain is a distributed ledger that operates as a closed database secured with cryptographic concepts and the organization's needs. Only those with permission can run a full node, make transactions, or validate/authenticate the blockchain changes. The changes needed to be made in our blockchain are related to the creation of insurance, which will require reshaping of the assets. After each transaction is recorded, in order to change any record, 51% of the participants must reach consensus. The farm's hash from IPFS is stored so that the insurer can access this information at any time and update as needed. These updates can also be verified by the asset manager to make sure all the alterations are correct. Transactions are additionally employed to store the farm's IPFS hash, permitting the farm owners to access their farm's data. When a user performs a transaction, the asset ID is associated with the transaction's transaction ID. Owing to the immutability of blockchain transactions, asset managers are given access to an unalterable log that they can audit at any time in future. Asset managers utilize this transactional data to securely look at information, including earlier iterations of asset records and what modifications have been made to the assets.

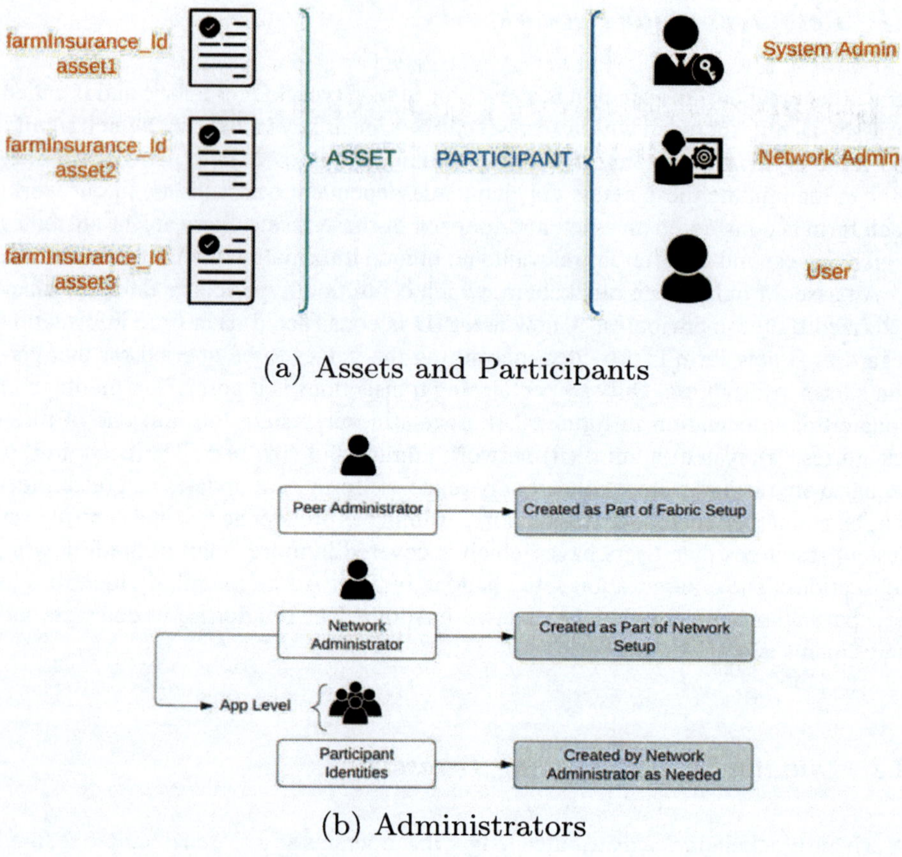

Fig. 2 Resources required for the proposed system

3.3 Deployment of Transactions Inside Historian Registry

The registry carrying the records of transactions is known as the Historian Registry. The Historian Registry is updated after each transaction is submitted and verified to be correct. And thus, a complete history of every transaction made within the network is created. In this record, along with the transaction history, information about who made the transaction and their identities are also preserved. A report of this Historian Registry is created in hyperledger fabric that includes transaction information such as timestamp, initiator ID, details of the changes made and a list of successful transactions that occurred. The Historian Registry cannot be viewed by anyone; only people with permissioned access and proper roles can inspect it. The declaration and definition of the queries are contained in the query document. Both untampered information and assets are subjected to queries. The Historian Registry and Asset Registry demonstrate the modifications recorded about the assets when a transaction is reported, along with the participant information and their identities

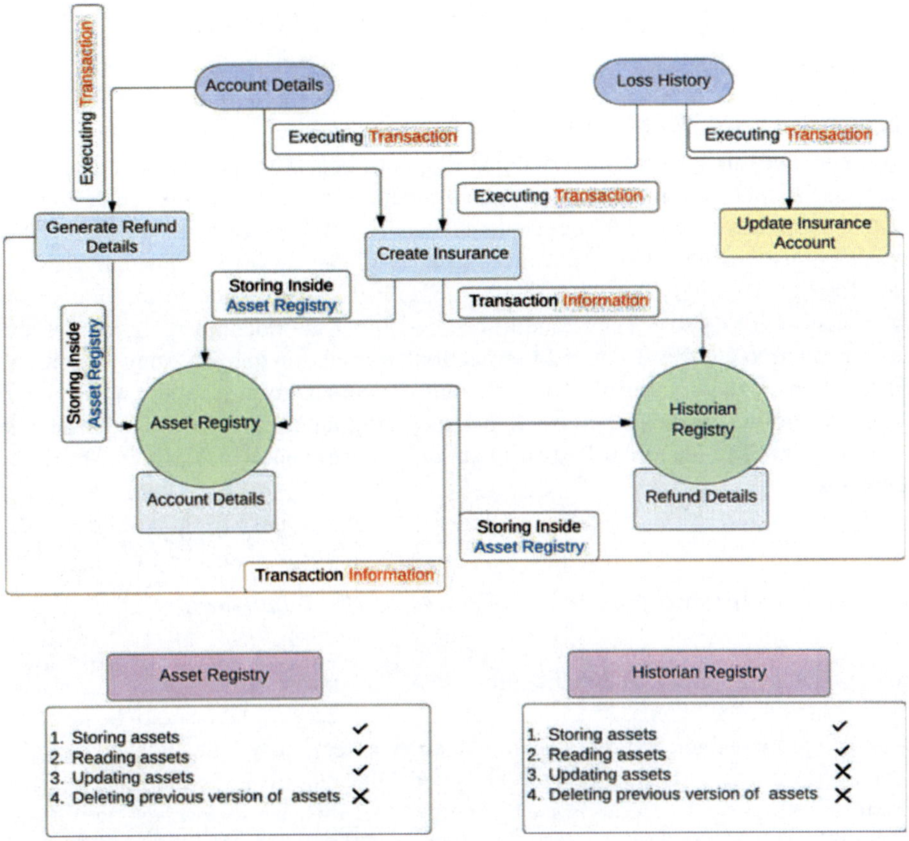

Fig. 3 Functionalities of asset and historian registries

in reporting such transactions. The Historian Registry's listing does not include the rejected transactions. Correspondingly, the Fabric runtime also refers to a number of configuration-related tasks it performs as transactions. The transactions which we identified and sought to conduct in our system can be divided into the following, (i) CreateAsset, (ii) ReadAsset, (iii) UpdateAsset, (iv) DeleteAsset, (v) AssetExists, (vi) TransferAsset, (vii) GetAllAssets, (viii) SearchAssets. Additionally, IPFS hash is linked to the assets. We may access the Historian Registry to obtain all of our private blockchain transactions, which are all preserved in blocks. In conjunction with asset ID, the Historian Registry is also categorized by asset ID, which is responsible for carrying out the transaction, and transaction genre. Like traditional databases, hyperledger fabric facilitates queries, by utilizing this feature inside the Historian Registry, any transaction's status may be confirmed. Figure 3 shows the implementation of queries to acquire asset, transaction, and heavyweight data from IPFS.

3.4 Regulating Access Using Access Control Module

The hyperledger fabric is a type of private blockchain that has access control as a unique feature. It refers to a script that lists the policies of a specific company. This feature allows our system to restrict a participant's access to the system's resources according to their designation and assigned function. We can limit the use of diverse systems utilizing create, read, update and delete (CRUD) procedures because of the extent of customizability. For instance, an insured can retrieve an insurance cover note from Asset Registry and view what alterations were made to the document from Historian Registry. They cannot, however, alter the documents directly. They are permitted to use the "RequestModification" transaction to update any inaccurate information in the document. The peer and network administrators, on the other hand, are eligible to update the assets since they are granted permissions in the access control script. The Historian Registry cannot be destroyed, hence accountability is preserved.

3.5 IPFS Hosting Electronic Copies of the Documents

In Hyperledger Fabric, a constraint of the blockchain is that only textual data may be recorded inside the blocks. Images and other large pieces of data cannot be directly uploaded to the ledger. For our system, we need to keep image files like each applicant's NID photo, a GD copy, etc. Therefore, we require a data storage system where the files' electronic copies will be kept. For this, we took advantage of the decentralized, peer-to-peer interplanetary file system (IPFS). However, before we can connect our system to the IPFS network, we must first install and initialize the IPFS client. The client is ready to upload an image file to IPFS when the daemon has been started. Following that, we must launch a new Ubuntu terminal from the picture file location. To upload the document's pdf, perform the bash command "ipfs add [nameoftheimagefile]". As a result, a hash is produced. Following that, this hash is added to the asset it belongs to using chaincode and transactions. Any changes made to the uploaded soft copy will significantly alter the hash..The soft copy of the document is downloaded and saved where the terminal is open when the command "ipfs get [hash]" is issued.

4 Result Analysis

We had to create a real implementation using the Ubuntu 20.04 LTS operating system and a Docker container because hyperledger fabric lacks a testing environment similar to composers (Hyperledger Composer Playground). For carrying out the necessary duties, such as producing endorsement responses, creating blocks, and broadcasting

Fig. 4 Setting up hyperledger fabric

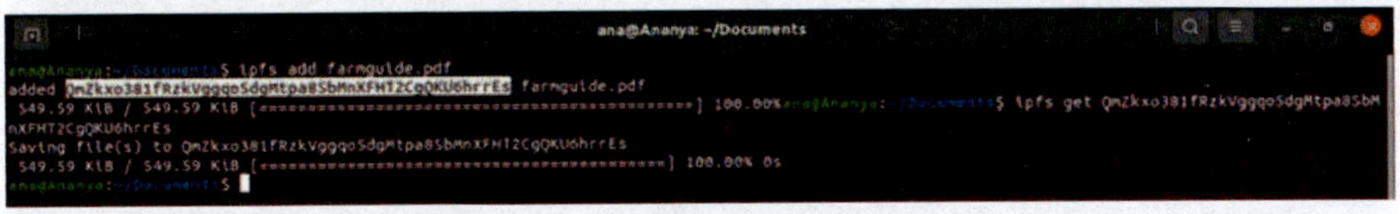

Fig. 5 Uploading and retrieving insurance information from IPFS

them throughout the network; the endorsers, peers, and orderers were created. The chaincodes are written in Javascript. In Fig. 4, the installation of prerequisites is shown, along with the fabric setup and activation.

After uploading the relevant electronic copy of the farm insurance, the IPFS generated a content identifier, which is what the hash in each specific asset for the farm insurance refers to. The IPFS daemon is started, followed by the initialization of IPFS, which is covered in the "methodology" section, in order to upload and receive the CID of farm insurance. The farm insurance's soft copy is uploaded using the command "ipfs add [fileName]," and it is retrieved using "ipfs get [hash]", as seen in Fig. 5.

Using the "CreateAsset" transaction listed in the chaincode, we generated numerous assets. The name of the specific transaction, together with any required parameters, must be supplied to the Fabric network in order for it to be executed. The process to initiate and execute the "CreateAsset" transaction has been shown in Fig. 6. We programmed this specific transaction, so that, after generating the asset, the user is given the success message through the asset details in the Ubuntu bash. Additionally, as shown in Fig. 6, the "AssetExists" transaction allows users to see if a certain asset is present in the Fabric network by providing an asset ID as an input. The output displays "true" if it exists and "false" otherwise.

Any existing asset may be updated using the "UpdateAsset" transaction, but in order to do so, the fabric network must receive the name of the specific transaction

Fig. 6 Execution of "CreateAsset" and "AssetExists" transactions

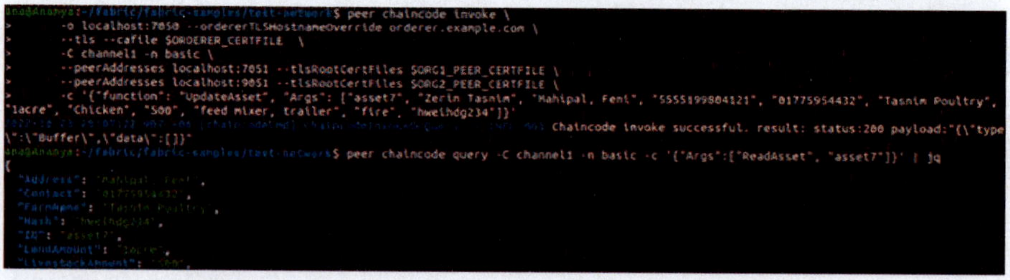

Fig. 7 Execution of "UpdateAsset" and "ReadAsset" transactions

as well as any necessary parameters as shown in Fig. 7. The "ReadAsset" transaction (Fig. 7), which accepts the asset ID as an input, and the "SearchAssets" transaction (Fig. 8a), allow users to get specific assets from our system. In "SearchAssets", any asset parameter from the assets can be used, instead of the NID. The asset ID must be entered when using the "DeleteAsset" operation to remove any asset, as shown using Fig. 8a and b. A demonstration of using the "GetAllAssets" transaction to get all the data associated with the farm insurance is displayed in Fig. 8b.

In addition, we developed our system using both Ethereum and Bitcoin. We used the Remix IDE and the solidity programming language to build our Ethereum-based solution. From GitHub, we downloaded an example Bitcoin project that uses the "Python-bitcoinlib" package to implement Bitcoin. All three solutions, including hyperledger fabric, had their throughput evaluated. For Ethereum, the "block.number" and "block.timestamp" functions from the solidity programming language were utilized; for Bitcoin implementation, the "time.time" and "time.now" functions from the "time" library of Python were used. Table 1 shows that fabric appears to be even better suited for getting farm insurance than the other two systems since it has the lowest throughput of the three blockchain-based systems.

(a) Initiating "SearchAssets" Transaction

(b) Initiating "DeleteAsset" and "GetAllAssets" Transactions

Fig. 8 Searching and deleting assets from asset registry

Table 1 Throughput comparison among the fabric-based system, Ethereum and Bitcoin

Amount of transactions (Txns)	Required time in hyperledger (s)	Required time in Ethereum (s)	Required time in Bitcoin (s)
620	18.6	40.4	89.3
550	16.7	39.7	86.4
470	13.4	37.6	84.2
365	12.5	35.3	83.5
240	10.8	33.8	81.1
170	8.9	32.5	79.6

Required time to execute per transaction on average

Hyperledger	Ethereum	Bitcoin
0.021	0.183	0.642

5 Conclusion and Future Recommendations

In our research, we developed a system that allows a person who owns a farm to secure farm insurance using the hyperledger fabric private blockchain, which is intended to provide protection to farm owners in the case of a catastrophe or loss that farm owners cannot afford. It also addresses all essential blockchain elements including tracking, verification, and validation. To verify that our chaincodes are accurate, a few contractual tests have been carried out. IPFS makes sure that all information saved on the blockchain network is encrypted, allowing only the insurance buyer to read and distribute it as they see fit.

We experienced a few complexities while developing the system. For example, each fabric version causes a change in the chaincode commands. As we used version 2.3.1, its chaincodes may not work in other versions. Also, there is no testing platform for fabric; the entire system can only be tested after implementation.

Moreover, we did not create any front end for our system which makes it hard to utilize the system. In future, we hope to develop a Web client for our system to make the system user-friendly. In addition, we hope to utilize the OCR functionality which recognizes texts of the digital image. By using such technology to scan images, users would be able to automatically add data or information from any image, eliminating the need for human entry.

References

1. Aleksieva V, Valchanov H, Huliyan A (2019) Application of smart contracts based on Ethereum blockchain for the purpose of insurance services. In: 2019 international conference on biomedical innovations and applications (BIA). IEEE, pp 1–4
2. Gatteschi V, Lamberti F, Demartini C, Pranteda C, Santamaría V (2018) Blockchain and smart contracts for insurance: is the technology mature enough? Future Internet 10(2):20
3. Hassan A, Ali M, Ahammed R, Khan MM, Alsufyani N, Alsufyani A et al (2021) Secured insurance framework using blockchain and smart contract. Sci Program 2021
4. Jha N, Prashar D, Khalaf OI, Alotaibi Y, Alsufyani A, Alghamdi S (2021) Blockchain based crop insurance: a decentralized insurance system for modernization of Indian farmers. Sustainability 13(16):8921
5. Lepoint T, Ciocarlie G, Eldefrawy K (2018) Blockcis—a blockchain-based cyber insurance system. In: 2018 IEEE international conference on cloud engineering (IC2E). IEEE, pp 378–384
6. Loukil F, Boukadi K, Hussain R, Abed M (2021) Ciosy: a collaborative blockchain-based insurance system. Electronics 10(11):1343
7. Nguyen T, Das A, Tran L (2019) Neo smart contract for drought-based insurance. In: 2019 IEEE Canadian conference of electrical and computer engineering (CCECE). IEEE, pp 1–4
8. Raikwar M, Mazumdar S, Ruj S, Gupta SS, Chattopadhyay A, Lam KY (2018) A blockchain framework for insurance processes. In: 2018 9th IFIP international conference on new technologies, mobility and security (NTMS). IEEE, pp 1–4
9. Reno S, Haque M (2023) Utilizing off-chain storage protocol for solving the trilemma issue of blockchain. In: Emerging technologies in data mining and information security. Springer, pp 169–179

10. Roriz R, Pereira JL (2019) Avoiding insurance fraud: a blockchain-based solution for the vehicle sector. Procedia Comput Sci 164:211–218
11. Vakilinia I, Badsha S, Sengupta S (2018) Crowdfunding the insurance of a cyber-product using blockchain. In: 2018 9th IEEE annual ubiquitous computing, electronics & mobile communication conference (UEMCON). IEEE, pp 964–970
12. Zhou L, Wang L, Sun Y (2018) Mistore: a blockchain-based medical insurance storage system. J Med Syst 42(8):1–17

Food-Health-Chain: A Food Supply Chain for Internet of Health Things Using Blockchain

Puja Das, Amrita Haldar, Moutushi Singh, Anil Audumbar Pise, and Deepsubhra Guha Roy

1 Introduction

AI, the Internet of things, and blockchain are all part of a recent change in thinking in engineering and technology. Mobile communication tools, IoT, intelligent networks, predictive analytics, AI, and blockchain, have given the healthcare sectors a new perspective and perspective. These digital interactions are merging to reshape the healthcare industry. Blockchain technology is a game-changing invention that is helping to design a new age. In 2008, a person named Satoshi Nakamoto created "Bitcoin", which is a decentralized digital money based on blockchain technology [1]. Blockchain is a distributed ledger that records digital transactions. If data is entered and stored in this network, it is visible and available to everybody, but it cannot be modified [2].

P. Das
Department of Computer Science, HMM College for Women, Dakshineswar, Kolkata 700035, India

A. Haldar
Department of Computer Science and Business Studies, IEM, Kolkata 700091, India

M. Singh
Department of Information Technology, IEM, Salt Lake, Kolkata 700091, India
e-mail: moutushi.singh@iemcal.com

A. A. Pise
Department of Data Science and Machine Learning Computer Science, University of the Witwatersrand, Johannesburg, South Africa
e-mail: anil@siatik.com

D. Guha Roy (✉)
IEM Centre of Excellence for Cloud Computing and IoT, Department of CSE (AIML), Institute of Engineering and Management, Kolkata 700091, India
e-mail: roysubhraguha@gmail.com

J. K. Mandal et al. (eds.), *Proceedings of International Conference on Network Security and Blockchain Technology*, Lecture Notes in Networks and Systems 738,
https://doi.org/10.1007/978-981-99-4433-0_30

Table 1 Comparative analysis of different distributed framework

	Transaction output	SC language	Database	Access	Consensus
Ethereum	Low	Solidity	Level DB	Using SC	PoW
Hyperledger	High	Java, Jscript, Python	Level DB	ACL	PoET, PBFT
Bitcoin	Low	Python	Web	Network	PoW

In other words, data remains unchanged and is temper resistant, once it is posted to the decentralized network. Because it is a distributed ledger, it has fewer possibilities of collapsing than a centralized system. From a commercial standpoint, blockchain is every product with its unique means of tracking, recording, and verifying information. Blockchain was first utilized as the core frame technology in cloud storage for any e-commerce [3]. This technology is eligible to be used in supply chain systems as a result of all of this. We'll go through some of the most important aspects of blockchain technology here:

- **Decentralized:** The primary characteristic of blockchain is that all data can be recorded, stored, and updated in a distributed manner, with no need for a single node.
- **Transparent**: Each node in the blockchain system transparently records data, and this transparency is maintained when the data is updated, which is why blockchain is trusted [4].
- **Immutable:** In the blockchain, all published data will be maintained indefinitely, unless and until someone can seize control of more than 51 percent of the nodes in the same amount of time.

This article will look at how supply chain companies may be benefited from this technology and how it can help them solve challenges (Table 1).

1.1 Motivation and Contribution

To make living easier, the globe has been transformed into a digital environment. The Internet of things (IoT) is the most excellent platform for connecting every tiny device or object to a large network. Because we are in an IoT network, everyone knows everything, and IoT is a little bit expensive; when a person transfers data to communicate, it becomes public. Every transaction is significant to a specific individual, yet this information, whether it is money, personal information, or health data, attracts hackers and intruders [5]. To address this, we are attempting to create a safe solution based on blockchain. The following are the contributions:

- We propose a recommended-based distributed, secure food supply chain for healthcare using the Internet of health things (IoHT).
- We presented a case study considering two different health issues and how they can be benefited from the proposed food supply chain plan.

- A smart contract-based approach is also applied to safeguard information and data.
- We offer proper food recommendations in a distributed, transparent, immutable, and secure environment.

1.2 Organization of Paper

This article is arranged as follows: Sect. 2 provides the literature review and a theoretical foundation of blockchain, among other topics. In Sect. 3, traditional supply chain model for farming is defined. The suggested system model is presented in Sect. 4. The experimentation outcomes for our suggested method in Sect. 5. Finally, Sect. 6 concludes this study.

2 Literature Survey

Satoshi Nakamoto published the first white paper on Bitcoin in 2008. The white paper described peer-to-peer electronic cash known as "Bitcoin" that allowed online payments can be made without the use of an intermediate such as a bank or a sovereign government [6]. Blockchain is the ground-breaking technology that underpins Bitcoin. Bitcoin is only one of the tens of thousands of apps that make use of distributed ledger technology. Blockchain implements a decentralized ledger with an infrastructure of emulated databases that are synchronized via the Internet and visible to anyone within the network, compared to existing databases with a central authority or administrations (like banks and public sector accountants) that handle all transactions. A blockchain network can be open to the public and available to anybody globally, or it can be private with limited membership (Fig. 1). Ethereum is another blockchain technology (edition 2.0). Vitalik Buterin was the one who came up with the idea [7].

It is built in such a manner that anybody with a basic understanding of computers may create and deploy decentralized apps on the blockchain. Its own currency, "Ether", and an Ethereum Virtual Machine (EVM). Ethereum has its own computer program, dubbed "solidity", that is used to code the apps. Aside from that, Ethereum allows you to create a "smart contract", which is a self-contained digital system with built in code that controls transactions between users on the blockchain. EVM and Ether are used to operate these transactions [8]. We can also use the blockchain platform that enables IoT devices. In a smart city, blockchain will also aid in the development of an authentication scheme that connects blockchain technology with smart devices to create a secure communication platform. Blockchain may be used in a variety of supply chain management systems to establish a decentralized network that provides transparency, security, neutrality, and dependability to all supply chain activities. Medium.io is a corporation that has developed a method for regulating the

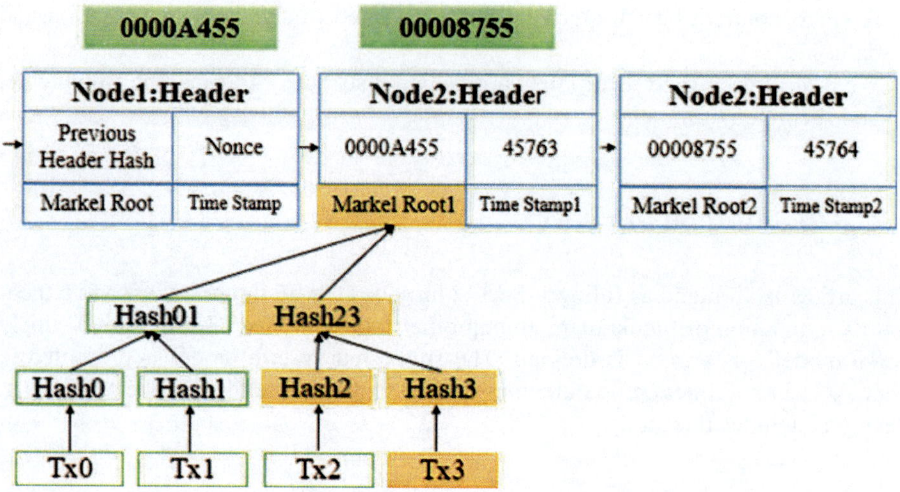

Fig. 1 Blockchain internal structure and its advantage

efficient operation of the medicine industry by integrating IoT sensor devices with blockchain technology [9]. The healthcare supply chain plays a critical role in getting food from farmers to people's plates as per patients' requirements. The major cause of inefficiency in the healthcare supply chain is a lack of mutuality and cooperation among the client or patient. As a result, a sound inventory management system is responsible for effective production, handling, distribution, and marketing, fulfilling consumer expectations, and delivering a high-quality healthcare industry [10].

3 Traditional Supply Chain Model for Healthcare

The conventional healthcare supply chain was simpler: It was continuous, and generally, it is used for business with firms that were quite close to you. The traditional healthcare supply chain consists of some units, namely develop, plan, medicine source, create, deliver, and support. Every module has its own importance, but this supply chain faces various challenges. On the other hand, we have different type of patients and their variations of food requirements. The same kind of food is provided to each patient in the general system, which is not recommended. For that healthcare industry needs a digital and secure healthcare system that must be flexible and linked.

Figure 2 shows a general supply chain at the association level within the framework of a whole supply chain setup. Every firm is located in a layering system and fits a minimum supply chain. It generally has various and changeable suppliers and customers at a similar point in time and over time [10]. The supply chain management monitors the multi-stage supply chain organization that contains farmers, brokers or middle man or agents, processing companies, and end users [11].

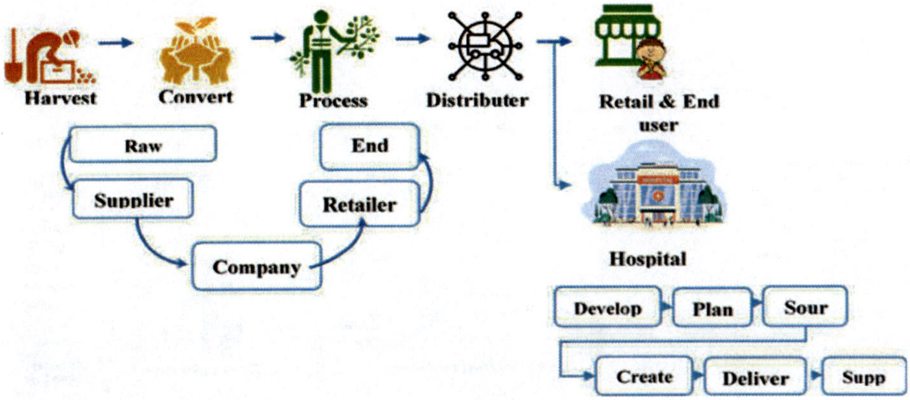

Fig. 2 Traditional healthcare system with food generation

4 System Design

In built-up systems, the supply chain is made up of a succession of schemes or modules that include people, existing resources, awareness, processes, financial agreements, and communications that make it easier to get a product from the manufacturer to the end-user or consumer. It isn't easy to get a comprehensive picture of all connections within an extensive health care-chain system because materials are stored in various locations in a typical supply chain system. As a result, the patients and end-user only have limited access to the entire system. As a result, there is a lack of openness and trust among the institutions' members. Presently, health care-chain, a blockchain-based supply chain for providing the best food as per requirement, is being announced.

In future, this object will be a decentralized distributed organization that employs blockchain technology to assemble, stock, and complete essential creation information for each product throughout its entire cycle. This dispersed block of data may potentially establish a secure, shared record of transactions for each production, as well as precise product data. Accumulating raw material, system communication mode, and user interaction mode are the three key components of the system model.

4.1 Raw Materials and Quality Assurance

Figure 3 depicts a possible future application of blockchain in the supply chain for industrial systems. The suggested solution involves a federated distributed organization that employs blockchain to collect, store, and retrieve crucial creation data throughout each invention's product lifecycle. This creates a secure, community record of each creation's dialog, as well as accurate product material. Many actors, including producers, suppliers, manufacturers, distributors, sellers, and eventually

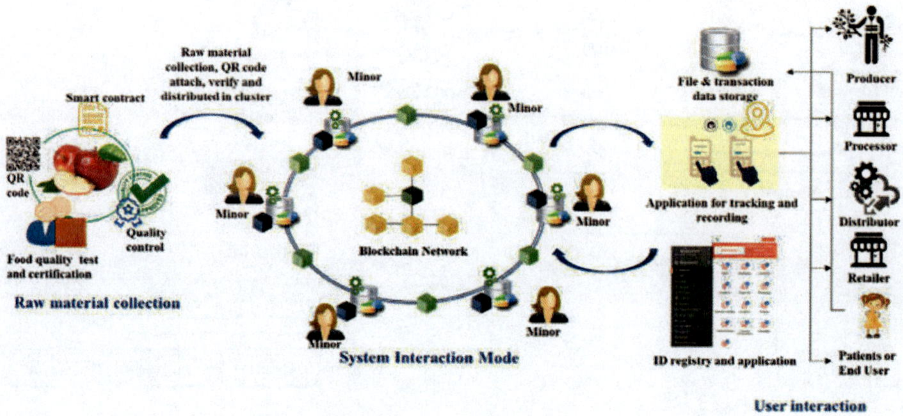

Fig. 3 Proposed blockchain-based food supply chain for healthcare system

the end customer, control how a production or raw material evolves during its life cycle.

4.1.1 Interaction With the User

The interactive query mode is divided into two parts: The first contains a store, trace, and application, while the second includes the conventional portion, which includes producers, processors, wholesalers, retailers or healthcare units, and customers or patients. The first component, on the other hand, is primarily concerned with system interaction. Those data are mostly captured and verified from the blockchain network, and they are stored and traced here. The second component is explained further down.

4.1.2 Production

These digital profiles store the most important information about the patients. This information includes things about which patients need what. Then, after completing a digital contract (smart contract) that is maintained on the blockchain, a new job is begun between the farmers and the intercessors or farming dispensation businesses, where the product is substituted. Figure 4 shows a blockchain framework.

To carry out the complete suggested process and distribution of health care—food goods from genesis to end customers, chain management systems entail many entities. As a result, tracking and tracing the entire process are complicated. To ensure total traceability, we keep track of each trade transaction from start to finish, adding the product's unique identity and lot number to each subsequent transaction and recording the hashes to keep hash network one going.

The procedural data is kept in the healthcare chain to maintain the hash chain. The data hashes are stored in the Ethereum blockchain, which circumvents the limitation.

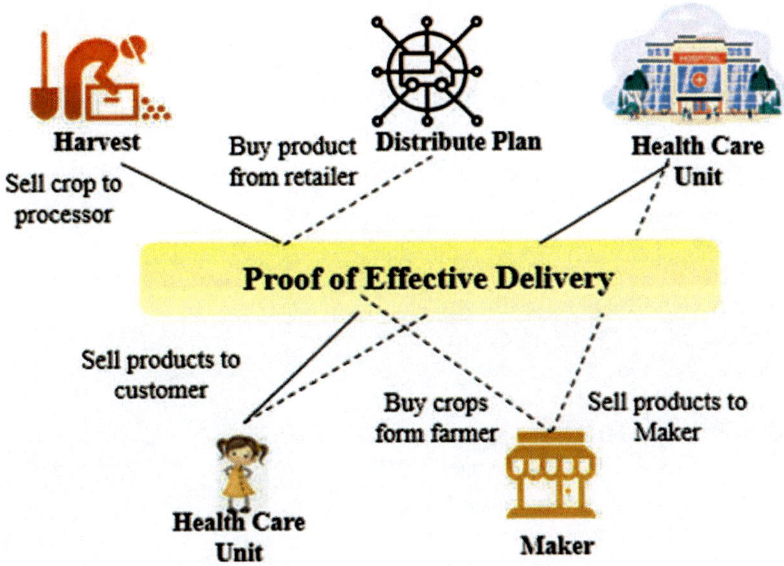

Fig. 4 Blockchain-based food transfer of patients

An access control method is used to create or decrypt messages from the blockchain, which guarantees the network's privacy and secrecy. The implementation of security techniques ensures that only the authorized user operates. Furthermore, each smart contract function may only be conducted by certain entities. Unauthorized individuals are not permitted to carry out any tasks. The entity registration process is represented by Algorithm 1, in which several supply chain entities are registered in the system and interact via smart contracts.

4.1.3 Procurement

The procurement center is now complete. To appreciate agricultural dispensation enterprises, the digital form of the development is efficient by supplying information linked to warehousing and seed transference from farmers.

4.1.4 Processing

Once the doctor has received the food routine, they will begin circulating.

Algorithm 1
Transparent Food Supply.

Input: *Add_node, msg.sender, node + +*
Output: Transparent Food Supply
INITIALIZATION
Add_node
Add Entity //**Start Identifying**
If msg.send() == **Authority** then entity initials
identity ← user_name
endif
REGISTRATION
If msg.send() = *APPROVED then*
Update node value node++
endif
ADD NEW NODE
New_node ← APPROVED user
if make_tran() = APPROVED then
Update node value node++
endif
endif
make_tran() // **Make Transaction**
REGISTER SUGGESTION
If msg_send++ *s an element from Approval list*
Addsuggestion() *Satisfied Ratings*
suggestion ++
else
Revert contract and presentation error
Savesuggestion++
Sendsuggestion()
Requggestion()
Find Suggestion()

4.1.5 Distribution as Food Requirements

Following the receipt of shipments from farming processing firms, information on the quality, loadings, logistics, and delivery is continuously updated on blockchain at precise breaks by the suppliers. It will keep track of all supplier actions when delivering products to retailers.

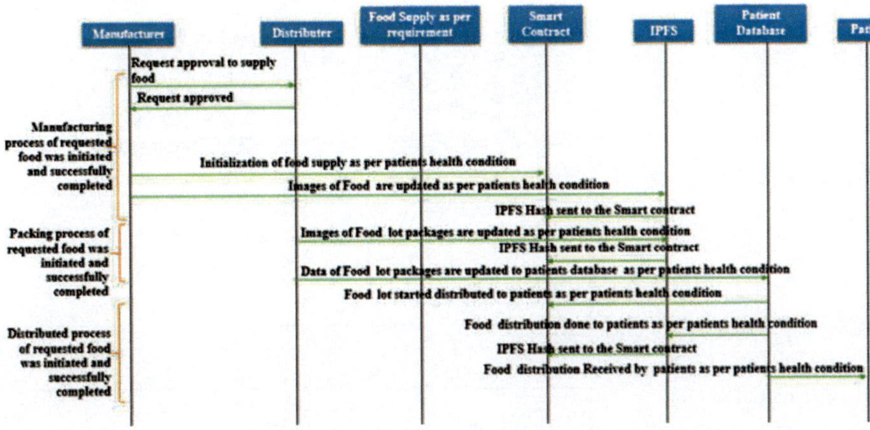

Fig. 5 Sequence diagram presentation interactions between the health entities with smart contract

4.1.6 Healthcare System

When merchants get farming bags, they roughly gather all of the information by reading the tags attached to the food packages. Because all of the information about food distribution is stored in a digital sketch on the blockchain, anyone with a blockchain-related programmer may access it.

4.1.7 The Patient or Client

At this point, the final user receives their product and provides comments. Figure 5 depicts sequence diagram that presentation interactions between the health entities with smart contract.

5 Experimental Analysis

We are trying to segregate the foods based on the health hazards the food contains, by implementing a QR code for every batch of product. The QR code is an effectual transmission medium for information. QR codes are very popular in applications like product traceability, advertising, mobile payment, passport verification, and a lot more. Figure 6 represents the QR code in which we have stored the necessary details of a particular batch of shipping cases which contains harvested agricultural products which are clustered into batches based on the patient's food types, for example, food like sugarcane, honey, sweet potato (potato family products) can be marked with its health benefits/hazards. The product details like "Product ID", "Product Code", "Product Manufactured Date", "Product Unit of Measure", "Product Price", and

Fig. 6 Represents QR code of a product

PID:1010, PCode:Potato,PMFGDate:30122021,PPrice:15RS,PUOM:KG,HHTYPE:DIAB	PID:4110, PCode: Red lentil, PMFDate:30122021,PPrice:40RS,PUOM:KG,HHTYLE:URIC

Fig. 7 Scan output of QR code

"Health Hazard Type" are embedded in the QR code, which can be tracked from any node in the supply chain. The QR code can be encrypted using AES-265 for further security, with the help of custom logic and keys (Fig. 7).

5.1 Test Run

We selected mainly two food items to test track our application: (1) potato and (2) red lentil. Both potatoes and red lentil are sourced from local farmers. The products are then shifted to a warehouse after gathering them from the farms and are sold at the farmer's market. For the simulations, an open-source blockchain technology, namely Ethereum, is employed. We executed this as a Web service (SaaS) on Remix Integrated Development Environment (IDE), Ganache, and Meta mask and generated the QR codes, which contains all the relevant information related to the product, to evaluate the performance of a blockchain-based supply chain network. Remix makes it easier to write, execute, and test smart contracts. Figure 8 shows the cost unit intake by the proposed system.

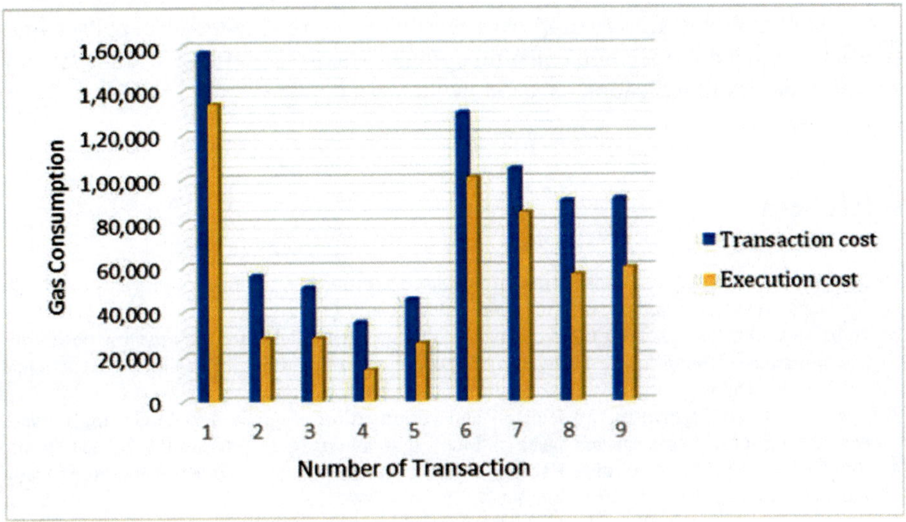

Fig. 8 Cost unit intake by the proposed system

We imitated the production in the above-mentioned farms in the decentralized application platform to reproduce the entire plot of the real-world food manufacturing system. For both potatoes and red lentils, each of them has hundred kilograms of product are accumulated into the same block and blockchain consisting of two main blocks "the genesis block and the hash of genesis block" which is generated for each individual batch. Since each batch of product made it to the farmer's market via the distribution center, each batch is processed in such a way that they have three blocks in the chain. The "genesis block" is the "hash of the genesis block" when it was sent to the distribution center and the "hash of the block" from the distribution center after it reaches the agriculturalist's marketplace.

6 Conclusion and Open Challenges

Blockchain technology can potentially change many existing traditional systems into more secure, decentralized, transparent, and collaborative systems while empowering individuals. We analyzed some of the critical aspects of blockchain technology and explored possible application fields in this paper. The article's attention was shifted to the applicability in healthcare specifically Internet of health things (IoHT)-based supply chains. A blockchain-based health supply chain was presented. This technique proves a smarter option for the health system for food suggestions. This degree of input may help this industry enhance its technology and food suggestions for individuals. Smart contracts may be integrated into this system to increase data integrity.

As part of ongoing efforts to improve the effectiveness of healthcare supply chains, we want to expand the recommendation system to achieve end-to-end transparency and verifiability of medication usage in the hereafter.

References

1. Nguyen DC et al (2021) Blockchain and AI-based solutions to combat coronavirus (COVID-19)-like epidemics: a survey. IEEE Access 9:95730–95753
2. Guha Roy D et al (2021) A blockchain-based cyber attack detection scheme for decentralized internet of things using software-defined network. Software: Practice and Experience 51(7):1540–1556
3. Liu P et al (2020) Investment decision and coordination of green agri-food supply chain considering information service based on blockchain and big data. J Clean Prod 277:123646
4. Roy DG et al (2019) QoS-aware secure transaction framework for internet of things using blockchain mechanism. J Netw Comput Appl 144:59–78
5. Sanal Kumar KP et al (2021) Security and privacy-aware artificial intrusion detection system using federated machine learning. Comput Electric Eng 96:107440
6. Barbosa MW (2021) Uncovering research streams on agri-food supply chain management: a bibliometric study. Global Food Secur 28:100517
7. Lee G-S et al (2020) A blockchain system for history management of agrifood. J Korea Acad Industr Cooper Soc 21(10):159–165
8. Das P et al (2021) Impact of blockchain-based cyber security implementing industry 4.0. Industry 4.0 Interoperability, Analytics, Security, and Case Studies. CRC Press, 13–34
9. Balasubramanian S et al (2021) A readiness assessment framework for Blockchain adoption: a healthcare case study. Technol Forecast Social Change 165:120536
10. Kuo T-T et al (2021) Benchmarking blockchain-based gene-drug interaction data sharing methods: a case study from the iDASH 2019 secure genome analysis competition blockchain track. Int J Med Inform 154:104559
11. Das P et al (2023) Block-a-city: an agricultural application framework using blockchain for next-generation smart cities. IETE J Res 2023:1–11

Sentimental Analysis for Social Media Topic Analysis Using Multi-tweet Sequential Summarization

A. Pandiaraj, R. Venkatesan, K. S. Chandru, and G. Vimalsubramanian

1 Introduction

Twitter fills in as a data conveying stage where it gathers a large number of tweets each day. In any case, a few clients, particularly new clients, regularly think that it is hard to comprehend moving points in Twitter when facing overpowering and sloppy tweets. Existing work has endeavored to give a short piece to clarify a theme; however, this just gives restricted advantages and can't fulfill the clients' assumptions.

Moving Topic Analysis framework gives a profound examination through point versatile slant arrangement and multi-tweet consecutive outline, which expects to give a sequential of sequentially, requested short sub-rundowns for a moving theme to give a total tale about the improvement of the subject while holding the request for data show. Correspondence specialists are utilizing Twitter information to gauge the beat of popular assessment and create business insight for which the Trending Topic Analysis framework will be of incredible worth [1].

A. Pandiaraj (✉)
Department of Computing Technologies, SRM Institute of Science and Technology, Kattankulathur, Chennai, India
e-mail: pandi.mnmjain@gmail.com

R. Venkatesan
Department of IT, Bannari Amman Institute of Technology, Sathyamangalam, India
e-mail: venkatesanr@bitsathy.ac.in

K. S. Chandru
Department of CSBS, Bannari Amman Institute of Technology, Sathyamangalam, India
e-mail: chandruks@bitsathy.ac.in

G. Vimalsubramanian
Department of CSE, Kalasalingam Academy of Research and Education, Krishnan Kovil, Srivilliputhur, India
e-mail: vimalsubramanian@klu.ac.in

J. K. Mandal et al. (eds.), *Proceedings of International Conference on Network Security and Blockchain Technology*, Lecture Notes in Networks and Systems 738,
https://doi.org/10.1007/978-981-99-4433-0_31

One of the uses of moving subject analyzer discovers its place in political spaces where an ideological group can follow the public's perspectives on subjects of their advantage which had been moving once and use it work on their political status. Financially content generators to play out a profound investigation of any moving subject under examination and use to further develop their business advantages could utilize it. As of late, the Trending Topic Analysis framework additionally assumes an urgent part in Research regions because of the expanded ubiquity of Twitter [2]. This investigation cum synopsis framework could fill in as a rich instrument for zeroing in just on critical spaces of subjects and beat the staggering idea of tweets [3].

2 Literature Survey

Kumar and Bala [4] proposed a consolidated methodology of passage reinforcer calculation along with WSD and Matter Inference strategies to produce a 1-line listing. Passage reinforcer calculation plan at creating a chart that support in distinguishing the frequently focus content by essentially looking heavily huge arrangement of ways in chart. This technique came up short on the mundane idea of rundowns and made realization problems produced in the outline.

Kim et al. [5] performed the summary using a non-invariable Bayesian model practical to HMM. It was expected that a sharp discriminant model would allow for situatedness as a function of selected farsighted features of individualistic tweets. A critical centering was the analysis of the attempt to use a short measure data model known as the HDP-HMM to deal with a flood of tweets identifying with an individual topic tweets. However, the compact provided didn't match the general character.

Zarrad et al. [6] proposed an idea-based improvement structure for substance report using ILP. Quarry data contained novel and tweets close by web content pertinent to the theme. The accentuation was not on underdeveloped new layout structures but rather commute and planning grouped text informant to make summations that are many helpful anyway there was a shortfall of broadcast of sub-circumstance conspicuous confirmation to demonstration the subject propelling cycle.

Wu et al. [7] explained Twitter topics by giving a straightforward overview of the news. An exploratory travel app for Twitter called Tweet Pattern composed communicate by common basic status, and the result is substance processed through an unfaceted query. The topic derivation group relied on grammar isolation, language representation, close twinned ranges and heuristics to bedding topics. Regardless, the structure required transient perspective in traces made and substance protrusion not noted.

Saravanan et al. [8] created successive outlines on a theme utilizing Stream and Semantic-based methodologies. In any case, the methodology missed the mark concerning the coherence of synopses produced. The framework neglected to catch the assessment communicated in the information consequently prompting clashing synopses.

3 Proposed System

The proposed framework "Moving Topic Analyzer" is cultivated through subject versatile opinion arrangement and multi-tweet synopsis. The proposed framework targets creating sequentially requested sub outlines, which tosses light in understanding the point advancement of the moving themes. It is expected that the item under investigation will contain some secret subtopics, which will be revealed with the proposed modeling.

The selection of final information is done by excerpt moving parts on a general. It is arrogated that the topic under investigation contains some secret substance which are uncovered using the planned framework. The client to absolutely analyze a moving subject to a striking degree of particular. A basic pre-preparing is carried out to establish the quarry aggregation that can be utilized. Near the pre-preparing phase, the proposed system also undertakes the translation of non-English tweets, which is imperative to prevent the discarding of the general evaluation of the moving topic.

Pre-handled information is subsequently dissected through theme versatile slant order to recognize the popular assessment. Assessment named information will then, at that point be prepared for the successive outlines. Subtopic acknowledgment is refined using a stream-based system and a semantic-based strategy. Finally, a sub-summation check is performed, which provides the top rated tweets that combine certain original segments of tweets. A repeat check is performed to remove duplicates from the selected tweets, followed by a threshold check to ensure a meaningful mix of customer perspectives.

3.1 Proposed Architecture

The proposed architecture is shown in Fig. 1. The planned hypothesis canvas an inclination substance based on a specific neighborhood. The carrying out of the proposal can be narrowly divided into tierce conception. The collection of reference point aggregation, which includes the extraction of trending topics and the filtering of the designated substance for encourage investigating. This is followed by the derivation of tweets supported on the substance.

Eventually, the successive summary stage in this task addresses the problem of theme development. Latent themes are invisible in the datasets, which requires an economic model for theme perception. In this module, course-based subtopic uncovering hypothesis and meaning-supported subtopic perception hypothesis help to detect subtopics within an inclination topic. The detailed system architecture is outlined in Fig. 1.

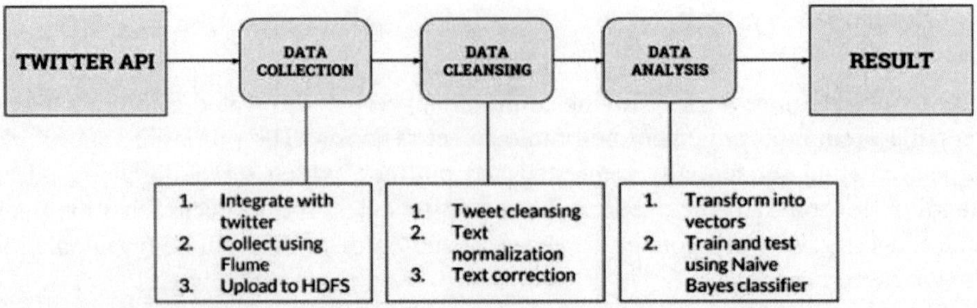

Fig. 1 System architecture of trending topic analyzer

3.2 *List of Modules*

The proposed system is composed of the following modules:

1. Data extraction module
2. Pre-processing module
3. Feature extraction module
4. Sentiment classification module.

3.2.1 Data Extraction Module

Moving focuses are removed based on the neighborhood "WOEID", which unusually perceives every domain in the universe. The Chitter API allows the derivation of examples from the neighborhood that are well suited. The eliminated example is scaled down to match the need for the topic to have a few subtopics tucked inside. Moving themes are included in the information index. When moving topics are added to the database, they are named with the month span in which they were moved. The tagCurrentMonthRange() method referenced in the previous estimation plays out the recently referenced assignment. The steps for extracting moving topics are shown in Fig. 2.

Following algorithm discusses the steps involved in Trending Topic Collection.

Algorithm Trending Topic Collection

Input: Target Place's WOEID

Output: Set of Trending Topics

Begin:

Connect to Twitter ()

for each trend do()

add to target topic database

end for

end

Function ()

Tags a month range for the set of trends extracted

Function ()

Gets top trending topics of that region

WOEID mentioned in method *getTrends()* decides the region of interest to extract top ten trends per request.

3.2.2 Pre-processing Module

Tweets basically contain a lot of droning placid and insufficiently outlined tidings controlled characters. Therefore, cleansing up and setting up the critical data is fundamental to any openhearted of investigation that will be done with the data. Treatment of tweets relate further processing of the data: removing URLs—URLs and associations are removed from the content of the tweet. Replacing slang words—slang words such as LOL, OMG are replaced with equivalent English words, e.g., "Laugh

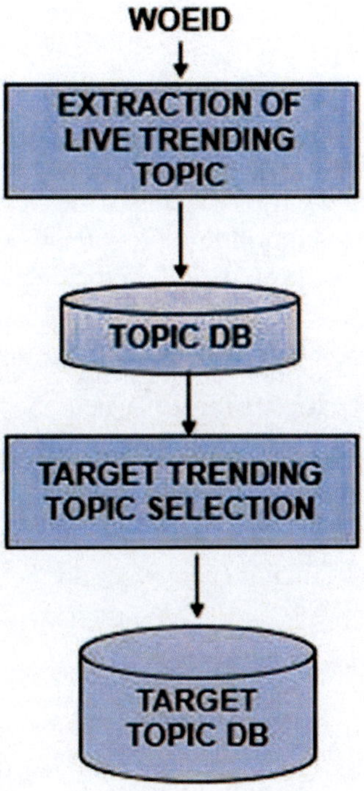

Fig. 2 Trending topic collection

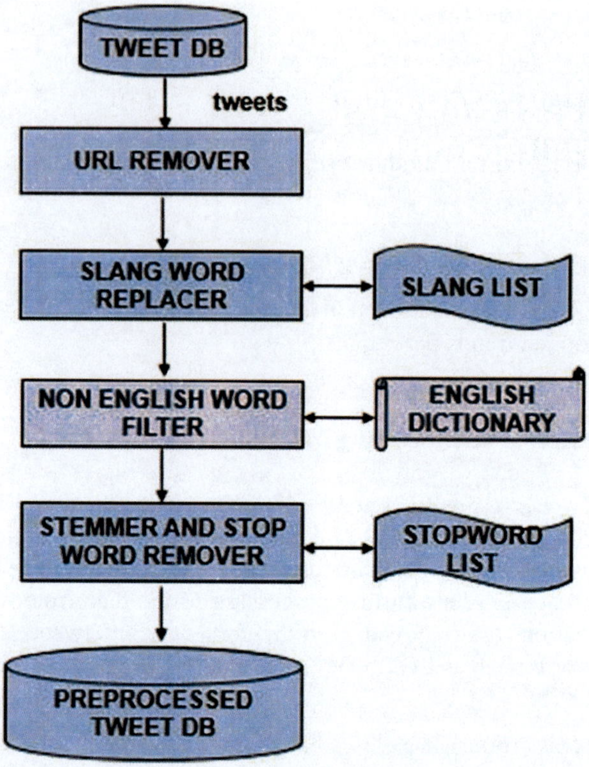

Fig. 3 Tweet pre-processing

wildly", "Oh My God", etc. Non-English words are filtered out using a word reference. Figure 3 shows Stemmer and stopword cleaning English words are stemmed and only the stems of the words are recorded. Stop words in the tweets are taken out.

Algorithm below details each of the above-mentioned pre-processing steps and results are stored in the database for further modules to act on.

Algorithm Pre-processing of Tweets

Input: Target Tweets

Output: Pre-processed target tweets

Begin

For each tweet do

If(.())

Remove

End if

For each word in tweet do ()

End for

For each word in tweet do ()

End for

For each word in tweet do ()

End for

End

3.2.3 Feature Extraction Module

Component extraction engaged with this possibility makes a huge outcome on the demonstration of the speculation grouping undertakings. Tweets have an uncommon nature not at all like typical English sentences like @ image used to allude to a client, emoji's communicated in the tweets, and so on fusing such components while performing highlight extraction upgrades the general interaction. Figure 4 shows the collaboration referred to.

Pre-processed tweets are selected for text feature where POS Tagging is applied as shown in Fig. 5. Each word will be tagged with its respective parts-of-speech.

Fig. 4 Text feature extraction

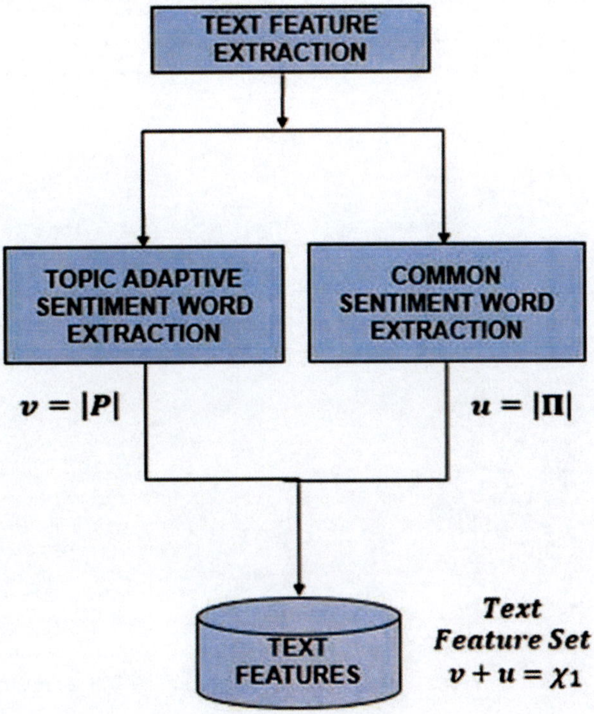

Most frequent verbs, nouns, and adjectives are taken as the text features. Figure 6 displays the verbs, nouns, adjectives with its count.

Words that satisfy a threshold will be filtered and treated as Topic Adaptive words for the set. Same procedure is repeated for the topics under study.

Non-Text Feature Extraction

Non-message features fuse the @-network features, customer assumption parts, and emoticons. These parts get the excellent thought of tweets as components. In Twitter, @ picture in a tweet is used to imply a customer whom an individual would tweet to for instance @abc suggests that the tweet is insinuating the customer ABC. Utilizing

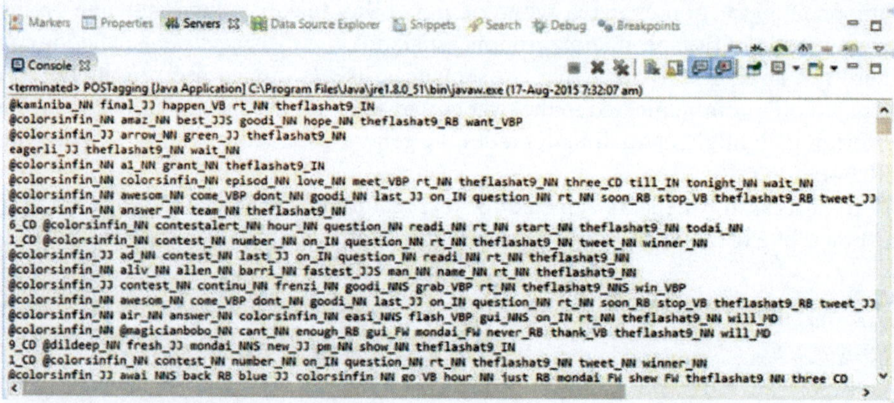

Fig. 5 Parts-of-speech tagged tweets

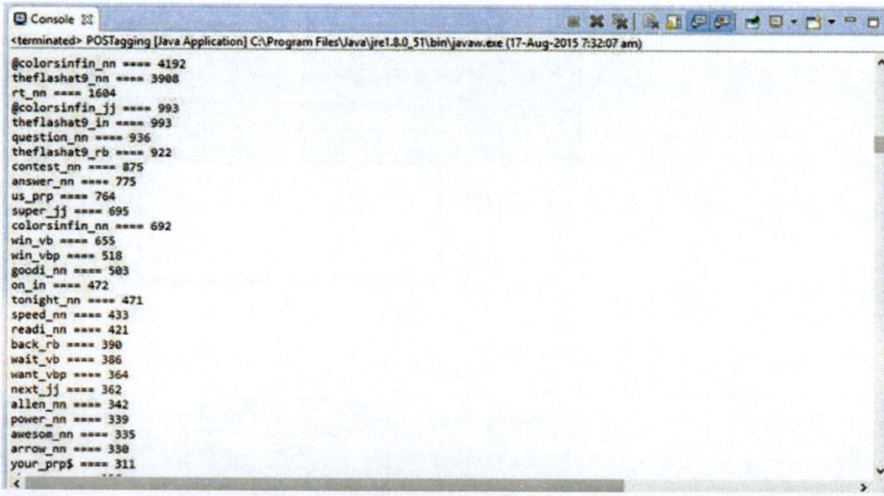

Fig. 6 Text features with frequency count

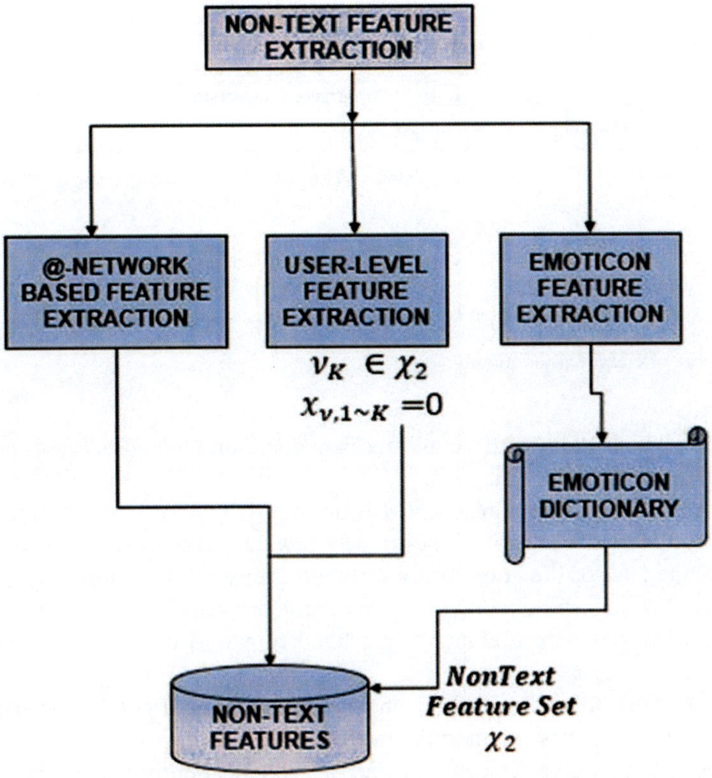

Fig. 7 Non-text feature extraction

this image helpful components can be removed consequently taking out the need to mark the whole dataset. Figure 7 shows the three kinds of non-text highlights extricated [9].

3.2.4 Sentiment Classification Module

In conventional systems, compact may contain contradictory placid that decrease the comprehensibility and overall striking of the compendious produced [3]. Idea request is a subject-irritable endeavor, i.e., a classifier arranged from one substance will accomplish more awful on some other. This is particularly an issue for the tweets notion examination. Since the themes in Twitter are exceptionally assorted, it is difficult to prepare a general classifier for all points. In addition, contrast with the item surveys, Twitter needs information naming and a rating instrument to obtain slant marks. The very scanty message of tweets additionally cuts down the presentation of an opinion classifier. In this paper, we propose a semi-managed point versatile slant

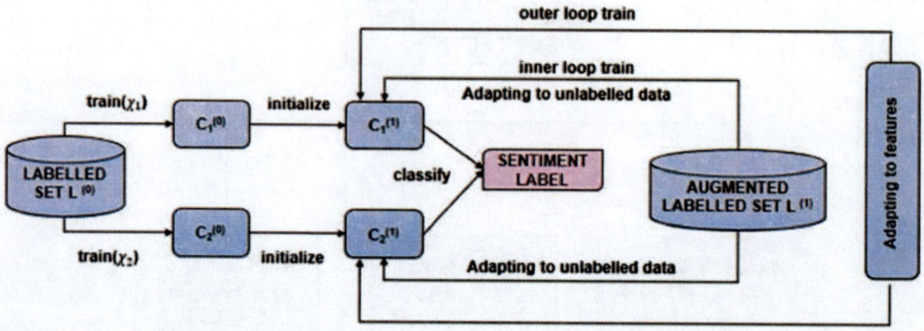

Fig. 8 Sentiment classification module

order model, which starts with a classifier dependent on typical segments and mixed stamped data from various subjects.

Figure 8 shows semi-managed SVM is quite possibly the fewest encouraging to use unlabeled information consolidating with few named ones since SVM limits the primary danger. Also, the community-oriented preparing (co-preparing) system is an elective covering and accomplishes a decent presentation, which is frequently utilized in situations whose elements are handily parted into various perspectives [10].

The framework takes the textual and non-textual highlights, $\times 1$ and $\times 2$ as autonomous perspectives for modify. In the co-preparing plan, two categorize C1 and C2 are created depending on $\times 1$ and $\times 2$ independently using labeled output information L. The comparative highlight values are atmosphere as x and x's separately for matter and non-matter include values. The non-labeled information is cooperatively selected to increase the labeled information index L used for the following cycle. Similarly, the classifier prepared on the consolidating marks using the increased marked information index L acquires the final estimation order result. Figure 9 portrays the interaction stream of the calculation followed by the TASC learning calculation is depicted where adjusting to unlabeled information and adjusting to highlights is performed iteratively returning the model, expanded marked information, and components [11].

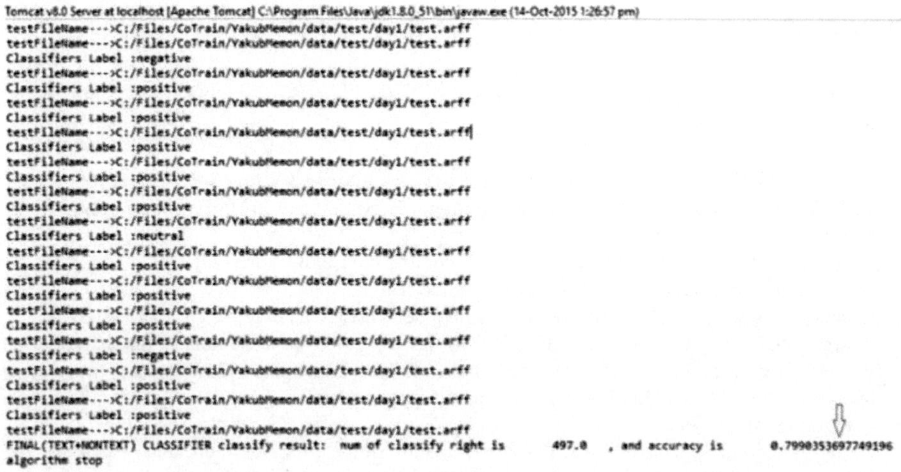

Fig. 9 Sentiment classification results

4 Conclusion

Web-based media is an essential piece of the world and it will just continue to assemble its impact later on. Examination of this humongous datasets can work on a huge number and enterprises. We have utilized one method of investigation called opinion examination; there are numerous approaches to work on its precision by conveying huge datasets thinking about the emoji's and internationalization. Numerous modern blower and channel calculations will assume a significant part in connecting the language obstruction and in taking care of loud substance and poorly shaped words. Supposition investigation of Twitter utilizing huge information assisted us with examining tremendous measure of datasets. Apache Spark is a quick and universally useful group registering framework. It gives significant level APIs in Java, Scala, Python and R, and an advanced motor that supports general execution charts. It additionally upholds a rich arrangement of more elevated level instruments including Spark SQL for SQL and organized information preparing, MLlib for AI, GraphX for diagram handling, and Spark Streaming. Another element we should investigate is whether the data on the relative location of words in a tweet has an impact on the exhibition of the classifier.

In this examination, we are zeroing in on broad notion investigation. There are ways of working in the field of investigating assumptions with reasonably well-known settings. For example, we have seen that customers largely use our site for certain types of keywords that can isolate two or three particular classes, namely: governmental issues/legislators, VIPs, items/brands, sports/athletes, and media/films/music. So we can endeavor to perform separate notion examination on tweets that just have a place with one of these classes (for example the preparation information would not be general however explicit to one of these classes) and analyze the outcomes we get in the event that we apply general opinion investigation on it all things being equal.

References

1. Sun B, Ng VT (2014) Analyzing sentimental influence of posts on socialnetworks. In: Proceedings of the 2014 IEEE 18th international conference on computer supported cooperative work in design
2. Ohmura M, Kakusho K, Okadome T (2014) Social mood extraction fromtwitter posts with document topic model. In: 2014 international conference on information science and applications (ICISA)
3. Sanjay R (2013) Big data and Hadoop with components like Flume, Pig, Hive and Jaql. In: International conference on cloud, big data and trust
4. Kumar M, Bala A (2016) Analyzing Twitter sentiments through big data. In: 2016 3rd international conference on computing for sustainable global development (INDIACom)
5. Kim JS, Yang MH, Hwang YJ, Jeon SH, Kim KY, Jung IS, et al (2012) Customer preference analysis based on SNS data. In: 2012 second internationalconference on cloud and green computing, pp 106–113
6. Zarrad A, Jaloud A, Alsmadi I (2014) Theevaluation of the public opinion. In: IEEE/ACM 7th international conference on utility cloud computing
7. Wu Y, Ren F (2011) Learning sentimental influence in twitter. In: International conference on future computer sciences and application
8. Saravanan M, Sundar D, Kumaresh S (2013) Probing of geospatial stream data to report disorientation. In: IEEE recent advances in intelligent computational systems (RAICS)
9. Pandiaraj A, Prakash SL, Kanna PR (2021) Effective heart disease prediction using hybridmachine learning. In: 2021 third international conference on intelligent communication technologies and virtual mobile networks (ICICV), Tirunelveli, India, pp 731–738. https://doi.org/10.1109/ICICV50876.2021.9388635
10. Pandiaraj A, Sundar C, Pavalarajan S (2021) Sentiment analysis on newspaper article reviews: contribution towards improved rider optimization-based hybrid classifier. Kybernetes 51(1):348–382. https://doi.org/10.1108/K-08-2020-0512
11. Pandiaraj A, Venkatesan R, Manochitra S, Lakshmanaprakash S (2022)Neural network based approach on sentimental analysis using herb. In: 2022 4th international conference on smart systems and inventive technology (ICSSIT), Tirunelveli, India, pp 1092–1100. https://doi.org/10.1109/ICSSIT53264.2022.9716505

Development of IoT-Based Biometric Attendance System Using Fingerprint Recognition

Prasun Chowdhury, Debnandan Bhattacharyya, Ritaban Das, Sourav Kr. Burnwal, and Asis Prasad

1 Introduction

Being in digital era, one of the popular and extensively used biometric techniques is fingerprint verification. It is a fact that every human being is born with unique pattern on their fingers and that is the sole motivation for developing fingerprint verification-based systems. Manual attendance system has many drawbacks. Manual attendance tracing and report generation are a tedious task, and chances of error are more. Backup of the record is even more costly and any attendance processing (leave calculation, attendance percentage calculation, etc.) need to be manually done. Because of cost effectiveness and simplicity, fingerprint-based biometric systems are widely used. It is a solution to get rid of boring and inefficient manual attendance system. However, biometric system ensures physical presence of the candidate much vividly than any other system. The objectives of this paper are

- To improve the existing attendance system process to fully-computerized and automated attendance system using Internet of things (IoT).
- To design a featured Webpage connected with remote database to keep track the attendance record. Both, the candidates as well as the admin can access the page and obtain required attributes and information.
- To build an automated leave management system (LMS) where the candidates can apply for leave and the admin can approve or reject the leave. The notice management system (NMS) simplifies the task of the admin to publish public notices.

P. Chowdhury (✉) · D. Bhattacharyya · R. Das · S. Kr. Burnwal · A. Prasad
Department of Electronics and Communication Engineering, St. Thomas' College of Engineering and Technology, Kolkata, India
e-mail: prasun.jucal@gmail.com

J. K. Mandal et al. (eds.), *Proceedings of International Conference on Network Security and Blockchain Technology*, Lecture Notes in Networks and Systems 738,
https://doi.org/10.1007/978-981-99-4433-0_32

2 Literature Survey

In [1], an embedded computer-based lecture attendance management system was proposed. Though it was an improved system with electronic card reader serially interfaced to a PC, the demerit of such system is that someone can take proxy attendance for another person if given the person's electronic card. In [2], authors used a wireless attendance management system that authenticates using the iris of the individual. This system uses an off-line iris recognition management system for image capturing, extracting precise details, storing and matching the captured image with the one stored in database. This system takes care of wrong clocking in or buddy punching. Buddy punching is when one worker or student inappropriately clocks in for another. The only problem of this biometric system is that people usually have the fear that the Iris scanner, which sometimes might contribute to the damage of their eye and so tend not to embrace it. The paper [3] proposed a model for implementation of an automated attendance management system for students of a class by making use of face recognition technique, using Eigen face values, principal component analysis (PCA), and convolutional neural network (CNN). After these, the connection of recognized faces is established by comparing with the student's faces from database. Problems with this system is that it may be failed to distinguish between similar or twins' faces. Aging effect of persons may lead to update of face-images for same individuals. The authors in [4] designed and implemented a system which authenticates the users based on passwords. But this system could not eliminate impersonation since the password can be shared or tampered. Passwords many times can be forgotten, or the system can be hacked, thereby preventing users to access the system. The attendance system proposed in [5] uses the STC89C52 microcontroller as the main control chip, 12,864 LCD as the man–machine interface, matrix keyboard to input student ID and fingerprint-identification module as sensors. Drawback of the system is that keyboard is required for data input as well as additional circuitry has to be implemented. Apart from the above-mentioned techniques, there are solutions such as RFID-based authentication system and GSM-GPRS-based authentication system. The GSM-GPRS-based systems use the location of class for attendance marking which is not dynamic. Thus, wrong attendance might be recorded if there is a change of venue. The authors of paper [6] introduce the concept of RFID-based communication; voice data-based analysis with biometric-based properties. The system improves payroll management system, leave management system, and performance appraisal also. But problems of RFID-based authentication systems are that RFID cards can get lost, stolen, and it requires the installation of RFID detectors. RFID cards cannot eliminate impersonation also. However, the fingerprint-based attendance system is a cost effective and simplified means of identification. The fingerprint is distinctive to each individual. Even identical twins do not share exactly similar fingerprint features, and it cannot be transferred, lost, or forgotten like the password. It allows employees to register for attendance with ease and eliminate

Table 1 Review of some related works

Technology	Description	Advantage	Disadvantage
RFID and android	This monitoring system uses RFID to verify the attendance, and the recorded data can be accessed using an application developed using Android [7]	1. System can be accessed remotely 2. Performance graph will be generated 3. RFID cards are difficult to tamper	RFID cards can be exchanged, and hence, the system does not ensure physical presence of the candidate strongly. Whereas fingerprint-based systems ensure physical presence
Internet of Things (IoT)	Hardware of the system includes ARM9 S3C2440 processor board, FPS200 solid state fingerprint sensor. Database is designed using SQLite database management tool. Additionally, vein recognition is also used [8]	1. Additional vein recognition is used 2. Information can be accessed anywhere	Software system architecture is very complex and hard to implement for a large network. This can be simplified
LabVIEW	The system is based upon 8051 microcontroller, R305 optical fingerprint sensor and LabVIEW [9]	1. User friendly 2. High speed 3. Efficient and low-cost embedded platform	The system is suitable for small database only. Remote database setup can eliminate this problem

errors because the system generates exports at the end of specific period. The advantage of this system is that it can work as a standalone system unlike other fingerprints identification systems already in existence.

Many researchers have implemented fingerprint-based attendance system which makes use of a fingerprint sensor along with other technologies. Reviews of some such techniques have been discussed in Table 1.

3 System Design

The proposed system has two distinguishable parts: the hardware circuitries (containing the fingerprint sensor module and the OLED display) and the software tools to maintain the database and user interface (UI).

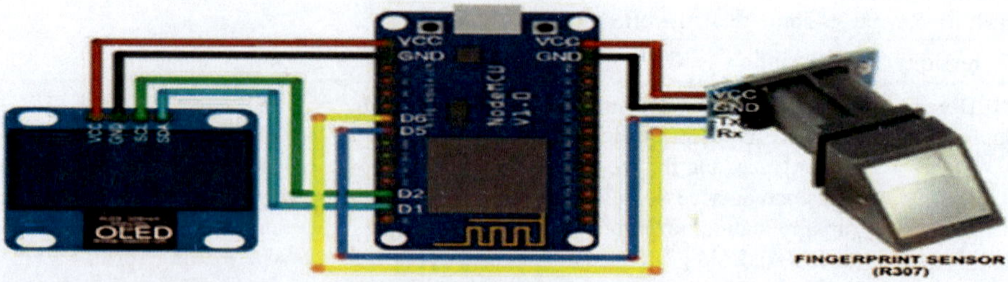

Fig. 1 Circuit diagram of the system

3.1 Hardware Requirements

Hardware implementation of the system consisting of central processing unit, fingerprint sensor module, and display module. Figure 1 shows the circuit diagram comprising of different hardware components like processing unit (Node MCU), fingerprint sensor module (R305), display module (OLED) [10].

3.2 Software Requirements

XAMPP is used as PHP development environment that contains MariaDB, PHP, and Perl. Frontend design of the Website is done using HTML5, CSS3, JavaScript, and Bootstrap4. Background running PHP codes connects frontend with the backend using structured query language (SQL).

3.3 System Operation

The PU communicates with the server machine using IP address which sets up a point-to-point network communication between these two. The interaction between server and remote database is carried out using php codes running in the server machine. Block diagram of the proposed system is given in Fig. 2. It should be noted that, the processing unit (NodeMCU) communicates with only one php code though there are multiple concurrent processes running simultaneously. The system runs on different stages: user enrollment, authentication, attendance processing, and leave management. Database interaction is a common process associated with every stage of the system.

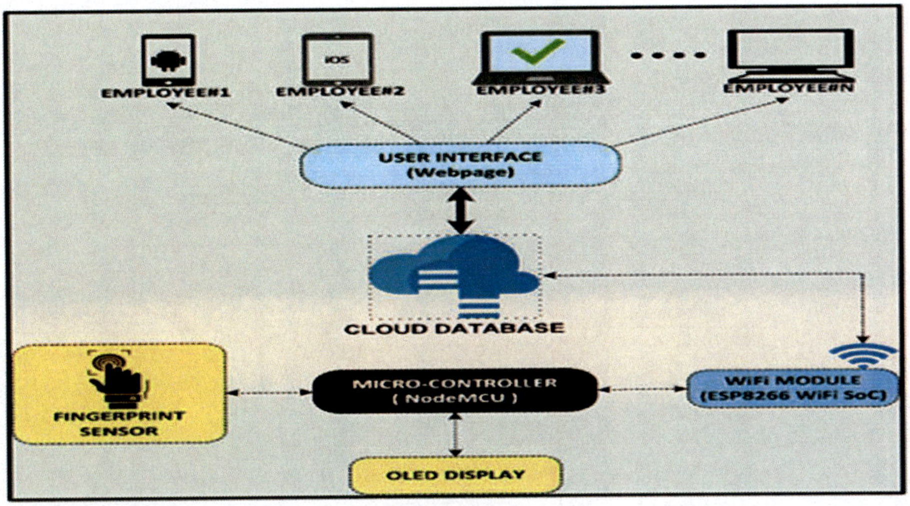

Fig. 2 Block diagram of IoT-based attendance system

3.4 Database Interaction

Database interaction is done by HTTP. The server has its IP address and the processing unit communicates with the server using that IP address. This is done by connecting the NodeMCU to Internet and initializing HTTP-client that establishes connection with predefined link containing the IP address of the server. This communication uses 'POST' method to transfer payloads. Payloads are the featured string that contains sensitive information while the system is running. All the data and flow controlling signals are transferred in the form of payloads.

3.5 User Enrollment, Modification, Deletion

Admin of the system has the authority to add new user or modify/delete pre-existing user data. To do so, admin first login into the system with admin credentials and navigate to 'manage user' section. Flowchart of the process is given in Fig. 3. User ID is a unique identification number for every user. To add user, admin need to enter new User ID and other information (name, date-of-birth, gender, email, role in organization hierarchy, designation, joining date, etc.) about the user. Server enters the information in database and the php code interacting PU, 'POST' the User ID to start fingerprint scanning. Thus, the enrollment process is done. Modification or deletion in user data is similar process in which the admin needs to choose the User ID to carryout modification or deletion.

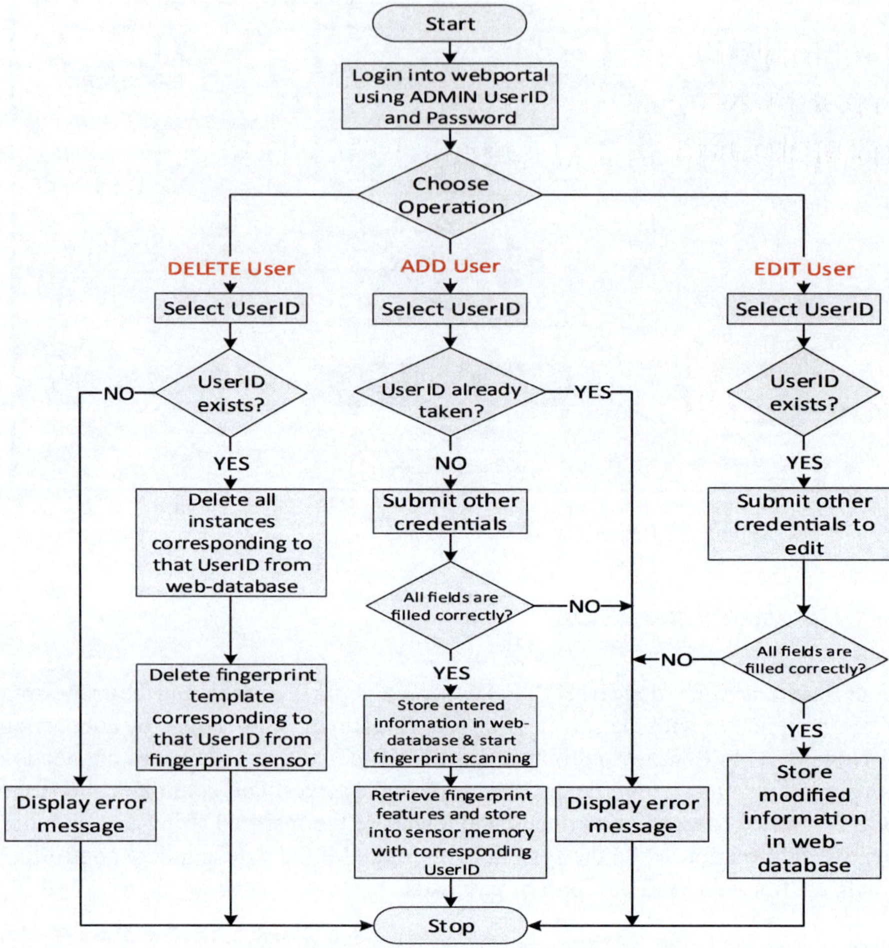

Fig. 3 Flowchart of registration process

3.6 *Authentication and Attendance Processing*

Authentication is done when a user places his/her finger in the fingerprint sensor to register attendance. Flowchart of the authentication process is given in Fig. 4. Initially, the fingerprint sensor captures the fingerprint of the user by illuminating the curves and ridges of the finger. Then, it tries to find the template in the local database. If any match is not found, the OLED displays error message that 'user is not registered'. If a match is found, the sensor sends the User ID of the matched user to the PU; and in turn, PU posts the User ID to the server. For a received User ID, the server looks for an entry with that User ID in the current date. If there is no such entry in Attendance Log table, the server adds an entry with the received User ID, current date and Time-IN attribute as timestamp and 'welcome' message is displayed

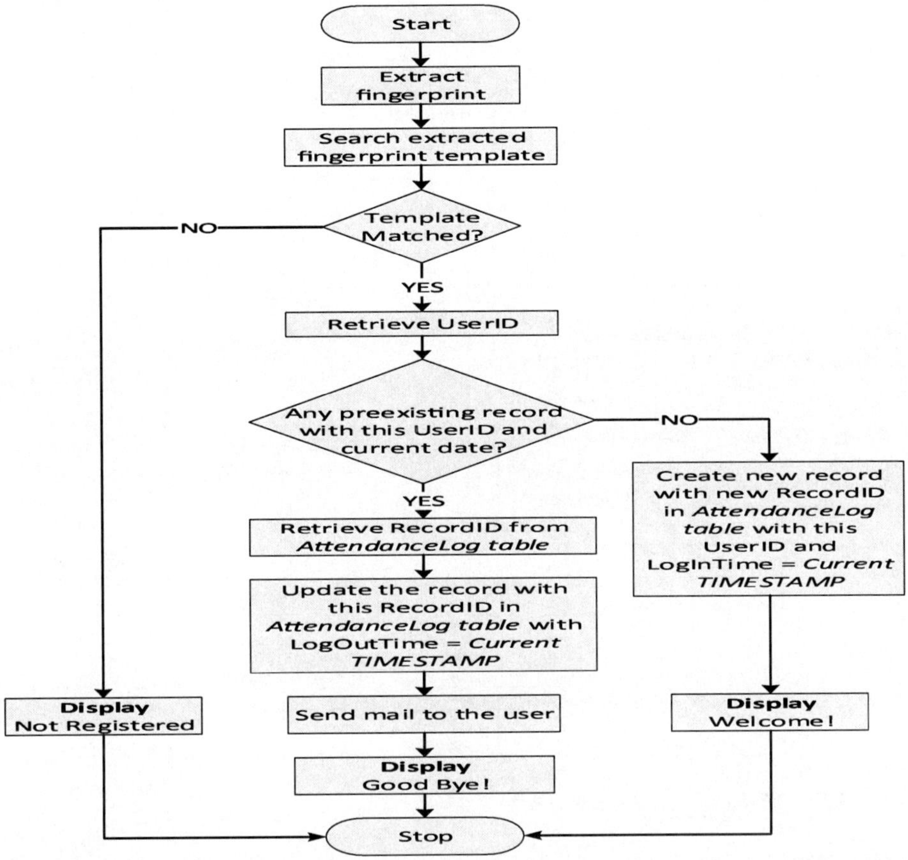

Fig. 4 Flowchart of authentication process

in OLED display. Otherwise, if such an entry is already found, the Time OUT is set to current timestamp, and 'Good Bye' message is displayed in OLED display. This system does not allow any duplicate entry of user attendance in Attendance Log table. A user can place his/her finger in fingerprint sensor as many times as he/she wants in a day; but the first occurrence is considered to be Time-IN, while the last occurrence is considered to be Time OUT. When a user registers his/her leaving attendance, a personalized mail is sent by triggering a PHP Mailer code by Windows Task Scheduler (in Windows-based system) or Cron Job (in Mac-based system).

3.7 *Leave Management System*

Leave management system (LMS) provides a platform for the users to submit pre-leave or post-leave applications. admin can view the live applications and either reject

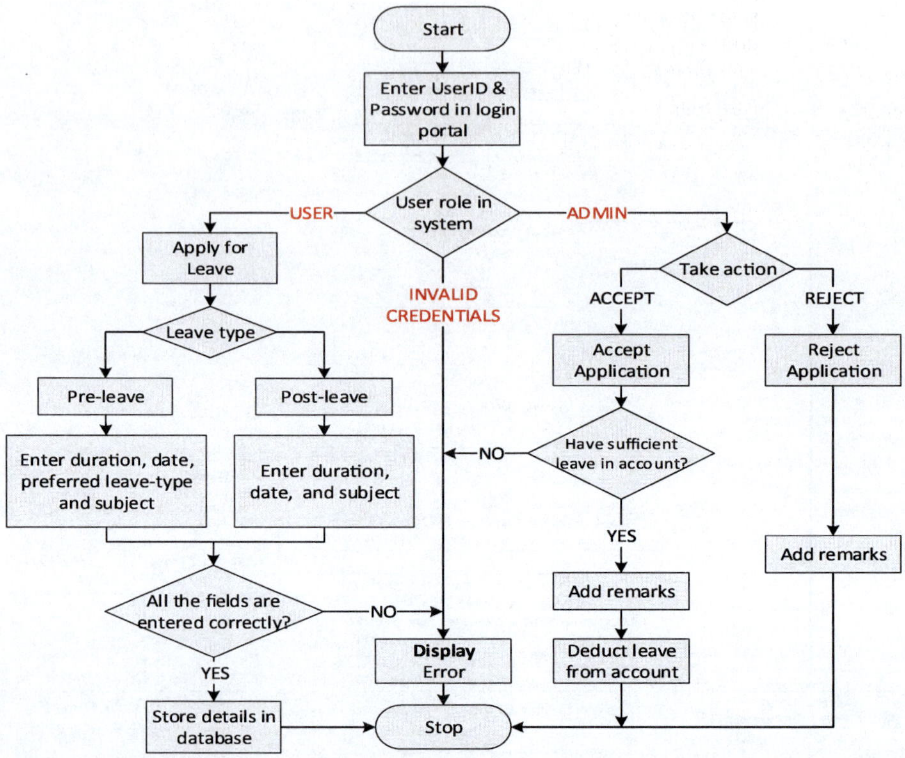

Fig. 5 Flowchart of leave management system

or accept. According to admin's action, leave deduction from user leave balance will take place. LMS is transparent and much faster than the conventional method of leave management. Flowchart of the LMS is given in Fig. 5. There are two different stages in leave management system (LMS):

Application from registered user: In this step, a registered user login into the portal and navigates to 'apply for pre-leave' or 'apply for post-leave' page based of the category he/she want to apply. For pre-leave applications, user can choose preferred leave type among Earned Leave (EL), Casual Leave (CL), and Medical Leave (ML) depending upon the availability in his/her account (system checks for availability automatically); but for post-leave applications, it is the responsibility of the admin to assign leave type.

Action taken by the admin: Admin can view all the live leave application by navigating to the 'live pre-leave application' and 'live post-leave application' pages. Admin can either ACCEPT or REJECT an application. For post-leave applications, admin need to assign a leave type if he/she accepts it. Remarks are mandatory to notify the reason if the application is rejected. Application status gets modified as soon as admin takes action. If an application is accepted, leave is deducted from the leave balance of the corresponding user. If a user has zero leave balance in either of EL, Cl, or ML account, he/she gets a remainder just after login.

3.8 Notice Management System

Notice management system (NMS) makes the notice updating task easy for the admin. Through a dedicated portal, admin will be able to add new notices and delete previous notices. Notices are to be displayed on the home page so that those can be accessed even without logging into the portal. This is shown in Figs. 6, 7, and 8.

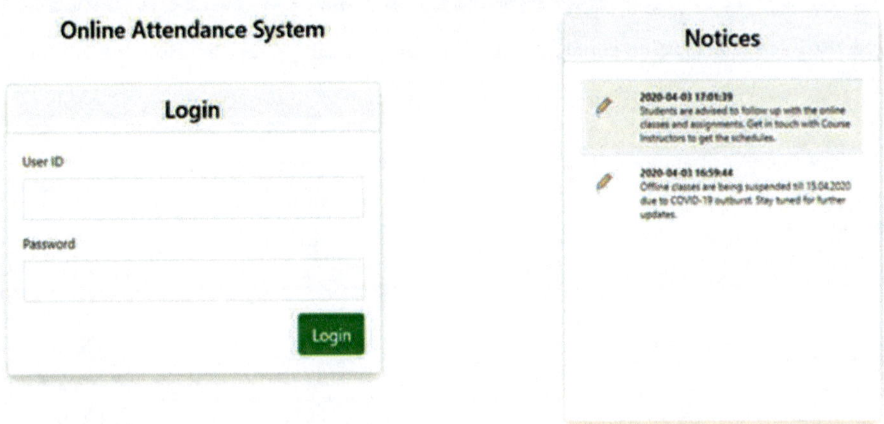

Fig. 6 Login portal with public notice display board

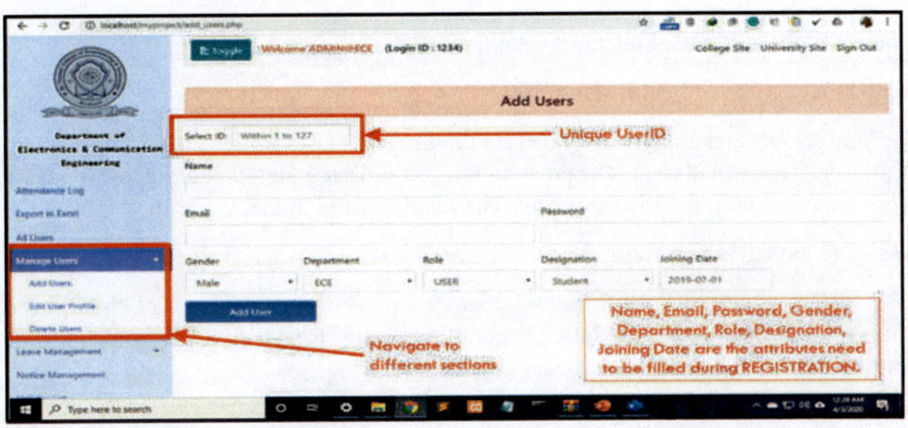

Fig. 7 Admin guided registration form

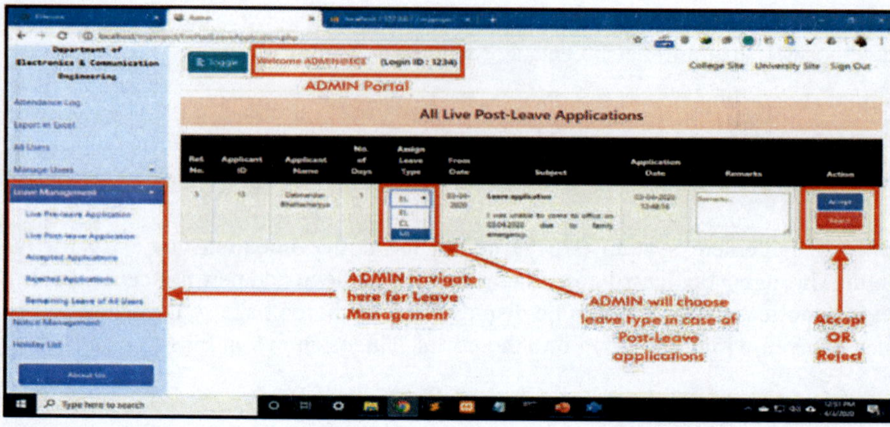

Fig. 8 Live leave application portal as seen by admin after successful submission of application by user

Table 2 Attendance monitoring test result

Test session	Attendance on the session (A)	Successful login (S_1)	Successful logout (S_2)	Unsuccessful login (U_1)	Unsuccessful logout (U_2)	Efficiency $(\frac{2A}{S_1+S_2})$ (%)
1	24	24	23	0	1	97.91
2	31	30	30	1	0	96.77
3	42	42	41	0	1	98.80
4	25	25	25	0	0	100.0

4 Result

There were two types of testing conducted on the proposed system:

Test-1 (Attendance Monitoring Test)

To conduct test and analyze the result under different circumstances, a sample space of 65 registered candidate was created. On those 65 registered candidates, the tests were experimented for four sessions. The results are in Table 2.

Test-2 (Leave Management Test)

The system is tested with 25 simultaneous connections, and each of them were successful to post-leave application simultaneously. Result of leave management test is given in Table 3.

Table 3 Leave management test result

Test session	No. of parallel login	No. of user tried to submit leave application (A)	Successful submission (S)	Unsuccessful submission	Efficiency ($\frac{S}{A} \times 100$) (%)
1	45	25	24	1	96.0
2	22	20	20	0	100

5 Conclusion

The proposed IoT-based biometric attendance system may help to reduce the possibilities of cheating or any human intervention in recording the attendance and helps to ease the enrolled candidates to keep track of their attendance. Attendance reports can be accessed directly by the candidate and the admin through featured Webpage, thereby providing transparency. This promotes efficient management of attendance system and improves organizational excellence.

6 Future Work

The robustness of the system can be improved by adding features which indicates the location of the candidate. Such a system will be helpful for the organizations where same employee works in multiple branches simultaneously.

Future enhancements that can be developed from this project are to expand the current database to store the complete bio-data of the candidate. In we think of from the point of view of an educational institution, a system can be designed to incorporate attendance, online assignment submission, online exam functionalities embedded into one platform.

References

1. Wu Q, Wang H, Li X (2017) Fingerprint attendance system based on embedded linux and WeChat official accounts. In: International Conference on Computer Technology, Electronics and Communication (ICCTEC) (2017)
2. Okokpujie KO, Noma-Osaghae E, Okesola OJ, John SN, Robert O (2017) Design and implementation of a student attendance system using iris biometric recognition. In: International conference on computational science and computational intelligence (CSCI) (2017).
3. Sawhney S, Kacker K, Jain S, Singh SN, Garg N (2019) Real-time smart attendance system using face recognition techniques. In: 9th international conference on cloud computing, data science and engineering (Confluence)

4. Koppikar U, Hiremath S, Shiralkar A, Rajoor A, Baligar VP (2019) IoT based smart attendance monitoring system using RFID. In: 1st international conference on advances in information technology (ICAIT).
5. Zhan H, Wang Q, Hu Y (2017) Fingerprint attendance machine design based on C51 single-chip microcomputer. In: International conference on computer technology, electronics and communication (ICCTEC)
6. Shukla VK, Bhandari N (2019) Conceptual framework for enhancing payroll management and attendance monitoring system through RFID and biometric. In: Amity international conference on artificial intelligence (AICAI)
7. Yadav DK, Singh S, Pujari S, Mishra P (2015) Fingerprint based attendance system using microcontroller and LabView. Int J Adv Res Electric Electron Instrument Eng 4(6):5111–5121
8. Wang J (2015) The design of teaching management system in universities based on biometrics identification and the internet of things technology. In: IEEE 10th International Conference on Computer Science and Education (ICCSE), Cambridge University, pp 979–982
9. Srinidhi MB, Roy R (2015) A web enabled secured system for attendance monitoring and real time location tracking using biometric and radio frequency identification (RFID) technology. In: IEEE international conference on computer communication and informatics (ICCCI), Coimbatore
10. https://www.rhydolabz.com/documents/finger-print-module.pdf

Performance Analysis of Public and Private Blockchains and Future Research Directions

Vemula Harish and R. Sridevi

1 Introduction

Blockchain is a modern technology that offers decentralized public ledger maintained by all network participants called peers, blockchain is one of the game changing technologies at present times, it provides an alternative solution for centralized systems, this is first introduced by Nakamoto et al. [1] in the year 2008 who is anonymous. Blockchain uses cryptographic algorithms to secure the transactions, all transactions in blockchain are immutable, i.e., nobody can change or alter the transactions once it is committed to the public ledger. Initially, blockchain is used in the cryptocurrencies, but later, it is attracted by many researchers and academia, the key features of the technology include security, transparency, availability, immutability, and confidentiality.

Since the inception of blockchain technology, many platforms have been developed, and source code is also available as open-source projects, but there are technical challenges and performance bottlenecks which must be addressed and optimized, as blockchain development is not fully developed for production environments. The researchers have conducted study of enterprise blockchain platforms and their performance analysis, still there are many aspects that need to be addressed which are not covered.

V. Harish (✉) · R. Sridevi
Jawaharlal Nehru Technological University, Hyderabad (JNTUH), Telangana, India
e-mail: vemula.harish31@gmail.com

R. Sridevi
e-mail: sridevirangu@jntuh.ac.in

J. K. Mandal et al. (eds.), *Proceedings of International Conference on Network Security and Blockchain Technology*, Lecture Notes in Networks and Systems 738,
https://doi.org/10.1007/978-981-99-4433-0_33

1.1 Smart Contracts

Smart contracts are introduced in blockchain 2.0, smart contract is the program which contains a set of conditions and logical statements used for transaction processing. It can also hold the assets so that it enhances the security and trust, all organizations must agree to the terms and conditions provided in the smart contract. With the help of smart contracts, enterprises can operate in a trustless environment so that they can expand their businesses globally. Smart contracts are immutable once deployed. We cannot modify any conditions that are written in it but can change the version by modifying the source code, and again verifying and validating are required by network participants such as organizations, etc.

1.2 Distributed Consensus

Blockchain resolved the distributed consensus problem through proof of work (PoW) consensus [2]. Distributed consensus is a common agreement between the network participants who are geographically separated over the globe and are connected through the peer-to-peer communication without centralized server, it is difficult to come to consensus due to several reasons such as nodes may go down or crash, peers may spend the same digital currency token twice, and nodes may behave maliciously as there is no trust among them. These issues are addressed by Nakamoto using proof of work consensus over the Bitcoin network. He added the transparency and immutability through which one cannot deny the transactions that he or she made, because it is globally available in the public ledger. Anybody can verify that the transaction is committed and final, subsequent transactions which are invalid are discarded in the network by mining. Miners will verify the transactions, and most of the time only one miner wins the puzzle and creates the block. If multiple miners win the puzzle at a time, then both of them create a block of transactions which result in the forks in the blockchain, later forks are resolved using the longest chain principle.

This study mainly focuses on the scalability issues and performance bottlenecks of various blockchain platforms that come under public and private categories, also identify, and analyze the previous work done in this aspect, provide complete study of existing solutions given by researchers and open issues that current practitioners are facing.

Section 2 provides the previous research done in this aspect by researchers, Sect. 3 discusses public and private blockchain platforms, their transaction processing along with pros and cons. Section 4 provides the comparison of consensus algorithms of both public and private blockchains. Section 5 provides the tools and techniques available for performance monitoring of blockchains. Section 6 provides experimental results and analysis; finally, Sect. 7 concludes the paper.

2 Related Work

The performance evaluation and benchmarking of blockchain platforms are an active area of research, and there are many platforms evolved as a result of research over public and private blockchains. But these are suffering from scalability and performance issues; however, researchers are modifying the underlying architecture, redesigning, and adopting the existing optimized solutions to improve the performance [3]. Wang et al. [4] discussed the distributed consensus issues in blockchain and suggested new ways of designing distributed mechanisms. Huang et al. [5] studied the raft consensus algorithm and proposed a simulation model for analyzing and optimizing raft consensus algorithm. Hao et al. [6] emphasized-on performance evaluation of consensus mechanisms of private Ethereum and Hyperledger fabric, also compared practical byzantine fault tolerance (PBFT) and proof of work (PoW). Cao et al. [7] studied and suggested the guidelines to choose appropriate consensus algorithms, they have considered, proof of stake, proof of work, direct acyclic graph consensus algorithms into consideration and evaluated research parameters such as block generation time, transactions per seconds (TPSs), confirmation latency. Thakkar et al. [8] conducted a performance study by considering HLF platform. Swanson [9] conducted an in-depth study of DLT consensus and crypto economics. In existing literature, most of the researchers focused on performance, consensus, and benchmarking of blockchain platforms they have considered either private or public blockchain platforms, there is a lack of performance study in comparing public and private blockchains, along with consensus algorithms, scalability issues.

3 Public and Private Blockchains

3.1 Public Blockchain Platforms

Initially, blockchain technology is used for Bitcoin (BTC). Which is the cryptocurrency introduced to eliminate the intermediate money exchanges, it is valid globally. In the beginning, many people didn't accept global currency and the technology behind it due to fear and lack of governance, but later, they realized the potential it has. Proof of work consensus solved double spending problem, it also offers immutability, so that once the transaction is committed, it cannot be modified or altered. Even though it is vulnerable to 51% attack which is infeasible in a real-time scenario where there are millions of peers participating in the network. Public blockchain platforms are open for everybody, there is no need for trust among the network participants. To circulate the enough coins in the network, new coins are minted and given to miners as incentive.

Miners fetch the pending transactions from mempool and validate them by solving cryptographic puzzles, by finding appropriate nonce and cutting them into blocks. Figure 2 shows the transaction processing in public blockchains. Nonce is a random number which is combined with the previous block hash and Merkle root for generating the current block hash. Data is maintained and replicated at all peers in the network and always maintain consistency, in public blockchains because of the PoW consensus, high computation power is required to mine the blocks, as public blockchains are transparent, most of the enterprises don't want their transaction information to be disclosed, as a result private blockchains are emerged. Public blockchains proved computationally expensive.

3.1.1 Scalability in Public Blockchains

Scalability in blockchains refers to the number of nodes that can participate in the network, and number of transactions that can be processed by the network. Public blockchains are not highly scalable in terms of transactions processed over a period. As it is crucial to achieve the distributed consensus over the large public network. Mining is the time taking process as miners require huge computational power to solve the cryptographic hash and finding the appropriate nonce to meet the difficulty of the proof of work. Miners will compete to solve the puzzle resulting in less scalability. There are other aspects also impacting scalability such as network latency and dynamic topology. These issues are resulting in taking around 10 min for creating a single block of transactions in the Bitcoin network. Block is not immediately final in public blockchains because there exists probability that more than one miner may solve the proof of work puzzle at a time, then both will create their own blocks with their own choice of transactions, and both blocks are appended to the previous blocks. This results in splitting the single chain into two parallel chains which is called as forking in blockchain. The longest chain principle is used to resolve the forks. Here, the transactions which are included in a small chain are not committed and then discarded. Hence in public blockchains, transactions are not final immediately. After certain no of blocks are added to the chain, they are committed to the public ledger permanently.

The Genesis block is known as the first block in the blockchain, it is created when a network is established and contains all network configuration details and doesn't have any previous block hash. Subsequent blocks are appended as a chain to one another, and block contains Merkle root hash which is a special data structure that represents all transactions included in the block. Figure 1 shows the block components, block contains header, previous block hash, Nonce, Merkle root, timestamp, and all transaction information. Block hash is a unique value generated using the combination of block components along with nonce. Cryptographically all block hash values are linked to each other and possess an avalanche effect (Fig. 2).

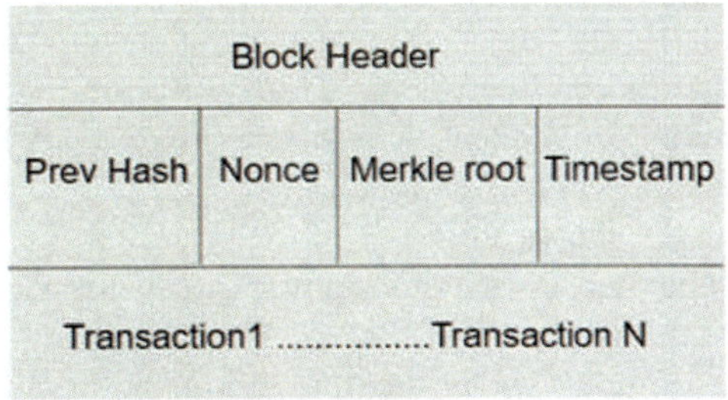

Fig. 1 Block in public blockchain

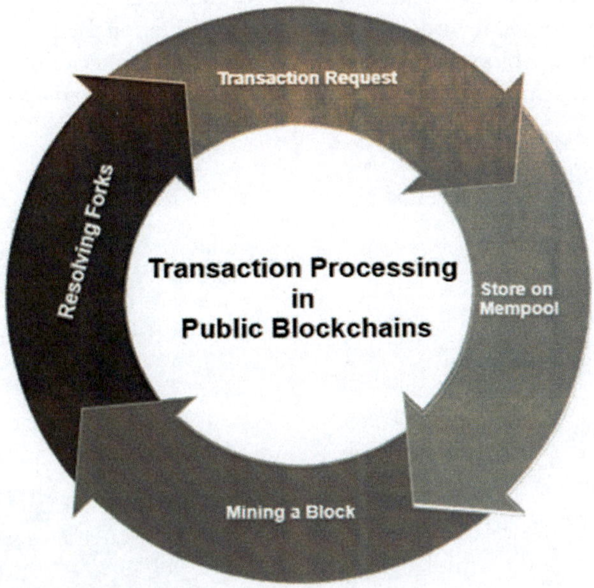

Fig. 2 Transaction processing in public blockchains

3.2 *Private Blockchain Platforms*

Private blockchains are also known as permission blockchains, they restrict the network participants, i.e., not open to everybody, so that only trusted members can join the network. Permission blockchain platforms use certificate authorities and membership service providers which provide the necessary public key infrastructure (PKI), public, and private key credentials are essential to prove their identities, authorized nodes can only make transactions. Communication between the network participants is encrypted and is available within the channel. Permission blockchains are best suitable for the enterprises that don't want to share their communication information publicly. Most of the organizations are moving to permission blockchain platforms.

Organizations can protect their privacy by using channels which are secure and reliable. Moreover, transaction finality is immediate in permission blockchain platforms when compared to public blockchain platforms, i.e., once the block is committed, there is no probability for forks in permission blockchains due to the nature of consensus, which is single leader based. Hence, it ensures that only one leader will cut the block.

As shown in Fig. 3 block in permission blockchain contains Header, Data, and Metadata, it contains signing information of its validators and endorsers along with timestamp.

Figure 4 shows the transaction processing in private blockchain, and it is different from public blockchains, majorly, there are three phases as shown in Fig. 5, while transaction is being processed. Validators are the trusted nodes that verify all transactions and their dependencies, after validation the transactions go to consensus phase where blocks are created by grouping the transactions together. Once the consensus is successful, then these blocks are committed to the ledgers of all peers permanently.

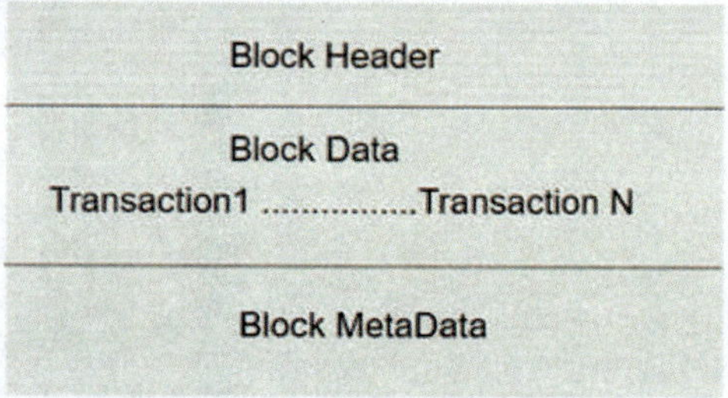

Fig. 3 Block in permissioned blockchains

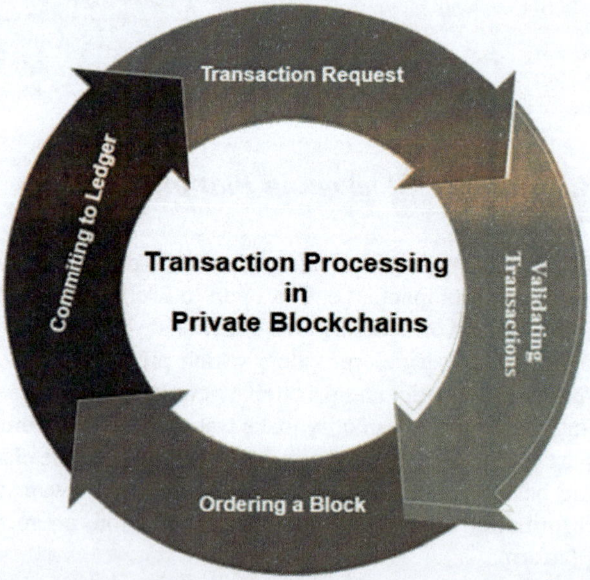

Fig. 4 Transaction processing in private blockchains

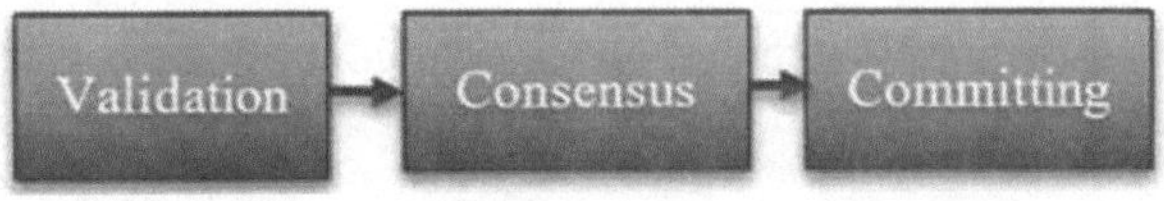

Fig. 5 Transaction processing phases

4 Consensus Comparison of Public Versus Permission Blockchains

Consensus is an agreement between the network participants or peers, in distributed systems such that all peers must have consistent information at any given time. Consensus mechanism varies based on the type of blockchain platform, i.e., in public blockchains consensus is achieved in a different way compared to permission blockchain platforms. There are many consensus algorithms available such as POW, POS, POA, PBFT, RAFT, and DAG.

Main goal of the consensus algorithm is to have a common agreement among all peers on valid transactions that are included in the block. Table 1 shows some of the consensus algorithms used in blockchain platforms.

4.1 Consensus in Public Blockchains

In public blockchains, miners or other special nodes create a block of transactions and validate them, while they execute the consensus protocol for publishing the block. Here, major performance bottleneck is that everybody cannot create the block at a time so that they will compete among themselves for block creation. Finally, one miner wins the race and publishes his block of transactions to the network and all other network participants see this block and they agree up on the transactions or block by considering this block hash as their previous block.

Fork is created when more than one miner solves the puzzle at a time and creates two valid blocks of transactions which result in multiple chains. Thus, they don't

Table 1 Examples of public and private consensus algorithms

Public blockchain platforms	Public consensus algorithms	Private blockchain platforms	Private consensus algorithms
Bitcoin	POW	Hyperledger	RAFT, SBFT
Ethereum	POW, POS	Ripple	RPCA
Solana	POH	R3corda	Valid and Unique
Cardano	POS	Quorum	DAG

POW Proof of Work, *POS* Proof of Stake, *DAG* Directed Acyclic Graph, *RPCA* Ripple, Ripple Protocol Consensus Algorithm, *POH* Proof of History, *RAFT* Reliable, Replicated, Redundant, and Fault-Tolerant [10]; R3 corda, valid consensus and unique consensus guarantee immediate finality, in Bitcoin after six successive blocks are added, then the transactions in the current block will be committed permanently to the ledger.

4.2 Consensus in Permission Blockchains

Consensus in permission blockchains is achieved by a trusted set of nodes, most of the permission platforms use leader nodes who are responsible for ordering the transactions into blocks, forward to other consensus nodes so that all of them must agree upon the block of transactions and maintain consistency. Further they broadcast this block along with signatures to all other peers in the channel, finally all network participants of the channel verify the integrity and authenticity then they commit to their individual ledgers.

Here in permission blockchains, nodes are trusted, hence we don't need to exercise the computationally expensive operations like proof of work, instead they use consensus algorithms such as POET [11], RAFT [12], and PBFT [13].

5 Performance Analysis of Blockchain Platforms

5.1 Performance Parameters

Performance of the blockchain platform refers to how fast and efficient transactions happen in the network. To measure the performance of blockchain platforms, most common metrics are throughput and latency.

5.1.1 Throughput

This is defined as a rate of the number of transactions committed to the ledger over a period, it is also known as transactions per second (TPS).

5.1.2 Latency

This is defined as the time taken for processing the transactions, i.e., time difference from transaction arrived and committed to the ledger. Depending on the blockchain platform, the latency can be observed at various transaction processing phases, reducing the latency by improving the underlying platforms architecture, or by using optimal techniques in processing, the transactions will help in improving the overall performance of the blockchain platforms.

5.2 *Performance of Public Blockchains*

Bitcoin and Ethereum are some of the public most popular blockchain platforms, transaction delay or latency is more in public blockchains compared to permissioned platforms because of its open nature and consensus. Depending on the difficulty of target transaction processing is a time-consuming process. Bitcoin uses 10 min for creating a single block of transactions. Reducing the block creation time impact, no transactions to be included in a block and difficulty level. Moreover, the transaction processing time is also affected by the network delay. Here to reduce the latency and to improve the performance of public blockchain platforms, reducing the block size becomes the performance bottleneck which increases the additional overhead because many blocks are created, and every block must be validated and processed.

In real-time scenarios, there are many network participants sending transactions to the blockchain platform, it should be able to handle a greater number of transactions within less time. Maximum throughput of the blockchain platform refers to the capability of handling maximum transactions. To enhance the performance of blockchain platforms, it is important to understand the performance barriers and remove them so that they are improved or enhanced. These performance enhancements must not affect any other components, nature of the blockchain.

5.3 *Performance of Permission Blockchains*

When performance is compared between public and private blockchain platforms, they are not identical by nature, permission blockchain platforms such as Hyperledger Fabric supports immediate transaction finality. Transaction latency is very less compared to public blockchains as they don't require algorithms like proof of work. Latency depends on block creation time or number of transactions included in block, network delay, etc.

Throughput varies from one permission platform to another platform depending on their core architecture, consensus algorithm and various components that process the transaction.

6 Results

In this experiment, the permission blockchain platform's performance is analyzed over Hyperledger fabric, considering the number of transactions ranging from 100 to 500.

Table 2 shows the transaction throughput and latency observed over Hyperledger fabric platform using open-source block bench [14] tool, using a single node with intel i3 7th Gen CPU, 8 GB RAM configuration. Average transaction processing

Table 2 Performance of private blockchain platform

Transactions sent	Transaction latency (ms)	Throughput
100	4,780	20.920
200	11,514	17.370
300	19,494	15.389
400	21,339	18.745
500	23,956	20.871

rate observed is 93.296 when transactions are sent in incremental fashion starting from 100 to 500 within 1 min of timeframe from five clients; here, the throughput is calculated based on the latency, i.e., total no. of transactions processed divided by time taken to process the transactions in seconds.

Example: Time in seconds = 4,780/1,000 = 4.78 s.

Then, the throughput = 100/4.78 = 20.920.

The performance also depends on the node's configuration, network bandwidth in case over local area network.

From the observations of the above Fig. 6, the permission platform's performance is better than public blockchains such as Bitcoin and Ethereum [15]. In Bitcoin, the transaction throughput is around 5 TPS [16], and current Ethereum throughput is around 25 transactions per second [17].

From Fig. 7, it is clear that the public blockchain platforms are struggling to scale in processing the numbers of transactions compared to permission platforms.

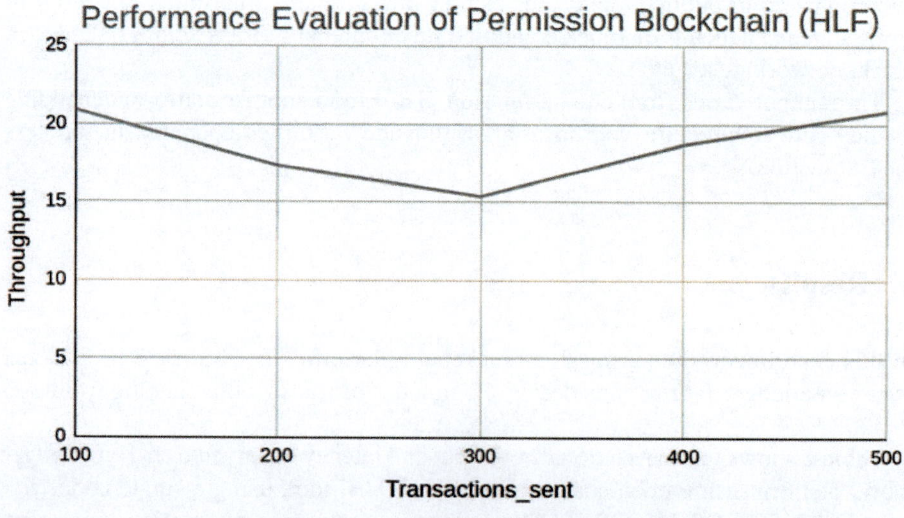

Fig. 6 Performance analysis of permission blockchain platform (HLF)

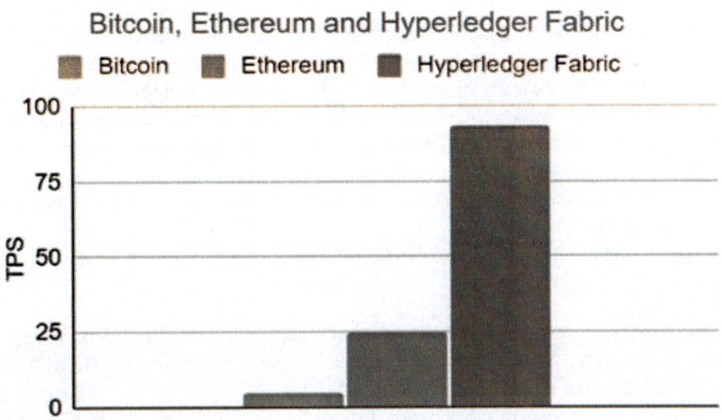

Fig. 7 Performance comparison of BTC, ETH, and HLF

7 Conclusion

Blockchain technology is offering a powerful and secure way of sending transactions over untrusted environments. This technology helps organizations to spread their businesses over the globe, but still, it is suffering from performance and scalability issues. In future to address these issues and make it more efficient and reliable, a detailed study of public and permission platforms is essential, the architectural components that process transactions are to be re-designed in a better way. Most of the researchers are focusing on either public or private platforms but we need scalable and hybrid platforms to meet the current needs. In our future work, we try to address the performance and scalability issues of the blockchain platforms by adopting best practices, efficient algorithms, and frameworks.

References

1. Nakamoto S et al (2008) Bitcoin: a peer-to-peer electronic cash system. http://bitcoin.org/bitcoin.pdf
2. Harish V, Sridevi R (2020) A brief survey on blockchain technology. In: Proceedings of the third international conference on computational intelligence and informatics. Springer, Singapore
3. Singh A et al (2022) A survey and taxonomy of consensus protocols for blockchains. J Syst Archit 127:102503
4. Wang W, Hoang DT, Hu P, Xiong Z, Niyato D, Wang P, Wen Y, Kim DI (2019) A survey on consensus mechanisms and mining strategy management in blockchain networks. IEEE Access 7:22328–22370
5. Huang D, Ma X, Zhang S (2019) Performance analysis of the raft consensus algorithm for private blockchains. IEEE Trans Syst Man Cybern Syst 50(1):172–181
6. Hao Y et al (2018) Performance analysis of consensus algorithm in private blockchain. In: 2018 IEEE Intelligent Vehicles Symposium (IV). IEEE

7. Cao B et al (2020) Performance analysis and comparison of PoW, PoS and DAG based blockchains. Digital Commun Netw 6(4):480–485
8. Thakkar P, Nathan S, Viswanathan B (2018) Performance benchmarking and optimizing hyperledger fabric blockchain platform. In: 2018 IEEE 26th international symposium on modeling, analysis, and simulation of computer and telecommunication systems (MASCOTS), pp 264–276. doi: https://doi.org/10.1109/MASCOTS.2018.00034
9. Swanson T (2015) Consensus-as-a-service: a brief report on the emergence of permissioned, distributed ledger systems. Report.
10. https://groups.google.com/g/raft-dev/c/95rZqptGpmU
11. https://www.investopedia.com/terms/p/proof-elapsed-time-cryptocurrency.asp
12. https://raft.github.io/
13. Barger A et al (2021) A byzantine fault-tolerant consensus library for hyperledger fabric. In: 2021 IEEE International Conference on Blockchain and Cryptocurrency (ICBC). IEEE
14. https://github.com/ooibc88/blockbench
15. Ethereum W (2014) Ethereum whitepaper. Ethereum. https://ethereum.org. Accessed 07 July 2020
16. https://academy.binance.com/en/glossary/transactions-per-second-tps
17. https://zebpay.com/in/blog/ethereum-100000-transactions-per-second-coming-soon

Protecting the Privacy of IoMT-Based Health Records Using Blockchain Technology

T. C. Swetha Priya and R. Sridevi

1 Introduction

Due to tremendous increase in the population of India, it has become a pressure on government to accommodate the healthcare facilities. The government is trying to accommodate the increasing urban population but could not reach the daily increase rate of population. So this has urged the need for latest technology to meet the increasing population health requirements. As the population increases, the need for medical facilities also will be more, and there will be more people with different health problems. So, there will be a need for remote healthcare in recent times. So that there will not be any need for more hospitals. Instead we can have latest technological advancements in terms of embedded devices, smart wearable gadgets, and many such low-cost devices used for healthcare of people. According to modern researchers, it is evident that these less expensive small smart devices have the ability to record health information of a patient and even monitors patient condition 24 × 7.

The IoT technology [1, 2] is one such technology that handles low cost, smart devices, or embedded devices which offers wireless connectivity between smart medical devices, patients, and doctors. IoT technology is based on wireless sensors that continuously records the signals and maps them with various parameters and will be communicated through the IoMT-based network. The received information is processed, stored, and examined with already present data. This data will be used by doctor to suggest appropriate treatment.

T. C. Swetha Priya (✉) · R. Sridevi
Department of Computer Science and Engineering, JNTUH University College of Engineering, Science and Technology, Hyderabad, Telangana, India
e-mail: tcswetha3552@gmail.com

R. Sridevi
e-mail: sridevirangu@jntuh.ac.in

J. K. Mandal et al. (eds.), *Proceedings of International Conference on Network Security and Blockchain Technology*, Lecture Notes in Networks and Systems 738,
https://doi.org/10.1007/978-981-99-4433-0_34

The IoMT [3, 4] comprises medical devices [5], software, and related hardware connected over the Internet for providing connectivity to health-related information. This concept was previously called as IoT for healthcare [6]. IoMT allows remote connectivity of wireless devices for communicating and analyzing the health-related information over the Internet. IoMT has evolved from the concept of IoT. IoT is an interconnection of computing devices that communicate data over the devices without human intervention. IoT supports connectivity between electronic devices providing communication of data between devices in various applications. IoT has made our lives easier when compared to the previous situation.

Figure 1 demonstrates how the communication between devices connected over the network in medical application will happen. Here, all the data will be stored in the central repository in digital format and will be made available to all the stakeholders like the medical lab staff, patients, and doctors involved in the network. So, a patient's complete information can be maintained with the help of this type of IoT-based systems automatically without human intervention. So, such patient's record can be helpful in analyzing the past and predicting the future status of a patient's health condition. So, with the help of such systems, we can automate the old medical devices to meet the present day needs to support real-time data by adding extra devices, sensors, converters, and modems.

IoT systems has complex architecture comprising of various components that interact with each other to support real-time data monitoring, gathering, transferring, and analyzing collected data. IoT includes various technologies that include

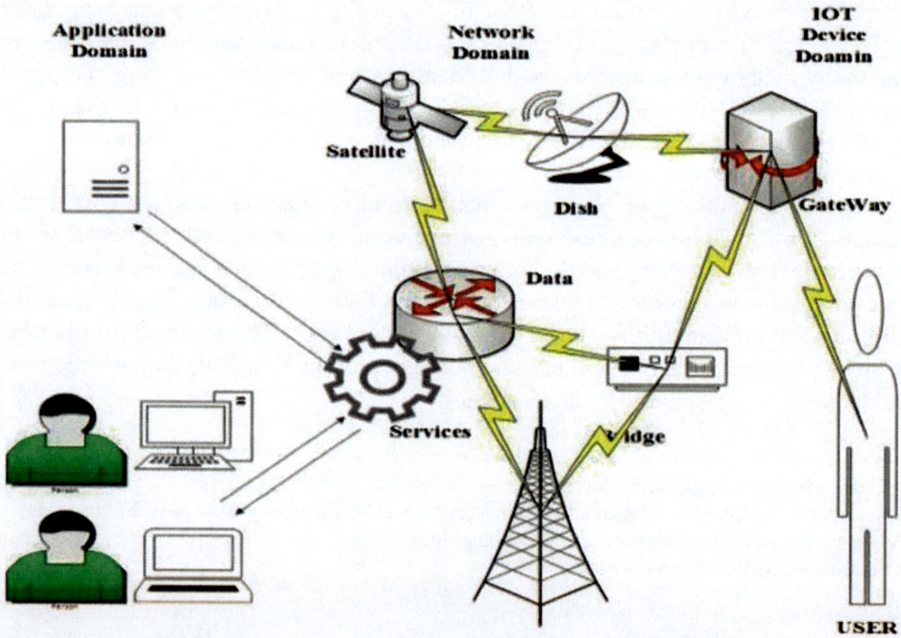

Fig. 1 Communication in IoT

smart homes, augmented reality, smart grids, and so on. IoT integrates the components with these technologies to support real-time data. Now, healthcare has also been transformed through IoT. So the term IoMT has evolved from healthcare application of IoT. IoMT devices can be embedded into various categories like depression monitoring, remote patient monitoring, hygiene monitoring, glucose monitoring, connected inhalers, smart contact lenses, robotic surgery, heart rate monitoring, mood monitoring, and ingestible sensors which has capability to collect data from remote locations. This IoMT can also be helpful in early recognition of various diseases, prevention of dangerous diseases, monitoring monthly status of a patient, and remote diagnosis and cure in critical situations.

But to store such huge data of increasing population, we need a technology that can handle large databases. But relational databases do not allow dynamic updates to the patient's data. So a technology that is similar to relational database called blockchain technology is introduced. Blockchain technology is one of the emerging computer protocol generally used for storing electronic records on multiple nodes of the network [7]. Blockchain stores the information in encrypted form or simply in the form of transactions called blocks that are connected in the form of chains. Each block will maintain a header for identification; block ID, the hashed link to corresponding chains. This feature does not allow alteration or deletion of data. Each node in the network has complete blockchain information giving data access publicly to the devices present in the IoMT system. The devices or nodes are patients, lab technicians, and doctors. This is disseminated data storage where data is available to all nodes. So even a small variation in data is immediately identified. So, one advantage of such technology is that we do not require any professional for maintaining security of the data in blockchain.

2 Related Work

Many of the authors have been into research for many years for providing security to the IoMT Network. They have proposed various solutions for protecting the privacy of IoMT [8]. The basic solution that was proposed is encryption of data. The data which is in plain text format is encrypted at the sender side, and it is transformed into cipher text. The cipher text that is communicated over public transmission medium is now sent to receiver which converts cipher text back into plaintext. One such method proposed is end-to-end key management in which keys are exchanged with less utilization of resources. Another solution is proposed by [9] which deal with privacy and protection of healthcare systems. This paper has proposed a lightweight encryption algorithm which is an extension of DES. Hu et al. [10] and Li et al. [11] proposed a scheme that reduces utilization of resources. This method is based on cloud-based IoT sensors [12] to monitor personal information of patient including digital signature and timestamp information. This method uses an improved version of data encryption standard which uses homomorphism algorithm. Gong et al. [9], Li et al. [11], and Sun et al. [13] proposed a key agreement-based secure authentication method for a IoT cloud system. The method secures medium when the participants register for network. This paper proposes that it addresses challenges in healthcare systems. Li et al. [11], Alasmari and Anwar [14], Esposito et al. [15] proposed cloud-based method for wireless networks in healthcare field. It supports various dynamic security policies that depend on attribute encryption and cipher text policy. Louni et al. [16] introduced an access control method to Patient E Health Records stored in trusted servers. This proposal could provide security at higher level for patient health information by providing attribute-based encryption for encrypting the health records for providing good access control to healthcare information. The servers or cloud [16, 17] environment that is storing the health records are not completely trusted. But the health records should be stored with consistency and integrity. But this data is lacking security, and data is altered or may be removed by unauthorized users. So we need security policies that restrict the unauthorized access. So one way to secure data is encrypting sensitive data before it is transmitted to the other party. Here, the information that has to be secured is patient attributes like disease and type of illness caused. So, these attributes must be protected by providing proper access to patient health records. Yeh et al. [18] have designed an advanced communication technique that is based on networks. They proposed an IoT system for body sensor network that provide effectiveness and security to IoT network. Hu et al. [10] have proposed a multi-communication standard-based IoT system for healthcare devices.

But the previous works have a drawback on data storage system connectivity among various data gathering devices that monitors data constantly at periodic time interval and analyzing data. So, to overcome this drawback of database connectivity and security to the healthcare information, this paper suggests a blockchain-based technology to secure the privacy of IoMT health records.

3 Proposed Work

Due to the widespread use of Internet of things which connects physical things through the means of Internet. Also with the increase of embedded devices technology, IoT technology has widespread demand in all fields including medical and healthcare technology [19]. It helps the future generation to have access to the information at any time and to become smarter and stronger in retrieving up to date information. IoT has capability to integrate real and virtual world. It covers a huge range of advancements technologically that comprise of wearable embedded devices, sensors, cloud computing, ICT, etc., so, IoT has been into one of the fast growing technology in healthcare fields [20]. In the modern wireless communication era, IoT has gained exponential growth. The aim of IoT is to connect every device or object at any instance.

The IoMT comprises a set of smart devices that are attached to Internet for providing medical service to any type of users. As time is passing, healthcare industry is slowly adapting IoT-related solutions leading to advancement of IoMT technology. IoMT technology works in this manner: A test report is sent from pathological lab to patient relatives mobile or the smart watch having tracker collects data and that is examined by doctor's smart phone. This technology involves reliable and an affordable cost handy devices that are embedded in the watch or smart phones that enable interconnection between patients, medical equipment, lab technicians, and doctors. The sensors present will record patient information and compare with existing patient information, and by implementing decision support systems, the doctor can give better treatment and can predict further health risks and warn patient of risks and can suggest better diet.

Because of the widespread and diverse nature of IoMT, several security problems arise. Because of increased utilization of smart devices, integrity of information sent over IoMT should be handled in an efficient manner. So, there is a need for securing the IoMT network from cyber attackers and other third parties through which transferring of patient data is possible. So, one of the possible solutions for securing this patient information is blockchain technology.

3.1 Overview of Blockchain Technology

Blockchain [21, 22] technology because of its unique characteristics such as security, distributed nature, decentralization, and data transparency offers a greater potential to promote various fields. A blockchain works without a centralized server. A blockchain is a network that makes use of databases distributed throughout to share and store information. A blockchain has the capability to maintain a large set of records. Here, there is no central authority to monitor data. It also allows unauthorized parties to do transactions in the network. The transactions that are performed in the network will be notified to each and every peer system present in the network so that all nodes will be aware of the transactions. If any unauthorized user tries to change the data, the other peers will know that some unauthorized activity is being done on the data records because of the variation of hash of the data records present in successive blocks. So, because of these advanced security features and hashing capabilities of blockchain [23, 24], it is an apt solution for securing IoMT health records.

The blockchain includes blocks that are linked to one another that have cryptographic hash of the previous blocks. The chain only stores an associated hash of data records. These blocks include the transactions performed between nodes, base stations, and users. Because of inclusion of hash, all the data records and transactions are immutable. A blockchain is depicted in Fig. 2.

An application programming interface enables communication as well as exchange of data between devices. With the help of API only, interoperability between various sensors and processes is possible. Also it is possible to perform hashing on data and store data directly from sensor nodes, processes, and algorithms.

Due to the advanced characteristics and properties of blockchain technology, it has become one of the mostly used technologies. Some of the relevant characteristics of the blockchain technology are summarized as follows:

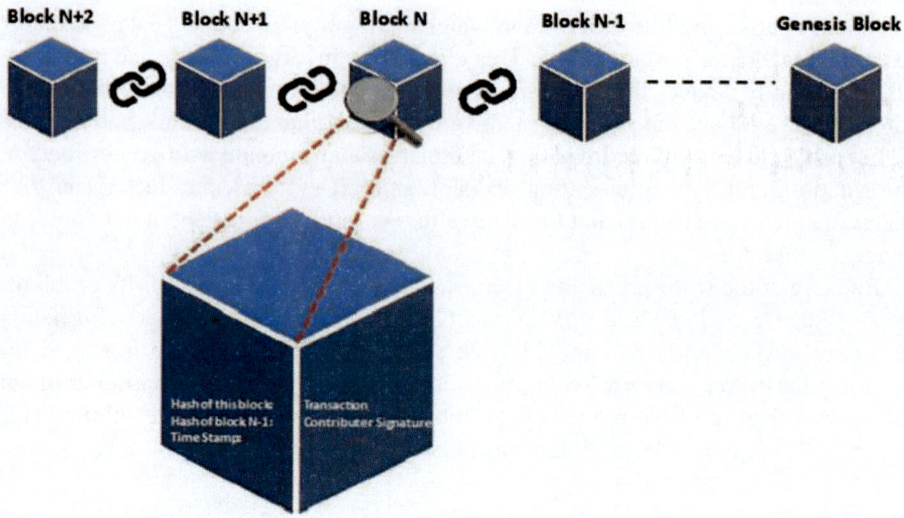

Fig. 2 Structure of blockchain

i. **Stability**

The patient health record once stored in blockchain will not get modified or removed by anybody. This feature enables users to choose this technology.

ii. **Access**

Each blockchain has two types of access: permission less and permission blockchains. Permission less blockchain does not require any permission for data access. It allows any user to access data in blockchain network. A permission blockchain requires access to network to view or alter data. This type of blockchain allows only a certain amount of nodes and gives access rights to only those nodes.

iii. **Cryptographic Hash**

Each block in blockchain is associated with hash of the previous block. This type of implementation of hashing ensures that any kind of changes made to information present in the block affects the subsequent hash values making the entire chain invalid.

iv. **Timestamp**

The records in blockchain will be time stamped. A combination of timestamp with cryptographic hash provides more security. Each record including block creation, transaction, and storage of data in blockchain will also be time stamped.

v. **Decentralized Nature**

This feature helps in eliminating the central authority dependency and removes the single point of failure. Here, a blockchain supports decentralized network where the entire blockchain or only a part of blockchain is distributed over the network. The decentralized versus centralized [25] nature of blockchain is illustrated in Fig. 3.

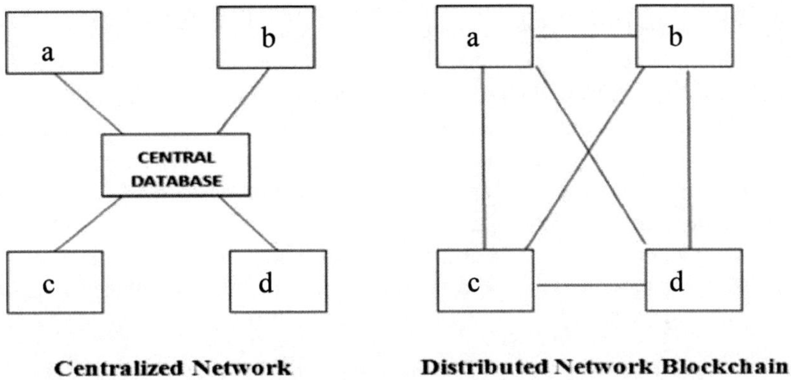

Fig. 3 Distributed versus centralized network

3.2 *Types of Blockchains*

Basically, three categories of blockchain exist. They are summarized as follows:

i. **Public**

This category of blockchains allows any type of users to access blockchain, and all users are having equal permissions to create, access, or modify the data in the blockchain. This category of blockchains comes under permission less type of blockchain.

ii. **Private**

This category of blockchains allows only a specific group of users who have permission rights to access blockchain. So this blockchain can be considered as permission blockchain. This category of blockchain is controlled by a central authority that has rights to give access to a specific set of authorized users.

iii. **Consortium Blockchain**

The private blockchain is controlled by an authority where as a consortium blockchain is controlled by a group of third party organizations rather than a single one. It also provides high levels of security compared to other type of blockchains.

Because of these advanced features and new opportunities provided by blockchain, blockchain technology has wide scope in healthcare field. It is used to store the patient health record in electronic form. It follows distributed architecture that is the database is stored in hundreds and thousands of computers and users. That means data is redundantly saved in encrypted form in chains. This type of concept helps in reducing the loss of information because this redundant data acts as a backup, i.e., even if data is lost, it can be retrieved from other users. This eliminates the distributed denial of service attacks making it impossible for the hackers to replace or destroy data. In this blockchain technology, new data can be easily added but we cannot modify or remove existing data from chain. Also this technology provides high level of encryption using private keys which hides the original information from malicious attackers. This has an advantage that if any malicious user tries to change data in record and want to save the data, it requires confirmation from other peer users. If there is any mismatch in any data from other users, that data record will be canceled. This makes the whole blockchain complete. This feature of blockchain makes it a suitable choice for IoMT technology. With this blockchain technology, the patient can be sure that his information is secure and will not be altered by any others. Also a patient can give access to his health-related information to the concerned doctor who is treating the patient. The doctor also can get data from any place in the world.

3.3 *Implementation of Blockchain Technology in Healthcare*

Blockchain helps in creation of patient record where high security can be provided to data from anywhere. Also the patient data will be synchronized from any place giving the doctor chance to review the history of patient and can suggest recommended treatment based on current and past history of patient only with prior permission of patient. This technology helps in getting the patient information at one place by providing security in a distributed environment.

Previously, creation of patient record, collection of patient information, storing data, and securing the data require more time and waste of space. Even the hardware cost is also more when manual work is more. But now with advancement in IoMT technology, the updating of data will be done on time automatically as that in real time and even the time is also reduced. With IoMT, the doctor can get updated patient information within a few seconds. This helps in identification of patient's health problem and provides diagnosis and medicate in an early stage without going to serious conditions.

Also previously, the data received from multiple sources is not in a common format. When data has to be consolidated and stored in single standardized format it used to take several years for consolidation. This posed a major problem to reliability of information. Also manual patient information monitoring [26] requires confirmation from the patient. But introduction of blockchain technology in healthcare has solved this problem by using standardized protocols and e-records of the patient. It also records each and every transaction between medicine dealer and hospital in the blockchain for providing authentic medicines to patient.

In future, blockchain technology will rule the IoMT. The IoT devices [27] whether may be smart watches, gadgets, etc., will capture patient information using embedded sensors in smart devices and can gather information like heartbeat, temperature, blood pressure, oxygen levels, etc., and send the data to e-record of the patient 24 h and 7 days. This information can be monitored by doctor at any time and can diagnose the disease within few hours and give appropriate treatment. Thus, blockchain technology will create a new revolution in medical field. Blockchain technology in healthcare can be illustrated as shown in Fig. 4.

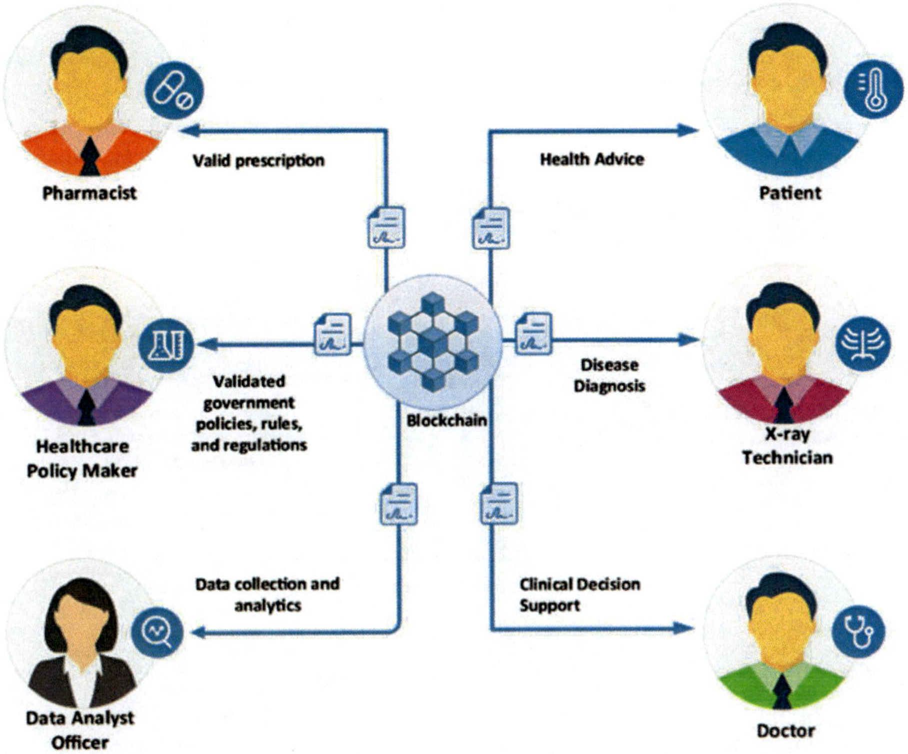

Fig. 4 Blockchain in healthcare

4 Discussion

Blockchain technology is an effective solution for providing solution in healthcare [28]. For secure transmission of patient information, blockchain technology is used in the proposed architecture to provide more security to data. The data structure used is blockchain which stores the important patient health related information in encrypted form. Figure 5 shows the proposed IoMT architecture based on blockchain.

In this type of architecture, the doctor will be in some distant or remote place monitoring the patient [29], and based on condition of patient, the doctor advises proper medication by analyzing the reports generated from laboratory. The reports received from diagnostic center which are in electronic form will be uploaded by practitioner and are updated to the existing patient's history. Along with electronic health records, the patient health information is also captured from the smart wearable devices. These electronic health records are maintained confidentially. By analyzing the tracking information along with the reports, the doctor who is present in distant location can suggest proper treatment at any time. This data will also be recorded in the blockchain in the form of new blocks. The diagnostic laboratory people who are part of IoMT architecture have access to add electronic records to blockchain. When

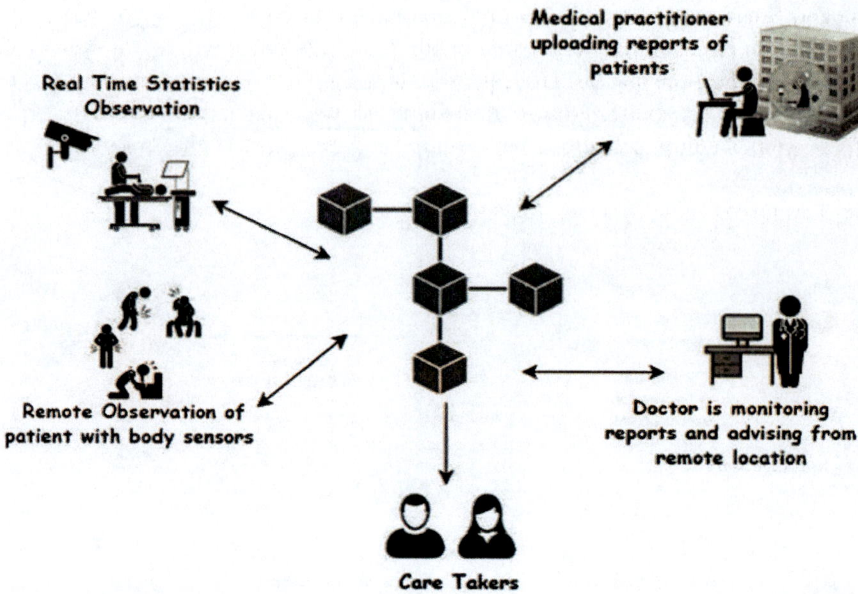

Fig. 5 Blockchain-based IoMT architecture

new patient information is added to IoMT network, a new block of data is appended to end of blockchain as depicted in Fig. 6.

Every block has the information related to patients, time at which patient record is created and the owner who has created a block. When a novel block is appended to the chain, this information is broadcasted into patient network. Every device in the IoMT system receives the block, and after getting approval from most of the peers, then only the block will be added to the end of chain. If this do not have any comparison with preceding block, then this block do not belong to this chain. But once any new block matches previous block, then it is added to chain. But once a block enters into the chain, it should not be altered or removed from chain. If any such alteration to data happens, then it will be immediately noticeable to every node in the network. That is how complete patients history is publicly visible to all peers in network in an authorized manner.

Appending a block to blockchain is depicted in Fig. 7. In this manner, blockchain provides high security to patient's data. The healthcare provider will take care of addition of new patient information. As and when a new patient enters into the IoMT network, a new block is created with the timestamp, patient data, and identity information. Now, the newly created block has to be broadcasted to all the peers present in the network. Now, the decision will be made by peers whether to approve the new block or not. If the approval is received from all the devices in the IoMT system, then the block is appended to existing blockchain. Otherwise, the block is rejected. The same process will be followed throughout the network upon addition of new patient.

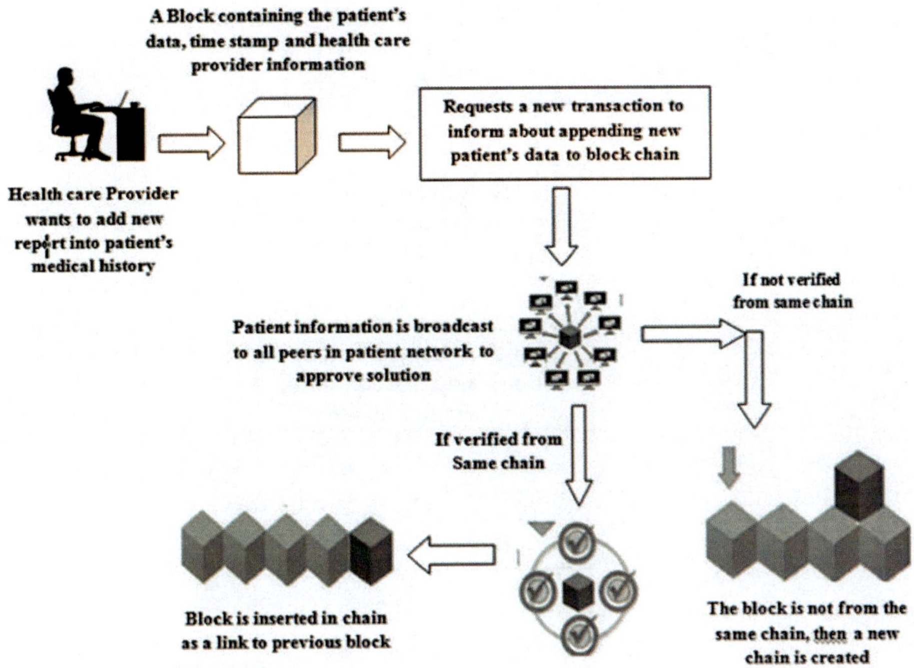

Fig. 6 Adding a block to blockchain

The blockchain technology considers the patient health records that goes to or comes from different devices connected in the IoMT network as transactions [30]. A transaction is illustrated as shown in Table 1.

Information that is exchanged among nodes, patients, lab assistants, and doctors is considered as transactions. The fields present in the table are explained as follows: The previous transaction field is a number that represents the transaction ID. The transaction number represents the transaction count. The node ID represents the number of node. The next field represents the transaction type. This particular field has 5 options or simply they are 5 different types of transactions. They are Start, Save, Retrieve, Examine, and Update. The **Start** transaction represents the first transaction. The **Save** option is used when the patient information has to be stored into the chain. The **Retrieve** option is used when the patient or doctor wants any crucial information for diagnosis. The **Examine** option is used to analyze the patient information that is retrieved from option 3. The **Update** option is used when any patient information has to be modified or any changes have to be updated to chain. The next field is SigReq that represents the node's unique signature. The last field is the actual data that is exchanged over network.

A block comprises of collection of **transactions**. The transaction is stored in the available space in block. If the block gets full, then the transaction is stored in the newly created block. A block header is maintained for every block so that the hash is

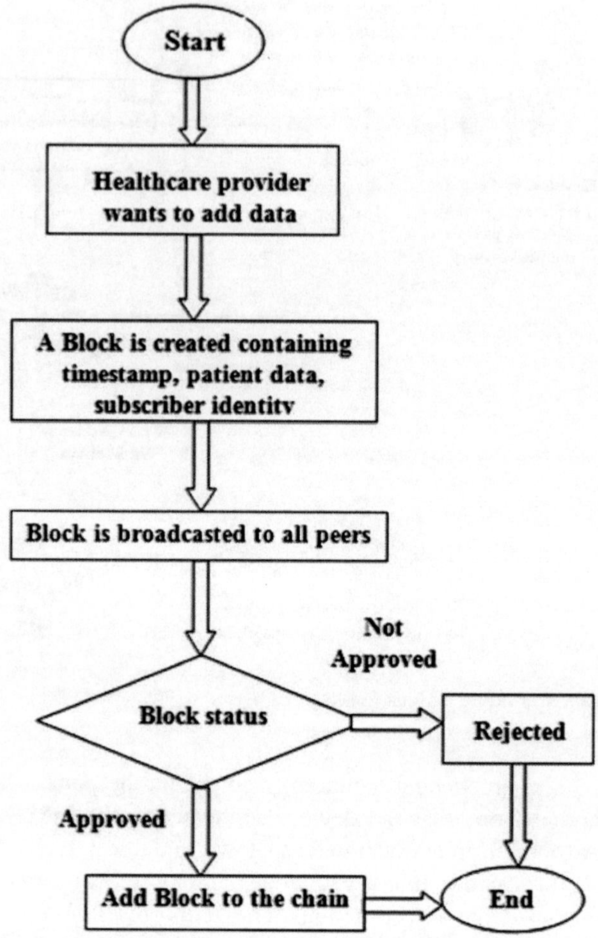

Fig. 7 Process of adding a block to blockchain

Table 1 Transaction

Preceding transaction	Transaction ID	Node number	Type	SigReq	Information
			0 = Start		
			1 = Save		
			2 = Retrieve		
			3 = Examine		
			4 = Update		

Table 2 Representation of policy

Requester	Request for	Node ID	Action
	…	……	
U_h	Retrieve	< Node IDs >	Allow
…	.	…	
N_1......N_h	Update	< BS IDs >	Allow

computed for preceding and subsequent blocks. So the computation of hash in this chain ensures consistency of information.

The access given to the patients, lab assistants, and doctors is named as policies [30]. These are stored in representation of block. The modifications made to the IoMT network result in the formation of novel policies. The structure of policy is presented in Table 2.

The patient health information update to all the nodes is done as follows: First, the patient nodes send the information to be updated to all peer nodes or to all the base station [31, 32] in the form of cipher text. Now, the policy lookup is done, and verification is done on whether the patient has access to update the requested information or not. Once the status in the policy lookup table is "allow", then the requested information is encrypted and updated to all the nodes. This updation has to be recorded as transaction. This updated record must be signed with timestamp and is encrypted and exchanged to patients in the IoMT system. So, this facilitates to have updated information with all the peer nodes. So if any person tries to change the data, then all the nodes will be notified about the changes made, and these updations will be canceled because there will be mismatch in the computed hash if the data is modified in only in single block. This helps in protecting the privacy of the patient's health records.

If the users of IoMT network, i.e., the lab assistants, doctors, patients, etc., requests access to the patient's updated information, then the policy check has to be done for that particular user. If the policy lookup has "retrieve" permission, then the user is allowed to access the requested information. Then, the patient health record is obtained and is signed with the key [33] of the base station that provides data. Then, the transaction has to be updated to all the users of IoMT network. This transaction data is encrypted and signed with random keys. Now, this information is communicated to all the nodes in the network. This is how patient health information is accessed from the blockchain providing secure access to the data.

5 Conclusion

Because of the revolutionary rise in IoT technology in various fields, many industries could utilize this opportunities very fastly. One such industry is the healthcare industry which could make use of these Internet of medical things-related things [34]. Because of this varied nature, security will be one of the important issues. So, blockchain technology promises privacy and security of health records in IoMT. It provides security to electronic health records of patient and provides access publicly to the users in IoMT network. Thus, blockchain provides privacy and security to the patient information. Blockchain technology will not replace advanced or ancient technologies. But blockchain can be a complementary application to other similar technologies. In future, this even may lead to development of new technologies that provide privacy to health records.

6 Future Work

As we keep on storing the new patient's health record information in the form of transactions, the size of blockchain will also increase. This leads to need of extra space. But the traditional method of using database systems may not be efficient method for storing the blockchain. So, to overcome this extra memory requirement issue, in future, we can extend the blockchain to be saved in an external cloud environment.

References

1. Atzori L et al (2010) The Internet of Things: a survey. Comput Netw 54:2787–2805
2. Cisco (2017) Enterprises are leading the internet of things innovation. Huffington Pos, Huffpost News
3. Rodrigues JJPC et al (2018) Enabling technologies for the Internet of Health thing. IEEE Access 6:13129–13141
4. Robert S et al (2012) Internet of M-health things. In: 2011, IET Seminar
5. Mohan A (2014) Cyber security for personal medical devices Internet of Things. In: IEEE international conference on distributed computing in sensor systems, Marina Del Rey, CA, pp 372–374
6. Porambage P et al (2015) Secure end-to-end communication for constrained devices in IoT-enabled ambient assisted living systems. In: IEEE 2nd World Forum on Internet of Things, Milan, pp 711–714
7. Qiao R et al (2018) Blockchain based secure storage scheme of dynamic data. Comput Sci 45:57–62
8. Halpin H, Piekarska M (2017) Introduction to security and privacy on the blockchain. In: European symposium on security and privacy workshops (EuroS & PW), IEEE Computer Society.
9. Gong T et al (2015) A medical healthcare system for privacy protection based on IoT. In: Proceedings of the 7th international symposium on parallel architectures, algorithms, and programming (PAAP), pp 217–222
10. Hu J-X et al (2017) An intelligent and secure health monitoring scheme using IoT sensor based on cloud computing. Journal of Sensors 2017:3734764
11. Li M et al (2013) Scalable and secure sharing of personal health records in cloud computing using attribute based encryption. IEEE Trans Parallel Distrib Syst 24(1):131–143
12. Hassan Alieragh M et al (2015) Health monitoring and management using internet-of-things (IoT) sensing with cloud-based processing: opportunities and challenges.

13. Sun W et al (2018) Security and Privacy in the medical internet of things: a review. Hindaw Secur Commun Netw 2018:1–9
14. Alasmari S, Anwar M (2016) Security and privacy challenges in IoT-based health cloud. In: International conference on computational science and computational intelligence. doi: https://doi.org/10.1109/CSCI.2016.43
15. Esposito C et al (2018) Blockchain: A panacea for healthcare cloud-based data security and privacy? IEEE Cloud Comput 5(1):31–37
16. Louni A et al (2016) Healing on the cloud: secure cloud architecture for medical wireless sensor networks. Futur Gener Comput Syst 55:266–277
17. Cubo J, Nieto A, Pimentel E (2014) A cloud-based internet of things platform for ambient assisted living. Sensors 14(8):14070–14105
18. Yeh K-H (2016) A secure IoT-based healthcare system with body sensor networks. IEEE Access 4:10288–10299
19. Páez DG et al (2014) Big data and IoT for chronic patients monitoring. Springer, pp 416–423
20. Yin Y et al (2016) The Internet of Things in healthcare: an overview. J Ind Inf Integr 1:3–13
21. Zyskind G, Nathan O, Pentland AS (2015) Decentralizing privacy: using block chain to protect personal data. In: Proceedings—2015 IEEE Security and Privacy Workshops, SPW 2015, pp 180–184
22. Yuan Y, Wang F (2016) Blockchain: the state of the art and future trends. Acta Autom Sinica 42(4):481–494
23. Hölbl M et al (2018) A systematic review of the use of blockchain in healthcare. Faculty of Electrical Engineering and Computer Science, University of Maribor, Slovenia
24. Darshan KR, Ananda Kumar KR (2015) A comprehensive review on usage of internet of things (IoT) in healthcare system. In: International conference on emerging research in electronics, computer science and technology
25. Krishnan B, Sai SS, Mohanthy SB (2015) Real time internet application with distributed flow environment for medical IoT. In: International conference on green computing and internet of things, Noida, pp 832–837
26. Khan SF (2017) Health care monitoring system in Internet of Things (loT) by using RFID. In: IEEE international conference on industrial technology and management, pp 198–204
27. Barro-Torres SJ et al (2012) Real-time personal protective equipment monitoring system. Comput Commun 36(1):42–50
28. Kumar DD, Venkateswarlu P (2016) Secured smart healthcare monitoring system based on IoT. Imperial J Interdiscip Res 2(10)
29. Saha HN, Auddy S, Pal S et al (2017) Health monitoring using Internet of Things (IoT). IEEE J 2017:69–73
30. Swetha Priya TC, Kanaka Durga A (2020) Clustering-based blockchain technique for securing wireless sensor networks. Data Engineering and Communication Technology. Springer, Singapore, pp 461–471
31. Deng R, He S, Chen J (2018) An online algorithm for data collection by multiple sinks in wireless sensor networks. IEEE Trans Control Netw Syst 5(1):93–104
32. Reddy NG, Chitare N, Sampalli S (2013) Deployment of multiple base-stations in clustering protocols of wireless sensor networks. In: 2013 international conference on advances in computing, communications and informatics (ICACCI), pp 1003–1006
33. Rahman M, Sampalli S (2015) An efficient pair wise and group key management protocol for wireless sensor network. Wireless Pers Commun 84(3):2035–2053
34. Yeole AS, Kalbande DR (2016) Use of Internet of Things (IoT) in healthcare: a survey. In: Proceedings of the ACM symposium on women in research 2016 (WIR '16). Association for computing machinery, New York, NY, USA, pp 71–76. https://doi.org/10.1145/2909067.2909079

Secured Covert Communication Through Blockchain Technology

Sharmistha Jana, Saraswati Dutta, Shovan Roy, Kousik Kundu, Alok Halder, Debkumar Bera, and Thanh Nhan Vo

1 Introduction

In order to avoid attackers, essential information is concealed in public media and transmitted through public channels using covert communication [1, 2]. By using covert communication technology, the important information is embedded into the redundant cover files' text, image, and video content, and the resulting stego files are transmitted. The transmission is then made using an open channel in order to meet the goal of safely transmitting essential information [3]. Technology for covert communication may both encrypt information being transferred and hide the fact

S. Jana (✉) · S. Dutta · S. Roy · K. Kundu
Department of Mathematics, Midnapore College [Aotonomus], Midnapore, West Bengal, India
e-mail: sharmistha.jana@midnaporecollege.ac.in

S. Dutta
e-mail: saraswati.dutta@midnaporecollege.ac.in

S. Roy
e-mail: shovan.roy@midnaporecollege.ac.in

K. Kundu
e-mail: kousik.kundu@midnaporecollege.ac.in

A. Halder
Department of Computer Science, Khragpur College, West Midnapore, West Bengal 721102, India

D. Bera
Department of Computer Science, Vidyasagar University, West Midnapore, West Bengal 721102, India

T. N. Vo
Department of Information Management, Chaoyang University of Technology, Taichung 41349, Taiwan, R. O. C.
e-mail: vtnhan@tdmu.edu.vn

J. K. Mandal et al. (eds.), *Proceedings of International Conference on Network Security and Blockchain Technology*, Lecture Notes in Networks and Systems 738,
https://doi.org/10.1007/978-981-99-4433-0_35

that it is occurring, making it one of the key methods for transmitting information secretly [4, 5].

Blockchain [6, 7] is a perfect platform for covert communication [8] due to its high channel robustness, strong participant anonymity, and simple group data transmission. Without using a centralized server, the data can be sent simultaneously between numerous nodes, lowering the possibility of covert communication leakage and tampering [9, 10]. Additionally, the blockchain's underlying network uses P2P (peer-to-peer) technology, and nodes record transactions by flooding mode. This creates the ideal environment for achieving covert communication. The shortcomings of conventional covert communication, in which the transmitter sends in one route and the recipient receives by leaving traces, can also be made up for by the aforementioned qualities of blockchain [10]. The first covert communication plan on the blockchain was put up by Partala [8]. The first covert communication technique that can be validated in terms of security is this one. By using the Bitcoin address as a cover and transmitting the secret message contained in the Bitcoin address by publicizing the Bitcoin transaction, Zhang et al. [7] devised a steganographic approach.

In an implicit steganography method based on Bitcoin addresses, Cao et al. [9] advised encrypting the secret message while making the public key so that it is not immediately contained in the Bitcoin address made by the public key. A covert communication technique based on Bitcoin transactions was proposed by Luo et al. [10]. It makes use of an index matrix of address interaction to reuse addresses for interaction relations and conceal secret messages through the interactive relationship between address and transaction amounts. The aforementioned techniques can decrease the number of transactions while increasing the embedding rate of information. There are still issues, such as the inability to reuse addresses or the ease with which transactional activities can be revealed by doing so. A blockchain-based application environment for covert communication is also made available by the quick development of digital currencies like Ethereum [11], Dogecoin [12], and Litecoin [13]. The majority of current covert communication techniques rely on the Bitcoin blockchain, which restricts the use cases for blockchain-based covert communication. Furthermore, when senders utilize transactions to deliver secret messages, recipients are unable to contact with negotiated addresses in the event of betrayal. It is vital to categories these receivers in such a situation.

In recent years, blockchain technology has witnessed significant advancements and diverse applications across various domains. One notable development is the Omni protocol, formerly known as Mastercoin, introduced in a comprehensive white paper by Willett et al. [14]. The Omni protocol presents an intriguing approach to blockchain by incorporating features like smart contracts and assets into the Bitcoin blockchain. Another pioneering concept that expanded the blockchain landscape is colored coins, outlined by Rosenfeld in 2012 [15]. Colored coins involve the association of real-world assets or information with specific tokens, showcasing the potential to revolutionize asset representation and transfer on the blockchain. Additionally, research has explored innovative techniques to embed covert communication within existing systems [16]. Roy and Changder propose a steganography method employing projection-aided payload dimension reduction and reconstruction for military covert communication [17]. Similarly, Hijaz and Frost delve into the exploitation of OFDM systems for covert communication [18]. Shifting focus to predictive modeling in the blockchain domain, Sridhar and Sanagavarapu introduce a Multi-Head Self-Attention Transformer for Dogecoin price prediction [19]. This work explores advanced machine learning techniques to forecast cryptocurrency prices, aiding decision-making for investors. On a technical front, Duong et al. present a hardware implementation for the efficient generation of blocks in the Litecoin blockchain system [20]. Their research focuses on enhancing the blockchain's performance through hardware optimization, demonstrating the continuous efforts to improve blockchain scalability and efficiency. These cited works collectively illustrate the diverse research directions and ongoing innovations within the blockchain space, emphasizing the growing potential and multidisciplinary nature of this transformative technology.

This study suggests a covert communication technique of secret data that is not restricted to Bitcoin and expands the use scenarios of the covert communication method based on blockchain technology based on the aforementioned current issues. In addition, the secret message recipients are split up into many groups, which lessens the likelihood that the secret message will be leaked and further restricts the extent of confidentiality. Here is the main contribution of the suggested work.

- The blockchain technology for virtual currency is the foundation for the group steganography method that is suggested.
- By accurately delivering secret messages to the specified sending group, the technique eliminates the constraint of one person sending numerous receivers receiving information indiscriminately and ensures the security of the group on covert communication.
- The address interaction index matrix is built to carry a hidden message. The matrix reduces resource usage, enhances information embedding rate, and prevents the development of random transactions by sensibly modulating the address.
- A method for modulating transactions by observing the moment of block formation is suggested. The method ensures secrecy by adapting the amount of addresses used in transactions to the block generation rules in order to avoid using identical addresses within the same block.

2 Related Works

The message or information is concealed in the locations as part of the covert communication strategy based on blockchain technology. Setting a unique code will allow the recipient to recognize the address. A flag to draw out the hidden content. A solution for clandestine communication via a blockchain channel, known as the BLOCCE system, was initially put up by Partala [8] in 2018. It is the first cryptographically safe steganographic communication system on blockchain that uses least significant bit (LSB) steganography on Bitcoin addresses. The V-BLOCCE system, which Zhang et al. [7] presented in 2019, uses vanitygen to generate specific Bitcoin addresses and circumvents BLOCCE's drawbacks, such as the system's low information embedding rate, high transaction volume, and poor management efficiency. To enable the receiver to recognize the unique transaction, the system must enter information in the OP RETURN field. The modified transaction is distinct from the standard transaction, which makes it simple for the attacker to suspect something is wrong. The steganographic method is not widespread since the system generates Bitcoin addresses using specialized software. A chain-based data embedding method was presented by Cao et al. [9] in 2020. The method disguises secret messages as random integers in the resultant public key to avoid the secret message from being directly stored in the Bitcoin address generated by the public key. The idea provides a safe covert communication channel; however, it still has problems with low embedding rates and waste.

2.1 *Objective of the Research Work*

To the extent that we know, there aren't many research papers on blockchain-based secure covert channels. There are, however, techniques and services that routinely add arbitrary data to a blockchain. The insertion might be based on either the output scripts that identify the recipient of the unlocked funds or the input scripts that unlock funds for the transaction. Deviations from the default payment templates give users the ability to insert a message into the blockchain, depending on the transaction type (coin base transactions that create new money or regular transactions that make payments).

2.2 *Contribution of the Paper*

- The sender sends the message securely through the blockchain, which cannot be accessible by third party because of its architecture.
- It is peer-to-peer (P2P) networking.
- Blockchain can also compensate for the defect traditional covert communication in which the sender sends directionally and the receiver receives by leaving trace.
- The data embedding method increases security by concealing secret messages as random numbers in the resulting private key.
- A covert communication channel cannot be easily identified because of the mechanisms it uses for illegal data transfers. However, it can be detected by monitoring system performance.

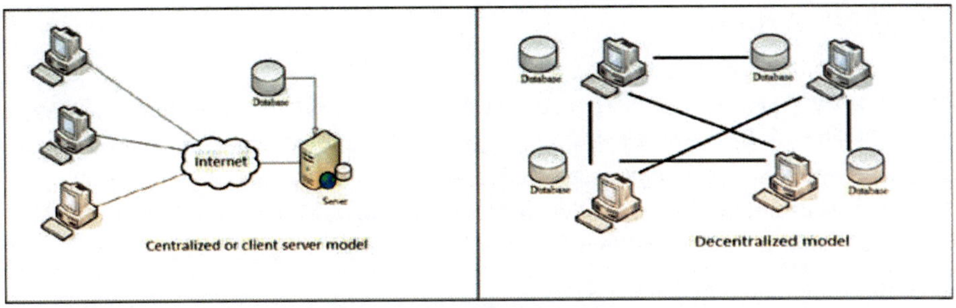

Fig. 1 Difference between centralized and decentralized model

3 System Implementation Methods

3.1 Centralized Application

A centralized application that operates on a single server or cluster of servers is managed and owned by a single company. The user interacts with a typical program, downloads it from the app store, and uses it by sending requests and information to a single server or group of servers. This request is processed by the server, and the proper response is given. Hence, the backend code of the centralized application runs in centralized servers. Figure 1 shows the difference between centralized and decentralized model.

3.2 Decentralized Application

The user communicates with a blockchain that uses smart contracts which is a group of code, logic, and data that can be utilized to operate the program, provides the user with the option to download the source code. User needs to pay the developer a predetermined sum (in crypto currency) in order to utilize these apps. Hence, the backend code of the decentralized program utilizes a peer-to-peer network and no need to run on a centralized server. Figure 2 shows example of decentralized model.

Fig. 2 Plan has been presented in a condensed version of our suggested plan

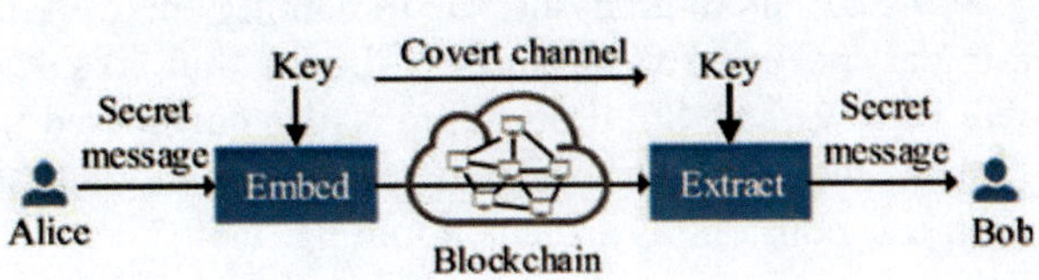

3.3 P2P Network

P2P stands for peer-to-peer network which is created when more than one device is connected to a single network directly and no additional server is needed. In this network, every connected device acts at least once as a server and client. For instance, User A must configure his computer to provide (sharing) access to the printer if User B want to use the printer that is linked to User A's PC.

3.4 Blockchain

In simple term, we can say blockchain is a transparent, immutable, and distributed ledger. Each block in a blockchain has the following elements: block number, node, time stamp, data, previous hash, and hash value. Simply put, block number is a number. It is kind of serial number of any record. Hash value has value is a group of 64 hexadecimal characters. Each characters contains 4 bits (because of hexadecimal number). So in total it has $64 \times 4 = 256$ bits. This hash value is generated by processing block number, nonce, time stamp, and data. There are five requirements of hashing algorithm are: One way (That means only encryption should possible not decryption) Deterministic (Always generated a particular hash for a specific data), Fast Computation, Withstand Collisions (Difficult to tamper data because decryption is not allowed.) and Avalanche Effect (If any data or content changed even by a single letter or symbol then whole hash will change, i.e., hash of "ABC" will be different from hash of "ABC.").

The term "nonce" refers to a single number which is a combination of "number-used-once." Nonce provides predictable variation and the demonstration of real achievement of a goal in a proof-of-work system. In accordance with the Bitcoin protocol, data chunks called "blocks" are constructed in a precise way and are inviolable. Each block in a blockchain is eternally immutable, remaining there frozen in the same state as it was initially stored. Nonce is a 32 bit number. Range of nonce is 0 to 232–1 (approx 4×109 numbers).

A Timestamp, represented as UNIX TIME, encapsulates the creation time of a block, serving a pivotal role in addressing specific challenges during the mining process. Because total valid hashes are 1077 and total generated nonce are approx 4×109. So, maximum hashes are remaining uncovered during mining. If we consider a miner speed is 108 hashes/sec, then within 40 s 4×109 these number of nonce will be traversed. So, time stamp is also considered with nonce to generate hash to find any block from transaction pool. When target hash value is smaller than miners hash it is consider as a success mining, that time is consider to store as time stamp value.

4 Proposed Method

The BLOCCE approach has its own drawbacks, including poor information embedding performance, excessive address utilization, sluggish communication performance, and exorbitant prices. On the contrary V-BLOCCE along with the usage of reused addresses generated by private key has much more increased performance. This work implements the system model of the V-BLOCCE model in Python programming platform to achieve the objective mentioned earlier.

The secret communication technique that makes use of vanitygen-generated addresses. The least significant bits of the addresses from the message from A are preserved when Alice converts the message to the blockchain. Both of those addresses and A's allow Bob to read the message bits. Encryption is also utilized in the real procedure, and the message's beginning is specified.

In our hypothetical situation, Alice tries to send Bob a secret message through the blockchain while the opponent looks for any such communication. Alice cannot change the "cover texts" (i.e., the blocks) stored in the blockchain, in contrast to traditional steganography because of the consensus mechanism. Alice, however, has total control over the message she sends to the chain.

Moreover, the consensus method will defend these messages from potential adversarial efforts at alteration. These payments will be used, and the payment addresses in particular, to send Bob a secret message. We will send one bit for each block that appears in the chain in order to keep things simple. The following gives a general outline of the plan.

i. Alice generates a number of private and public key pairs

$$\left(s_k^{(1)}, p_k^{(1)}\right), (s_k^{(1)}, p_k^{(1)}), \ldots, \left(s_k^{(n)}, p_k^{(n)}\right)$$

and generates the message addresses $a^{(1)}, a^{(2)}, \ldots, a^{(n)}$ corresponding to these keys.

ii. Alice creates messages from her own account to these addresses, ordering them according to the secret text message m such that the message addresses' least significant bits (LSBs) make up m.

iii. Alice sends the message to the blockchain in the proper order.

iv. Bob reads the concealed text message from the LSBs of the payment addresses and searches the blockchain for messages sent by Alice.

Because to her control over the created key pairs, it should be noted that Alice does not lose any messages when the scheme is run. While some proof-of-work implementations of blockchains may have substantial transaction costs, there are other blockchains with various consensus processes that do not require transaction fees. The following is an illustration of the suggested method's general process.

Step 1. Encryption of sender message to generate temporary cipher text.
Step 2. Encoding with base64 to generate the cipher text.

Step 3. Private key generation.
Step 4. Embedding of private key generated addresses with the cipher text.
Step 5. Transection of the address through blockchain.
Step 6. Address received in the receiver's end.
Step 7. Extraction of cipher text address.
Step 8. Base64 decoding to get temporary cipher text.
Step 9. Decryption to the original message.

The corresponding algorithms for embedding and extraction is presented in algorithm A and algorithm B.

4.1 Embedding Algorithm

procedure Embed$((k,\lambda),m,B)$

  $c \leftarrow Enc(k,m)$

  Concatenate $c^0=\lambda||c$

  Set $N=|c^0|$

  Interpret c^0 *as a bit representation* $c^0{}_1c^0{}_2 \ldots c^0{}_N \in \{0, 1\}^N$

  Set $i=1$

    while $I \leq N$ *do*

    Generate unseen $(s_{k,}, p_k) \leftarrow Gen\ \Sigma(1^s)$

    $a \leftarrow H(p^{(i)}{}_k)$

    Interpret ***a*** *as a bit representation* $a_1a_2 \ldots a_n \in \{0, 1\}^n$

      if $a_n=c^0{}_i$ *then*

      $\mu \leftarrow M_H$

      Generate a unique identifier ***t*** *for the message*

      $\sigma \leftarrow Sign(s^{(A)}{}_k, (p^{(A)}{}_k, a, \mu, t))$

      Submit $(p^{(A)}{}_k, a, \mu, t, \sigma)$

      Wait for the blockchain to publish a new block

      Update H

      $i \leftarrow i+1$

      end if

    end while

  end procedure

4.2 Retrieval Algorithm

```
procedure Retrieval ((λ,k),B)
    i=1
    j=1
        while have not yet discovered λ do . Take for λ
        C=Read (j)
        if C=⊥ then
        Wait until a block appears and read it: C=Read (j)
         end if
            for any message  P ε C do
                if P is from p^(A)_k  then
                Retrieve address a from P and get the LSB a_n
                 Scan if  found the entire λ ε {0, 1}^n_λ
                end if
            end for
            j←j+1
        end while
        i=1
        while I ≤ N−n_λ do
            C=Read (j)
            if C=⊥ then
            Wait for a block to emerge before reading it.: C=Read (j)
            end if
            for any message  P ε C do
                if P is from p^(A)_k then
                Retrieve address a from P and get the LSB a_n
                c_i←a_n
                i←i+1
                end if
            end for
            j←j+1
        end while
    Compile c=c_1c_2. . . c_N−nλ
    m←Dec(k,c)
    print  m
    end procedure
```

5 Security of the Proposed Work

In this section, we keep BLOCCE's security in account. By simulating a specific concealed text attack by a probabilistic polynomial time adversary on BLOCCE, we are able to construct a security definition for the block chain-based covert channel, adhering to the ideas of security for a stegosystem. We begin by enumerating our presumptions. The development of a security definition known as payment in distinguishability follows, which calls for the adversary to be able to tell the difference between the payload carrying payments and random data. Lastly, we demonstrate how BLOCCE conforms to this definition.

Based on a condensed version of the ideal blockchain B, our security proofs. The applied cryptographic hash function is characterized as a random oracle, and we assume that digital signatures are existentially unforgeable. In our case, there are three participants.

1. Our scheme's transmitter, Alice, is identified by her (payee identification) public key $p^{(A)}{}_i$. She and Bob, the recipient, have already agreed on a private key (λ, i) $\in \{0, 1\}^{m\lambda} \times \{0, 1\}^{\text{mi}}$ that is only known to the two of them. She tries to communicate with Bob privately via the streamlined ideal blockchain B. The total number of embedded bits, N, is known to both Alice and Bob. Lastly, Alice is able to sample from her "typical" distribution of payment amounts MH given her payment history H.
2. Bob is a proxy for the target of our plan. He understands Alice's public key $p^{(A)}{}_i$, secret key (λ, i), total number of embedded bits N, and expects a private message from Alice over the blockchain.
3. The opposition takes an effort to identify any hidden communication occurring on the blockchain. We suppose that the opponent is aware of Alice's public key $p^{(A)i}$. With the oracles, the warden also has full access to the blockchain B. Read and send. The opponent's task is to discern between ordinary payments and payments for covert communications.

6 Experimental Result

We utilize "Hello Rimpa" as the covert message in the experiment. We use is an Intel i7 (CPU: Intel(R) Core(TM) i7-8750H CPU @ 2.20GHz 2.21GHz, and RAM is 32.0GB) processor, and the operating system is Windows 10. The proposed method is tested using Python, and Python 3.6 is the platform used for compilation (Fig. 3). We encode the secret information in the 7 bits of the transaction amount to be compatible with the comparison approach.

Currently, blockchain-based clandestine communication is only theoretical, and it has a lower embedding capacity than conventional covert communication. Researchers are driven to examine covert communication on blockchain because of its qualities, which could compensate for some of the drawbacks of traditional covert communication.

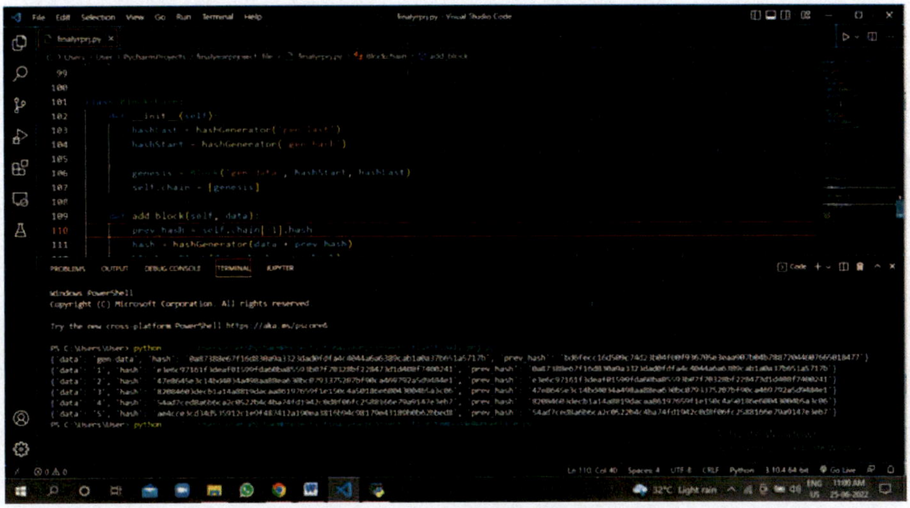

Fig. 3 Blockchain created using python programming applying our proposed algorithm

The secret message is sent using n addresses, indicated as, to enable comparison of the capacity for existing techniques to transmit secret messages. $A = (a_0, a_1, \ldots, a_{n-1})$, and the number of transactions generated is denoted as T_r, $T_r = (\text{tr}_0, \text{tr}_{1,\ldots,} \text{tr}_i)$. The secret message is denoted as M, $M = (m_0, m_1, \ldots, m_{|M|-1})$. The embedded capacity of each transaction is 1 bit and 8 bits respectively in [8, 9]. The average embedding capacity of each transaction is Eq. (1) and (2) respectively for [10] and our new method.

$$C = (|S|/|S_{si=1}|) + |B|, \tag{1}$$

$$C^{/} = (|M|/|M_{\text{mi}-1}) + |B|, \tag{2}$$

where S is the sequences of address interaction. Transaction number for various embedded capacities in byte are shown in (Fig. 4).

7 Future Scope

We limited ourselves to embedding just one bit for each block in order to make our investigations more straightforward. The time it takes to publish a new block is measured in minutes rather than seconds in many applications, including Bitcoin. As a result, our plan has a poor throughput over time. Nonetheless, it is simple to raise the embedded bit count per block. Several LSBs of the address could be matched as one method of doing it. Since the bits are chosen at random from an oracle, the

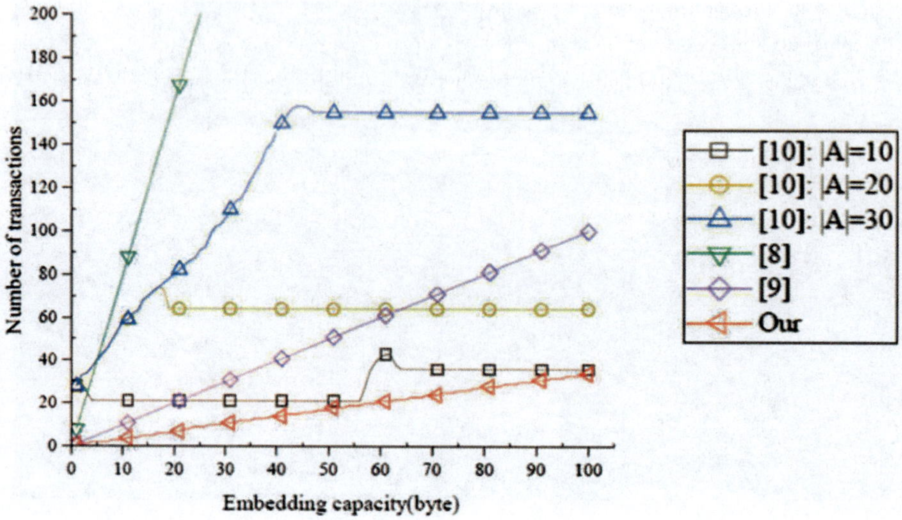

Fig. 4 Transaction number for various embedded capacities in byte

predicted computation time for embedding increases exponentially with the number of bits. A block of payments that includes multiple installments is more effective. In this situation, it is necessary to guarantee the sequencing of the message bits, for instance by utilizing the public payer keys or the unique payment identifier.

8 Conclusion

We offer the first implementation of a covert communication channel over a blockchain that can be proven to be secure. Our arguments are demonstrated using the random oracle model, in which the blockchain's cryptographic hash function is represented as a random oracle. We create a condensed version of the ideal blockchain that simulates the blockchain implementations that underlie real-world cryptocurrencies. Based on this approach, we propose a technique for embedding a single bit for each block via blockchain payments. We demonstrate the method's dependability and expected polynomial time of execution. We formulate the idea of payment in distinguishability, where the transmitted concealed message should be computationally indistinguishable from random payments, to describe the security of covert channels on a blockchain. Our suggested methods are helpful for healthcare organizations, enterprises in the legal and financial sectors, tax advisors, private banking, pharmaceutical, automotive, and industrial sectors, among others, to protect consumer data.

Acknowledgements The authors are thankful to the Vidyasagar University for providing infrastructural facilities required for carrying out the project.

References

1. Filler T, Judas J, Fridrich J (2011) Minimizing additive distortion in steganography using syndrome-trellis codes. IEEE Trans Inf Forensics Secur 6(3):920–935
2. Boroumand M, Chen M, Fridrich J (2019) Deep residual network for steganalysis of digital images. IEEE Trans Inf Forensics Secur 14(5):1181–1193
3. Zhang Y, Luo XY, Wang JW, Guo YQ, Liu FL (2021) Image robust adaptive steganography adapted to lossy channels in open social networks. Inf Sci 564(5):306–326
4. Holub V, Fridrich J, Denemark T (2014) Universal distortion function for steganography in an arbitrary domain. EURASIP J Inf Secur 1:1–13
5. Zhang Y, Qin C, Zhang WM, Liu FL, Luo XY (2018) On the fault tolerant performance for a class of robust image steganography. Sig Proc 146:99–111
6. Liang W, Fan Y, Li KC, Zhang D, Gaudiot JL (2020) Secure data storage and recovery in industrial blockchain network environments. IEEE Trans Industr Inf 16(10):6543–6552
7. Zhang LJ, Zhang ZJ, Wang WZ, Waqas R, Zhao CH, Kim S, Chen HL (2020) A covert communication method using special bitcoin addresses generated by vanitygen. Comput Mater Continua 65(1):597–616
8. Partala J (2018) Provably secure covert communication on blockchain. Cryptography 2(3):1–18
9. Cao HT, Yin H, Gao F, Zhang ZJ, Khoussainov B, Xu SB, Zhu LH (2020) Chain-based covert data embedding schemes in blockchain. IEEE IoT J online. https://doi.org/10.1109/JIOT.2020.3040389
10. Luo XY, Zhang P, Zhang ML, Li H, Cheng QF (2021) A novel covert communication method based on bitcoin transaction. IEEE Trans Ind Inf. Online. https://doi.org/10.1109/TII.2021.3100480
11. Wang X, He J, Xie Z, Zhao G, Cheung S (2020) Contract guard: defend ethereum smart contracts with embedded intrusion detection. IEEE Trans Serv Comput 13(2):314–328
12. Simmons GJ (1984) The prisoners' problem and the subliminal channel. Crypto, Plenum Press, New York.View at: Publisher Site | Google Scholar
13. Nakamoto S (2008) Bitcoin: a peer-to-peer electronic cash system. Decentralized Bus Rev 21260.View at: Google Scholar
14. Willett J, Hidskes M, Johnston D, Gross R, Schneider M (2016) Omni protocol specification (formerly mastercoin), vol 28. White paper
15. Rosenfeld M (2012) Overview of colored coins. White paper, bitcoil.co.il, Coinprism, vol 41, p 94. View at: Google Scholar
16. Helen D, Arivazhagan D (2014) Applications, advantages and challenges of ad hoc networks. J Acad Ind Res 2(8):453–457
17. Roy R, Changder S, Steganography with projection aided payload dimension reduction and reconstruction for military covert communication. Int J Comput Appl 139
18. Hijaz Z, Frost VS (2010) Exploiting OFDM systems for covert communication. In: 2010-MILCOM 2010 military communications conference. San Jose, CA, USA, pp 2149–2155.View at: Google Scholar
19. Sridhar S, Sanagavarapu S (2021) Multi-head self-attention transformer for dogecoin price prediction. In: Processing 14th international conference on human system interaction (HSI), pp 1–6
20. Trung Duong LV, Van Hieu D, Luan PH, Hong TT, Duc Khai L (2021) Hardware implementation for fast block generator of litecoin blockchain system. In: Proceeding 2021 international symposium on electrical and electronics engineering (ISEE), pp 9–14; vol 6, no 3, pp 920–935, 2011

MetaFund: Blockchain Based Crowdfunding Platform

Rohan Shinde, Keval Dhanani, Sahil Chorghe, and Anand Godbole

1 Introduction

Crowdfunding platforms are charging some fees for every project they have listed. Well-known crowdfunding platforms require real projects and usually reject start-up ideas. The primary feature of blockchain is its decentralized character. It may have an impact on crowdfunding campaigns by lowering processing expenses. Because blockchain eliminates the need for intermediaries or third parties in financial transactions, it has the potential to make crowdfunding considerably more inexpensive.

The major motivation behind using blockchain instead of traditional centralized methods is that it makes transaction transparent. A contributor might track their contribution from them all the way through a charity to the recipient and beyond. Every transaction can be recorded using blockchain technology. The immutability and tamper-resistance of data provided by the blockchain itself will further boost transparency and accountability. The implemented system benefits from the use of Polygon's proof-of-stake (PoS) consensus algorithm and thereby provides faster transaction speeds and lower gas fees.

This paper is organized as follows: in Sect. 2, we have summarized some related works on using blockchain to create a crowdfunding platform. Section 3 includes the

R. Shinde (✉) · K. Dhanani · S. Chorghe · A. Godbole
Sardar Patel Institute of Technology, Mumbai 400 058, India
e-mail: rohan.shinde@spit.ac.in

K. Dhanani
e-mail: keval.dhanani@spit.ac.in

S. Chorghe
e-mail: sahil.chorghe@spit.ac.in

A. Godbole
e-mail: anand_godbole@spit.ac.in

J. K. Mandal et al. (eds.), *Proceedings of International Conference on Network Security and Blockchain Technology*, Lecture Notes in Networks and Systems 738,
https://doi.org/10.1007/978-981-99-4433-0_36

proposed system that includes the architecture of the system using Polygon sidechain. In Sect. 4, we have included the implementation details. Section 5 discusses our results. Section 6 discusses conclusion and future work.

2 Literature Survey

In [1], LikeStarter, a blockchain-based decentralized platform combines social interactions with crowdfunding mechanisms. LikeStarter assigns Likoins to users that fund a given project. These tokens can be used to buy artifacts as well as token that enables voting. Tokens can be staked on projects that a user supports. Additionally, users own a share on the project they are donating to. The value project determines value of the token. Any users possessing Likoins gain voting rights. Price is suggested based on votes and the process is time consuming.

In [2], the authors have used the smart contract as a transaction protocol which automatically executes, controls, and documents actions of the transactions according to the agreement on behalf of project creators and investors. This paper proposes a method which includes two contracts, one stores the projects and other handles the transactions for all projects. The acceptance of the fundraising request is based on voting of majority of contributors. The application developed is still in its exploratory stage and various law specific issues are not addressed.

In [3], the paper discusses the issue of the lack of transparency in the transactions involving contributions supplied by the government or other contributors. Authors state that allowing contributors to trace their contributions is necessary to increase openness in social funding. The goal of their work was to maintain funds security and ensure donation traceability. Charity chain uses blockchain to record every transaction. The method in the paper would track donations and inform the donor when their money has successfully reached the beneficiary using smart contracts. The computational efficiency and scalability of the byzantine consensus algorithm are utilized. Because it is a public platform, the Ethereum platform is used by the authors. Authors used Ethereum which uses proof-of-work consensus mechanism which is more power consuming and relatively slow process when compared with proof-of-stake.

In [4], using blockchain helps regulate the process of crowdfunding and makes the process more secure. Author has researched about how using blockchain is better than traditional centralized platforms. Blockchain-based crowdfunding is still a new concept and there is not a lot of research available. Authors have only provided results of literature surveys and a comparison between them. No application has been developed.

In [5], authors have provided an architecture for public welfare crowdfunding. Open consensus is used to provide transparency and establish a trust model. Information disclosure module is used to display information about a funding request. The application then tracks the source of funds to the usage. Also, a point-based system is used to reward trustworthiness. The architecture described is very vague and is based on a literature survey. Although it is not a complete one-stop solution, it provides a starting point for a blockchain-based application for fundraising.

In [6], the authors have suggested a framework applying the concepts of blockchain to insure cyber products. An auction-based system is suggested where winners will be selected to insure the product. The major goal here is to ensure a cyber product which could be a software or hardware. The proposed model counters risk by using blockchain as a trust factor to reduce risk tendency. Blockchain also provides transparency allowing the donors to monitor the work of the product. The auction function that was created is based on Ethereum blockchain and provides slower performance in terms of transaction speed. The auction function also had significant gas cost. This paper provides a good background on how investors can be convinced to invest into a product given transparency and incentive, both of which are properties that blockchain can provide.

In [7], the authors presented a unique resource tokenization technique based on blockchain to crowdfund wireless network deployment. Load on a particular base station can be verified using the proposed proof-of-load algorithm. When a particular percentage of a base station's resources is allotted to the mining process, BCR increases as miner base stations can only spend these resources to build a block in blockchain.

In [8], the authors have proposed a system using Ethereum blockchain, they have written a smart contract in solidity which is used for investing and tracking the donations. Investors have control over whether a project is funded or not. Proof-of-work consensus algorithm is used to run the contract which makes the entire process slow. The project cannot get funds without approval, so malicious users can donate a small amount and get a vote in the project. Approval system is redundant and adds delay to the process.

In [9], the authors have suggested an Ethereum-based system which uses the proof-of-authority consensus algorithm. They have three types of users, non-government organizations (NGOs) which can request for funds, government body which approves the request of NGO, and the donors who donate the money. Even though the system of donating and receiving money is transparent, the proposed system adds a dependency on government body to approve the requests which makes it partially centralized.

In [10], the authors have proposed a system which can efficiently and reliably track charity donations with the use of blockchain technology. The authors have made comparison between blockchains and the impact the public and private factor of blockchain can have on the reliability of the system. The authors have also detailed the blockchain used and the working of their version of smart contract to create a system for reliable donations.

In [11], the authors have suggested the advantages blockchain has over other conventional centralized systems. The security, transparency, and immutability of blockchain are discussed. Also, some other key features that can benefit conventional systems include decentralization which removes the use of central third-party authority, persistency meaning once data is added it cannot be tampered, and anonymity meaning user can maintain their privacy while interacting with systems. The types of blockchains, i.e., public, private, and consortium are discussed. Consensus algorithms and their efficiency are also discussed.

In [12], the authors have discussed blockchain as an efficient way to implement trust management systems and to increase security in authentication infrastructures. The authors have discussed a theoretical model for trust management and how it can be used in a blockchain. Further, they have tested this model under certain attacks and displayed the efficiency of blockchain to secure the system.

3 Proposed System

This section describes the architecture of the platform, the proposed methodology, and the algorithms used. The user follows a series of steps for creating campaigns. Campaign creators will connect to the MetaMask wallet and select their accounts that they want to create a campaign with. The account address for the MetaMask wallet will be fetched onto the server. The server will then create a campaign along with the creator's account address and send the details to smart contract. A smart contract will get the campaign details and mine a block on the Polygon network with the account address and campaign details. Once a block is mined, the campaign is successfully created. Block-mined and campaign-created status will be sent back to the server, which in turn will update the campaign creator. A list of created campaigns can be viewed by the campaign donor from the server. To donate, the donor must connect to their MetaMask wallet and select an account. The account address of the donor will be sent to the server, and the server will map the address to the amount of donations that the donor will make. These donation details will be sent to the smart contract. A smart contract will then mine a block with donation details and update the status of the campaign on the Polygon network. The updated status of the campaign will be returned to the server and can be viewed by the donor (Fig. 1).

3.1 Smart Contracts

Smart contracts are rules that are executed whenever we want to perform any action on the blockchain. They eliminate the need for a third party to execute the functionality. Instead, we can write a smart contract to perform that functionality. They are deployed on the blockchain and are constantly running, waiting for input.

(1) CreateCampaign(): This function is used to create a new campaign on the platform. Details such as address of creator, title, description, donation goal and deadline are required.
(2) Contribute(): This function is called whenever a person wants to donate to a particular campaign. There are certain conditions on contributing such as:

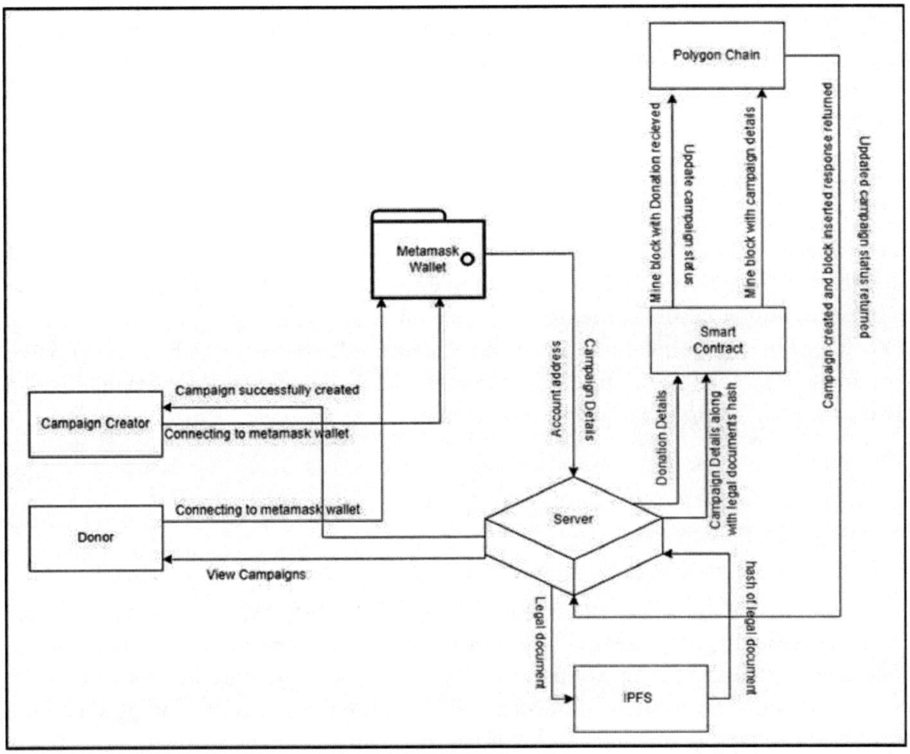

Fig. 1 Block architecture of the system

you can only donate if the campaign is active, you cannot donate to your own campaign, and you can donate to a campaign only once.

(3) Withdraw(): This function is used by a campaign creator, the conditions applied are: the campaign has to pass its deadline, the campaign should have reached its goal, and the campaign should have some balance left to withdraw.

(4) Refund(): This function can be used by donor in case they change their mind and want to refund their donation. The conditions for a refund are: the campaign goal and deadline should not be complete, someone who has not contributed to the campaign should not be able to refund any amount and a refund cannot be made if the campaign balance is zero.

3.2 Hardhat

Hardhat is used as a development environment for Ethereum blockchain. It facilitates the compilation of contracts and their execution on a development network. We installed the hardhat node package and used it to compile and deploy our smart contract to the Polygon Mumbai test net. Once the smart contract is deployed using hardhat, it returns an address, and this address can be used to interact with the functions of the smart contract using JavaScript.

3.3 IPFS

Interplanetary file system (IPFS) is a peer-to-peer, distributed file system. IPFS is used to store files remotely in a distributed environment. We used IPFS to store the legal documents that the campaign creator will share while creating the campaign. IPFS will return a unique hash of this document to the server, which will be further sent to our smart contract. The smart contract will then add the campaign details and the document hash to the polygon blockchain.

4 Implementation

The implemented system has two major components: the donor component and the campaign creator component. The MetaMask wallet is required for accessing the system and donating to the campaigns. The donor, upon visiting the website of the system, can connect to their MetaMask wallet through the given prompt. Once the user has connected their MetaMask wallet, they can access the platform. The donor can view ongoing campaigns and select the ones they want to donate to. The donor must also have enough Matic coins in their MetaMask wallet to cover both the donation amount and the gas fee displayed during wallet confirmation. If the user has enough Matic coins, they can donate to the campaign by inputting the amount and completing the transaction in their MetaMask wallet. The donation can be seen as progressing the campaign (Figs. 2 and 3).

When the campaign creator visits MetaFund for the first time, he needs to connect to his MetaMask account in order to sign up for MetaFund. If this user does not have a MetaMask account, he must create one, and then only he will be able to use MetaFund. If the user has a MetaMask account, he will need to sign in to his account, and then MetaMask will automatically get signed in for this user. The MetaMask wallet will be connected to MetaFund for transactional and account purposes. Now the user can create a campaign for a specific category, where he needs to provide all the necessary details that will be mandatory to create a campaign. After filling in the details, the user can now publish the campaign on MetaFund. The user (campaign creator) can withdraw the funds collected only if the campaign goal is reached. If the campaign goal is yet to be reached, the user cannot withdraw the current collected funds.

Polygonscan is a block explorer platform that allows users to access any Polygon blockchain transaction. As Polygon is a public blockchain, developers have access to various features like viewing list of top accounts, statistical graph, gas tracker, each transaction details, and list of pending transaction on the Polygonscan platform.

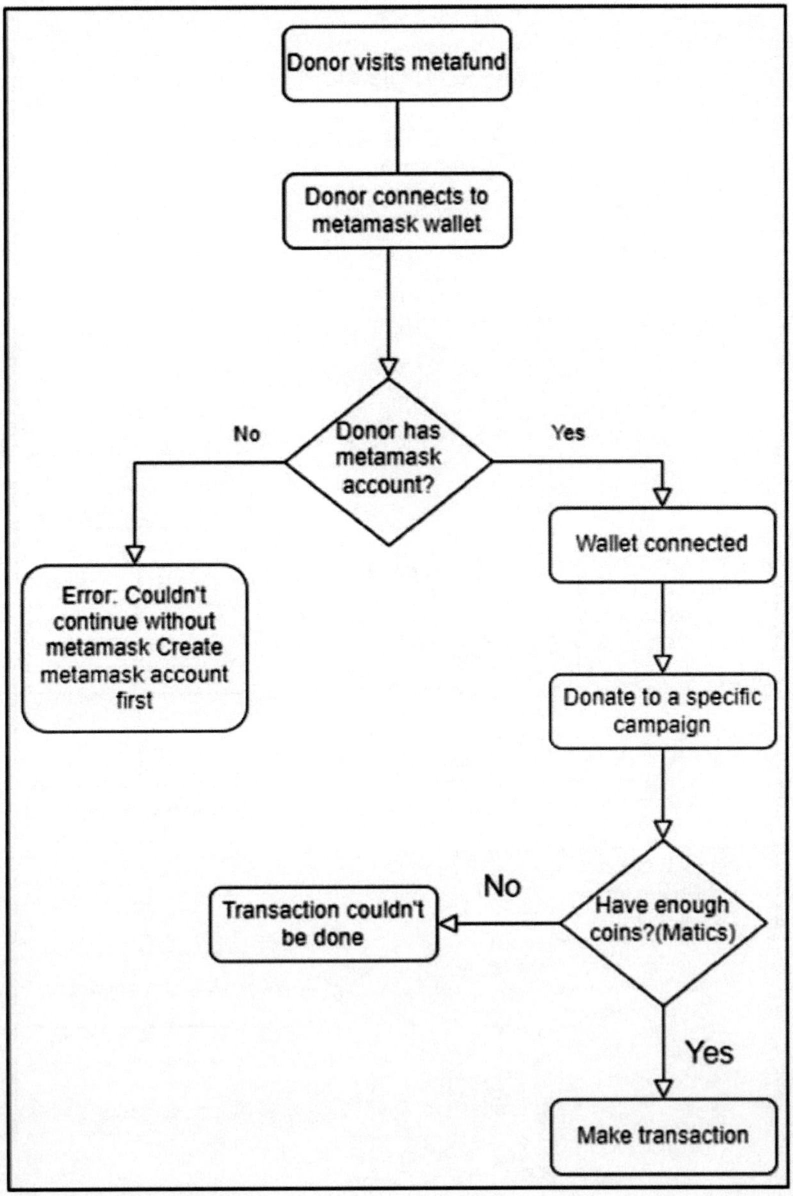

Fig. 2 Donor process

Our transaction details can be viewed on the Polygonscan platform [in Fig. 4] as we deployed our smart contracts on the Polygon Mumbai test net. The metadata of the block that was added can be noted and it contains details such as transaction hash, block number, gas fees, etc.

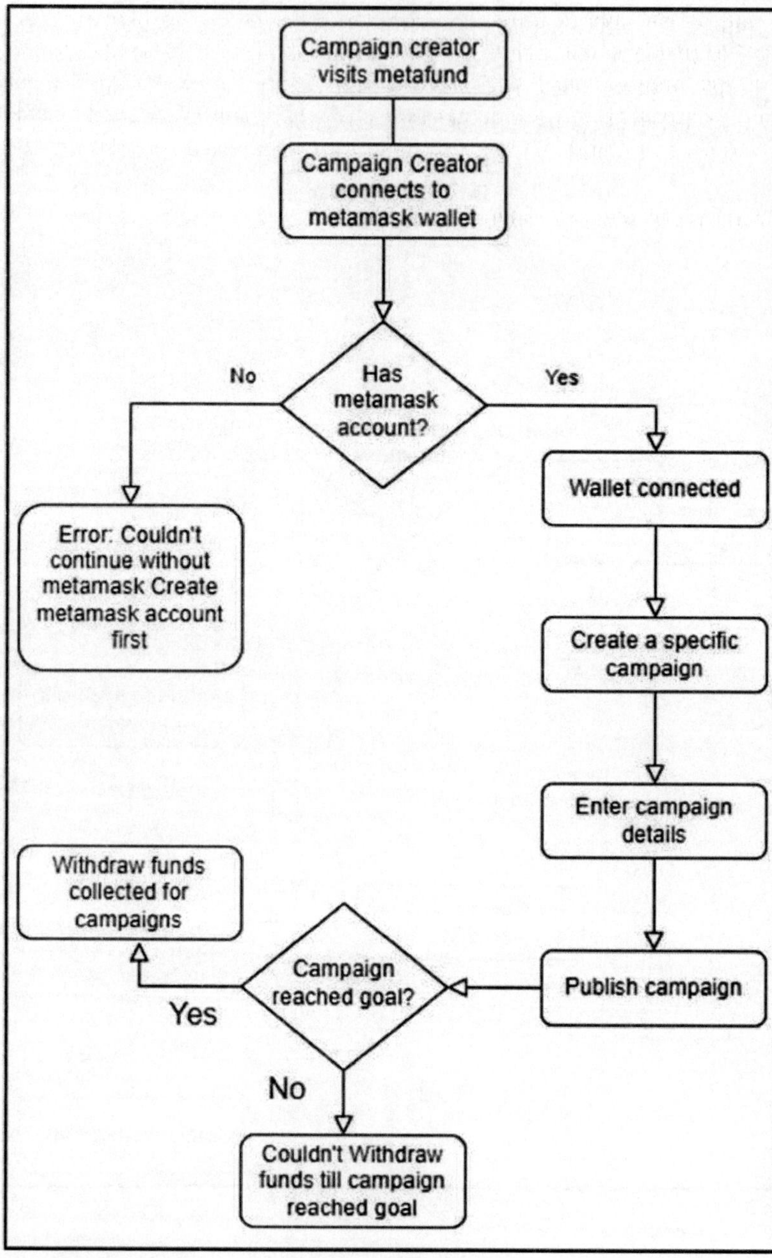

Fig. 3 Campaign creator process

Fig. 4 Polygon scan transactions

Table 1 Gas cost for campaign creation and donation

Function	Gas cost
Creating a campaign	0.00239 MATIC
Donating to a campaign	0.0000528 MATIC

5 Results and Discussion

We have proposed the use of Polygon blockchain in our solution as Polygon has low gas price and lower transaction fee as the network uses the proof-of-stake algorithm, so the energy cost and consumption are much lesser when compared with Ethereum (Table 1).

This can be seen in the results, as the gas fee is lower when using our proposed system. For creating a new campaign, the user needs to pay a gas fee of 0.00239 MATIC. Similarly, while donating to any campaign, the user needs to pay a gas fee of 0.0000528 MATIC. So, the total cost of that transaction would be 0.0000528 MATIC plus the amount he/she wants to donate.

6 Conclusion

Blockchain is an upcoming technology, and a lot of applications can be done securely and efficiently using blockchain. We implemented one such application of crowd-funding for start-ups, medical requirements, and charities using blockchain. We used the Polygon sidechain because of its faster transaction speeds and proof-of-stake consensus algorithm. The smart contract we wrote allowed us to ensure transparency and make sure that the donor had control over their donations. We also provided an incentive for donors in the form of MetaCoin, which is a platform-specific coin that can be earned by donating to campaigns. Although our platform has various features and improvements over past research, there is still room for improvement. Smart contracts could be improved to support trust building by tracking the status of the campaign. They could also be improved to allow donors to approve goal-based withdrawals of funds.

References

1. Zichichi M, Contu M, Ferretti S, D'Angelo G (2019) LikeStarter: a smart-contract based social DAO for crowdfunding. In: IEEE INFOCOM 2019—IEEE conference on computer communications workshops (INFOCOM WKSHPS), pp 313–318. https://doi.org/10.1109/INFCOMW.2019.8845133
2. Yadav N, Sarasvathi S (2020) Venturing crowdfunding using smart contracts in blockchain. In: 2020 third international conference on smart systems and inventive technology (ICSSIT), pp 192–197. https://doi.org/10.1109/ICSSIT48917.2020.9214295
3. Sirisha NS, Agarwal T, Monde R, Yadav R, Hande R (2019) Proposed solution for trackable donations using blockchain. Int Conf Nascent Technol Eng (ICNTE) 2019:1–5. https://doi.org/10.1109/IC-NTE44896.2019.8946019
4. Hartmann F, Grottolo G, Wang X, Lunesu MI (2019) Alternative fundraising: Success factors for blockchain-based versus conventional crowdfunding. In: 2019 IEEE international workshop on blockchain oriented software engineering (IWBOSE), pp 38–43. https://doi.org/10.1109/IWBOSE.2019.8666515
5. Fan Y, Ao C, Jingren L (2021) Research on the application model of public welfare crowdfunding based on blockchain technology. In: 2021 2nd international conference on E-commerce and internet technology (ECIT), pp 441–445. https://doi.org/10.1109/ECIT52743.2021.00098
6. Vakilinia I, Badsha S, Sengupta S (2018) Crowdfunding the insurance of a cyber- product using blockchain. In: 2018 9th IEEE annual ubiquitous computing, electronics and mobile communication conference (UEMCON), pp 964-970. https://doi.org/10.1109/UEMCON.2018.8796515
7. Sevindik V (2021) Blockchain based resource tokenization for crowdfunding of wireless network investment. In: 2021 11th IFIP international conference on new technologies, mobility and security (NTMS), pp 1–50. https://doi.org/10.1109/NTMS49979.2021.9432653
8. Benila MS, Ajay V, Hrishikesh K, Karthick R (2019) Crowd funding using blockchain. Glob Res Dev J Eng 4(4):19–24
9. Singh A, Rajak R, Mistry H, Raut P (2020) Aid, charity and donation tracking system using blockchain. In: 2020 4th international conference on trends in electronics and informatics (ICOEI) (48184), 457–462. https://doi.org/10.1109/ICOEI48184.2020.9143001
10. Wu H, Zhu X (2020) Developing a reliable service system of charity donation during the Covid-19 outbreak. IEEE Access 8:154848–154860. https://doi.org/10.1109/ACCESS.2020.3017654
11. Zheng Z, Xie S, Dai H, Chen X, Wang H (2017) An overview of blockchain technology: architecture, consensus, and future trends. IEEE Int Congress Big Data (BigData Congress) 2017:557–564. https://doi.org/10.1109/Big-DataCongress.2017.85
12. Alexopoulos N, Daubert J, M¨uhlh¨auser M, Habib SM (2017) Beyond the hype: On using blockchains in trust management for authentication. In: 2017 IEEE Trustcom/BigDataSE/ICESS, pp 546–553. https://doi.org/10.1109/Trustcom/BigDataSE/ICESS.2017.283

Blockchain Technology Adoption in Small and Medium Enterprises: Indian Perspective

D. Divya and O. N. Arunkumar

1 Introduction

Industries have been going through a digital era in which only those companies that are ready to accept modern technologies like artificial intelligence (AI), the Internet of Things (IoT), and blockchain technologies can survive [1]. Blockchain technology has become popular in the last few years. In the case of technological adaptation, SMEs struggle more than major sector organizations because they lack the technology infrastructure necessary to deploy blockchain technologies. Yet, it is very important to create cutting-edge business models that emphasize performance, sustainability, and greater product and service quality in the SME sector [2].

Adopting blockchain technology (BCT) could be a starting point for developing a better business model for SMEs. BCT is a revolutionary financial technology that offers high levels of confidentiality, irreversibility, delivery, transparency, and precision [3]. Due to these advantages, blockchain technology has emerged as a popular technique for money transactions in various industries. Blockchain technology replaces software in which a single entity is in charge of and maintains track of all data [4]. This technology has redefined various perspectives of the industry with new prospects for improved security, efficiency, and cost-cutting [5]. These advancements in financial transactions aid Indian companies to enter into global markets which impacts the economic growth of Indian industries. For constant performance

D. Divya (✉) · O. N. Arunkumar
Symbiosis Centre for Management Studies (SCMS), Symbiosis International (Deemed University) (SIU), Electronics City, Hosur Road, Bengaluru, Karantaka, India
e-mail: divya.d@scmsbengaluru.siu.edu.in

O. N. Arunkumar
e-mail: arunkumar@sibm.edu.in

O. N. Arunkumar
Symbiosis Institute of Business Management (SIBM), Bengaluru, India

J. K. Mandal et al. (eds.), *Proceedings of International Conference on Network Security and Blockchain Technology*, Lecture Notes in Networks and Systems 738,
https://doi.org/10.1007/978-981-99-4433-0_37

and long-term sustainability, SMEs in India must update their business models to incorporate BCT [6]. Identifying the barriers in implementing blockchain technology is an important task as it helps SMEs to equip themselves with the requirements.

This paper discusses the difficulties in implementing blockchain technology in SMEs. Section 2 discusses the related works, and Sect. 3 describes data and methodology, Sect. 4 presents the results and discussion, and Sect. 5 conclusion.

2 Related Works

Several researchers have contributed to the literature on blockchain technology in which the researchers are trying to explore the various perspectives of blockchain technology implementation. Rakshit et al. [6] studied the impact of blockchain technology on the internationalization of SMEs. In his study, he highlighted the inefficiency of blockchain technology in dealing with previously tampered datasets. He discussed SME auditors' degree of acceptability and tolerance to various financial technology forms, such as blockchains and cryptocurrencies, and the impact of blockchain technology on the quality of financial reporting.

In addition to that, identifying the cost benefits of blockchain technology and the impact of perceived cost on organizational behavior has to be studied in order to understand an organization's attitude toward blockchain implementation [1]. From an organizational perspective, it is very important to understand the possibilities of blockchain integration with other technologies such as AI and IoT [1]. Adopting blockchain necessitates spending money on employee training, technology deployment, and technology purchases. Yet, reduced cost and efficiency and security influence are the driving forces for blockchain adoption [5]. However, in order to utilize this network's benefits, global and national trade unions or organizations should motivate more SMEs to join this global network. Moreover, international trade organizations should establish a worldwide regulatory framework as soon as feasible to reduce uncertainty regarding the blockchain and to encourage more businesses to use it safely [7]. Organizations must carry out a cost–benefit analysis through which the SMEs can decide upon the departments in which blockchain technology can be adopted in a cost-effective manner [8].

Kaur et al. [9] discussed the barriers to adopting blockchain technology in SCF of SMEs. This study highlighted technological, organizational, and security barriers to implementing blockchain technology in SMEs in India [9]. This paper frames its theoretical framework based on force filed theory, the resource-based view (RBV), and information processing theory (IPT). However, an in-depth analysis of organizational readiness is needed to understand the technical and financial readiness of the organization [10]. Another major factor that has to be studied in this area is the momentum with which consumers will adopt blockchain technology [12].

Hence, we are trying to find answers to the following questions.

RQ1: What is the impact of organizational readiness in implementing blockchain technology?
RQ2: What is the momentum with which consumers will adopt blockchain technology?

To address these two research questions, it is very important to understand the technical and organizational feasibility of blockchain technology implementation in SMEs and to understand the diffusion of blockchain technology among consumers. Hence, it is very important to understand various factors that affect the technical feasibility of blockchain technology implementation in SMEs.

3 Methodology

Based on the literature review, we identified variables as barriers to the implementation of blockchain technology which is given in Table 1.

After identifying the variables, another important step is to measure the impact of each factor on the implementation of blockchain technology. Interpretive structural modeling (ISM) is a well-known technique for determining links between particular elements that characterize a problem or a concern. Hence, a questionnaire survey was conducted to get expert opinion regarding the various factors that affect the organizational readiness of a company and to identify the variables that affect the momentum at which companies adopt blockchain technology. This questionnaire survey covers various perspectives on technology and organization. The survey was conducted

Table 1 Factors affecting the implementation of blockchain

Sl. No	Factors	Variables	Notation
1	Technological readiness	Implementation cost of blockchain technology	V1
		Software revision of blockchain technology	V2
		Maintenance cost of blockchain technology	V3
2	Financial readiness	Level of sophistication of IT usage	V4
		IT project management	V5
3	Customer adaptability	Complexity of the innovation	V6
		Compatibility with the organization	V7
		The benefit compared with other existing technological choices	V8
		Observability (the ability to measure the internal states of a system by examining its outputs)	V9
		Trialability (verified by means of a trial)	V10

among five respondents out of which two of them were academic experts and three of them were blockchain technology experts. The study used an organized methodology that included the following interpretive structural modeling (ISM) techniques.

The subsequent actions were taken:

- Identify and note the main impediments affecting blockchain technology adoption.
- Establish how the variables relate to one another in the context, then create a structural self-interaction matrix (SSIM) to show how the difficulties relate to one another in pairs.
- Create a reachability matrix that is evaluated for contextual relation transitivity before creating the final matrix.
- Create a hierarchical relationship model by level partitioning based on the final reachability matrix.

The "leads to" contextual relationships are used to assess the relationship dynamics between each pair of implementation obstacles in order to examine the difficulties associated with blockchain technology (*i* and *j*). The relationship's direction between the variables (*I* and *j*) has been shown by the usage of the following four symbols:

(1) V is used to show how variables *i* and *j* are related (i.e., whether variable *i* "will help accomplish" variable *j*).
(2) If variable *j* "will be accomplished" through variable *i* then the relation between *j* and *i* is represented by the letter A.
(3) If variables *i* and *j* "help achieve each other," *X* is used for both direction relations.
(4) O is used when there is no connection between *i* and *j*. (i.e., if variables *i* and *j* are not related).

The next section describes results obtained by using the ISM methodology.

4 Results and Discussion

After collecting the responses from five participants a VAXO matrix has been designed which represents the abovementioned relationship in a matrix format (Table 2).

After developing the VAXO matrix the next step is to develop the final reachability matrix that also evaluates the measure of influence (MI), referred to as driving power, and the measure of being influenced (MBI), also known as "dependence," for all the blockchain implementation challenges (Table 3).

The final reachability matrix is used to guide the level partitioning process. The reachability and antecedent set for each barrier are first established. The element itself, as well as any additional elements it might influence, make up the reachability set. The element itself as well as additional elements that might be influencing it

Table 2 VAXO matrix

Barriers	V10	V9	V8	V7	V6	V5	V4	V3	V2	V1
V1	A	V	A	A	O	O	A	X	V	X
V2	O	X	O	O	A	O	X	V	X	
V3	O	O	O	X	X	O	O	X		
V4	O	X	O	X	X	O	X			
V5	X	O	X	X	X	X				
V6	O	X	X	X	X					
V7	O	X	X	X						
V8	X	X	X							
V9	X	X								
V10	X									

Table 3 Final reachability matrix

Barriers	V1	V2	V3	V4	V5	V6	V7	V8	V9	V10
V1	1	1	1	1*	0	1*	1*	1*	1	1*
V2	1*	1	1	1	0	1*	1*	1*	1	1*
V3	1	1*	1	1*	1*	1	1	1*	1*	0
V4	1	1	1*	1	1*	1	1	1*	1	1*
V5	1*	1*	1*	1*	1	1	1	1	1*	1
V6	1*	1	1	1	1	1	1	1	1	1*
V7	1	1*	1	1	1	1	1	1	1	1*
V8	1	1*	1*	1*	1	1	1	1	1	1
V9	1*	1	1*	1	1*	1	1	1	1	1
V10	1	1*	1*	0	1	1*	1*	1	1	1

make up the antecedent set. Each driving variable's intersecting sets are determined after the reachability and antecedent sets are derived.

The top level of the ISM hierarchical model is occupied by the intersection sets and reachability sets for driving variables that are identical. The challenges for the second level are determined using the same procedure, and so on, until each challenge's level is determined (Tables 4 and 5).

Results obtained from the ISM model help to classify the variables into different levels which are given in Fig. 1.

A structural model has been developed after the identification of level variables which is illustrated in Fig. 1. From the analysis, it is found that the level of sophistication of IT usage is the primary requirement for SMEs to adopt BCT. IT project management, complexity of innovation, compatibility with the organization, benefit compared with other technological choices, and trialability are the level 2 variables in which small-scale industries have to focus before the implementation of BCT. Once

Table 4 Set of variables identified for the first level

Variables	Reachability set	Antecedent set	RS ∩ AS	Level
V1	1, 2, 3, 4, 6, 7, 8, 9, 10	1, 2, 3, 4, 5, 6, 7, 8, 9, 10	1, 2, 3, 4, 6, 7, 8, 9, 10	L1
V2	1, 2, 3, 4, 6, 7, 8, 9, 10	1, 2, 3, 4, 5, 6, 7, 8, 9, 10	1, 2, 3, 4, 6, 7, 8, 9, 10	L1
V3	1, 2, 3, 4, 5, 6, 7, 8, 9	1, 2, 3, 4, 5, 6, 7, 8, 9, 10	1, 2, 3, 4, 5, 6, 7, 8, 9	L1
V4	1, 2, 3, 4, 5, 6, 7, 8, 9, 10	2, 3, 4, 5, 6, 7, 8, 9	2, 3, 4, 5, 6, 7, 8, 9	
V5	1, 2, 3, 4, 5, 6, 7, 8, 9, 10	3, 4, 5, 6, 7, 8, 9, 10	3, 4, 5, 6, 7, 8, 9, 10	
V6	1, 2, 3, 4, 5, 6, 7, 8, 9, 10	2, 3, 4, 5, 6, 7, 8, 9, 10	2, 3, 4, 5, 6, 7, 8, 9, 10	
V7	1, 2, 3, 4, 5, 6, 7, 8, 9, 10	2, 3, 4, 5, 6, 7, 8, 9, 10	2, 3, 4, 5, 6, 7, 8, 9, 10	
V8	1, 2, 3, 4, 5, 6, 7, 8, 9, 10	2, 3, 4, 5, 6, 7, 8, 9, 10	2, 3, 4, 5, 6, 7, 8, 9, 10	
V9	1, 2, 3, 4, 5, 6, 7, 8, 9, 10	1, 2, 3, 4, 5, 6, 7, 8, 9, 10	1, 2, 3, 4, 5, 6, 7, 8, 9, 10	L1
V10	1, 2, 3, 5, 6, 7, 8, 9, 10	2, 4, 5, 6, 7, 8, 9, 10	2, 5, 6, 7, 8, 9, 10	

Table 5 Level 2 and level 3 variables identified

Variables	Reachability set	Antecedent set	RS ∩ AS	Level
V4	4, 5, 6, 7, 8, 10	4, 5, 6, 7, 8	4, 5, 6, 7, 8	L3
V5	4, 5, 6, 7, 8, 10	4, 5, 6, 7, 8, 10	4, 5, 6, 7, 8, 10	L2
V6	4, 5, 6, 7, 8, 10	4, 5, 6, 7, 8, 10	4, 5, 6, 7, 8, 10	L2
V7	4, 5, 6, 7, 8, 10	4, 5, 6, 7, 8, 10	4, 5, 6, 7, 8, 10	L2
V8	4, 5, 6, 7, 8, 10	4, 5, 6, 7, 8, 10	4, 5, 6, 7, 8, 10	L2
V10	5, 6, 7, 8, 10	4, 5, 6, 7, 8, 10	5, 6, 7, 8, 10	L2

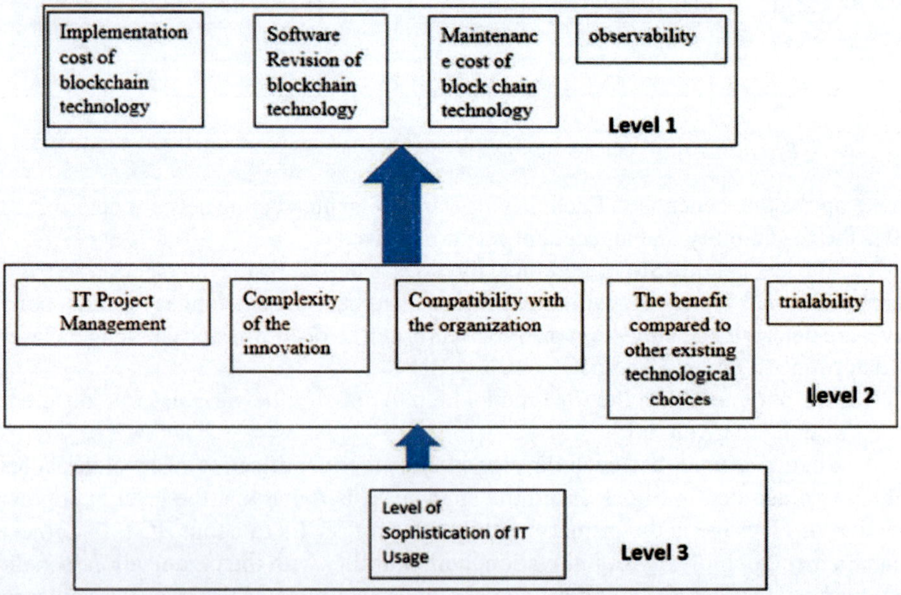

Fig. 1 Level diagram

the company is equipped with first and second-level variables, SME should focus on variables such as the implementation cost of BCT, software revision of BCT, the maintenance cost of BCT, and observability. From the analysis, it can be concluded that the blockchain adoption process starts from level 1 variables, continues with level 2 variables, and ends at level 3 variables. Once the SME is equipped with all the level variables, BCT can be adopted in SMEs.

5 Conclusion

This paper discusses the difficulties in implementing blockchain technology in SMEs. In order to understand the relationship between variables, ISM model has been used. ISM model helps to identify the cause-effect relationship between variables. A comparison of the existing literatures with the current study is given in Table 6.

Results obtained from the study show that the level of sophistication of IT usage leads to level 2 variables which are IT project management, complexity of the innovation, compatibility with the organization, benefit compared with other existing technological choices, and trialability. Finally, the level 2 variables lead to level 1 variables which are the implementation cost of blockchain technology, software revision of blockchain technology, the maintenance cost of blockchain technology, and observability. In a conclusion, financial readiness is at the bottom level (level 1), customer adaptability is at the middle level (level 2), and technological readiness is at the top level (level 3).

Table 6 Comparison of the current study with existing systems

	Current study	Existing literature
Factors considered in the analysis	Technological readiness, financial readiness, customer adaptability	Technology barriers, organizational barriers, external barriers, knowledge barriers, security barriers (Kaur et al. [9])
Methodology	Interpretive structural modeling (ISM)	TAM approach [5] Fuzzy-AHP, Fuzzy DEMATEL [9]
Results	For implementing blockchain technology, a company's financial readiness is the primary factor, and the second factor is customer adaptability. Technological readiness comes as a third factor for blockchain adaptability	Technology barriers have the highest priority, followed by organizational barriers and security barriers

References

1. Abu-salim T, El Barachi M, Mohamed A, Halstead S, Babreak N (2022) The mediator and moderator roles of perceived cost on the relationship between organizational readiness and the intention to adopt blockchain technology. Technol Soc 71:102108. https://doi.org/10.1016/j.techsoc.2022.102108
2. O'Dwyer M, Gilmore A (2018) Value and alliance capability and the formation of strategic alliances in SMEs: the impact of customer orientation and resource optimisation. J Bus Res 87:58–68. https://doi.org/10.1016/j.jbusres.2018.02.020
3. Ahluwalia S, Mahto RV, Guerrero M (2020) Blockchain technology and startup financing: a transaction cost economics perspective. Technol Forecast Soc Change 151:119854. https://doi.org/10.1016/j.techfore.2019.119854
4. Chowdhury MJM, Colman A, Kabir MA, Han J, Sarda P (2018) Blockchain versus database: a critical analysis. In: Paper presented at the 2018 17th IEEE international conference on trust, security and privacy in computing and communications/12th IEEE international conference on big data science and engineering (TrustCom/BigDataSE)
5. Sciarelli M, Prisco A, Gheith M, Muto V (2021) Factors affecting the adoption of blockchain technology in innovative Italian companies: an extended TAM approach. J Strategy Manag Aahead-of-print. https://doi.org/10.1108/JSMA-02-2021-0054
6. Rakshit S, Islam N, Mondal S, Paul T (2022) Influence of blockchain technology in SME internationalization: evidence from high-tech SMEs in India. Technovation 115:102518. https://doi.org/10.1016/j.technovation.2022.102518
7. Ilbiz E, Durst S (2019). The appropriation of blockchain for small and medium-sized enterprises. J Innov Manage 7:26. https://doi.org/10.24840/2183-0606_007.001_0004
8. Khan SAR, Godil DI, Jabbour CJC et al (2021) Green data analytics, blockchain technology for sustainable development, and sustainable supply chain practices: evidence from small and medium enterprises. Ann Oper Res. https://doi.org/10.1007/s10479-021-04275-x
9. Kaur J, Kumar S, Narkhede BE et al (2022) Barriers to blockchain adoption for supply chain finance: the case of Indian SMEs. Electron Commer Res. https://doi.org/10.1007/s10660-022-09566-4
10. Charalambos LI, Benbasat I, Albert S Electronic data interchange and small organizations: adoption and impact of technology. DexterSource: MIS Q 19(4):465–485, (Dec 1995)
11. Dubey R, Singh T (2015) Understanding complex relationship among JIT, lean behaviour, TQM and their antecedents using interpretive structural modelling and fuzzy MICMAC analysis. TQM J, 27(1):42–62
12. Rogers R (2010) Structured interview of reported symptoms. The Corsini Encyclopedia of Psychology, 1–2

An E-Coupon Service Based on Blockchain

S. Deepika, K. P. Vijayakumar, and Vijayan Sugumaran

1 Introduction

Electronic coupons (e-coupons) are becoming a popular marketing strategy as the market for electronic commerce expands [1, 2]. E-coupons are in digital form which makes them convenient for users as well as an effective management tool for coupon suppliers like merchants and marketers. Since e-coupons are computerized codes, coupon providers can disseminate the e-coupons effortlessly online to their clients and gain insights concerning e-coupon utilization and downloads. Moreover, clients can manage their e-coupons on their computers or versatile gadgets. Because of these advantages of e-coupons, amid the COVID-19 crisis, the global market for mobile coupons estimated at US$393.1 Billion in the year 2020 is projected to reach a revised size of US$14.8 Trillion by 2027, growing at a CAGR of 67.9% over the period 2020–2027. Even while the e-coupon business is growing and offers several advantages, there are still some difficulties. For the convenience of management, the majority of e-coupon providers maintain e-coupon data in a centralized system. The data in the centralized database system is utilized to validate an e-coupon when it is used. Due to the centralization of the data, an administrator might, however, readily change it, which may cause e-coupon fraud and forgeries. For instance, a malicious attacker may modify the discount rate or an e-coupon could be used more than once (doubling up on purchases). Real e-coupon crime costs in the US are estimated by

S. Deepika (✉) · K. P. Vijayakumar
School of Computer Science and Engineering, Vellore Institute of Technology, Chennai, India
e-mail: deepika.s2021@vitstudent.ac.in

K. P. Vijayakumar
e-mail: vijayakumar.kp@vit.ac.in

V. Sugumaran
School of Business Administration, Oakland University, Rochester, MI 48309, USA
e-mail: sugumara@oakland.edu

J. K. Mandal et al. (eds.), *Proceedings of International Conference on Network Security and Blockchain Technology*, Lecture Notes in Networks and Systems 738,
https://doi.org/10.1007/978-981-99-4433-0_38

PennLive to be between $300 and $600 million annually [3]. Numerous applications and services are now offered in a mobile environment as a result of the development of short message services, multimedia message services, and wireless application protocols. As a result, if an e-coupon system enables a consumer to get an e-coupon through a mobile device, the user cannot only redeem the discount online but also forward it to other customers or businesses via Bluetooth or short message service (SMS). As a result, the coupon issuer can work with the telecom sector to conduct marketing initiatives. Due to the portability and convenience of this type of e-coupon, it may be used widely in e-commerce for things like gift certificates, the promotion of new products, marketing, and more. It is predicted that mobile consumers will soon adopt e-coupons as a way of life. Security, effectiveness, and manageability must all be considered in the creation of a functional e-coupon system [4].

An e-coupon framework proposed by Hsueh et al. [5] combines blockchain innovation with a hash chain to extend the security of e-coupons. Our current research work is in line with prior activities to utilize blockchain innovation to ensure the accuracy of e-coupon information. By developing a secure smart contract, we concentrate on the operational integrity as well as the data integrity of e-coupons in our work. To strengthen the security of the service in this article, we suggest an electronic coupon service built on a blockchain system. We achieve this by creating smart contracts, which ensure the accuracy of the operations and information, and by constructing a server to support the electronic coupon service and interact with the blockchain system. Additionally, the smart contract is automated for the advantage of the client. The following are the contributions made by our work:

- Novel smart contracts and HMAC algorithm are proposed to achieve better performance in terms of speed and security.
- The performance of the proposed system is evaluated by comparing it to the existing system [1].

The rest of the paper is organized as follows. The motivation and foundation are provided in Sect. 2, and the related work is discussed in Sect. 3. The proposed framework is described in Sect. 4. The prototype implementation is explained in Sect. 5. The test discoveries and their examination are given in Sect. 6. The paper is concluded in Sect 7.

2 Motivation and Background

The use of e-coupons is growing as a result of the growth of smartphones and e-commerce [6, 7]. E-coupons are in contrast to conventional paper coupons and allow coupon suppliers to quickly gather and handle coupon data. Additionally, e-coupons provide options to the clients for using and managing them online or through a smartphone [4]. Verifying an e-coupon is the most crucial step in using e-coupon services since fraudulent attacks can change e-coupons and cause a financial disaster. Extant research [2, 8, 9] provided various techniques such as MD5, MAC, and the one-way hash function to authenticate the e-coupons and to prevent the forging of e-coupons. Despite that, they do not, offer any way to stop information from being falsified on a centralized system. The forging of e-coupons cannot happen during data transmission, and it is possible to forge e-coupon data when employing the aforementioned methods. A server administrator for the e-coupons has the authority to alter any e-coupon data for their benefit. In this paper, a blockchain-based e-coupon service is built in order to prevent fraudulent e-coupon forgery and modification of data on the server.

Implementation of blockchain technology is an alluring alternative to handle security concerns (such as information accuracy) in disseminated frameworks [10, 11]. The majority of blockchain frameworks maintain a time-stamped chain of components with participation from each client to get around the problems. The header and body of a block are its two components. Exchanges are incorporated within the block body. The Merkle tree's root [6], which is composed of the exchanges in the block body, is found in the block header along with other metadata and the hash of a prior block. The pieces are associated by the hash of the previous block, and an unused block will be included after the conclusion of the chain. These characteristics deny blockchain exchanges from being changed or erased. As a result, a blockchain framework can offer inter-individual exchanges and Byzantine blame resilience (BFT) [12] without the requirement for an intermediary. Smart contracts can work effectively without the involvement of a reliable third party because the blockchain maintains the business logic code and status esteem [4, 13]. Through the utilization of the smart contract, one can make distributed apps (DApps) with tall levels of security, promptness, straightforwardness, and anonymity [14, 15]. Therefore, we create exceedingly secure and user-friendly e-coupon service by exploiting the high level of security offered by blockchain and naturally executing an e-coupon smart contract.

3 Related Work

The previous studies [2, 4, 7, 8, 12] have been carried out to provide secure e-coupons. Using message authentication code (MAC) for e-coupon security, new e-coupon models and e-coupon protocols are presented by Blundo et al. in [2]. Agarwal et al. [7] provide a method predicated on centralized third-party coupon checking for double-spending. Hsueh et al. [4] use digital signatures (i.e., PKI) to sign the e-coupon, and hash functions are employed to check the veracity of the data and validate each digital signature. Chang et al. [8] use MAC and one-way hash functions to enable e-coupon providers to protect users from using their coupons more than once without incurring additional processing costs on devices. A user can check whether or not a foreign actor has altered an e-coupon using the approaches discussed in [2, 4, 7, 8]. They effectively handle e-coupon issues and their use as a result, mitigating e-coupon fraud and counterfeiting. These strategies are ineffective, though, because the e-coupon server database can be maliciously altered by an attacker. Additionally, these methods are unable to stop an administrator's nefarious actions. In terms of strengthening the security of e-coupons, our study is comparable to these works [2, 4, 7, 8]. On the other hand, we concentrate on enhancing the security of the database that stores e-coupons and combating e-coupon forgeries.

To verify the forging of e-coupons, the blockchain technique is combined with a hash chain provided by Hsueh et al. [5]. They guarantee the precision of the data contained in e-coupons by deploying blockchain innovation. In terms of using blockchain to ensure the smartness of electronic coupons, our research is consistent with the work [5]. We employ smart contracts to ensure the cleverness of the e-coupon business logic, including downloading, using, and giving an e-coupon. Podda et al. [12] compared several blockchain-based coupon frameworks. They moreover give a standard format for advanced coupons and record the fundamental affirmations that a coupon framework ought to provide. Hsu et al. [9] consider how to leverage blockchain innovation and cryptography to form a secure e-voucher framework and evaluate the system's security necessities. They also released a rational application model that helps the campus welfare nutrition voucher system by utilizing blockchain innovation in the context of vouchers. In the case of the proposed service, instead of looking into a system for a particular use case (such as a campus welfare lunch voucher framework), it is a nonspecific e-coupon framework with an e-coupon-smart contract format. Hinarejos et al. [16] proposed a multi-merchant, blockchain-based promotional point system that protects client privacy and enables point transfers between customers.

From the above-discussed related works, it is observed that these studies only provide e-coupon security features by using a blockchain system for a specific use case to determine whether a modified e-coupon has been done so maliciously. But the rapid growth in the need for power to run the blockchain and the third-party intervention for coupon verification makes the system slow and vulnerable. Thus, we provide a technique for using e-coupons that avoids these issues by applying smart contracts and signing algorithms in the blockchain.

4 Proposed System

Using a safe blockchain and smart contracts can improve the security and efficiency of an e-coupon business. By incorporating blockchain into our system, we strengthen the reliability of business logic and e-coupon data. The main arrangement is to sign a contract before utilizing the e-coupon service, including giving, receiving, gifting, and using the e-coupon. We briefly go over these actions below.

4.1 Register Contract

Before the business logic of a smart contract can be executed, each member must develop a wallet. The wallet is configured to be linked to the blockchain and maintains public and private key sets. To authorize the use of an e-coupon or to issue one for a product, for instance, an e-coupon provider could use the wallet. A customer can download, spend, or give an e-coupon using the wallet. After generating the wallet, the e-coupon server saves the wallet address and UID. The discoveries are then sent via the e-coupon server.

4.2 E-coupon Issue

While issuing an e-coupon, a supplier describes the e-finer coupon's details, such as the price, the number of e-coupons, the validity date, and so on. The electronic coupon is classified as a free coupon when the cost is set to zero. How many e-coupons will be issued is indicated by the quantity of e-coupons. If the number is 0, customers cannot download the electronic voucher. In the validation date, the starting date demonstrates the day on which clients can start downloading e-coupons. The expiration date indicates how long the e-coupon can be downloaded. All connected operations will halt right away if a coupon expires (such as downloading, gifting, and utilizing the e-coupon). If the transaction is significant, the information connected to the e-coupon is kept on the blockchain, and the smart contract is then carried out (that checks certain conditions like provider Id, etc.). At that point, the e-coupon provider requests that the e-coupon server executes the smart contract. In order to synchronize the blockchain and e-coupon server, the e-coupon server already collects e-coupon data from the blockchain and stores it in its database. Customers will finally get access to the e-coupon list.

4.3 E-coupon Download

Before clients begin downloading e-coupons, the e-coupon server gives them a list of e-coupon details. The client at that point creates a transaction to download the desired e-coupon utilizing the wallet address and the coupon address. Then, the customer employs their private key to sign the transaction. The accompanying e-coupon smart contract examines the validity of the transaction signature, the length of the coupon, and the number of coupons that will be available after the transaction has been added to the blockchain in order to validate the authenticity of the downloaded e-coupon. The blockchain's smart contract completes the transaction and updates the client's download status if the transaction is valid (i.e., the e-coupon is downloaded). Otherwise, the blockchain generates a meager outcome. The e-coupon server accumulates and logs the occurrence of downloading an e-coupon from the blockchain in its database.

4.4 Gifting an e-coupon

To offer client 'B' an e-coupon, client 'A' begins by making a transaction utilizing the following components: client B's wallet address, client A's wallet address, and the address of the e-coupon to be gifted. The transaction is made by client A, signed with their private key, and sent to the blockchain. To verify the transaction and check whether client A has sufficient coupons, on the off chance that the e-coupon hasn't been terminated, etc., the e-coupon smart contract executes a gifting operation, as shown by the address. The execution result, which demonstrates whether the operation was successful or not, is returned by the blockchain. The server at the time gets and holds the data regarding the e-coupon that has been gifted, and it informs client B of this information.

4.5 Purchasing with e-coupon

One transaction is completed, when a client utilizes a coupon. The utilization of the e-coupon in another transaction must be confirmed by the source of the e-coupon. To avoid clients from utilizing electronic coupons without the provider's consent, the smart contract changes the status after a substantial transaction. The client is at that point informed by the blockchain regarding the result of their request to utilize the e-coupon. The server synchronizes its status after accepting a request to utilize an e-coupon. The request to utilize it is communicated to the provider. A vendor at that point creates a transaction to approve its utilization and signs it using their private key to give consent to utilize it the merchant will upload the marked transaction to the blockchain, where it will be forwarded to the smart contract that validates the use of the coupon. The server informs the client and the provider when the utilization of the e-coupon is being affirmed and logs the event.

4.6 Rsa and Hmac

From prior works, it is discovered that RSA has the most potential when developing blockchain-based coupon services combined with hash functions. Three steps make up the RSA scheme: key generation, encoding or encryption, and decoding or decryption. According to the RSA algorithm, plain text is encoded with the sender's public key and then added to the general public file. Later, the private key of the sender is required to complete the decryption procedure. So, to develop a system that is more secure and faster than the previous ones, the algorithm used is required to have more such potential than RSA.

Blockchain's smart contracts in conjunction with HMAC and hash functions are utilized to increase the proposed system's speed and security because it is well-known that HMAC is significantly faster than RSA. The network is protected from numerous dangers by a hash-based message authentication (HMAC) on blockchain, which comprises various components including an access control mechanism, a signature, routing controls, notarization, etc. Details of the e-coupon transaction, including the user request, the ID that was sought, and the business logic contract, the HMAC is used to sign and validate those details. The transaction will either be accepted into the block if the signature is legitimate or it will be removed.

4.7 Malicious Block Avoiding Contract

A contract is created in the model to prevent malicious block additions to the blockchain. We are validating the user's (the coupon provider) most recent month of activity based on this smart contract, and if the user is approved (contract satisfaction), they can add the coupon to the blockchain. Within a month, a provider can only issue coupons 5 times with 100 total coupons within a price limit of Rs. 500 or lower, and the validity of the coupon must be less than or equal to one week only. The frequency and quantity of coupons issued during the current month, the overall number of coupons sold during the current month, and other variables all affect how valid a coupon provider is, which makes the system more trustable.

4.8 Coupon Suggestion Contract

There are three categories of users in this system: normal, prime, and elite. This contract makes coupon recommendations based on the user type. The free coupons and deals with 10% or lower are accessible to normal users. Users of prime have access to discounts and deals with a 30% or lower. All coupons are accessible to elite users. Without this contract, every user would receive every coupon that is currently accessible, which could occasionally cause the system to lag or force users to look at more coupons than they need. By decreasing the number of coupons, users will see the loading time, and using this contract makes the system simpler and faster to use.

Figure 1 depicts the proposed service for easier understanding, which includes all the above-mentioned levels. The user starts the process by registering either as a provider or a user and logs in using the registered information as a provider/user. If the login information is varied from the registered information (username, password), it

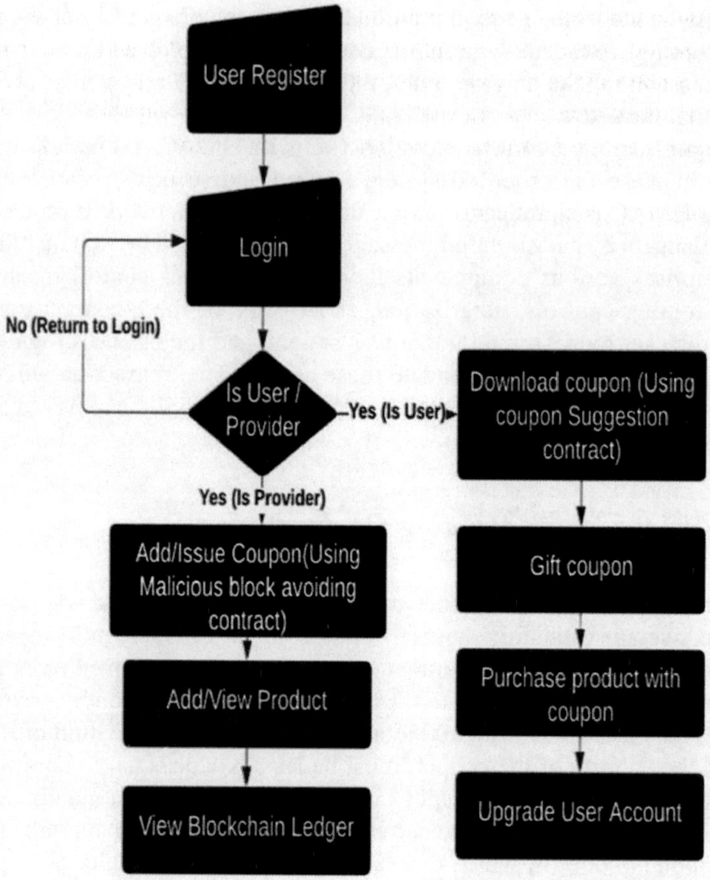

Fig. 1 Flow diagram

returns to the login page. If logged in as a valid user, they are provided with options to download e-coupons, gift, and purchase products using e-coupons and to upgrade their user type. If logged in as a provider, they can issue an e-coupon or a product, view the issued coupon/products, and also view the block chain ledger (which has the purchase details).

5 Prototype Implementation

An API has been developed in the proposed system that consists of Menus such as dashboard, coupon, and upgrade. The dashboard displays the results of each process. The coupon menu includes several options such as download coupon, gift coupon, product purchase, and cart item. The upgrade menu provides upgrade user, account amount credit, and account information. An e-coupon service is supported by smart contract techniques as shown in Fig. 2. The dashboard and other options are provided on the main screen as shown in Fig. 2a to download the coupon and utilize the coupon to make a purchase or gift. The available coupons are displayed to a customer according to their user type as shown in Fig. 2b.

After purchasing/downloading the coupon, the customer can gift it to others if their wallet address is known, as depicted in Fig. 2c. The customer can use the options under upgrade to upgrade their user type, credit the amount to their account, and view their account information as shown in Fig. 2.

In addition to this, the API offers extra functionality such as creating an account and logging in as a consumer or provider, issuing an e-coupon, releasing a product, and examining the purchase ledger from the provider's perspective, which is not shown in the figures due to space constraints.

6 Result and Discussion

The performance assessment metrics such as contract verification time, signature signing time, and data accessing time (in milliseconds) are considered for the proposed system in evaluating the performance of the HMAC algorithm with smart contracts (2 contracts) and compared with the RSA algorithm. A transaction happens only when all the conditions in the smart contract are satisfied. Every time, a transaction happens (when a customer purchases a product using a downloaded/purchased coupon). Figure 3a shows the amount of time consumed for contract verification with 10, 100, and 1000 transactions by employing RSA and HMAC. The x-axis denotes the number of transactions (10, 100, 1000), and the y-axis denotes the amount of time in milliseconds for contract verification. The RSA consumes 25,583, 187,399, and 1,971,734 ms for contract verification with 10, 100, and 1000 transactions, respectively. The HMAC method consumes 664, 4858, and 42,429 ms for contract verification with 10, 100, and 1000 transactions, respectively.

Figure 3 shows the comparison of time consumption for signature signing for a single transaction by using RSA and HMAC algorithms. The x-axis denotes algorithms (RSA, HMAC), and the y-axis denotes the amount of time in milliseconds for signature signing. The amount of time taken by RSA and HMAC are 636 and 6, respectively.

Figure 3c represents the comparison between RSA and HMAC algorithms in terms of data accessing time. The x-axis denotes the number of transactions (10, 100, 1000), and the y-axis denotes the amount of time in milliseconds for accessing data. The RSA consumes 19,223, 123,799, and 1,335,734 ms for data accessing with 10, 100, and 1000 transactions, respectively. The HMAC method consumes 604, 4258, and 36,429 ms for data accessing with 10, 100, and 1000 transactions, respectively.

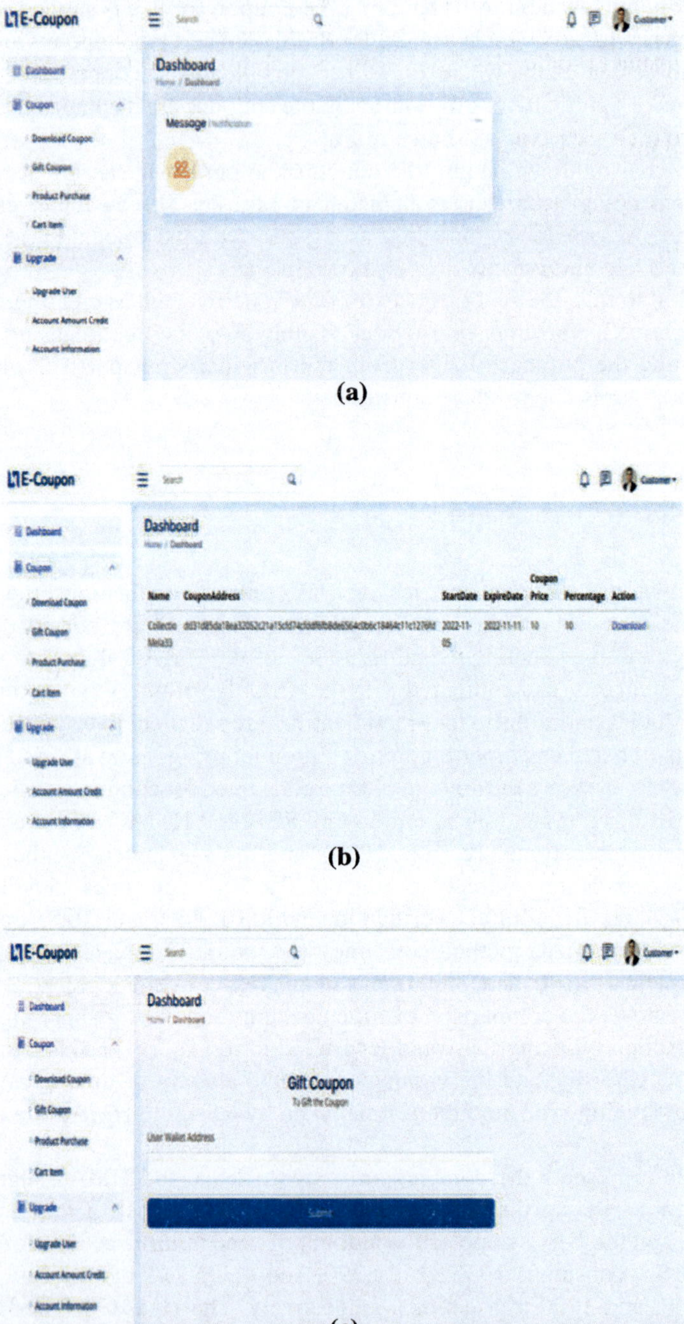

(a)

(b)

(c)

Fig 2 **a** Main screen, **b** coupon download screen, **c** gift coupon screen

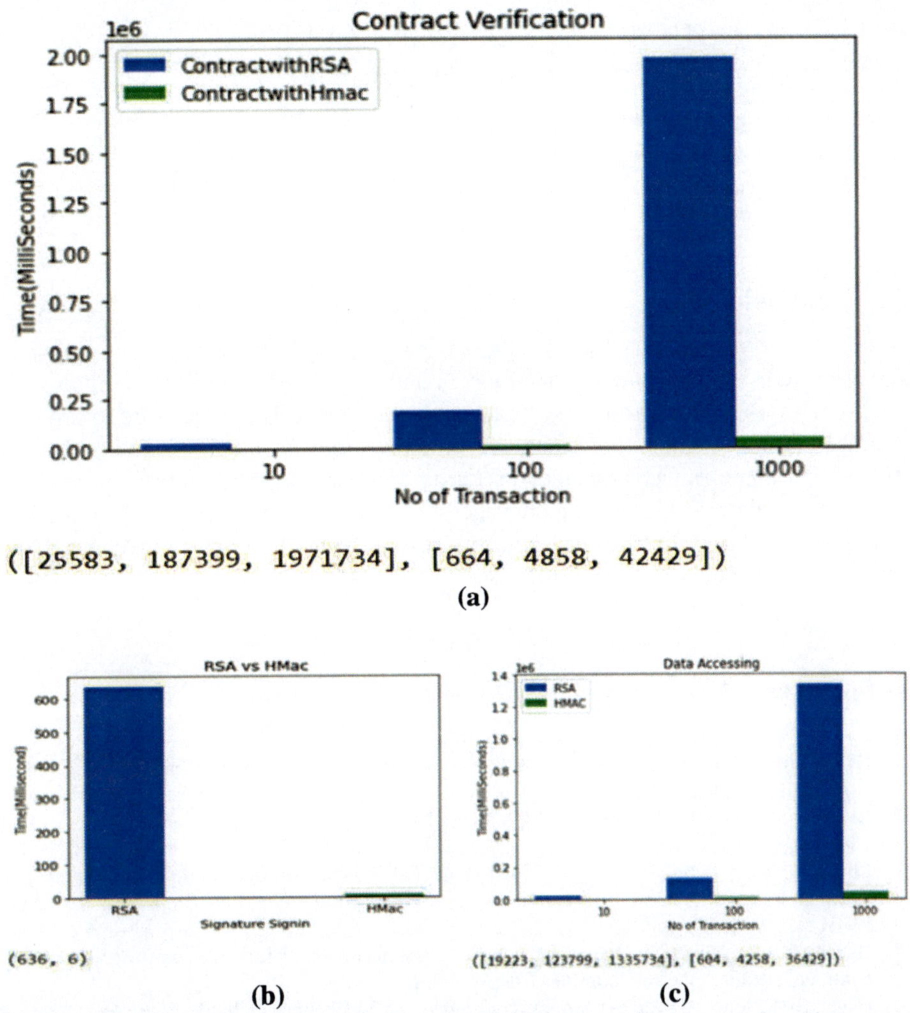

Fig. 3 **a** Contract verification time (RSA vs HMAC), **b** signature signing time (RSA vs HMAC), and **c** data accessing time (RSA vs HMAC)

In Fig. 3a and c, the y-axis of the contract verification and data accessing is denoted using scientific notation ($1e^6$) where 2.00 ms are equivalent to 2 million milliseconds in order to plot the graph with ease and large numbers. From Fig. 3 and Table 1, it is evident that the HMAC algorithm with smart contracts achieves better performance in terms of contract verification time (CVT), signature signing time (SST), and data accessing time (DAT).

Table 1 Comparison of proposed HMAC with existing RSA [1]

Metrics algorithms	CVT (ms)			SST (ms)	DAT (ms)		
	Iterations			Iteration	Iterations		
	10	100	1000	1	10	100	1000
RSA [1]	25,583	187,399	1,971,734	636	19,223	123,799	1,335,734
HMAC	664	4858	42,429	6	604	4258	36,429

7 Conclusion

The proposed system introduces a unique e-coupon service that is built by employing smart contracts in a blockchain system along with the HMAC algorithm to provide increased security. The experimental result shows that the smart contract with the HMAC approach provides enhanced security and performance in terms of contract validation, signature signing, and data accessing. Future work can be in the direction of improving the system to avoid the phishing attacks that the blockchain itself undergoes nowadays.

References

1. Han J, Son Y, Eom H (2022) A secure E-coupon service based on blockchain systems. IEEE Access, 10:21836–21846, Feb 2022
2. Blundo C, Cimato S, De Bonis A (2005) Secure E-coupons. Electron Commerce Res 5(1):117–139
3. Zhang B, Teng J, Bai X, Yang Z, Xuan D (2011) P 3-coupon distribution. In: 2011 IEEE international conference on pervasive computing and communications (PerCom). IEEE, pp 93–101
4. Hsueh SC, Chen J-M (2010) Sharing secure m-coupons for peer generated targeting via eWOM communications. Electron Commer Res Appl 9(4):283–293
5. Hsueh S-C, Zeng J-H (2018) Mobile coupons using blockchain technology. In: Proceedings international conference on intelligent information hiding multimedia signal process. Springer, Berlin, pp 249–255
6. Jiang J, Zheng Y, Yuan X, Shi Z, Gui X, Wang C, Yao J (2016) Towards secure and accurate targeted mobile coupon delivery. IEEE Access 4:8116–8126
7. Agarwal RG-PM-V, Modani N (2001) An architecture for secure generation and verification of electronic coupons. In Proceeding USENIX annual technology conference, Boston, Jun 2001, p 51
8. Chang CC, Wu C-C, Lin I-C (2006) A secure e-coupon system for mobile users. Int J Comput Sci Netw Secur 6(1):273
9. Hsu C-S, Tu S-F, Huang Z-J (2020) Design of an E-voucher system for supporting social welfare using blockchain technology. Sustainability 12(8):3362
10. Hu C, Lee T-T, Chatzopoulos D, Hui P (2018) Hierarchical interactions between Ethereum smart contracts across testnets. In: Proceedings of the 1st workshop on cryptocurrencies and blockchains for distributed systems, pp 7–12
11. Partridge K, Pathak MA, Uzun E, Wang C (2012) 'Picoda: privacy-preserving smart coupon delivery architecture. In: Proceeding of the HotPETs, pp 94–108

12. Podda AS, Pompianu L (2020) An overview of blockchain-based systems and smart contracts for digital coupons. In: Proceeding IEEE/ACM 42nd international conference software engineering workshops, Jun 2020, pp 770–778
13. Hewa, Tharaka (2021) Survey on blockchain-based smart contracts: Applications, opportunities and challenges. J Netw Comput Appl 177, Article No. 102857.pp 1–55, Jan 2021
14. Bartoletti M, Bellomy B, Pompianu L (2019) A journey into bitcoin metadata. J Grid Comput 17(1):3–22
15. Rane S, Uzun E (2014) A fuzzy commitment approach to privacy-preserving behavioral targeting. In Proceeding of the ACM MobiCom workshop secure privacy mobile environment, pp 31–36

A Blockchain Model to Uplift Solvency by Creating Credit Proof

C. K. Gomathy, V. Geetha, G. Lakshman, and K. Bharadwaj

1 Introduction

Blockchain technology guarantees the integrity and transparency of digital assets through decentralized hashing. Employing a distributed digital ledger, transactions are tracked on a blockchain by all network participants. Peer-to-peer transactions on a de-centralized network are made possible by blockchain. Establishing confidence between unidentified peers while documenting the transaction during a distributed, immutable system. Based on their credit history, an individual's solvency situation is represented by their credit score [1]. Giving credit for every transaction allows them to demonstrate that they are taking ownership of their own personal growth and uses. An individual's participation in the financial system, like eligibility for loans or mortgages [2], interest rates, and insurance premiums, is often determined by their credit ratings. It is going to also affect a person's ability to obtain credit or rent an apartment.

C. K. Gomathy (✉) · V. Geetha · G. Lakshman · K. Bharadwaj
Department of CSE, SCSVMV (Deemed to be University), Tamilnadu, India
e-mail: ckgomathy@kanchiuniv.ac.in

V. Geetha
e-mail: vgeetha@kanchiuniv.ac.in

G. Lakshman
e-mail: gidugulakshman02@gmail.com

K. Bharadwaj
e-mail: bharajchinnu@gmail.com

J. K. Mandal et al. (eds.), *Proceedings of International Conference on Network Security and Blockchain Technology*, Lecture Notes in Networks and Systems 738,
https://doi.org/10.1007/978-981-99-4433-0_39

2 Problem Statement

Everyone cannot access their credit score as it is difficult to record every transaction and know their true solvency. People having centralized and systemized accountability can use banks for this process as they pay. But a community who does not mainly depend on banks cannot know their true financial potential forever. The main problem lies here as there is no such system that provides credit proof for such people as they are not completely digitalized.

3 Existing System

Currently, our credit scoring systems are highly centralized, exclusive, and vulnerable to cyber-attacks. Technology has advanced to an extent that every employee who can access his credit score has a digitalized mode of transacting and recording for future credit score improvement. In the case of several communities who do not have that type of financial freedom, the existing system does not provide credit potential scores to them.

- As the existing system is centralized, there would be some security compromises.
- Not everyone is able to know their credit score
- Worthiness of solvency is being wasted
- True financial ability of an individual will not be known forever.

4 Proposed System

Credit will be given for each transaction. By addressing these issues, blockchain-based credit scoring [3, 4] has the potential to transform the banking industry [5] as we know it. Even the ability to obtain a credit card or rent an apartment may be affected. With a credit, you have more financial flexibility since you can use the lent amount as needed at any given time.

- Everyone who has an ability to earn will have a perfect credit score.
- Every transaction is being monitored so it is easy to predict credit score.
- As everything is recorded perfectly, it is difficult to vulnerability and cyber-attack.

5 System Architecture

See Fig. 1.

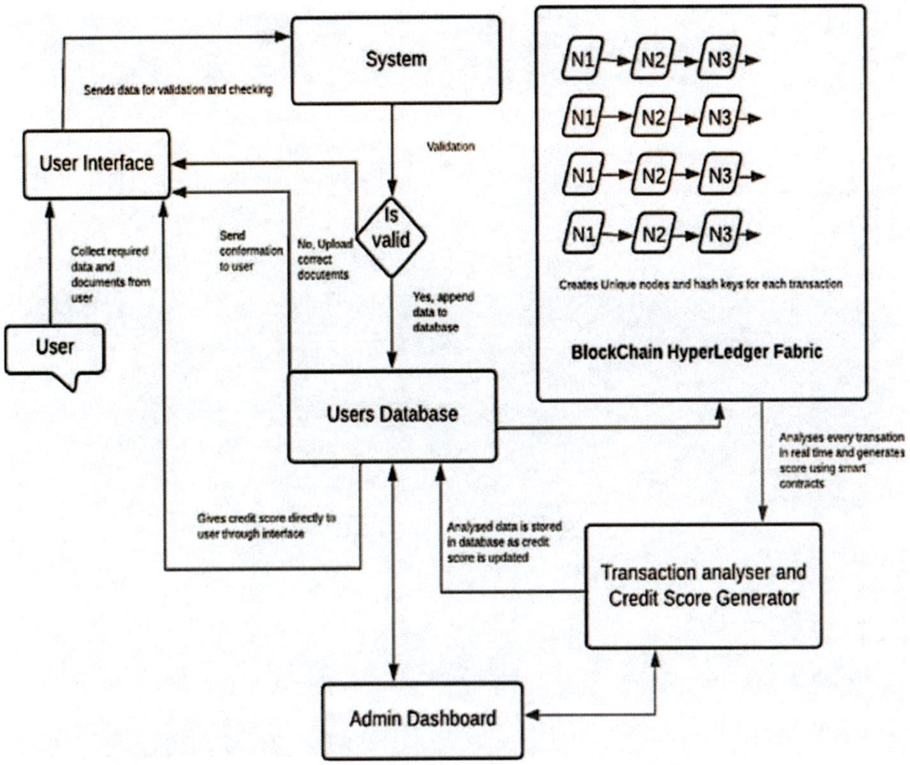

Fig. 1 System architecture

6 Results and Discussion

This work has processed the transaction and verified a transaction. This transaction has been included in a block. A list of blocks containing transactions has been displayed (Figs. 2, 3, and 4).

7 Conclusion and Future Enhancements

A blockchain ledger makes transaction histories more transparent than ever before. Due to the fact that it is a distributed ledger, all nodes in the network have access to the data. Like the Internet, transparency of transaction history has never been greater than it's with a blockchain ledger. All nodes within the network have access to the data since it is a distributed ledger. With speedier cross-border payments, identity management, smart contracts, cryptocurrency, and provide chain, blockchain technology, just like the Internet, is here to remain and will soon overtake all other innovations. Your credit-worthiness is shown by your credit score, which may be a

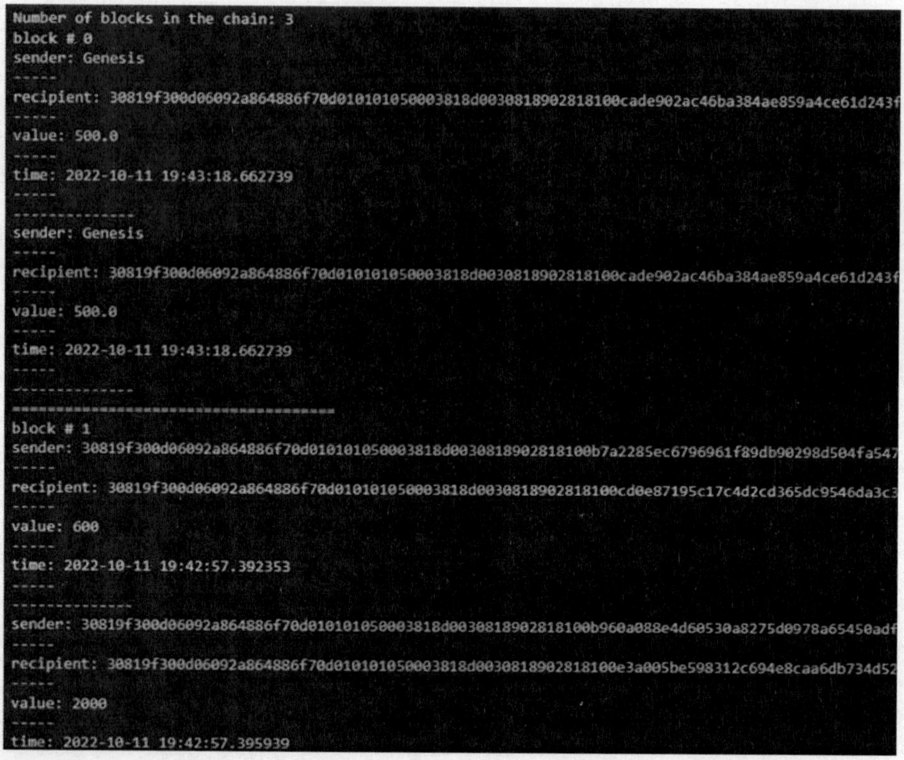

Fig. 2 Transactions added to a block

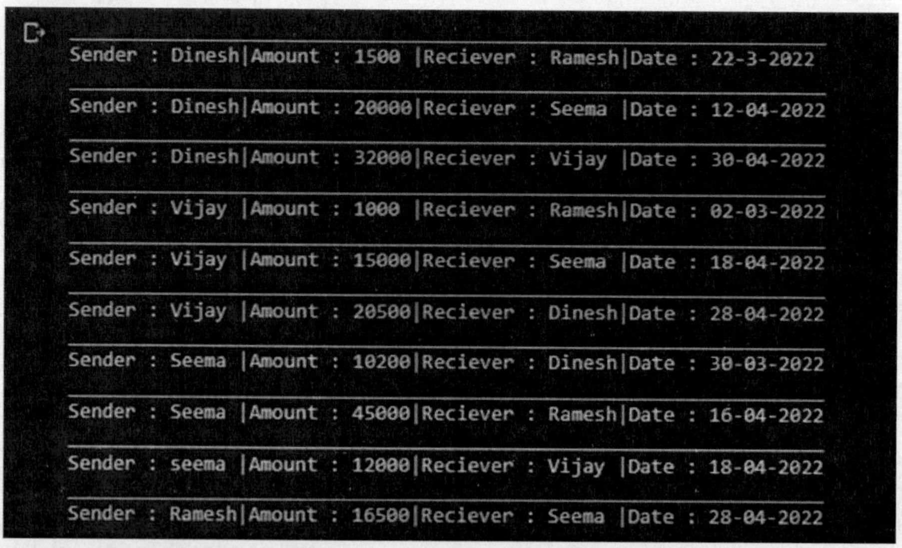

Fig. 3 Transactions stored in database

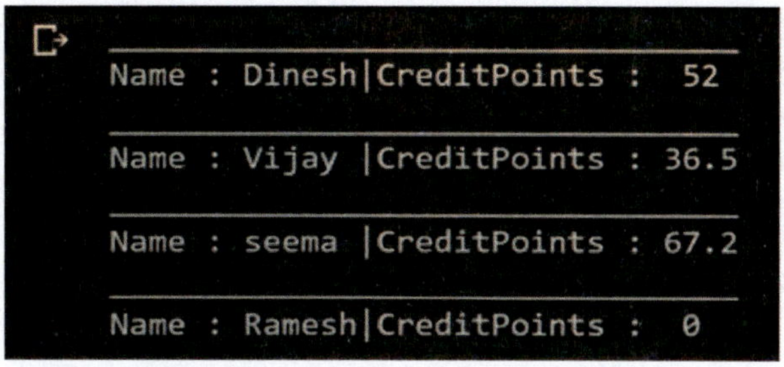

Fig. 4 Credit points assigned

three-digit figure. Every transaction that a customer processes receives a credit score from us. The likelihood of getting your loan accepted increases with a higher credit score. Additional advantages like reduced interest rates, improved payback terms, and a speedy authorization procedure are also probably going to be provided for you. Credit score transparency is often improved using a blockchain ledger. Transparent transaction records make it simpler to identify recurring patterns of activity that can indicate a high or low credit score. This might make credit scoring more accurate and make it more difficult for people to lie about their credit history to get a loan.

References

1. Kalpana Devi S, Sainadh B, Hariharan K, Hemanth T (2020) An unimpeachable system for providing credit scores using blockchain in educational institution (IJRTE), vol 8, no 6. ISSN: 2277-3878
2. Deer M, Mejibli I (2020) Money transfer system using blockchain technology: a case study of banks in Iraq. J Xi'an Univ Architect Technol
3. Du M, Chen Q (2020) Supply chain finance innovation using blockchain. IEEE Trans Eng Manage 67:1045–1058
4. Chang S, Luo H, Chen Y (2019) Blockchain-enabled trade finance innovation: a potential paradigm shift on using letter of credit. Sustainability 12:188 (MDPI)
5. Guo Y, Liang C (2016) Blockchain application and outlook in the banking industry. Finan Innov (Article no. 24, Springer). https://doi.org/10.1186/s40854-016-0034-9

CRYPTOLIGATION: An Offbeat Blueprint of Crypto Contract in the Decentralized Administration

Subhalaxmi Chakraborty, Subha Ghosh, Rajarshi Das, and Pritam Kundu

1 Introduction

Information systems and information technology must be able to adapt to social change and technological improvements in order to meet the needs of information and societal use of communication technologies [1]. Blockchain technology is an example of a disruptive technology that is still being developed in response to societal requirements [2–4]. This served as the foundation for the creation of additional technologies, including smart contracts [5]. Smart contract is a system for electronic transactions and can automatically carry out a contract's conditions. Blockchain is essentially a condensed form of payment verification. It aims to enforce a contract made between several parties [5]. Because they are disinter-mediated and generally transparent, smart contracts promise economic efficiency, a decrease in transaction and legal costs, and anonymous transactions [6]. It has many desirable characteristics that make it one of the most in-demand technologies, particularly in the financial industry [7] because it lowers the risk of fraud and non-payment, enhances the quality of financial contracts with a certain level of confidence, and is not influenced by intermediaries [8]. The technology supporting smart contracts has made tremendous strides recently, but there is still little knowledge about how to use them in different businesses.

This research explores and presents an offbeat methodology for creating intelligent contract blueprints. This study also discusses several methods and blueprints. Businesses have tried with and imitated this application of smart contract technology in several different sectors or as prototypes. This paper enhances current theoretical research and offers instances of smart contracts in several industries among other things [9]. This document also offers a thorough explanation of how smart contract

S. Chakraborty · S. Ghosh (✉) · R. Das · P. Kundu
University of Engineering and Management, Kolkata, India
e-mail: subhaghoshssk@gmail.com

J. K. Mandal et al. (eds.), *Proceedings of International Conference on Network Security and Blockchain Technology*, Lecture Notes in Networks and Systems 738,
https://doi.org/10.1007/978-981-99-4433-0_40

implementation works. These techniques allow for the regulation of dynamic identification using information. This document also offers a thorough explanation of how smart contract implementation works. These techniques allow for the regulation of dynamic identification using information. As the body of knowledge expands, blockchain will develop in a few sectors. As this study paper indicates, researchers are seeking to adapt it to a range of contexts, including smart contracts, supply chains, and healthcare (PHI) [10]. The security, privacy, and scalability of blockchain are the current research focuses [11]. Applications for blockchain integration in cognitive computing, IoT, supply chains, and healthcare look promising [12]. An overview of how smart contracts are applicable in the decentralized environment finishes this study work.

But for smart contracts to succeed in the commercial sector, they need to be deployed well. This necessitates a broad framework, including legal protections supplied by technological protection mechanisms and rights management data more often known as Digital Rights Management (DRM), to support the functioning and development of blockchains [11].

2 Related Works

Blockchains are shared and distributed ledgers that provide transactions with persistence and verifiability. An instruction on the blockchain is referred to as a transaction because it is cryptographically signed by the user [13]. The core concepts of blockchain are cryptography, hashing, digital signatures, open source systems, and distributed architectures [14].

As of today, there are 1639 different cryptocurrencies available on the digital currency market. In order to validate transactions, the blockchain uses several proof-of-concept, proof-of-work, and proof-of-stake concepts. A set of rules is included in a smart contract, a piece of software that operates on the blockchain. The development of blockchain technology over the past ten years demonstrates how widely used it is. A smart contract is a computer program that has self-driven, self-auditing, and security properties. Nick Szabo first proposed the concept of a smart contract in 1994 [15]. Heho defines it as a computerized transaction protocol that executes the provisions of a contract. In 2008, the year it first used blockchain data, Satoshi Nakamoto [16] published Bitcoin without a centralized command structure. They created an operation-based smart contract within a permission blockchain and carried it out in a unified and synchronized manner [17]. A consensus approach for the use of smart contracts in applications for digital rights management is put out by Hiroki and Ahmed [17] introduced a Hawk, which offers a novel method for creating cryptographically safe smart contracts. In an Internet of things application, Joshua Ellul and Gordon JPace [18] describe how to establish a smart contract with AlkylVM. The transformational nature of blockchain is suggested by Michael Harte, CTO of Barclays [18]. SCM challenges are characterized by issues with supply chain finance as well as issues with lead-time and throughput. Hurlburt [19] discussed the need

Table 1 Comparison of features with existing approaches and CRYPTOLIGATION

SI No.	Liquidity and token handling	Error handling	Atomicity	Security	Sustainability	Economic challenges
[1]	✘	✔	✘	✔	✘	✘
[4]	✔	✔	✘	✔	✘	✔
[7]	✔	✔	✘	✘	✔	✘
[9]	✔	✘	✔	✔	✘	✔
[11]	✘	✘	✔	✔	✔	✘
[14]	✘	✔	✔	✘	✔	✔
[16]	✘	✔	✘	✔	✘	✘
[18]	✔	✔	✘	✔	✘	✔
[20]	✔	✘	✔	✘	✔	✔
CRYPTOLIGATION	✔	✔	✔	✔	✔	✔

for ethical standards and operational direction in his article from April 2016. Before blockchains are widely used to replace conventional transaction databases, he notes that stringent rules, including acceptable behavioral norms, must be established [20].

However, the use of blockchain technology and smart contracts can offer advantages beyond cost savings and supply chain flow optimization [21]. Since the existing IT systems are interconnected, the suggested framework can make it easier for SMEs to be integrated into cross-organizational business activities by establishing an open and unified IT environment [22]. This would ensure that SMEs participate fairly in supply chains. The study's foundations include expert interviews, surveys, and case studies that took place in the context of international business ventures involving smart contracts. We have described the state of art comparison between existing approaches and proposed approach in Table 1.

3 Proposed Methodology

3.1 *CRANQ Analysis*

CRANQ which is a low-code integrated development environment has a high degree of reusability and allows component building. Its emphasis on defined data types and ports makes it simple to verify intent. This program is a graphical, user-friendly IDE that enables the compilation and deployment of the smart contract. This third-party application helps to understand how WEB 3.0 and the smart contract work. In a regular code editor, it cannot be understood the flow of code. Therefore, CRANQ makes it extremely simple to track down the smart contract and its working.

The fundamental connection between CRANQ and this research is the desire to create a smart contract with an unconventional design. It allows for the creation of a smart contract's blueprint using a drag-and-drop interface. Additionally, it aids a daily programmer in better time management. CRANQ has a huge collection of nodes that are used to generate smart contracts. Every node in CRANQ can have an output and input gateway. It is also flexible, scalable, and a tricky swift tool for every smart contract developer. This third-party application provides huge advantages of reusability and debugging. This application helps by providing the types of input to use factory deploy image (FDI). Figure 1 shows the process of getting access from CRANQ to use it as IDE in our research work.

With the use of smart contracts, which can be executed on top of blockchains using blockchain technology, parties can codify business rules like those found in legally binding contracts. A smart contract can be viewed as an electronic transaction protocol that permits the digital enforcement of the terms of an underlying legal contract in order to satisfy demands like payment, compliance with legal obligations, and enforcement without the need for a third party. Blockchain is basically a publicly available ledger where participants enter data and certify their acceptance of the transaction via an elliptic curve digital signature algorithm (ECDSA). The equation of an elliptical curve is calculated as

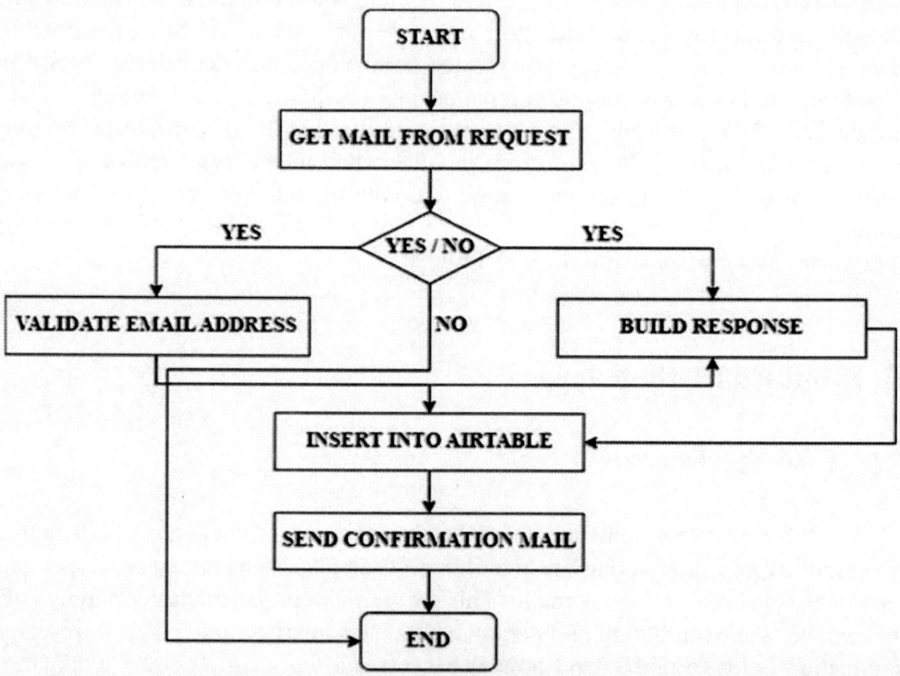

Fig. 1 Flowchart of getting access from CRANQ for generating smart contract

$$y^2 = x^3 + ax + b \tag{1}$$

In Bitcoin and most other implementations, $a = 0$ and $b = 7$, so this is simply

$$y^2 = x^3 + 7 \tag{2}$$

The financial sector has discovered significant applications for smart contract that is executed digitally. Ethereum, Corda, Ripple, Hyper Ledger, and Bitcoin are just a few examples of the blockchain networks where smart contracts are used. Many companies that operate blockchain technologies support smart contracts. To explain the importance of the smart contract, the report is analyzed the difference between traditional security, market-based contracts, and their smart contract counterpart. Blockchain will mature in the number of areas as the body of research grows. A blockchain is a network of computers that must all concur that a transaction has occurred before it is recorded in a chain of computer code, according to a description provided by the financial times in 2016.

The blockchain and smart contract technology are used to bring an offbeat up gradation in the decentralized market. The mathematical solutions are discussed for clarifying the theoretical expression of this ledger's cutting-edge technology. The IDE and its components are described in Sect. (3.1). For generating a smart contract, blockchain technology needs to maintain the various types of protocols. In later parts, these ideas as well as the algorithmic component of creating smart contracts are covered.

3.2 *Schematic Algorithmic Structure of CRYPTOLIGATION*

'CRYPTOLIGATION' word is discovered from 'crypto' and 'obligation'. 'Obligation' means one type of contract that creates and pays transaction fees to anonymous miners. An offbeat schematic algorithmic structure (Fig. 2) of CRYPTOLIGATION is designed and implemented on a specific scalable platform that can swiftly contribute work pressure in blockchain technology. This structure is an architect by keeping the means of security, scalability, flexibility, and maintenance of smart contracts in blockchain technology.

The starting position of generating a smart contract in CRANQ is used by getting help from the node of 'start' (as shown in Fig. 2). Mostly, the 'start' node is used to kick start the entire program. As per peer reviews, it is already found that there is no solidity language in CRANQ, but CRANQ only supports the nodal architectural view because of its nodal connections. 'Store' (known as 'Config' in Fig. 2) is used for reading the user inputs and passing the information to its neighbor nodes. Suppose somehow the input of the 'store' node is empty, then any attempt to read the content will result in a signal sent out via not found. These (private key, crypto wallet account address, a kind of network, provider URL) are validified by 'store' and then passed to its successive nodes. Therefore, an extra 'store' node is added

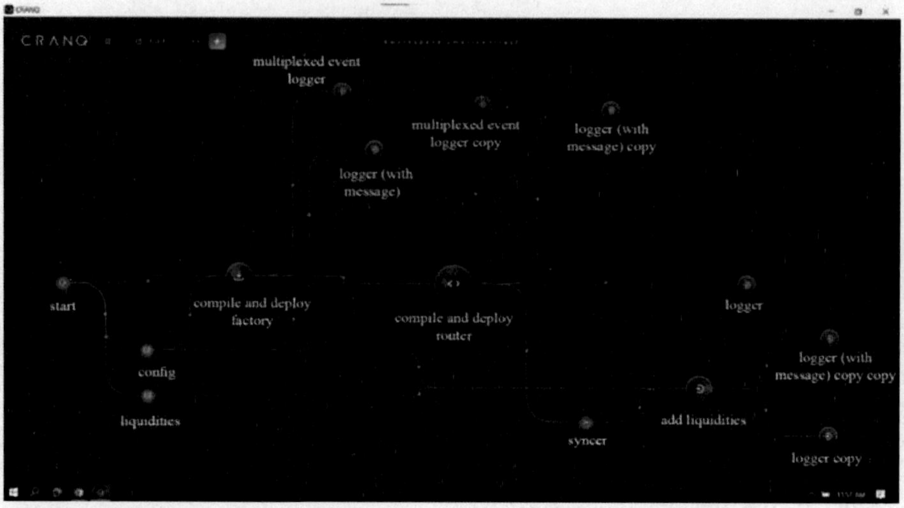

Fig. 2 Offbeat blueprint of smart contract

and named it 'liquidities'. This node is used mainly for getting the token addresses as inputs. Also, exchange rates have been given to this node. The following program code is used to create and connect the nodes to each other in this research work.

```
program Inflation (Output of START and CONFIG).
begin
    SetTimeOut(equals to or greater than start(NULL, 'START'), 0)
#fetch data
    State->data = data
    Outputs->written(NULL, tag)
#read instances (provider URL, private key, account address, network)
    constant value = state->data is not equals to undefined state->data: params-
    >data
    if value is not equals to undefined
        outputs->data(value, tag)
    else
        outputs["NOT FOUND"](NULL, tag)
#read token address and exchange rate
    State->data = data
    Outputs->written(NULL, tag)
end
```

Inside ‘compile and deploy factory’, a collection of various functioned nodes is gathered. This node helps to provide the authentication based on the inputs which have been collected from its previous node and also compiles smart contract. ‘Multiplexed event logger’ and ‘logger with message’ are used for debugging the ‘factory’ and ‘router’ nodes. These exception handling nodes are connected for getting green signal from its neighbor connections. The following program code has been provided for better experience.

program Inflation (Output of COMPILE AND DEPLOY FACTORY).

```
begin
#item getter
    constant {} = state
    let bundle = get bundle tag
    if bundle is not present
        bundle = static bundle
        tag and bundle
    let ports = get ports tag
    if ports are not present
        ports = set dynamic fields
        set tag and ports
    bundle[input] = data
    delete input of ports
    if size of ports is equals to 0
        delete bundle tag
        delete ports tag
    read instances()
    factory ABI()
#send factory address
    contract deployment waiter()
#find error and passes signal to a logger node
    factory deployer()
end
```

‘Compile and deploy factory’ node generates ‘factory address’ and passes it to ‘compile and deploy router’. This ‘router’ node helps to generating actual smart contract which is implemented in blockchain technology. The following pseudo-program code is attached here.

program Inflation (Output of COMPILE AND DEPLOY ROUTER).

```
begin
```

```
#get network
    get network()
#passes w-eth address
    while (router deployer assign data in chain)
        contract deployment waiter()
            return router address, router ABI
end
```

'Compile and deploy router' provides the router address to 'Syncer' (follows the below program code). 'Syncer' node is bundled up the same types of data or information and passes these for the next phase. These two nodes are different in functionalities. That is why an interface (called as 'application binary interface') is used to communicate in this particular architecture. In Fig. 2, the transaction is completed in 'add liquidities' node. This node is taking the output of 'Syncer' as input and helping to complete the transaction. Other side, this node fetches the data from user end with router contract inputs, and if any error occurs, then the 'logger (with message)' node gives the red signal, otherwise it will be green in whole procedure.

```
program Inflation (Output of SYNCER and LIQUIDITIES)
    begin
    #read router address and ABI
        constant {} = state
        let bundle = get bundle tag
        if bundle is not present
            bundle = static bundle
            set tag and bundle
        let ports = get ports tag
        if ports are not present
            ports = set dynamic fields
            set tag and ports
        bundle[input] = data
        delete input of ports
        if the size of the ports equals to 0
            delete bundle tag
            delete ports tag

#add in liquidities
Address = string,
ABI = {string: any}()
```

```
config()
#calculate deadline

    timestamp(new Date(), tag);

item getter()
#adder

    sum(data->a + data->b, tag);

liquidities adder()
end
```

4 Result and Discussions

There is already widespread agreement that the blockchain's influence will extend well beyond crypto currencies and might have disruptive consequences on the creation of distributed applications. Smart contracts able to support a new type of distributed computing are the primary enabler for this impact. The quantity and variety of smart contract platforms are constantly expanding; this article examines one of them. As a result, the ensuing technical landscape is becoming more complex and heterogeneous. The conceptual foundations of this new landscape are more integrated than one might anticipate from an application standpoint, as this article demonstrates, and smart contracts can in fact be seen as the fundamental building blocks-or services-of a blockchain-based, service-oriented computing paradigm. This proposed approach also demonstrates how we are putting into practice a smart contract approach that values interoperability and reusability as positive characteristics. Several obstacles must be overcome in order to implement the smart contract protocol in blockchain and fully utilize smart contracts. It is startling how little attention has been given to making it possible for developers to reuse contracts that have previously been deployed, especially given that it is generally thought that deploying a new contract is more expensive than just calling a job that has already been done. The idea of resource consumption and the cost of invocations are organically incorporated into smart contracts. While generic contracts that may require transaction processing may result in greater, unpredictable response times, libraries, and data contracts are processed locally inside of each node and have minimal response times. It makes sense that platforms focus on their own technologies these days as their unique selling points. Developing common protocols, interface styles, data formats, and, of course, mechanisms for authentication and certification is a problem. Finally, composition solutions must be designed and implemented in order to abstract away from technical details and give developers the tools and infrastructures they need to increase productivity. Only then will smart contracts' full potential be realized.

Whatever is envisioned is, in essence, a progression from the current technology silos to an abstract, reuse-oriented contract system that can maintain the assurances unique to blockchain technology. Figure 3 demonstrates that this eccentric smart

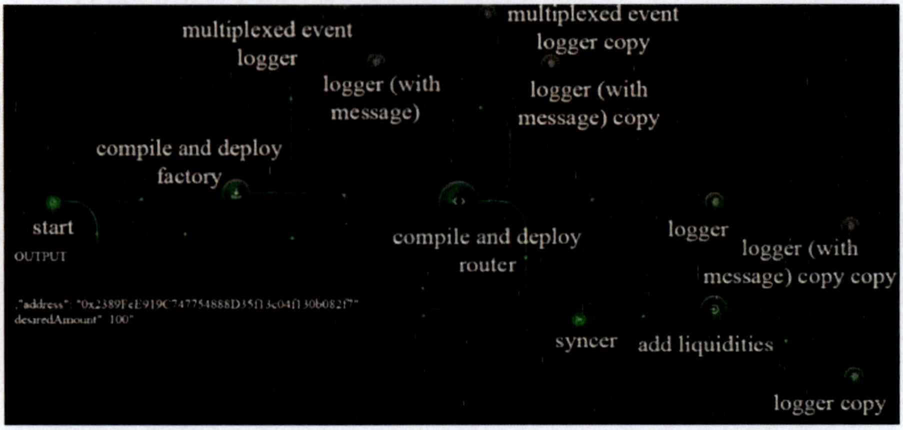

Fig. 3 Execution status of smart contract

Table 2 Comparison with other research papers

SI No	Research papers	Complexity	Transparency	Security
1	Prause, G. (2019). Smart contracts for smart supply chains [6]	O(n^2)	Moderate	Good
2	A triplicate smart contract model using blockchain technology [9]	O(n)	High	Moderate
3	CRYTOLIGATION	O(n)	Good	Best

contract has been successfully run and that all the nodes have received the go-ahead. So, our eccentric blueprint is now good to go and ready to be applied to a crypto currency online application. We have compared the complexity, transparence, and security of the existing approaches [6, 9] with proposed approach in Table 2.

5 Conclusion

The information technology has emerged as a crucial innovation. Here, we have discussed blockchain technology as a catalyst for new applications in the financial and non-financial sectors of healthcare, supply chain, and industrial manufacturing. According to the report, by providing secure trust frameworks, fostering agile value chain production, and fostering closer interaction with technologies like cloud computing and IoT, smart contracts can play a crucial part in revolutionizing the currencies of various sectors and applications. This smart contract's eccentric design aims to eliminate the necessity for an existing, unreliable third party by self-enforcing contractual requirements like payments and legal duties. In addition to self-organizing and self-optimizing, the study emphasizes how industry 4.0, blockchains, and smart contracts concepts are connected to each other and fit structurally well together. Blockchain has built-in, highly effective cyber security protections that enable instant contracts, engagements, and agreements. This paper explains how, why, and from what angle this smart contract is bringing about a seismic shift in the blockchain industry.

Acknowledgements I would like to convey my keen gratitude to Subhalaxmi Chakraborty, my research supervisor, for giving me the chance to do research and for her helpful advice during this process. Her energy, vision, genuineness, and drive have really motivated me. Additionally, I want to thank Subha Ghosh for his insightful comments and ideas.

References

1. Negara ES, Hidayanto AN, Andryani R, Syaputra R (2021) Survey of smart contract framework and its application. Information 12(7):257
2. Palit SK, Chakraborty M, Chakraborty S (2022) Performance analysis of 5GMAKA: lightweight mutual authentication and key agreement scheme for 5G network. J Supercomput 1–34
3. Palit SK, Chakraborty M, Chakraborty S (2022) MASKA: mutual authentication and session key agreement protocol in global mobility networks. In: Applications of machine intelligence in engineering. CRC Press, pp 309–321
4. Daniel F, Guida L (2019) A service-oriented perspective on blockchain smart contracts. IEEE Internet Comput 23(1):46–53
5. Gupta R, Shukla VK, Rao SS, Anwar S, Sharma P, Bathla R (2020) Enhancing privacy through “smart contract” using blockchain-based dynamic access control. In: 2020 international conference on computation, automation and knowledge management (ICCAKM). IEEE, pp 338–343
6. Prause G (2019) Smart contracts for smart supply chains. IFAC-PapersOnLine 52(13):2501–2506
7. Mohanta BK, Panda SS, Jena D (2018) An overview of smart contract and use cases in blockchain technology. In: 2018 9th international conference on computing, communication and networking technologies (ICCCNT). IEEE, pp 1–4
8. Younus D, Muayad A (2021) Supply chain using smart contract blockchain technology in organizational business. Eur J Res Dev Sustain
9. Eze P, Eziokwu T, Okpara C (2017) A triplicate smart contract model using blockchain technology. Circ Comput Sci-Spec Issue 1–10
10. Azka LI, Firdaus R (2023) Blockchain technology based smart contract model in Indonesia. Blockchain Frontier Technol 2(2):58–63
11. William ADJ, Rajendran S, Pranam P, Berry Y, Sreedharan A, Gul J, Paul A (2023) Blockchain technologies: smart contracts for consumer electronics data sharing and secure payment. Electronics 12(1):208
12. Kumar NM, Chopra SS (2022) Leveraging blockchain and smart contract technologies to overcome circular economy implementation challenges. Sustainability 14(15):9492
13. Lin SY, Zhang L, Li J, Ji LL, Sun Y (2022) A survey of application research based on blockchain smart contract. Wireless Netw 28(2):635–690

14. Verma M (2021) Smart contract model for trust based agriculture using blockchain technology. Int J Res Anal Rev 8(2):354–355
15. Dolgui A, Ivanov D, Potryasaev S, Sokolov B, Ivanova M, Werner F (2020) Blockchain-oriented dynamic modelling of smart contract design and execution in the supply chain. Int J Prod Res 58(7):2184–2199
16. Li X, Jiang P, Chen T, Luo X, Wen Q (2020) A survey on the security of blockchain systems. Futur Gener Comput Syst 107:841–853
17. Golosova J, Romanovs A (2018) The advantages and disadvantages of the blockchain technology. In 2018 IEEE 6th workshop on advances in information, electronic and electrical engineering (AIEEE). IEEE, pp 1–6
18. Halaburda H (2018) Blockchain revolution without the blockchain? Commun ACM 61(7):27–29
19. Meng W, Wang J, Wang X, Liu J, Yu Z, Li J, Chow SS (2018) Position paper on blockchain technology: Smart contract and applications. In: International conference on network and system security. Springer, Cham, pp 474–483
20. Tasatanattakool P, Techapanupreeda C (2018) Blockchain: challenges and applications. In: 2018 international conference on information networking (ICOIN). IEEE, pp 473–475
21. Sharma P, Jindal R, Borah MD (2022) A review of smart contract-based platforms, applications, and challenges. Cluster Comput 1–27
22. Zou W, Lo D, Kochhar PS, Le XBD, Xia X, Feng Y, Xu B (2019) Smart contract development: challenges and opportunities. IEEE Trans Softw Eng 47(10):2084–2106

Progression Analysis and Facial Emotion Recognition in Dementia Patients Using Machine Learning

Afrin Siddiqui, Pooja Khanna, Sachin Kumar, and Pragya

1 Introduction

Dementia is a term for loss of memory, and other mental abilities severe enough to interfere with daily life [1]. Dementia is not a specific disease; rather, it is an umbrella term that describes a group of symptoms generally related with a decline in memory or other thinking abilities severe enough to hinder a person's ability to perform simple everyday activities [2]. Studies have suggested that MRI features can predict rate of decline of Alzheimer's disease and may help suitable therapy in the future. However, to reach that stage clinicians and researchers will have to make use of machine learning techniques that can accurately predict the progress of a patient from mild cognitive impairment to dementia.

Alzheimer's is the most well-known common type of dementia, though there are numerous kinds that do exist but are in theory, incurable. Brain imaging via magnetic resonance imaging (MRI) is used for the evaluation of patients with suspected AD. MRI findings include both local and generalized shrinkage of brain tissue [3–6].

Although current Alzheimer's treatments cannot stop Alzheimer's from progressing, they can briefly slow down the worsening of dementia symptoms and improve the quality of life for those with Alzheimer's and their caregivers. Today, there is an overall exertion under approach to discover better ways to treat the disease. This paper aims to include collective research and practical implementation which can delay its onset and prevent it from developing [7–10].

A. Siddiqui · P. Khanna · S. Kumar (✉)
Amity University Uttar Pradesh, Lucknow Campus, Lucknow, India
e-mail: skumar3@lko.amity.edu

P. Khanna
e-mail: pkhanna@lo.amity.edu

Pragya
MVD College, Lucknow, India

J. K. Mandal et al. (eds.), *Proceedings of International Conference on Network Security and Blockchain Technology*, Lecture Notes in Networks and Systems 738,
https://doi.org/10.1007/978-981-99-4433-0_41

In this paper, a sound machine learning web application is proposed that can help clinicians predict early dementia. Section 1 describes the basic underlying concept of dementia disease and is elaborated with the help of clinical information, Sect. 2 presents progression analysis using machine learning and is performed on the dataset to resolve important conclusions. Exploratory data analysis (EDA) is done on data which is a longitudinal collection of 150 subjects aged 60 to 96. Sect. 3 discusses the concept of facial emotion recognition which is explained and how it can diagnose dementia at an early stage which saves the patient's cognitive abilities to deteriorate further finally in Sect. 4, evaluating the goal of the machine learning web application and how it can help clinicians in determining dementia disease in patients and its scope for battling key challenges in the future of neurological illnesses such as dementia and AD affecting people of all ages. Discussion about further useful functionalities that can be added to the web application which can, in turn, enhance its capability to provide even more promising results in early medical diagnosis.

2 Literature Survey

Alzheimer's is a type of dementia that causes problems with memory, thinking, and behavior. Alzheimer's is certainly not a normal part of aging; a number of ML techniques have been designed to extract features and perform operations on MRI images [11]. To identify affected people, Kloppelet [12] developed a dynamic technique employing weighted MRI-based SVM. Gray et al. [13] employed NB classification for determining a multidimensional classification the AD category. Morra et al. [14] designed a distinction between diversified platform for detecting AD with SVM and AdaBoost hierarchy. An improved DKPCA for AD MRI images was designed [15] by Neffen et. al. New platform design was verified on OASIS samples, and 92.5% accuracy was achieved employing an MSVM. Wang et al. [16] employed equilibrium transformation function to determine features in MRI image, and Ding et al. [17] technique enhanced function recovery. They employed grayscale matrix to differentiate ADNI datasets. Dashan et al. [18] used extract and discount technique of Harvard Medical School's, technique achieved an accuracy from 97 to 98%. Images from the ADNI databases were verified by Hinrich et al. [13]. Yue et al. [19] established a connected removal algorithm on voxels which exposes the relationship among objects. Ahmed et al. [20] designed an easy-to-implement CNN model. The model lowered the cost of computing and vastly increased accuracy of about 91%. Most of the experiments gave results, based on how well features were employed, one potential way to overcome the constraint is by employing deep learning techniques, since these techniques perform automatic selection of features [21].

3 Proposed Methodology

Recognizing dementia in the earlier stages will allow time for carers and those affected by dementia to understand what is happening, plan for future and establish links with support services, thus hopefully preventing crisis situations [4]. To find out the symptoms in which dementia is mainly isolated, we mainly use comprehensive assessment to analyze the symptoms, but the sensitivity of these assessment tools varies depending on age, education, social class, and living situation [5].

Process flow for early identification of dementia comprises of following steps.

i. Medical History Analysis of Subject (Image Processing)

   a. We can add an automated cognitive test which analyzes the facial emotion of subject based on some intuitive questions and closely examine the facial expressions of subject, whether there are any facial expression deficits present.
   b. For example, if the subject is being probed about a particular question, how is the subject reacting with respect to his facial expressions. What facial emotions is the subject conveying?
   c. A ML model is built keeping the aspects of image processing. The model takes facial portraits of the subjects (young, middle-aged, old men, or women) as input and analyzes those images by predicting the emotion displayed by subject. The model can classify the images into seven emotions—anger, disgust, fear, happy, sad, surprise.

ii. Computer-Based Test

   a. In this step, we include a system that can perform computer-based tests, based on medical technique of FCSRT. We conduct something like "Active recall" for the patients, i.e., Free and Cued Selective Reminding Test (FCSRT)
   b. The subjects(patients) search for items (e.g., apple—could be any object) in response to cues (fruits) and then further used to recall more similar items.
   c. Performance on the FCRST distinguishes dementia from normal aging with accuracy.
   d. We can build a system that requires the subject to draw/imitate an item from memory given the cues. After that, the subject's imitation and the actual item can be compared.

iii. Prescribed Tests/Screening Tests

   a. A standard medical workup for dementia often includes structural imaging with magnetic resonance imaging (MRI) or computed tomography (CT).
   b. A system can be built using ML that takes the subjects' MRI and after analyzing it, labels it as normal or abnormal.

c. It can be an interface where the subjects would be required to upload their MRI scans, and the model would look for abnormalities by comparing it to a normal brain imaging subject.

4 Implementation and Result: Progression Analysis Using Machine Learning

Progression analysis employs machine learning to comprehend the parameters that work is being implemented. Analysis main aim is to find hidden details that affect the slope of deterioration of dementia. Starting with the data, which is fundamental in implementing any machine learning algorithm, further Exploratory Data Analysis (EDA) is done alongside with some data preprocessing to ensure the machine learning model that will perform well. Further, discussion is done regarding the classification algorithm used for the model that is random forest classifier along with the result obtained [6] (Table 1).

Dataset consists of a longitudinal collection of 150 subjects aged 60 to 96. Each subject was scanned on two or more visits, separated by at least one year for a total of 373 imaging sessions. For each subject, three or individual T1-weighted MRI scans obtained in single scan sessions are included. The subjects are all right-handed and include both men and women. Seventy-two of the subjects were characterized as non-demented throughout the study. Sixty-four of the included subjects were characterized as demented at the time of their initial visits and remained so for subsequent scans, including 51 individuals with mild to moderate Alzheimer's disease. Another 14 subjects were sorted as non-demented at the time of their initial visit and were subsequently characterized as demented at a later visit.

Table 1 Column descriptors

Column name	Description
EDUC	Years of education
SES	Socioeconomic status
MMSE	Mini-mental state examination
CDR	Clinical dementia rating
eTIV	Estimated total intracranial volume
nWBV	Normalize whole brain volume
ASF	Atlas scaling factor

Table 2 Cognitive impairment scale

Method	Score	interpretation
Single cutoff	< 24	Abnormal
Range	< 21	Increased odds of dementia
	< 25	Decreased odds of dementia
Education	21	Abnormal for 8th-grade education
	< 23	Abnormal for high school education
	< 24	Abnormal for college education
Severity	24–30	No cognitive impairment
	18–30	Mild cognitive impairment
	0–17	Severe cognitive impairment

4.1 Mini-Mental State Examination (MMSE)

The Mini-Mental State Examination (MMSE) or Folstein test is a 30-point questionnaire that is used extensively in clinical and research settings to measure cognitive impairment [7]. Any score greater than or equal to 24 points (out of 30) indicates a normal cognition. Below this, scores can indicate severe ($\leq$9 points), moderate (10–18 points), or mild (19–23 points) cognitive impairment (Table 2).

4.2 Clinical Dementia Rating (CDR)

The CDR is a 5-point scale used to characterize six domains of cognitive and practical performance appropriate to Alzheimer. The CDR table provides descriptive secures that help the clinician in making fitting evaluations dependent on questionnaire data and clinical judgment. This score is useful for characterizing and tracking a patient's level of impairment/dementia (Table 3).

Table 3 Clinical dementia rating (CDR)

Score	Description
0	Normal
0.5	Very mild dementia
1	Mild dementia
2	Moderate dementia
3	Severe dementia

4.3 Estimated Total Intracranial Volume (eTIV)

Total intracranial volume (TIV/ICV) is a significant covariate for volumetric investigation of the brain and cerebrum regions, particularly in the investigation of neurodegenerative diseases. Unlike brain decay in the patients with AD, TIV did not change over time. The only huge predictor of TIV was gender. Men showed an approximately ~12% larger eTIV than women [8].

4.4 Atlas Scaling Factor (ASF)

ASF is a normalization technique to measure the standardized total intracranial volume for comparison, classification and predication in benign and infected cases for all age groups [9].

4.5 Exploratory Data Analysis (EDA)

Dataset is further investigated for the relationship between each feature of MRI tests and dementia of the patient. It might help us to understand the nature of the data and to select the appropriate analysis strategy method for the model later (Table 4; Fig. 1).

This chart indicates that there is a higher concentration of 70–80 years old in the demented patient group than those in the non-demented patients. Patients who suffered from that kind of disease have lower survival rate so that there are a few of 90 years old (Fig. 2).

This chart shows that non-demented group has much higher MMSE scores than demented group (Figs. 3, 4, and 5).

From the above charts, we can conclude that non-demented group has a higher brain volume ratio than demented group based on the assumption that neurological diseases affect the brain to be shrinking its tissue.

Table 4 Minimum, maximum, and average values of each feature

	Min	Max	Mean
Educ	6	23	14.6
SES	1	5	2.34
MMSE	17	30	27.2
CDR	0	1	0.29
eTIV	1123	1989	1490
nWBV	0.66	0.837	0.73
ASF	0.883	1.563	1.2

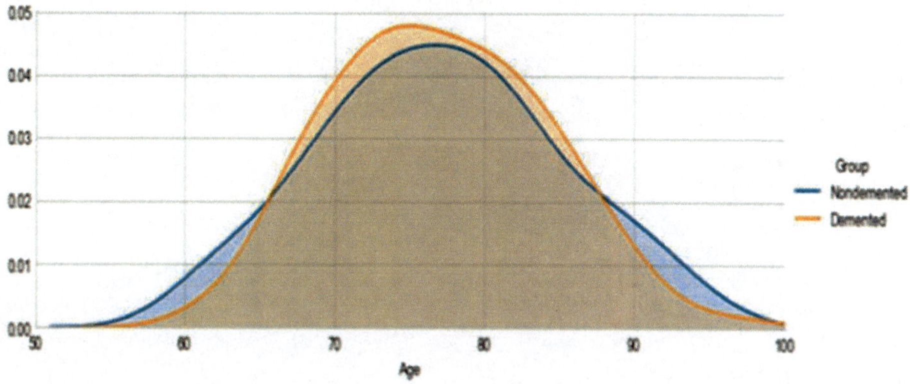

Fig. 1 Age versus non-demented and demented

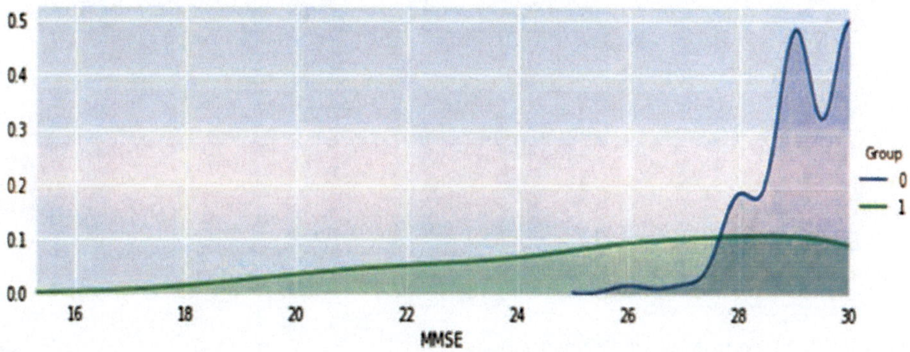

Fig. 2 Mini-mental state examination

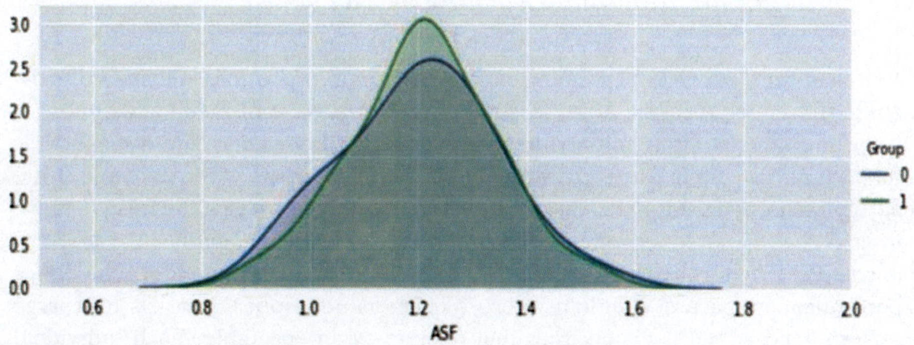

Fig. 3 ASF: Atlas scaling factor

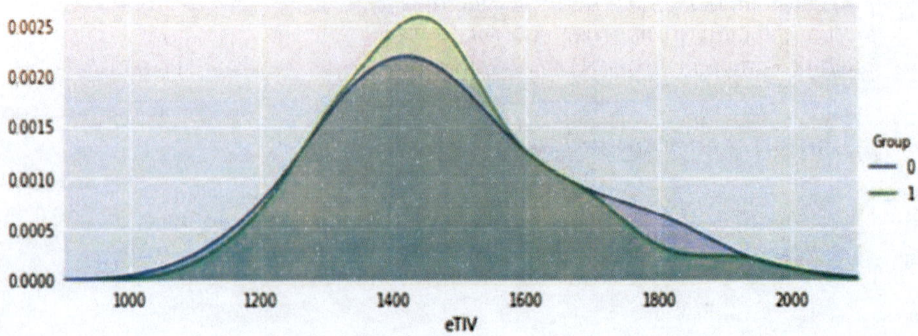

Fig. 4 eTIV: Estimated total intracranial volume

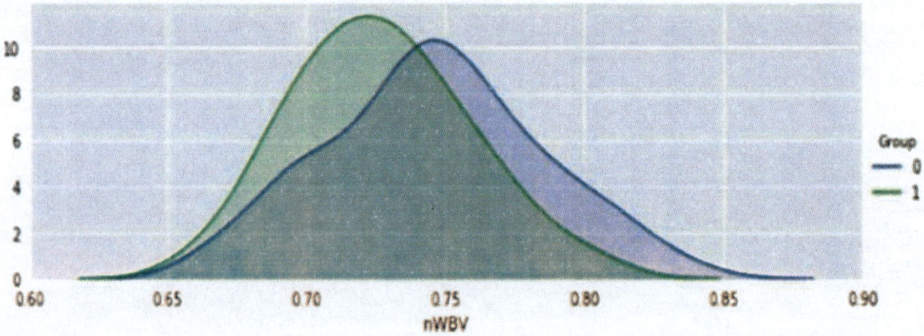

Fig. 5 nWBW: Normalized whole brain volume

4.6 Building the Machine Learning Model

The area under the receiver operating characteristic curve (AUC) is considered as the main performance measure here. It is assumed that in case of medical diagnostics for non-life-threatening terminal diseases like most neurodegenerative diseases, it is important to have a high true positive rate so that all patients with Alzheimer's are identified as early as possible. But we also want to make sure that the false positive rate is as low as possible since we do not want to misdiagnose a healthy adult as demented and begin medical therapy. Hence, AUC seemed like an ideal choice for a performance measure. Random forest, like its name implies, consists of a large number of individual decision trees that operate as an ensemble. Each individual tree in the random forest spits out a class prediction and the class with the most votes becomes our model's prediction [10]. Result obtained is depicted in figure, and accuracy for the model achieved from the proposed technique is 0.8393 or 84% (Figs. 6 and 7).

Paper also incorporates facial emotion recognition in dementia patients. As discussed, delayed diagnosis of dementia is one of the main challenges that keeps patients from having proper healthcare aid at the correct time. Most dementia patients

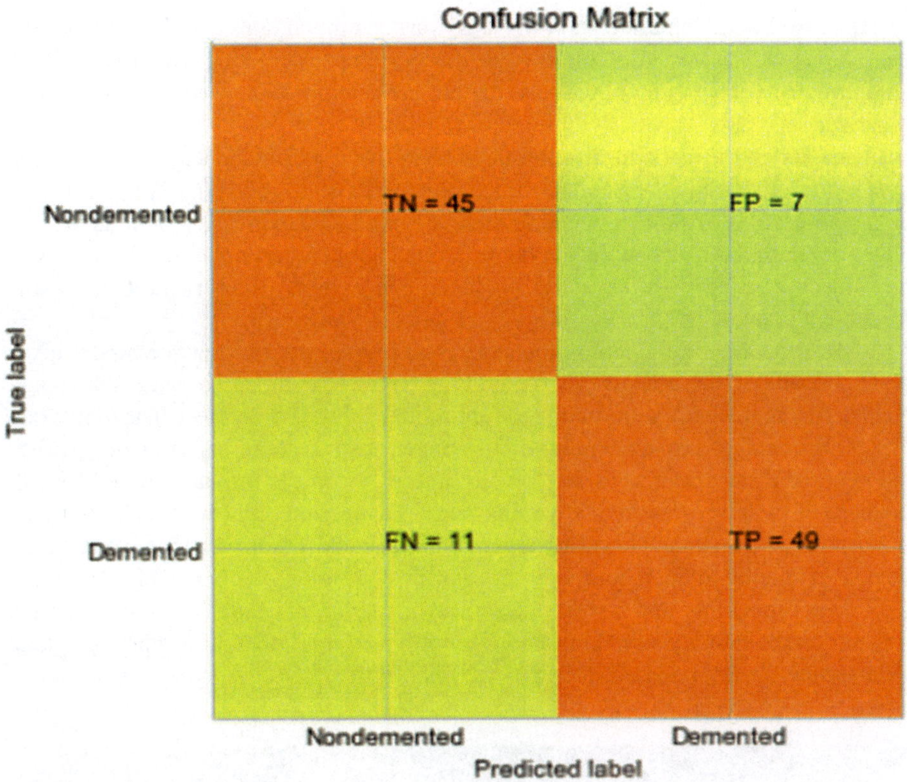

Fig. 6 Classification report and confusion matrix for the model

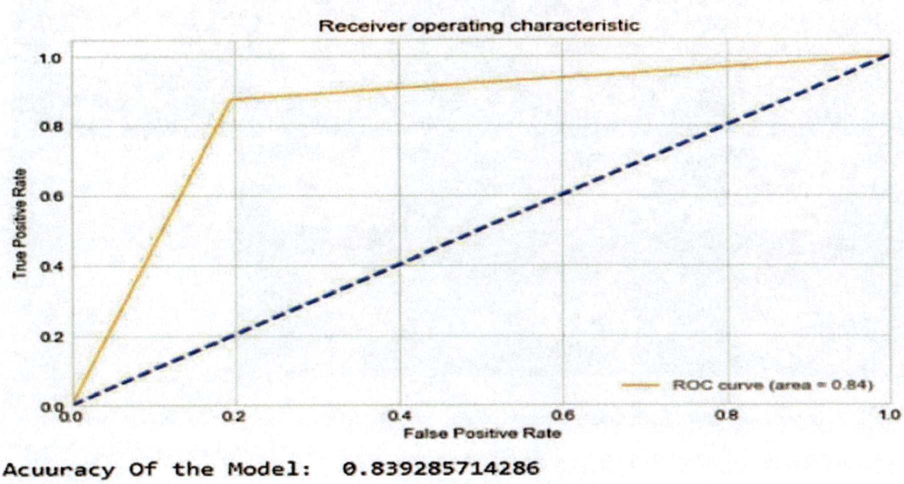

Fig. 7 Receiver operating characteristics for the proposed model

are diagnosed with dementia and its related variants much later than required which leads them to lead a very difficult lifestyle that pertains to depleting cognitive abilities which in turn leads to complications that put these underdiagnosed patients in a lot of danger.

In this part of the webapp, an automated cognitive test is added which analyzes the facial emotion of subject based on some intuitive questions and closely examine the facial expressions of subject, whether there are any facial expression deficits present [11]. For example, if the subject is being probed about a particular question, how is the subject reacting with respect to his facial expressions. What facial emotions is the subject conveying?

A ML model is built keeping the aspects of image processing. The model takes facial portraits of the subjects (young, middle-aged, old men, or women) as input and analyzes those images by predicting the emotion displayed by subject. The model can classify the images into seven emotions—anger, disgust, fear, happy, sad, surprise, and neutral. The dataset used was mainly imaged (in jpeg format) from FACES (a database of facial expressions in young, middle-aged, and older women and men).

Work used two main libraries OpenCV and facial emotion recognition 0.3.4. OpenCV is the huge open-source library for the computer vision, machine learning, and image processing and now it plays a major role in real-time operation which is very important in today's systems. Facial emotion recognition 0.3.4 is a pre-weighted open-source library released in October 2020 (Fig. 8).

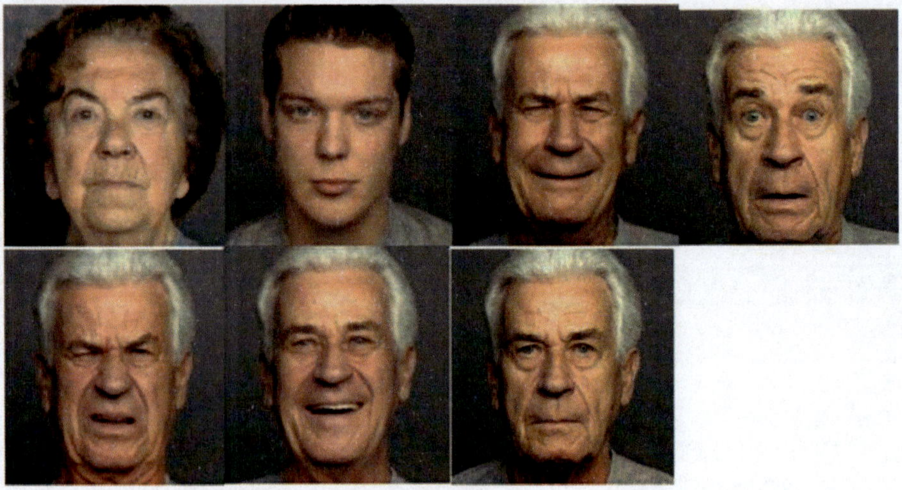

Fig. 8 Images from sample dataset

5 Conclusion

Work proposed employs simple but powerful machine learning techniques to create valuable advancements in the medical domain. Mental illnesses such as dementia and AD are often overlooked, when they can be diagnosed using such unique functionalities which are flexible enough to be used by one regarding their programming background. Clinicians can be highly benefitted from such creative intelligent applications which can assist them to identify dementia and similar neurological diseases that can be easily identified using MRI scans and collaborative cognitive tests. Developing the current webapp into an efficient robust machine learning system that can be efficiently used in various other functionalities that can help to determine dementia at an early stage. Adding more complex functionalities such as a replication interface of the medical technique called FCSRT. We can conduct something like "Active recall" for the patients, i.e., Free and Cued Selective Reminding Test (FCSRT). In this technique, the subjects(patients) search for items (e.g., apple—could be any object) in response to cues (fruits) and then further used to recall more similar items. Performance on the FCRST distinguishes dementia from normal aging with accuracy. According to World Health Organization (WHO), more than 50 million people have dementia worldwide and approximately 10 million new cases are reported every year. Thus, there is need to spread awareness which will help the community to know about this disease. Model designed achieves an accuracy of 84% which is at par with existing models.

References

1. Khan DM, Yahya N, Kamel N, Faye I (2021) Automated diagnosis of major depressive disorder using brain effective connectivity and 3D convolutional neural network. IEEE Access 9:8835–8846. https://doi.org/10.1109/ACCESS.2021.3049427
2. Kim C-M, Alvarado RL, Stephens K, Wey H-Y, Wang DJJ, Leritz EC, Salat DH (2020) Associations between cerebral blood flow and structural and functional brain imaging measures in individuals with neuropsychologically defined mild cognitive impairment. Neurobiol Aging 86:64–74
3. Association A (2019) 2019 Alzheimer's disease facts and figures. Alzheimer's Dement 15(3):321–387
4. Johnson KA, Fox NC, Sperling RA, Klunk WE (2012) 'Brain imaging in Alzheimer disease. Cold Spring Harb Perspect Med 2(4):a006213
5. Hanyu H, Sato T, Hirao K, Kanetaka H, Iwamoto T, Koizumi K (2010) The progression of cognitive deterioration and regional cerebral blood flow patterns in Alzheimer's disease: a longitudinal SPECT study. J Neurol Sci 290(1–2):96–101
6. Buvaneswari PR, Gayathri R (2021) Deep learning-based segmentation in classification of alzheimer's disease. Arab J Sci Eng 1–11
7. Barthel H, Zeisig V, Nitzsche B, Patt M, Patt J, Becker G, Dreyer A, Boltze J, Sabri O (2021) In: PET and SPECT of neurobiological systems. Springer International Publishing, Cham, pp 127–152, https://doi.org/10.1007/978-3-030 -53176-8_5

8. Zhang Y, Wang S, Xia K, Jiang Y, Qian P (2021) Alzheimer's disease neuroimaging initiative. "Alzheimer's disease multiclass diagnosis via multimodal neuroimaging embedding feature selection and fusion." Inf Fusion 66:170–183
9. Park TJ, Kanda N, DimitriosDimitriadis KJ, Han SW, Narayanan S (eds) A review of speaker diarization: recent advances with deep learning. arXiv preprint arXiv:2101.09624
10. Yuan S, Li H, Wu J, Sun X (2021) Classification of mild cognitive impairment with multimodal data using both labeled and unlabeled samples. In: IEEE/ACM transactions on computational biology and bioinformatics
11. Wang L, Li RC (2020) Multi-view orthonormalized partial least squares: regularizations and deep extensions. arXiv preprint arXiv:2007.05028
12. Thushara A, Amma CU, John A, Saju R (2020) In: Multimodal MRI based classification and prediction of Alzheimer's disease using random forest ensemble. IEEE, pp 249–256
13. Abed MT, Fatema U, Nabil SA, Alam MA, Reza MT (2020) Alzheimer's disease prediction using convolutional neural network models leveraging pre-existing architecture and transfer learning. In: 2020 Joint 9th international conference on informatics, electronics & vision (ICIEV) and 2020 4th international conference on imaging, vision & pattern recognition (icIVPR). IEEE, pp 1–6
14. Thushara A, Amma CUD, John A, Saju R (2020) Multimodal MRI based classification and prediction of Alzheimer's disease using random forest ensemble. In: 2020 advanced computing and communication technologies for high performance applications (ACCTHPA). IEEE, pp 249–256
15. Billeci L, Badolato A, Bachi L, Tonacci A (2020) Machine learning for the classification of alzheimer's disease and its prodromal stage using brain diffusion tensor imaging data: a systematic review. Processes 8(9):1071
16. Chakraborty I, Roy D, Garg I, Ankit A, Roy K (2020) Constructing energy-efficient mixed-precision neural networks through principal component analysis for edge intelligence. Nature Mach Intell 2(1):43–55
17. Xie L, Wisse LEM, Das SR, Vergnet N, Dong M, RanjitIttyerah, de Flores R, Paul A. Yushkevich, Wolk DA, Alzheimer's Disease Neuroimaging Initiative (2020) Longitudinal atrophy in early braak regions in preclinical Alzheimer's disease. Hum Brain Mapp 41(16):4704–4717
18. Khan RU, Tanveer M, Pachori RB, Alzheimer's Disease Neuroimaging Initiative (ADNI) (2021) A novel method for the classification of Alzheimer's disease from normal controls using magnetic resonance imaging. Expert Syst 38(1):e12566
19. Fang C, Li C, Forouzannezhad P, Cabrerizo M, Curiel RE, Loewenstein D, Duara R, Adjouadi M (2020) Gaussian discriminative component analysis for early detection of Alzheimer's disease: a supervised dimensionality reduction algorithm. J Neurosci Methods 344:108856. https://doi.org/10.1016/j.jneumeth.2020.108856
20. Altinkaya E, Polat K, Barakli B (2020) Detection of alzheimer's disease and dementia states based on deep learning from MRI images: a comprehensive review. J Instit Electron Comput 1(1):39–53
21. Raju M, Sudila TV, Gopi VP, Anitha VS (2020) Classification of Mild Cognitive Impairment And Alzheimer's disease from magnetic resonance images using deep learning. In: 2020 international conference on recent trends on electronics, information, communication and technology (RTEICT). IEEE, pp 52–57

Blockchain and Flutter-Based Quiz Mobile DApp Toward Decentralized Continuous Assessment

Priyanshu Kapadia, Megh Naik, Raaj Anand Mishra, and Anshuman Kalla

1 Introduction

Education is indispensable to our lives, and lately, we have experienced many new modes of acquiring it. Countries all around the globe are putting dedicated efforts, and building education reforms to educate people. For a given education system, irrespective of modes, assessment is an integral process. It provides insights at various levels such as students' performance, efficacy of the teaching learning methodologies used, efficacy of a course, and interest in a program. There are different types of assessment such as portfolio assessment, peer assessment, project-based assessment, and continuous assessment [1]. However, continuous assessment has been widely adopted.

Continuous assessment aims at measuring students' performance more frequently during a course of learning. Some of the ways to carry out continuous assessment are assignment, quizzes, oral presentation, and group tasks. The summative function of the continuous assessment is to prepare final transcripts and certificates which are issued to students at the end of a module, course, semester, or a program [2, 3]. One of the challenges is to make this functionality transparent and verifiable for not just teachers and school authorities, but also for students and their parents. This implies that various involved entities should be able see and verify anytime the computation of final grades from the continuous assessments. Any intentional or unintentional changes should be easily traceable.

Another issue is that a student shares only the final transcripts and certificates with entities such as companies (for recruitment) or other schools (for admission). However, these final transcripts do not presents the continuous learning logs of the

P. Kapadia · M. Naik · A. Kalla (✉)
Department of Computer Engineering, CGPIT, Uka Tarsadia University, Bardoli, Gujarat, India
e-mail: anshuman.kalla@ieee.org

R. A. Mishra
Dell EMC, Bangalore, India

J. K. Mandal et al. (eds.), *Proceedings of International Conference on Network Security and Blockchain Technology*, Lecture Notes in Networks and Systems 738,
https://doi.org/10.1007/978-981-99-4433-0_42

student. In other words, if someone wants to see how a student performed throughout the semester or program then that is either not possible or challenging with the use of state-of-the-art (centralized) technologies. Hence, the challenge is how to enable sharing of students' learning logs along with the final transcripts and certificates.

Furthermore, there has been increase in e-learning and blended learning which goes beyond the formal regular (face-to-face) mode of education [4]. Thus, students can acquire knowledge from various sources and using various modes (online, distance, regular). Many of these e-learning platforms have online quizzes as means of assessment. In this context, the challenge is how to ensure security and verifiability for the online quiz-based assessment to keep the trust intact.

In essence, there are numerous challenges associated with the continuous assessment such as transparent and secure conduction of assessment, verifiability of assessment process and computation of grades, and trusted sharing of learning or assessment logs along with final transcripts. The existing web or mobile applications are not right fit to overcome these challenges since they are centralized in nature. Thus, an interesting solution could be blockchain-based decentralized continuous assessment which allows transparency, security, verifiability, and trusted sharing.

The paper aims to contribute toward the development of decentralized continuous assessment by building a Blockchain and Flutter-based Quiz Mobile-Decentralized Application (BFQM-DApp). The front-end of the BFMQ-DApp is developed using flutter, while the back-end is made decentralized using blockchain technology along with smart contracts.

The main contributions of this paper are as follows.

- To propose a blockchain-based architecture for conduction of online quiz in an educational settings.
- To off-load the blockchain by securely storing the data in an off-chain distributed storage and pushing only the metadata on the blockchain.
- To implement the proposed architecture as a mobile application (BFMQ-DApp) using Flutter, Ganache (Ethereum) blockchain, smart contracts, and InterPlanetary File System (IPFS).
- To compute the cost of deploying various smart contracts designed to check the economic viability of the designed solution.

The demonstration video of the build BFMQ-DApp is available here[1].

Rest of the paper is organized as follows. Section 2 studies all the existing related works and distinguishes the current work. The proposed architecture and the overall flow is presented in Sect. 3. Section 4 provides the implementation details and discusses various smart contracts designed. The results are discussed in Sect. 5. Finally, Sect. 6 concludes this work.

[1] https://sites.google.com/view/blockchain-flutter-quiz-dapp/home.

2 Related Work

In recent years, blockchain has been considered an important technology for managing aspects of education ecosystem. Some of the applications of blockchain for education are (i) secure and privacy-protected sharing of degree and transcripts [5], (ii) secure and transparent admission process [6], (iii) scholarship management [7], (iv) prevention against degree certificate [8], and accreditation process [9]. One of the important aspects of the education is conduction of exams or quizzes toward continuous assessment. Some of the existing related works that make use of blockchain technology for secure, transparent, verifiable, and auditable conduction of examination are discussed below.

Shen et al. [10] emphasized on making the assessment process transparent and verifiable. More specifically, authors proposed the use of double-layer consortium blockchain for the conduction of quiz-based assessment. Here, students' answers for the questions rolled out during a quiz are stored of blockchain so that these can be publicly verified later. To overcome the issue of throughput and storage, authors used sharding technique where there is one main chain named as prime-chain and multiple sub-chains (shards) one sub-chain for each course. All the answer records are stored in prime-chain and sub-chains store only the summary. Moreover, use of group signature allows a teacher to completely trace any student inspite of pseudonymity. The paper not provides implementation details and experiment-based analysis.

Mitchell et al. [11] leveraged permissioned blockchain to make the examination reviewing process transparent and auditable. The examination reviewing process includes question paper creation, moderation, modification(s) if required, verification, and final submission. The authors proposed to shift this examination reviewing process on blockchain so that system can be secure, verifiable, and trustworthy. Authors used Hyperledger Fabric and Composer for implementation.

Tentea et al. [12] proposed a blockchain-enabled web application for online quiz that provides strong security against tampering of results. Authors used angular for front-end to develop interface. For back-end, they used firebase for real-time database and blockchain to store results. In addition to strong protection against result tampering, their web-based system provides single sign-in, quiz creation, and conduction. The work does not make use of any real blockchain platform and also does not provide cost estimation.

In [13], the authors aim at the issue of non-transparent and thus unreliable conduction of examinations. In particular, their focus is on making the assessment criteria (i.e., decision logic which is used to determine if the answer is correct or not) transparent with the use of blockchain. Authors implement their proposed solution using Bitcoin Core Testnet (public blockchain) and scoring server as back-end and web browser with a wallet enabled with ID and password as front-end.

Table 1 summarizes the related works and shows the clear distinction between these and our work. To the best of our knowledge, this is first work toward implementing blockchain and flutter-based mobile decentralized application for secure, transparent, and verifiable conduction of quiz.

Table 1 Comparison with the key related works

References	Aim	Implementation	Blockchain platform	Cost computed
[10]	Proposed use of double-layer sharding-based consortium blockchain for transparent and verifiable assessment process during conduction of quiz	No	–	No
[11]	Presented use of permissioned blockchain for secure, verifiable and auditable examination reviewing process	Yes	Hyperledger fabric and composer	No
[12]	Proposed a blockchain-enabled secure and tamper-proof web application for online quiz	Yes (web-based)	–	No
[13]	Proposed use of blockchain to make assessment criteria transparent and verifiable to avoid falsification of results and misjudgment	Yes (web-based)	Bitcoin core testnet	No
Our work	Propose and develop blockchain and flutter-based mobile quiz application for secure, transparent and verifiable conduction of quiz	Yes (mobile flutter-based)	Ganache (Ethereum)	Yes

3 Proposed Architecture

Figure 1 presents the proposed architecture for BFQM-DApp. There are four entities in the architecture: faculty member, student, blockchain platform, and distributed storage. Faculty members and students along with their mobile devices are the users and form the front-end of the DApp, whereas blockchain platform and distributed storage are at the back-end. Since storing all the quizzes directly on the blockchain would be expensive in terms of storage and transaction fee, thus the proposed architecture uses distributed storage. All the quiz-related data is stored in the distributed storage, and only the digital fingerprints are stored on the blockchain. In other words, we use distributed storage as off-chain storage to off-load the blockchain platform.

Faculty member, as a user, is allowed to perform following operations.

- **Approve Registration Request**: When a student registers for a course the request goes to the concerned faculty member (through blockchain platform). The faculty

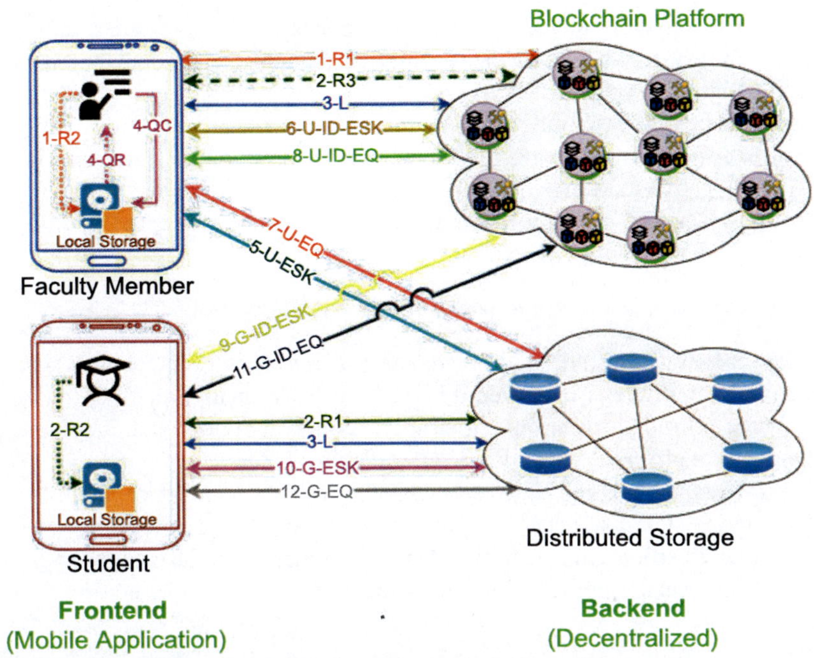

1-R1: Registration (sign-up) of faculty member on blockchain platform.
1-R2: Encrypt the private key of faculty member with the password entered during previous step (1-R1) and store it in the local storage of phone. This private key is used to sign all the transactions.
2-R1: Registration (sign-up) of student on blockchain platform.
2-R2: Encrypt the private key of student with the password entered during previous step (2-R1) and store it in the local storage of phone.
2-R3: Send the registration request of student to the faculty member for approval.
3-L: Login of both type of users (faculty member and student).
4-QC: Quiz creation by faculty member (question-wise) and storing in the local storage of phone.
4-QR: Retrieve the quiz (JSON format) from the local storage.
5-U-ESK: Upload the Encrypted Session Key. A session key (created for every quiz) is encrypted using public keys of all the students and is uploaded on distributed file storage. An ID is returned from distributed file storage.
6-U-ID-ESK: Upload the ID received in previous step on blockchain.
7-U-EQ: Upload the Encrypted Quiz. The quiz is encrypted using the session key and then uploaded on distributed file storage. An ID is returned.
8-U-ID-EQ: Upload the ID received in the previous step on blockchain.
9-G-ID-ESK: Student (after login) gets ID of encrypted session key from blockchain.
10-G-ESK: Using ID received in previous step, get the encrypted session key from distributed file storage.
11-G-ID-EQ: Get ID of encrypted quiz from blockchain.
12-G-EQ: Using ID, get encrypted quiz from distributed file storage.

Fig. 1 Proposed architecture for BFQM-DApp

member after viewing the request and crosschecking the relevant details of the student can approve (or reject) the registration request. Rejection happens in case of fake registration request.

- **Create Quiz**: Faculty member can create a quiz by entering questions along with the correct answers and marking scheme (such as marks per question, time per question, negative marking).
- **Viewing of Results**: Once the students have attempted the quiz, the results can be viewed by the faculty members.

Student type user is allowed to perform following operations:

- **Request for Registration**: When a student wants to enroll for a course, s/he does the registration process. However, it is the concerned faculty member (delivering that course) who approves the registration request. On approval, a student gets successfully registered.
- **Attempt the Quiz**: At the time of assessment, faculty member rolls out the quiz and all the registered students are allowed to attempt the quiz.
- **Viewing of Results**: Like faculty members, students are also allowed to view the result. However, an individual can view only his/her marks and not of other students.

Note that faculty members also need to register and their registration can be approved by head of the department or institute. Furthermore, the registration of the head of the institute needs to be approved by some governmental apex body. However, the proposed architecture assumes that these steps are already in place.

The overall flow and brief explanation of all the steps involved is shown at the bottom of the same figure. The first number in the notation used for any step signifies the step number. Step 1 (1-R1 and 1-R2) is the registration of a faculty member. During this step, all the required details such as name, email address, and password (for login) are entered. Furthermore, the private key of the user is encrypted using the same password and stored in the local storage of the user's device. This private key is used to put digital signature every time a user performs a transaction. Step 2 is registration of student type user. Compared to the registration of a faculty member (i.e., step 1), there is one additional sub-step involved here (i.e., 2-R3). In this sub-step, the registration request of student is forwarded to the concerned faculty member for approval. Step 3 (i.e., 3-L) allows both faculty member and student type users to login by entering their password.

Step 4 is quiz creation by a faulty member. Since a faculty member may create a large pool of questions over a period of time, thus these questions are first stored in local storage of faculty member's device. Once the faculty member is done with all the questions, the final quiz is retrieved from the local storage as one single file. Now, if this quiz file is upload on distributed storage in plaintext, any attacker can apply brute-force attack and search entire distributed storage to retrieve the quiz. So in order to ensure strong security this quiz is first encrypted and then uploaded on distributed storage. To do so, a *unique session key* is created for every quiz. Upon creation of a session key, this session key is encrypted with every student's public key and the final set of encrypted session keys are uploaded on the distributed storage (i.e., Step 5-U-ESK). On successful uploading, a unique ID is returned by distributed storage to faculty member, and this ID is then uploaded on the blockchain platform (i.e., Step 6-U-ID-ESK). In the step 7-U-EQ, the session key is used to encrypt the single quiz file and then upload the encrypted quiz on distributed storage. The ID returned is uploaded on the blockchain platform, i.e., step 8-U-ID-EQ.

In step 9-G-ID-ESK, student logins and gets the ID of encrypted session key from blockchain platform. Then in step 10-G-ESK, student retrieves the set of encrypted session keys and decrypts the session key using her/her private key. In step 11-G-ID-EQ and step 12-G-EQ, student gets the ID of encrypted quiz from blockchain and then retrieves the encrypted quiz from distributes storage, respectively. Finally, using the session key, student decrypts and attempts the quiz. The result of the quiz is uploaded on the blockchain by every student.

4 Implementation

To implement the proposed architecture, we have developed a decentralized application (BFQM-DApp) using different technologies and platforms. Table 2 summarized various technologies and platforms used. The back-end of BFQM-DApp consists of Ganache (Ethereum) blockchain plus InterPlanetary File System (IPFS). The former serves as blockchain platform, whereas the latter works as distributed storage. When any data or content is uploaded on IPFS in return the sender gets a Content Identifier (CID) which can be later used to retrieve the data. The front-end of the developed DApp comprises flutter-based mobile application which makes use of Web3 to interact with decentralized back-end.

IPFS is used for storing large quiz related data to reduce the cost. However, if we store the data on IPFS in plaintext, then attacker can apply brute-force attack to retrieve the quiz. Thus to ensure the overall security, BFQM-DApp encrypts all the data before storing on IPFS. To do so, our implementation generates an additional pair of public private keys since the encryption and decryption is not possible with key pair provided by Ganache blockchain. This additional key pair is generated at the time of registration using OpenPGP library. Thus, every user has two set of public private key pair, one provided by Ganache platform and second additionally generated. This second key pair is used for encryption and decryption of session key of every quiz.

4.1 Designed Smart Contracts

Four different smart contracts are designed for the developed BFMQ-DApp. Brief description of these smart contracts are as follows.

User Contract: This smart contract defines the structure of users' data and consists of member function for registering a new user. In particular, it has the logic

Table 2 Summary of technologies and platforms used

Technology/platform	Description
Flutter	Flutter [14] is a mobile app development framework that allows for the creation of apps for Android and iOS platforms. It was developed by Google and uses the Dart programming language
Ganache	Ganache [15] is a tool that allows developers to quickly set up a virtual blockchain for testing and development purposes
Web3	Web3 [16] is a library that allows developers to interact with the Ethereum blockchain from within a web application. It provides an API for reading and writing smart contract data, as well as for sending transactions on the network
IPFS	InterPlanetary File System [17] (IPFS) is a protocol and network designed to create a peer-to-peer method of storing and sharing hypermedia in a distributed file system. In IPFS, files are identified by their content, rather than by their location, which allows for a more robust and resilient file sharing system
Truffle	Truffle Suite [18] is a set of development tools for Ethereum blockchain development. It includes Truffle, a development environment, testing framework, and asset pipeline for Ethereum
OpenPGP library	OpenPGP [19] is a widely used library that provides cryptographic functions such as encryption and signing, to secure communication and data files. It is based on the OpenPGP standard and uses public key cryptography, allowing users to encrypt messages using the recipient's public key and decrypt them using their own private key

of (i) creating the student and faculty member user, (ii) retrieving the user details, (iii) updating the user details, and also the logic for verifying a student registration request by a faculty member.

Quiz Contract: As the name implies, this contract defines the structure used of quiz related data. The contract provides all the functionalities required for creating, encrypting, and uploading a quiz. More specifically, this contract consists of logic for (i) creating a quiz in JavaScript Object Notation (JSON) format, (ii) encrypting and uploading a quiz on IPFS, and (iii) pushing the CID (of the encrypted quiz) received from IPFS on the Ganache blockchain.

PublicKey Contract: This contract defines the structure of public key data which is mapped uniquely to user address. Various functionalities provided by this contract are (i) adding the public key to the blockchain according to each user enrollment address and (ii) retrieving the public key arrays for the encryption and decryption process of quiz data.

Result Contract: This contract maps each student's result for every quiz according to the quiz ID and user address. The mapping helps in easy retrieval of result data and verification as and when required.

Further details of the smart contract can be found at[2].

---

[2] https://sites.google.com/view/blockchain-flutter-quiz-dapp/home.

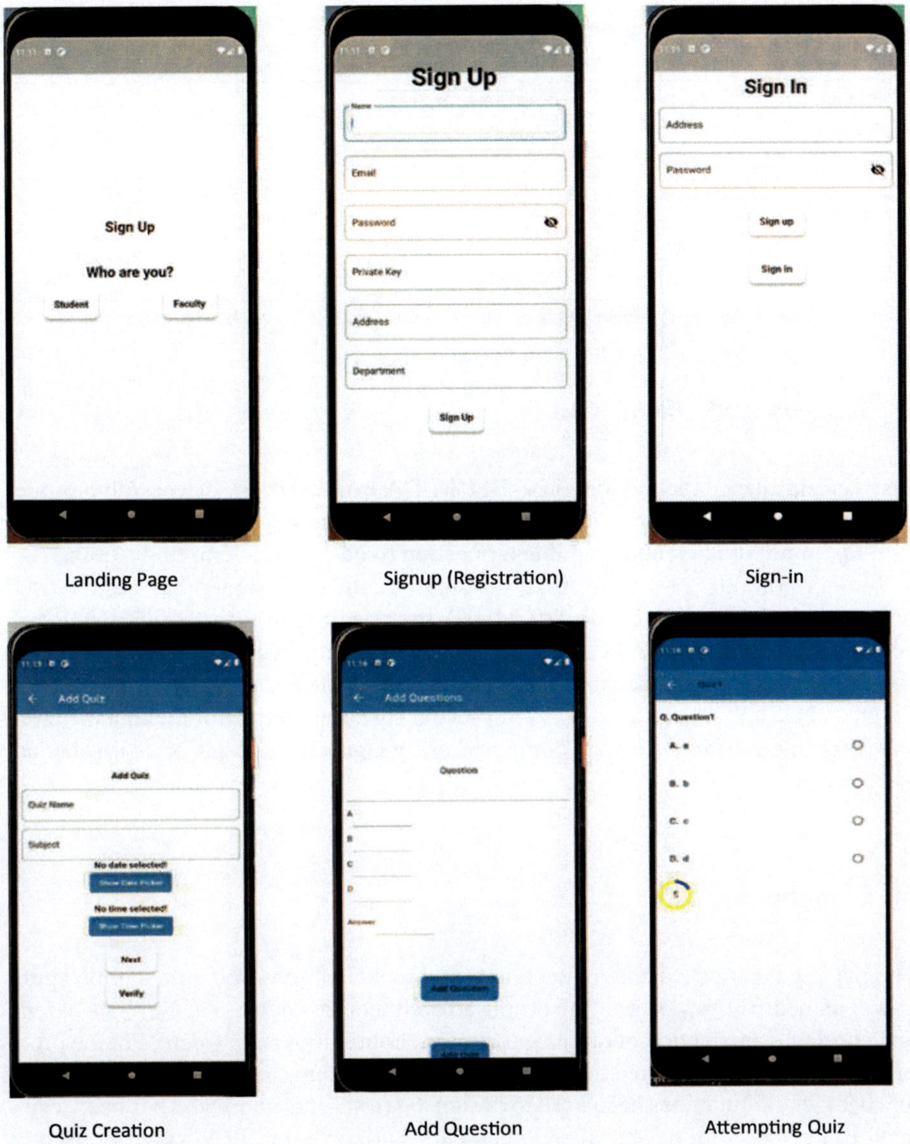

Fig. 2 Various snapshots of BFQM-DApp

Table 3 Various costs of smart contracts developed

Contract name	Cost (Gwei)	Cost (Ether)	Cost (USD)	Gas used	Gas limit
User contract	27,738,160	0.02773816	36.94	1386908	2,000,000
Quizzes contract	16,804,838	0.016804838	22.38	840242	2,000,000
PublicKeys contract	13,022,700	0.0130227	17.34	651135	2,000,000
Results contract	7,185,920	0.00718592	9.57	359296	2,000,000

1 Gwei = $1/10^9$ Ether, gas price = 20 Gwei, and 1 Ether = $ 1331.75 on Jan 10, 2023

5 Results and Discussion

The decentralized application (i.e., BFQM-DApp) has been successfully implemented and tested. Figure 2 shows snapshots of various steps such as landing page, sign-up, login, quiz creation, adding a question to quiz, and attempting a quiz.

Furthermore, the cost analysis of the smart contracts designed has been carried out to check the economic viability. Miners participating in Ethereum P2P network need to be incentivized for the resources they are offering. Thus, deployment of smart contracts on top of blockchain incurs some cost. Table 3 shows the cost (in Gwei, Ether, and USD), gas used, and gas limit set during computation for the four different smart contracts. These costs are computed using Ganache, and gas price is set equal to 20 Gwei.

6 Conclusion

The paper has proposed an architecture for a blockchain-enabled quiz mobile application named BFQM-DApp. The proposed architecture enables secure, transparent, and verifiable conduction of online quiz toward continuous assessment. For the proof of concept, we implemented the BFQM-DApp using Ganache blockchain which acts as a back-end. Flutter has been used to design the front-end, and Web3 has been leveraged to interact with decentralized back-end. Furthermore, IPFS has been used as decentralized (off-chain) storage. Four different smart contracts have been designed to encode various functionalities. The deployment cost of the smart contracts is also computed to check the economic viability. The future scope of the work includes latency computation, detailed cost computation of various functionalities (in addition to the costs of the designed smart contracts), and security analysis of overall applications. Since latency is an important factor, and it is also blockchain (type and) platform dependent, thus in the future we plan to implement the proposed architecture on hyperledger fabric blockchain as well and provide a comparative results.

References

1. Oli G, Olkaba T (2020) Practices and challenges of continuous assessment in colleges of teachers education in west Oromia region of Ethiopia. J Educ Teach Learn 5(1):8–20
2. Hernández R (2012) Does continuous assessment in higher education support student learning? Higher Educ 64(4):489–502
3. Peytcheva-Forsyth R, Saev S, Yovkova B (2021) Integrated continuing assessment in an online course as a mechanism for a smoother transition from face-to-face to distance learning. In: AIP conference proceedings, vol 2333. AIP Publishing LLC, p 050014
4. Commission UG (2023) Blended mode of teaching and learning: concept note. Available at https://www.ugc.ac.in/pdfnews/6100340_Concept-Note-Blended-Mode-of-Teaching-and-Learning.pdf. Accessed on 15 Jan 2023
5. Mishra RA, Kalla A, Braeken A, Liyanage M (2021) Privacy protected blockchain based architecture and implementation for sharing of students' credentials. Inf Proc Manag 58(3):102512
6. Kutty RJ, Javed N (2021) Secure blockchain for admission processing in educational institutions. In: 2021 international conference on computer communication and informatics (ICCCI). IEEE, pp 1–4
7. Tekguc U, Adalier A, Yurtkan K (2020) Scholarchain: the scholarship management platform with blockchain and smart contracts technology. Eurasia Proc Educ Soc Sci 18:86–91
8. Tariq A, Haq HB, Ali ST (2022) Cerberus: a blockchain-based accreditation and degree verification system. IEEE Trans Comput Soc Syst
9. Cahyadi D, Faturahman A, Haryani H, Dolan E, Millah S (2021) BCS: Blockchain smart curriculum system for verification student accreditation. Int J Cyber IT Service Manag 1(1):65–83
10. Shen H, Xiao Y (2018) Research on online quiz scheme based on double-layer consortium blockchain. In: 2018 9th international conference on information technology in medicine and education (ITME). IEEE, pp 956–960
11. Mitchell I, Hara S, Sheriff M (2019) Dapper: decentralised application for examination review. In: 2019 IEEE 12th international conference on global security, safety and sustainability (ICGS3). IEEE, pp 1–14
12. Tentea EC, Ionescu VM (2019) Online quiz implementation using blockchain technology for result tampering prevention. In: 2019 11th international conference on electronics, computers and artificial intelligence (ECAI). IEEE, pp 1–6
13. Kaneko Y, Tanaka S, Kimura T, Okumura J, Azuchi S, Osada S (2021) Deexam: a decentralized exam administration model using public blockchain. In: 2021 3rd blockchain and internet of things conference. pp 1–7
14. Flutter. Available at https://flutter.dev/. Accessed on 15 Jan 2023
15. Ganache. Available at https://trufflesuite.com/ganache/. Accessed on 15 Jan 2023
16. Web3. Available at https://web3js.readthedocs.io/. Accessed on 15 Jan 2023
17. IPFS. Available at https://docs.ipfs.tech/. Accessed on 15 Jan 2023
18. Truffle. Available at https://trufflesuite.com/docs/truffle/. Accessed on 15 Jan 2023
19. Openpgp. Available at https://www.openpgp.org/software/developer/. Accessed on 15 Jan 2023

Data Receiving Analysis for Secure Routing from Blackhole Attack in a Spontaneous Network Using Blockchain Method

Gaurav Soni, Kamlesh Chandravanshi, Nilesh Kunhare, and Medhavi Bhargava

1 Introduction

MANET is a network of mobile nodes with no infrastructure, which means it can change their geographic location. The result is a dynamic topology and unrestricted mobility scenario with limited resources. In MANET, due to its dynamic nature, the network is portioned into small networks. MANET is a multi-hop self-organizing system having mobile wireless nodes with restricted communication capacity [1, 2]. To send messages between two nodes that are not in direct communication range, intermediary nodes are required and these nodes are the malicious nodes. Due to the open for all, it has been easily affected from attackers and only reliable schemes can detect and prevent network from attacks. The blockchain is a novel and secure method by which we can secure the routing data from attackers. The blockchain method is only popular because decryption of the code is impossible and any changes to the code are broadcast to all connected nodes [3, 4]. Once hash code has been generated, decryption of the hash code is not possible.

The attacker is not interested in forwarding of data packets, and it will only focus on dropping the data of the sender. The attacker initially behaves normally, but once

G. Soni (✉) · K. Chandravanshi · N. Kunhare
School of Computing Science and Engineering, VIT Bhopal University, Sehore, Madhya Pradesh 466114, India
e-mail: gaurav.soni@vitbhopal.ac.in

K. Chandravanshi
e-mail: kamlesh.chandravanshi@vitbhopal.ac.in

N. Kunhare
e-mail: nilesh.kunhare@vitbhopal.ac.in

M. Bhargava
School of Engineering and Technology, SAGE University, Bhopal, India

J. K. Mandal et al. (eds.), *Proceedings of International Conference on Network Security and Blockchain Technology*, Lecture Notes in Networks and Systems 738,
https://doi.org/10.1007/978-981-99-4433-0_43

the route is established and the sender begins sending data, the attacker begins to misbehave or drops data in the network. MANET is subject to assaults by selfish or malevolent nodes due to multi-step routing and an open working environment, such as packet loss attacks, also known as blackhole attack, and selective redirect attacks, also known as gray hole attack [2]. All assurances of MANET security for solving real-world problems continue to pique the interest of corporate and university research initiatives. The primary goal of peer-to-peer mobile network research is to provide a dependable environment and secure communication [2].

There are various network applications that necessitate secure communication. MANET is inactive when the infrastructure cannot be installed, either because it is too expensive or too susceptible. A military base station on the battlefield is an example of vulnerable infrastructure. If the network infrastructure is damaged as a result of a disaster, a specialized wireless network can be deployed to coordinate rescue efforts. The location-based information system can store the location of nodes by that node movement prediction detection is possible [5]. MANET is extremely beneficial in a variety of situations, including emergency services, tactical networks, education, entertainment, natural disasters, and so on [6]. MANETs can be set up in a matter of meters or kilometers, and they can also be linked to the Internet, allowing information to be transferred all over the world. The Internet is only used to convey data to a certain destination via the sender. All senders may be sending data within or outside of the MANET.

This paper's significant contributions are securing the communication of ad hoc network by applying blockchain approach. The blockchain implementation in MANETs is outlined in a series of processes, which include transaction validation, block configuration, block validation, block chaining, and block maintenance. Furthermore, a suitable consensus mechanism for MANETs known as BSSB has been successfully implemented for blockchain validation. The BSSB scheme is applying rules to influence, if nodes can collaborate and accept a specific security strategy to prevent threats for improving network performance.

2 Attacks in MANET

The use of wireless links in MANET produces an open medium that makes it impossible to define a clear line of defense and exposes the network to attacks like passive eavesdropping to active eavesdropping. Each node should be prepared to contact with attacker directly or indirectly [7–9]. MANET attacks are classified into two different types: first is passive attacks and second is active attacks.

2.1 Passive Attack

Passive attacker or malicious nodes are those in which the assailant does not disrupt the routing protocol but instead uses traffic analysis to uncover valuable data. Spying, traffic analysis, and location disclosure are examples of such attacker.

2.2 Active Attack

Active attacker performing invasive actions such as changing, injecting, tampering, producing or malicious information, or fake control packets, result is multiple network distractions. These attacks and malicious activities are continued from beginning to end of simulation. Internal and external attacks are the two types of active attacks.

- **Blackhole Attack**

When a node requires a route to its destination, attackers will endeavor the susceptibility in route discovery methods for table-driven routing protocols such as AODV [10] and DSR [10]. The malicious node declares that it has a route to destination, when the node transmits the RREQ. When data packets are being sent in the network, blackhole nodes drop these packets, and these packets have not reached their destination.

- **The Gray Hole Attack**

A gray hole (GH) assault is a subset of a BH attack in which the initial attacker detentions routes; i.e., he is part of the network routes through which he drops packets of his choice (as in a BH attack).

- **Rushing Attack**

In a rushing attack, the attacker exceeds the packet's service data limit; the protocol on-demand only entails nodes to return the main RREQ, which is received for each road discovery. The malicious node will exploit this trait by rapidly sending RREQ packets across the network. All legal RREQ packets that follow will be dropped as a result.

- **Sybil Attack**

A Sybil attacker behaves like a malicious node that steals two or more other nodes' identities for participating in routing. Sybil nodes use fake identities to participate in routing. These other node identifiers will be generated by the nodes or physical devices.

3 Literature Review

In the relevant section, we discuss the work done in the field of the attacking blackhole to avoid, detect, and effect the attack in routing protocols.

Noguchi and Hayakawa [11] proposed a way of protecting against blackhole attacks. You can use the proposed way to send several RREPs for the same route request (RREQ) packet. Several RREPs aid both the source and intermediate nodes in the prospective path, allowing the destination node to acquire a large amount of route information. Multiple RREQ is normal, and how it is different presence attacker is not mentioned. Soni and Chandrawanshi [12] proposed a security scheme for blackhole attacker in high-speed 6G network. Without security guarantees, some disobedient or malicious cars make the system vulnerable to provide low-quality services. The security mechanism for locating an attacker's car is based on network traffic statistics. The approach detects the presence of the attacker and estimates the total number of packets lost by the blackhole attacker in the network. Sharndip Kaur, Anuj Gupta [13] proposed a new method for detecting and stopping blackhole attacks in MANET. To accomplish this, a metaheuristic search technique based on the AODV routing protocol has been created, which combines the minimum and maximum ACO choices with DRPI control tables. Soni and Sudhakr [14] proposed a L-IDS scheme for dropping attack. In this research, apply the integration on hop count for measures the exact value of forwarding. The correct integrated value means no attacker exist in network otherwise starts to prevent network from attacker. Abdelshafi and King [15] "Resistance to blackhole attacks in MANET." This section introduces the novel idea of trust in the protocol itself (SPT), which is used to detect intruders. Watching the regular behavior of the protocol and engaging a malicious node is to make an implicit statement of its harmful activity. Kaur et al. [16] "A comparative study based on the modeling of routing protocols during a wormhole assault in a MANET." In this part, we investigate and compare the performance of AODV, DSR, and ZRP. Only comparison is mentioned.

4 BSSB Security Scheme

The proposed **BSSB** security scheme detects the attacker via its dropping behavior and completely disables the attacker's network communication. After employing BSSB, the attacker node(s) are completely barred from the network, and no node in the network communicates with attackers for the purpose of establishing connections or sending data. The purpose of BSSB is to monitor the incoming and outgoing packets between senders and receivers. There is no work is done before to detect and prevent the MANET using blockchain method. The sender sends data, but it has not reached its destination because of the attacker's presence on route.

4.1 Algorithm: Data Receiving and Hop Count-Based Blackhole Detection and Prevention

```
Input:
Q_i: mobile nodes
I_j: intermediate node ε Q_i
S: Source node ε Q_i
R: Receiver node ε Q_i
H_c : hop count initial 0
H_seq: higher sequence no.
d_rcv: data receive
B: black hole node or Malicious  // m_1,m_2.....m_n
BS: preventer node// where apply BSSB
Radio Range Ψ550m^2
Proto: AODV
Output: PDR, Throughput, NRL, Delay, attack percentage
Procedure:
Network form by Q_i & Ψ
S sense medium call AODV
AODV execute (S, R, R_req)
    If I_j in ψ & I_j receive AODV R_req Then
        I_j ← create route table (insert I_j-1)
        I_j forward AODV(S, R, R_req) to next hop
        If I_j receive AODV R_req & not forward Then
            I_j generate H_seq
            I_j ← set H_c ∞
            Send R_rep to S
            S send data to I_j
            I_j drop data
                End if
        Else If I_j in Ψ & I_j == R Then
            Send R_rep to S
            S Send data to R by I_j link
        Else
            Not out of range
        End If
```

```
#Prevention Module by Blockchain (BSSB)
P watch I_j node
If I_j in P range & participate in R_req Then
    P watch the route behavior of I_j node
If I_j receives R_req & I_j != R & R_req not forward Then
        I_j ← generate H_seq & H_c ∞
        P send feedback to Q_i-1
        I_j ← B
        I_j block by BS node
        S Call AODV(S,R, R_req)
        Broadcast R_req to Q_i-1 where I_j not in route
        Else If I_j receives data & I_j != R & Data not
    forward Then
        P calculate d_rcv = forward/receives
        count++
        If count > 10  & d_rcv == 0 Then
        P check I_j routing table
            If I_j modified routing table Then
              I_j ← B
              I_j block by BS node /BSSB
              // number of "0" count is more
               up to hop count D (loop)
              S Call AODV(S,R, R_req)
              Broadcast R_req to Q_i-1
          where I_j not in rout
            Else
              I_j congested S decrease
        Else
              I_j congested
              S decrease data rate
                        End if
                  End if
            End if
        End if
```

The steps of detection and prevention of attacker are also mentioned clearly. If routing table is not matched, it means some misbehavior activity occur in the network through malicious node, then BSSB scheme is applied and block that hops and changes the path and forwarding data packet. In normal routing "B" node behaves as normal node because data sending is not started (Fig. 1).

If routing is matched then forward data packet until send all data packets are reached to the destination. Here we represent the routing information of source node S to destination node D, where each node will forward data packets to their next neighbor till destination is not found, i.e., mentioned in Table 1.

The malicious activity is shown in Fig. 2. The BSSB scheme is clearly differed from normal and susceptible condition. The nodes forward data packets to the destination but attacker dropped all data packets. Table 2 represents the data information of each node in case of malicious node or blackhole node. Here node is a malicious node and it will not forward data packets to their next node, then it consumes all data packets.

After that, forward data packets until all data packets have arrived at their destination. When it will set up the authentication type on wireless devices, then it can choose from a number of different options. It will likely originate from a collection

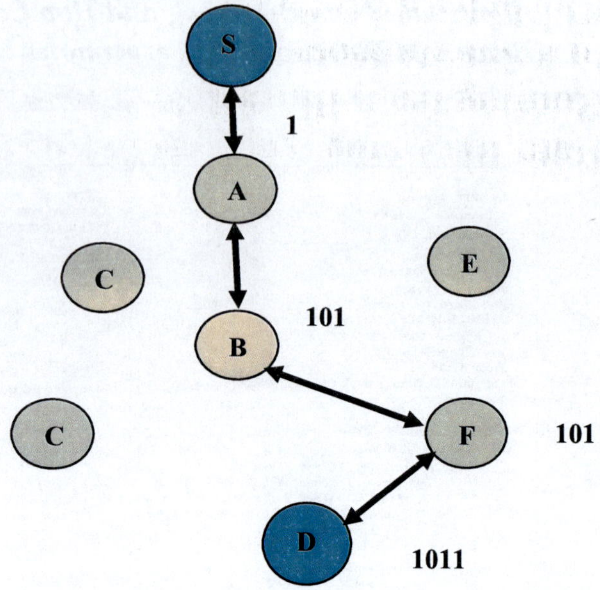

Fig. 1 Normal routing in S and D

Table 1 Record of source node S

Data information		Hash code for D
From	Through	
S to D	A	1
	B	10
	F	101
	D	1011

or cluster of packets that corroborate the occurrence of malicious behavior. It is standard practice to verify the identification of a device or user before granting access to a network and its resources. The use of binary digits to denote levels of access is the norm. Based on the user's affiliation with a given group, access is granted or denied.

5 Simulation Results and Description

The simulation has been performed with different node densities and more than one malicious node. The number of nodes is taken into account as mobile detector nodes. All node has random mobility speed to move in network. The simulator version is employed for simulation is NS-2.31[17]. The final parameters are listed in Table 3.

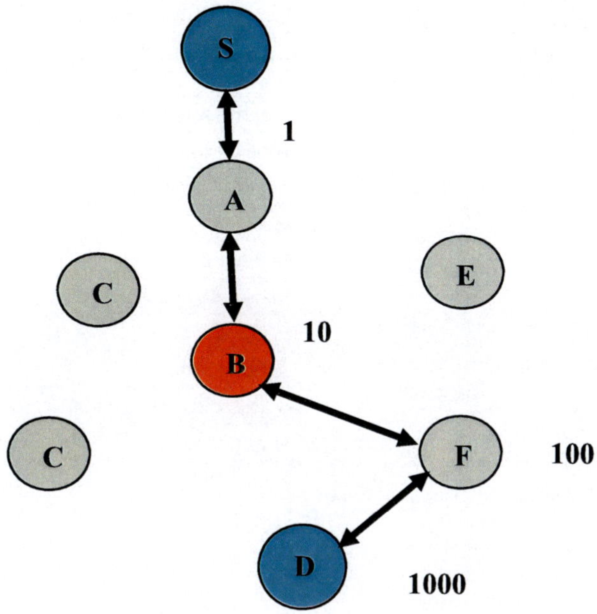

Fig. 2 Blackhole malicious actions are identified

Table 2 Malicious node identification

Data ınformation		Hash code for D
From	Through	
S to D	A	1
	B	10
	F	100
	D	1000 **(loop)**

6 Results Analysis

In this section, mention the results analysis of proposed PSSB, previous MRREP, and the normal blackhole AODV. Simulation parameters are common from all scenarios.

6.1 Packet Delivery Ratio Analysis

The performance of the proposed security scheme is providing better results as compared to the old prevention scheme in MANET. The number of mobile nodes in the network is continuously sending, reviving, or forwarding data, but the data success ratio is evaluated by PDR at the destination. In this graph, the node density scenario of 35 nodes shows a lower PDR value because nodes are densely populated in limited area. The PDR value of a blackhole attack is about 20% in 35 nodes and

Table 3 Simulation parameters

Parameters	Value
Attacker	Blackhole
Security scheme	BSSB
Time (seconds)	100
Area for simulation (meter Square)	800 * 800
Mobility model	Random waypoint
Number of nodes	20, 25, 30, 35
Blackhole nodes	1, 2, 3, 4
Medium	Wireless (802.11)
Routing protocol	AODV
Data packet size (bytes)	512
Traffic type	CBR, FTP
Nodes motion	Random
Propagation radio model	Two-ray ground
Rate	10 to 30 Packet/s

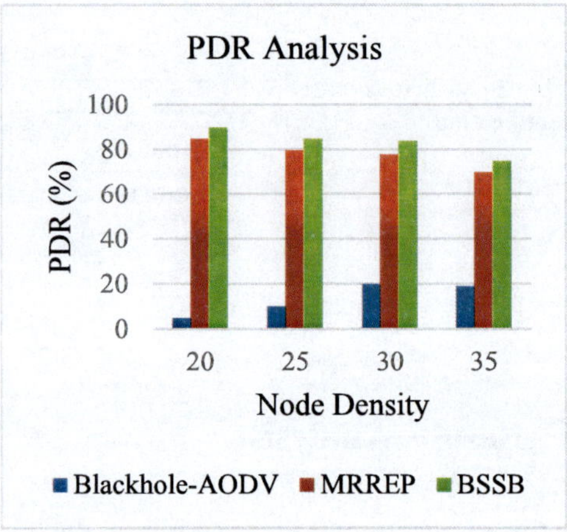

Fig. 3 PDR performance analysis

about less than 20% in the rest of the scenario, but after applying security, it reaches more than 90% in the rest of the protocol (Fig. 3).

6.2 Normal Routing Overhead Analysis

Because of the retransmission of control packets, data loss improves overhead control performance. Because of the loss of nearly complete data packets, the number of attacker nodes is increasing, increasing the overhead. The existing security scheme

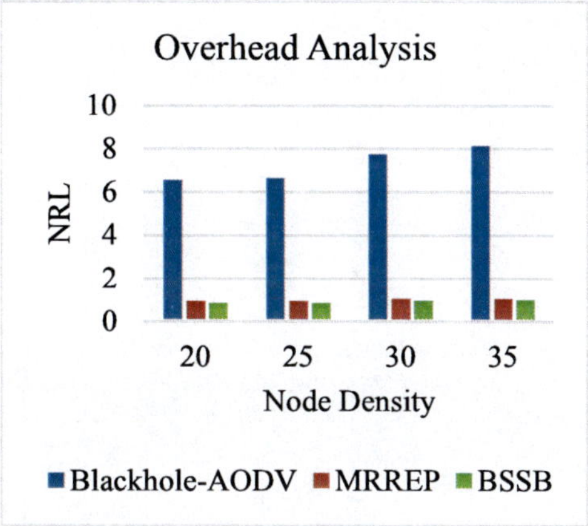

Fig. 4 Routing overhead analysis

has an overhead of less than one in the network. The proposed security scheme improves performance, which explains why the overhead is minimal. In the presence of an attacker, the overhead is also at its maximum or relayed nearly six times more than in the proposed scheme (Fig. 4).

6.3 *End-to-End Delay Analysis*

At the receiver end of the network, the delay is measured. The sender sends data packets after confirming the destination. Some packets in any network are lost, and a duplicate copy is sent to the receiver, or data is sometimes retransmitted in the network due to time expiration. In this graph, the performance of all three protocols in the network is measured, and it is observed that the performance of the proposed scheme is better because it provides less network delay. The delay in blackhole presence is more, which shows the degradation in network performance (Fig. 5).

6.4 *Throughput Analysis*

The throughput is actually evaluated to measure the amount of data received or sent per second in the form of packets, or Kbps (kilobits per second). In this graph, due to poor data reception in the presence of an attacker, the throughput of blackhole is really showing degradation in performance. The throughput improves with node density, and the attacker's performance provides a throughput of 22 Mbps only but proposed scheme produces better results. The throughput performance of the proposed scheme reaches up to 90 Mbps and provides better security as compared to previous work (Fig. 6).

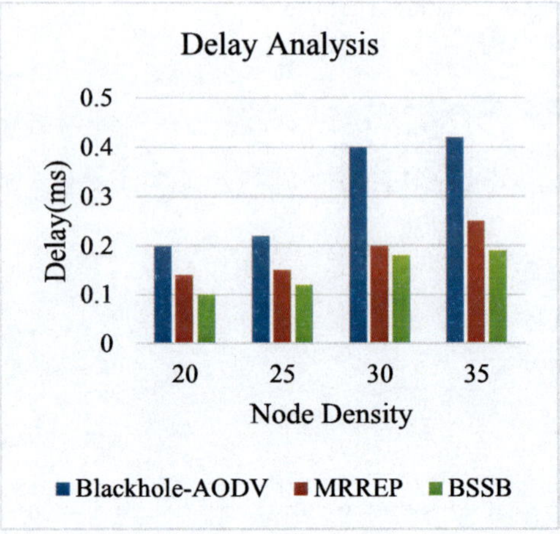

Fig. 5 End-to-end delay analysis

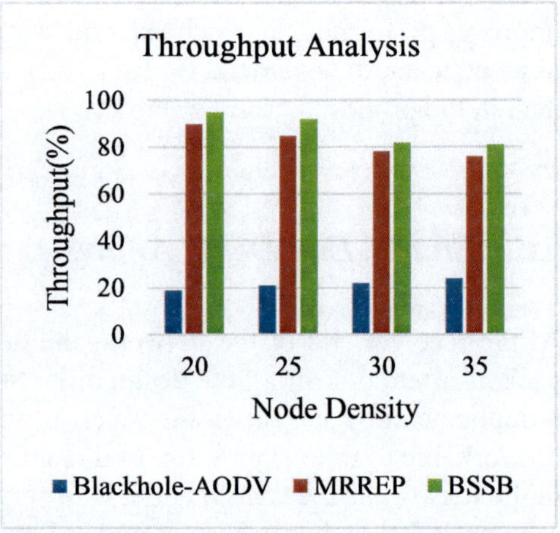

Fig. 6 Throughput performance analysis

6.5 Blackhole Attack Percentage Analysis

The number of attacker nodes in MANET only drops packets in the network. In this graph, the attacker's packet capture is evaluated, and it is seen that the percentage of loss in the network reaches a maximum of 40%. The attacker's packet capture causes data packets on the network to be lost. In this case, data loss in the network is only counted in the presence of an attacker, but the proposed prevention improves routing performance in the network (Fig. 7).

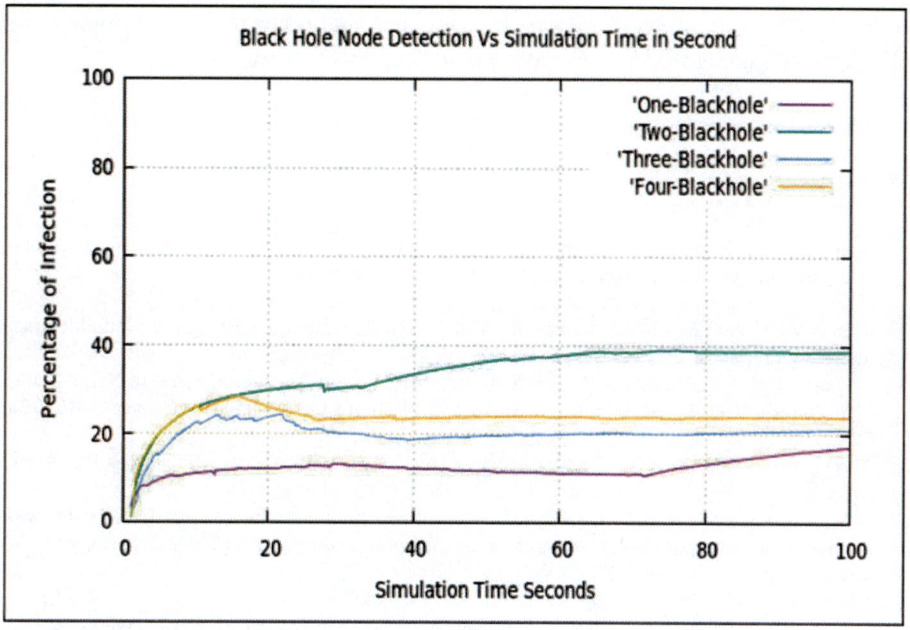

Fig. 7 Blackhole drop percentage analysis

7 Conclusion and Future Work

The blackhole attacker is a very dangerous attacker because it is not easily detectable in network because of the attacker nodes movements and the dropping of packets due to other reasons in network. According to the proposed method, if the number of dropped packets exceeds the limit, either the network's performance is degrading or an attacker is present in the network. The routing protocols don't know the attacker is there, or they can't do anything against them because the attacker is participating as a normal node and when data forwarding has been started, it drops data. In this paper, we propose a BSSB approach for malicious activity detection and prevention in MANET. The hash code for normal routing has been different as compared to routing done in the presence of an attacker. The performance of the MRREP scheme is trust-based, and trust is only dependent on better packet receiving. The proposed blockchain-based security is better than the existing scheme in terms of better routing performance. The proposed BSSB scheme shows a 5% improvement in PDR and an 5% improvement in throughput performance. The overhead and delay are also less than the previous MRREP approach. The performance of all protocols is measured through performance metrics, and the performance of the proposed scheme is better than the existing MRREP.

In the future, we will work on the difference between a packet being dropped because of an attacker and other reasons like congestion, collision, request timeout, and so on. The symptoms of an attacker dropping and other types of dropping are different, and updated-BSSB security scheme separates them.

References

1. Chandravanshi K, Soni G, Mishra DK (2022) Design and analysis of an energy-efficient load balancing and bandwidth aware adaptive multipath N-channel routing approach in MANET. IEEE Access 110003–110025
2. Kumar M, Mishra M (2012) An overview of MANET: history, challenges and applications. Indian J Comput Sci Eng (IJCSE) 3(1):121–125
3. Abdel-Sattar, A.S, Marianne A. Azer., :Using Blockchain Technology in MANETs Security," 2nd International Mobile, Intelligent, and Ubiquitous Computing Conference (MIUCC), (2022).
4. Guo H, Yu X (2020) A survey on blockchain technology and its security. Blockchain Res Appli 1–15
5. Soni G, Jhariya MK (2020) Quadrant base location tracking technique in MANET. In: 2nd ınternational conference on data, engineering and applications (IDEA), Bhopal, India, pp 191–196
6. Sivalingam KM (2003) Tutorial on mobile Ad Hoc networks
7. Soni G, Chandrawanshi K (2013) A novel defence scheme agaınst selfısh node attack in MANET. Int J Comput Sci Appl (IJCSA) 3(3)
8. Manmohan S, Mamoon R (2020) Security attacks In MANET—A comprehensive study. In: Proceedings of the ınternational conference on ınnovative computing and communications (ICICC), pp 1–6
9. Soni G, Chandrawanshi K, Verma R, Saraswat D (2022) High data priority endorsement and profile overhaul using block chain against remapping attack in MANET-IoT CRC Book Chapter 8, Dec 2022 (preprint)
10. Kannhavong B, Nakayama H, Nemoto Y, Kato N (2007) A survey of routing attacks in mobile Ad Hoc networks. IEEE Wireless Commun, Oct (2007)
11. Noguchi T, Hayakawa M (2018) Black hole attack prevention method using multiple RREPs in mobile Ad Hoc networks. 17th IEEE ınternational conference on trust, security and privacy ın computing and communications/12th IEEE ınternational conference on big data science and engineering
12. Soni, G., Chandravanshi, K., "A Novel Privacy-Preserving and Denser Traffic Management System in 6G-VANET Routing Against Black Hole Attack," Sustainable Communication Networks and Application pp 649–663(2022).
13. Kaur S, Gupta A (2015) A novel technique to detect and prevent black hole attack in MANET. IJIRSET 4(6)
14. Soni G, Sudhakar R (2020) A L-IDS against dropping attack to secure and ımprove RPL performance in WSN aided IoT. In: IEEE ınternational conference on signal processing and ıntegrated networks (SPIN), pp 377–383
15. Abdelshafy MA, King PJB (2016) Resisting blackhole attacks on MANETs. In: 13th IEEE annual consumer communications and networking conference (CCNC)
16. Kaur P, Kaur Ḍ, Mahajn R (2017) Simulation based comparative study of routing protocols under wormhole attack in MANET. Springer Science Business Media, New York, 28 Apr 2017
17. Content Available on link https://www.isi.edu/nsnam/ns/ access from 15 Oct (2022)

A Blockchain-Based Transparent Solution for Achieving Investment for Farming

Ayushya Chitransh and Barnali Gupta Banik

1 Introduction

Farmers are the backbone of any economy. They put all their time, effort, and assets into growing crops without certainty of investment return. Most of the time, they do not have enough cash flow for many things. Farmers must invest in their farming operations, including purchasing seeds, organic fertilizers, boosters, root hormones, and new equipment, modernizing their irrigation systems, investing in modern technology and data analysis tools, and expanding their operations by buying additional land or acquiring new farms [1]. Farmers may also invest in new or improved infrastructure, such as building new barns, greenhouses, or storage facilities.

Additionally, farmers may choose to invest in marketing and branding efforts to increase the value of their products and reach new customers [2]. Therefore, they need capital funding. There are several challenges that farmers may face when seeking capital funding:

Limited access to credit: Many small and medium-sized farmers may have difficulty accessing credit from traditional financial institutions due to a lack of collateral or poor credit history.

High-interest rates: Farmers may be charged higher interest rates on loans due to the perceived risk of farming as a business.

Short repayment periods: Farmers may be required to repay loans within a fleeting period, which can be challenging if the farm is not generating sufficient income.

Lack of information: Farmers may not be aware of the various funding options or may not have the necessary information to apply for funding.

A. Chitransh (✉) · B. Gupta Banik
DL Unify, DLT Labs, Hyderabad, India
e-mail: ayushyachitransh@gmail.com

B. Gupta Banik
e-mail: barnali.guptabanik@ieee.org

J. K. Mandal et al. (eds.), *Proceedings of International Conference on Network Security and Blockchain Technology*, Lecture Notes in Networks and Systems 738,
https://doi.org/10.1007/978-981-99-4433-0_44

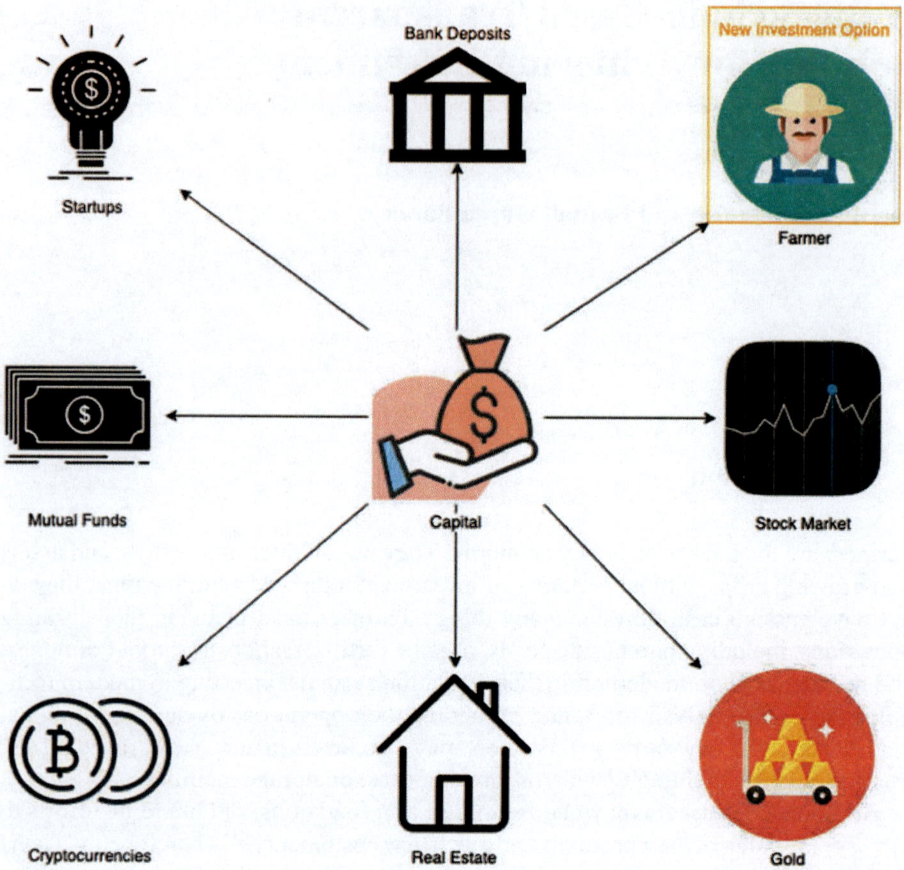

Fig. 1 Farmers as new investment opportunity for investors

The complex application process of applying for funding can be complex and time-consuming, which can be a barrier for farmers with busy schedules.

Limited options: There may be limited funding options available to farmers, particularly in rural areas. These challenges can make it difficult for farmers to access the capital they need to operate and grow their businesses.

Figure 1 demonstrates many opportunities for investors to invest in traditional options, and one new option can be farming

Therefore, in this article, a blockchain-based architecture has been proposed that will bring venture capitalists and farmers on the same page with the flexibility of choosing either one as per the requirement match. Since it is built on blockchain, it brings along some of the inherent benefits of blockchain. One of them is transparency for investors. They would be more aware of their investment. It also facilitates farmers in getting a better line of credit [3, 4].

2 Background and Motivation

The capital market can be a source of funding for farmers, either through issuing securities such as stocks and bonds or through loans and other forms of credit. Farmers may access the capital market through intermediaries such as investment banks, brokerage firms, and lending institutions. Some farmers may also be able to access capital directly by issuing securities on a public exchange. In addition to traditional forms of capital, farmers may also be able to access alternative funding sources such as crowdfunding platforms, peer-to-peer lending platforms, and microfinance institutions.

Small-scale farmers may have difficulty providing collateral due to a lack of assets, such as land or equipment, which can be used as collateral. This can make it difficult for them to access credit and other financial services, as lenders may view them as high-risk clients. There are a few options available to small-scale farmers who are unable to provide collateral [5]:

- Microfinance institutions: These institutions specialize in providing financial services to underserved communities, including small-scale farmers. They often have more flexible lending requirements and may be more willing to provide loans without collateral.
- Government programs: Some governments have programs to provide credit and other financial services to small-scale farmers. These programs may have relaxed collateral requirements or may not require collateral at all.
- Alternative collateral: In some cases, small-scale farmers may use alternative forms of collateral, such as livestock or crops, to secure a loan.

It's important to note that while these options may be available, they may not always be the most affordable or convenient options for small-scale farmers. It's also worth considering ways to improve access to credit and financial services for small-scale farmers, such as by developing more inclusive financial systems or implementing policies that support small-scale agriculture.

Blockchain is a distributed ledger technology that allows for secure, transparent, and tamper-proof record-keeping. It is a decentralized system, meaning that any single entity does not control it. Instead, it relies on a network of computers to validate and record transactions [6].

With the use of blockchain for this purpose, we can bring the following benefits to the system:

- Immutability: Once data has been recorded on a blockchain, it cannot be altered. This means that the terms of a contract stored on a blockchain would be tamper-proof and could be relied upon as a reliable source of information.
- Decentralization: Blockchain technology is decentralized, meaning that any single entity does not control it. This could be useful for contracts between investors and farmers, as it would allow both parties equal access to and control over the contract.

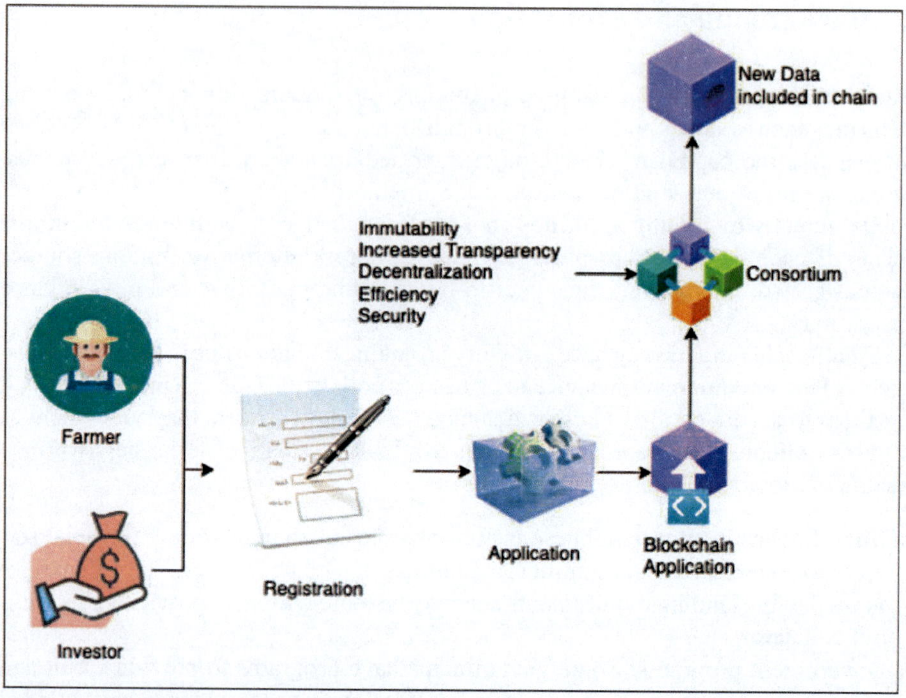

Fig. 2 Overview of Blockchain-based solution

- Smart contracts are self-executing contracts with the terms of the agreement between buyer and seller being directly written into lines of code. The code and the agreed-upon terms are stored and replicated on a blockchain network.
- Increased transparency: Blockchain technology allows for increased transparency, as all parties have access to the same information. This could be useful for contract-making between investors and farmers, as it would give both parties a clear understanding of the terms of the agreement.
- Efficiency: The use of blockchain technology can streamline and automate various contract-related processes, making the contract-making process more efficient.
- Security: Blockchain technology is secure, using encryption to protect the recorded data. This could be useful for protecting the terms of a contract between an investor and a farmer.

Figure 2 demonstrates the use of blockchain to gain benefits of the blockchain system and bring them to the agricultural industry. This would allow the growth of farmers and bring them closer to not only technology but also to better opportunities for farmers to get initial investment for farming. It also demonstrates the use of the platform by both farmers and investors to interact with each other.

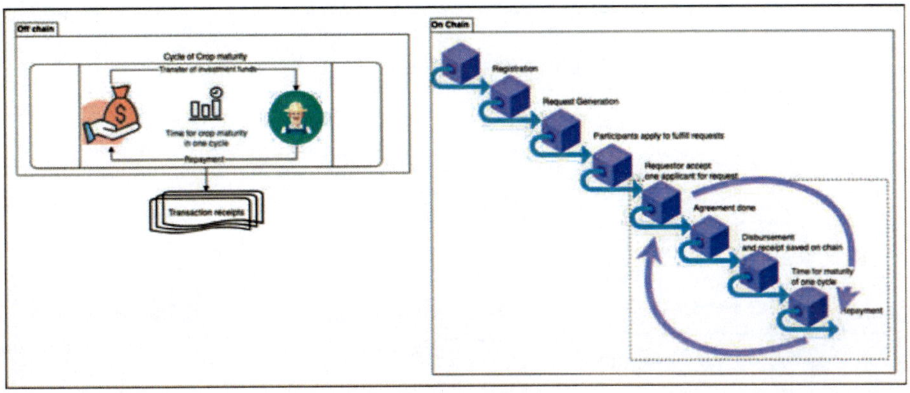

Fig. 3 Architecture of the solution

3 Proposed Solution

We bring a platform where investors can invest in farming, and farmers can find investors for the planned agricultural venture. The architecture of the proposed system has been demonstrated in Fig. 3. The proposed solution brings together farmers and investors by creating a distributed application. Some parts, which are essential to promote transparency and trust, are stored on the blockchain, while other components, which involve currency, are done off-chain, and their completion details are stored on the chain. This distributed application will be accompanied by another android application, which facilitates ease in accessing the interface to the blockchain ease of fingertips.

The process which involves blockchain includes the following steps:

1. Registration
2. Request generation
3. Application submission
4. Request acceptance
5. Agreement creation
6. Disbursement of funds
7. Repayment of the fund back to investors.

3.1 Registration

Registration of business entities, farmers, and investors will be done by a decentralized application (dApp). This will verify the identity of the entity. The data is private and should not be exposed to the public. Hence, proper security controls will be placed to keep the privacy of registered entities. This process is depicted in Fig. 4.

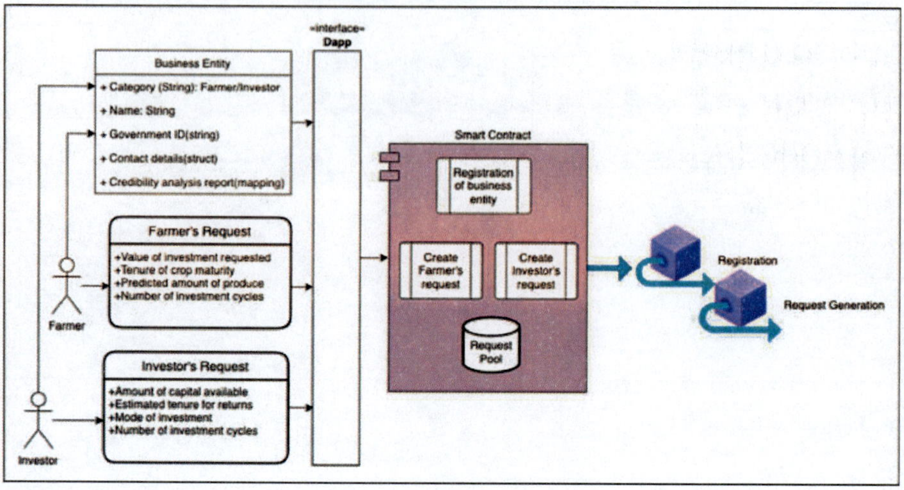

Fig. 4 Registration and request generation

3.2 *Request Generation*

Both business entities can generate the request. Requests generated by farmers will provide details about the crop the farmer intends to produce and the expected amount of investment the requester is looking for. The format of a farmer's request shall include the following key data items:

- Amount of investment required
- Tenure of maturity
- The Predicted amount of produce
- Number of investment cycles.

Similarly, an investor can also put forward his invitation for investment. An investor's request shall have the following key data items:

- Amount of capital available
- Expected maturity (estimated tenure of returns)
- Mode of investment-cash, online
- Number of investment cycles.

3.3 *Application Submission*

Farmer's requests are available for investors who are looking for investment opportunities. They can review the request of farmers and apply for investment.

Similarly, investors' request for an invitation to farmers is available. Farmers can easily find available investors and submit a proposal to get investment. This ease of finding and transparency of the process is the main advantage of blockchain here.

3.4 Request Acceptance

The request is subject to receiving multiple applications. The requestor, whether a farmer or investor, can understand, and study each application, review the credibility, and establish one-to-one connections to understand better and accept the application. This would lead to the finalization of terms of the agreement as per the request submitted by the requestor and as applied by the applicant.

3.5 Agreement

Once the requestor accepts the request from the selected participant, the request will be noted in the smart contract as an agreement. It would contain all the terms and conditions that both parties agreed to arrive at the settlement.

Once requests are available online, interested parties can apply for those requests. The request-generating entity will review the applicants, and the applicant will be accepted based on the criteria registered during the application submission process. Once accepted, this will generate an agreement between both parties. This agreement is valid for multiple cycles of crop production. This will allow flexibility to farmers in case one of their cycles gets affected by any natural calamities leading to loss of productivity.

The process of application, acceptance, and agreement has been demonstrated in Fig. 5.

3.6 The First Phase of the Cycle: Disbursement

In Fig. 6, the disbursement process is depicted. This is an offline process. The monetary transactions, though capable of being carried out on blockchain, pose the risk of market volatility and crypto laws, therefore all the processes for disbursement would be carried out off-chain, where both parties will meet physically, exchange the cash/digital disbursement of the pre-planned amount and physical receipt will be signed by both parties. The receipts would be stored on the blockchain to ensure the transactions were made, and both parties agreed to avoid the non-repudiation conflict.

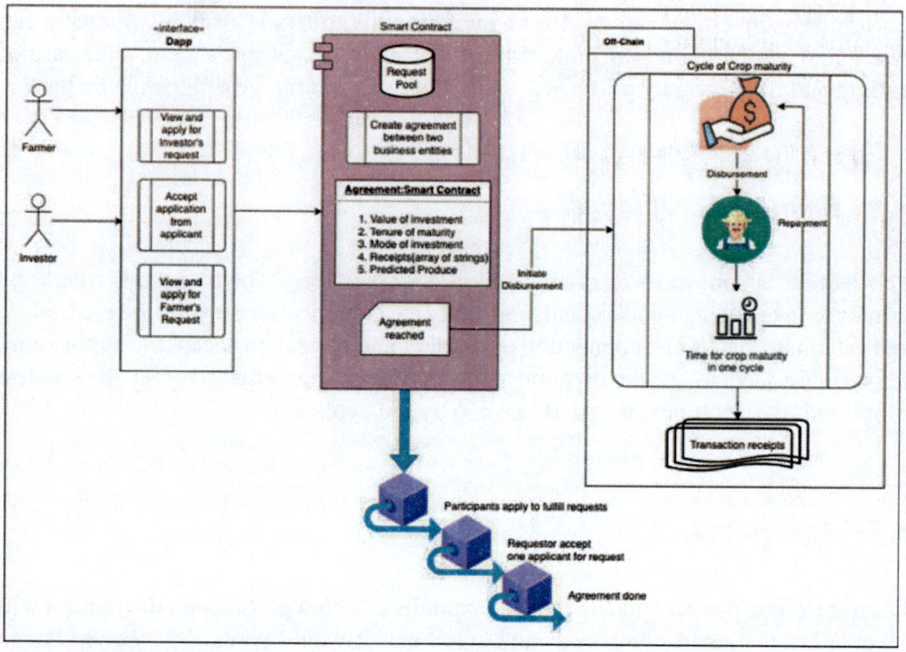

Fig. 5 Application, acceptance, and agreement

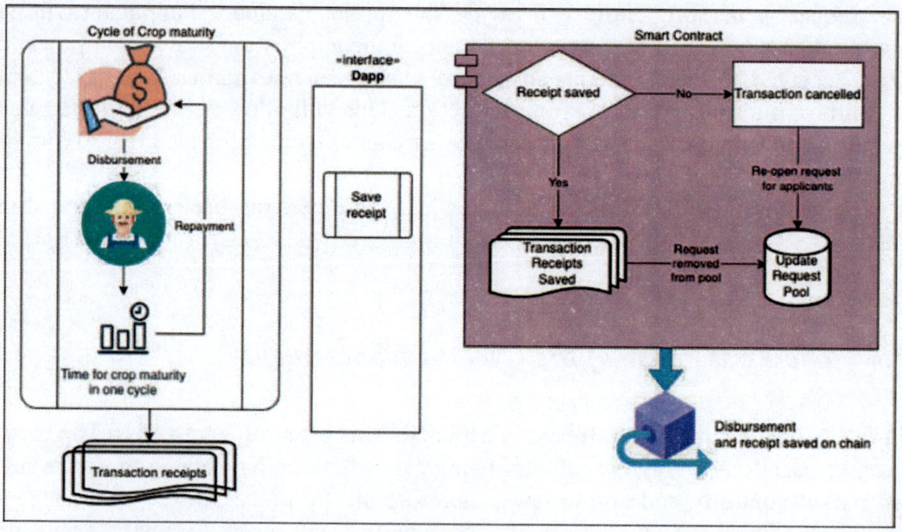

Fig. 6 Disbursement

If the disbursement is completed, the "request" must be deleted from the "Request Pool." If not completed, the request must be re-opened and updated in the "request pool" for future acceptance by any other farmer or investors.

3.7 *The Second Phase of the Cycle: Repayment*

This is also an offline process. In this step, farmers can return the borrowed amount to the investors. The time of repayment can be pre-decided because the agreement can be for multiple cycles of crop maturity. Once the repayment is done, the details of the transactions can be stored on the blockchain to ensure both parties agree to the transactions made. This cycle of disbursement and repayment can be repeated multiple times as agreed by both parties during the agreement.

4 Future Scope

This work has been ideated and demonstrated here. In future, this application will be built as a prototype and tested on agricultural entities. At that time, the following concerns also will be addressed:

- Expanding this process to other farm produce
- Farm market produce: This opens the scope for creating new markets for farmers to sell their produce, ensuring a better post-harvesting experience for farmers.

5 Conclusion

Blockchain technology can be used to create funding solutions for farmers. In this article, farmers can get digital identity for themselves on blockchain. This could help them to access the credit and insurance at right amount. In this proposed solution, smart contract has been used to ensure that farmers are able to receive transparent and efficient funding based on specific conditions and triggers, and it will allow the funders to disburse the funds to farmers. It involves smart contract which can ensure compliance with law and regulations, reducing risk of corruption. This proposed solution aims at increasing transparency, efficiency, accessibility, and automating processes also. In this solution, farmers can have direct transactions with the funders, reducing the transaction costs and time consumption compared to the traditional way of transactions through intermediaries.

References

1. LEI Innovation, Risk and Information Management, et al (2017) Blockchain for agriculture and food: findings from the pilot Study. Wageningen Econ Res. https://doi.org/10.18174/426747
2. Xiong H et al (2020) Blockchain technology for agriculture: applications and rationale. Frontiers Blockchain 3:7. https://doi.org/10.3389/fbloc.2020.00007
3. Tripoli M, Schmidhuber J (2018) Emerging opportunities for the application of blockchain in the agri-food industry. FAO and ICTSD: Rome and Geneva. Licence: CC BY-NC-SA 3.0 IGO
4. Liu P et al (2022) Investment decision of blockchain-based traceability service input for a competitive agri-food supply chain. Foods 11(19):2981. https://doi.org/10.3390/foods11192981
5. Garg A, Shivam AK (2017) Funding to growing start-ups. Res J Soc Sci 10(2):22–31
6. Sathya AR, Gupta B (2020) A comprehensive study of blockchain services: future of cryptography. Int J Adv Comput Sci Appl 11(10). https://doi.org/10.14569/IJACSA.2020.0111037

Author Index

J. K. Mandal et al. (eds.), *Proceedings of International Conference on Network Security and Blockchain Technology*, Lecture Notes in Networks and Systems 738,
https://doi.org/10.1007/978-981-99-4433-0